Jacek Koronacki, Zbigniew W. Raś, Sławomir T. Wierzchoń, and
Janusz Kacprzyk (Eds.)

Advances in Machine Learning II

Studies in Computational Intelligence, Volume 263

Editor-in-Chief

Prof. Janusz Kacprzyk
Systems Research Institute
Polish Academy of Sciences
ul. Newelska 6
01-447 Warsaw
Poland
E-mail: kacprzyk@ibspan.waw.pl

Jacek Koronacki, Zbigniew W. Raś,
Sławomir T. Wierzchoń, and Janusz Kacprzyk (Eds.)

Advances in Machine Learning II

Dedicated to the Memory of Professor
Ryszard S. Michalski

 Springer

Prof. Jacek Koronacki
Institute of Computer Science
Polish Academy of Sciences
ul. Ordona 21
01-237 Warsaw
Poland
E-mail: korona@ipipan.waw.pl

Prof. Zbigniew W. Raś
Woodward Hall 430C
University of North Carolina
9201 University City Blvd.
Charlotte, N.C. 28223
USA
E-mail: ras@uncc.edu,
          ras@pjwstk.edu.pl

Prof. Sławomir T. Wierzchoń
Institute of Computer Science
Polish Academy of Sciences
ul. Ordona 21
01-237 Warsaw
Poland
E-mail: stw@ipipan.waw.pl

Prof. Janusz Kacprzyk
Systems Research Institute
Polish Academy of Sciences
ul. Newelska 6
01-447 Warsaw
Poland
E-mail: kacprzyk@ibspan.waw.pl

ISBN 978-3-642-05178-4   e-ISBN 978-3-642-05179-1

DOI 10.1007/978-3-642-05179-1

Studies in Computational Intelligence   ISSN 1860-949X

Library of Congress Control Number: 2009940119

© 2010 Springer-Verlag Berlin Heidelberg

Typeset & Cover Design: Scientific Publishing Services Pvt. Ltd., Chennai, India.

Printed in acid-free paper

9 8 7 6 5 4 3 2 1

springer.com

Foreword

Professor Richard S. Michalski passed away on September 20, 2007. Once we learned about his untimely death we immediately realized that we would no longer have with us a truly exceptional scholar and researcher who for several decades had been influencing the work of numerous scientists all over the world - not only in his area of expertise, notably machine learning, but also in the broadly understood areas of data analysis, data mining, knowledge discovery and many others. In fact, his influence was even much broader due to his creative vision, integrity, scientific excellence and exceptionally wide intellectual horizons which extended to history, political science and arts.

Professor Michalski's death was a particularly deep loss to the whole Polish scientific community and the Polish Academy of Sciences in particular. After graduation, he began his research career at the Institute of Automatic Control, Polish Academy of Science in Warsaw. In 1970 he left his native country and hold various prestigious positions at top US universities. His research gained impetus and he soon established himself as a world authority in his areas of interest – notably, he was widely considered a father of machine learning.

His contacts with the Polish scientific community were very close over all the years; in the last couple of years he was an affiliate scientist at the Institute of Computer Science, Polish Academy of Sciences, Warsaw. This relation culminated some years ago with his election to the rank of Foreign Member of the Polish Academy of Sciences, a distinction granted to only a small number of world-wide best scientists, including numerous Nobel Prize and other prestigious awards winners.

Professor Michalski was one of those active members of the Polish Academy of Sciences who were always interested in solving whatever problems we had, always ready to help us in shaping the research policy of the Academy and discuss with us all difficult issues that are these days unavoidable in any large and prestigious research organization with so many strong links to science worldwide. He was always ready to offer us his deep understanding and scholarly vision of the future of the human scientific endeavor. As President of the Polish Academy of Sciences I sense very personally an enormous loss coming from no longer being able to ask for his opinion and advice.

I wish to congratulate the editors of these scholarly volumes, Professors Jacek Koronacki, Zbigniew Raś, Sławomir T. Wierzchoń and Janusz Kacprzyk, for their initiative to pay the tribute to the memory of Professor Michalski. Having known him for many years they realized that the best way to honor his life achievements would be to prepare a collection of high quality papers on topics broadly perceived as Professor Michalski's main interest and to present *in memoriam* volumes of the contributions written by those who had the luck to be his friends or, at least, to meet him on various occasions. I am really impressed that so many prominent authors have accepted the invitation and I thank all of them most deeply.

I believe the memory of Professor Richard S. Michalski should remain with us for ever. Hopefully, these volumes will contribute to reaching this objective in the most appropriate and substantial way.

Professor Michał Kleiber
President
Polish Academy of Sciences

Preface

This is the second volume of a large two-volume editorial project we wish to dedicate to the memory of the late Professor Ryszard S. Michalski who passed away in 2007. He was one of the fathers of machine learning, an exciting and relevant, both from the practical and theoretical points of view, area in modern computer science and information technology. His research career started in the mid-1960s in Poland, in the Institute of Automation, Polish Academy of Sciences in Warsaw, Poland. He left for the USA in 1970, and since then had worked there at various universities, notably, at the University of Illinois at Urbana – Champaign and finally, until his untimely death, at George Mason University. We, the editors, had been lucky to be able to meet and collaborate with Ryszard for years, indeed some of us knew him when he was still in Poland. After he started working in the USA, he was a frequent visitor to Poland, taking part at many conferences until his death. We had also witnessed with a great personal pleasure honors and awards he had received over the years, notably when some years ago he was elected Foreign Member of the Polish Academy of Sciences among some top scientists and scholars from all over the world, including Nobel prize winners.

Professor Michalski's research results influenced very strongly the development of machine learning, data mining, and related areas. Also, he inspired many established and younger scholars and scientists all over the world.

We feel very happy that so many top scientists from all over the world agreed to pay the last tribute to Professor Michalski by writing papers in their areas of research. These papers will constitute the most appropriate tribute to Professor Michalski, a devoted scholar and researcher. Moreover, we believe that they will inspire many newcomers and younger researchers in the area of broadly perceived machine learning, data analysis and data mining.

The papers included in the two volumes, Machine Learning I and Machine Learning II, cover diverse topics, and various aspects of the fields involved. For convenience of the potential readers, we will now briefly summarize the contents of the particular chapters.

Part I, "General Issues" is concerned with some more general issues and problems that are relevant in various areas, notably in machine learning, data mining, knowledge discovery. and their applications in a multitude of domains.

- **Witold Pedrycz** ("Knowledge-Oriented and Distributed Unsupervised Learning for Concept Elicitation") discusses a new direction of unsupervised learning and concept formation in which both domain knowledge and experimental evidence (data) are considered together. This is a reflection of a certain paradigm which could be referred to as knowledge-oriented clustering or knowledge mining (as opposed to data mining). The author presents the main concepts and algorithmic details. The distributed way of forming information granules which is

realized at the level of individual locally available data gives rise to higher order information granules (type-2 fuzzy sets, in particular).

- **Andrzej Skowron and Marcin Szczuka** ("Toward Interactive Computations:
  A Rough-Granular Approach") present an overview of Rough Granular Computing (RGC) approach to modeling complex systems and processes. They discuss the granular methodology in conjunction with paradigms originating in
  rough sets theory, such as approximation spaces. The authors attempt to show
  the methodology aimed at the construction of complex concepts from raw data
  in ahierarchical manner. They illustrate, how the inclusion of domain knowledge, relevant ontologies, and interactive consensus finding leads to more poer-
  ful granular models for processes and systems.
- **Stan Matwin and Tomasz Szapiro** ("Data Privacy: From Technology to Economics") deal with an omnipresent and increasingly relevant problem of data
  privacy. They attempt to relate two different approaches to data privacy: the
  technological approach, embodied in the current privacy-preserving data mining
  work, and the economic regulations approach. The authors claim that none of
  these two approaches alone will be able to address the increasingly important
  data privacy issues. They advocate a hybrid system, combining both approaches
  in a complementary manner. A view of privacy is presented in the context of an
  accepted taxonomy of economic goods, stating the question: if privacy is exchanged and traded, then what kind of good is it? The authors also show that,
  viewed in the light of an established economic growth theory, the involvement
  of privacy in the growth process leads to a steady state growth.
- **Phillipa M. Avery and Zbigniew Michalewicz** ("Adapting to Human Gamers
  using Coevolution") consider a challenging task of how to mimic a human ability to adapt, and create a computer player that can adapt to its opponent's strategy. Without this ability to adapt, no matter how good a computer player is,
  given enough time human players may learn to adapt to the strategy used, and
  routinely defeat the computer player. However, by having an adaptive strategy
  for a computer player, the challenge it provides is ongoing. Additionally, a
  computer player that adapts specifically to an individual human provides a more
  personal and tailored game play experience. To address specifically this last
  need, the authors investigate the creation of such a computer player. By creating
  a computer player that changes its strategy with influence from the human strategy, it is shown that the holy grail of gaming, an individually tailored gaming
  experience, is indeed possible. A computer player for the game of TEMPO, a
  zero sum military planning game, is designed. The player was created through a
  process that reversely engineers the human strategy and uses it to co-evolve the
  computer player.
- **Mirsad Hadzikadic and Min Sun** ("Wisdom of Crowds in the Prisoner's Dilemma Context") provide a new way of making decisions by using the wisdom
  of crowds (collective wisdom) to handle continuous decision making problems,
  especially in a complex and rapidly changing world. By simulating the Prisoner's Dilemma as a complex adaptive system, key criteria that separate a wise
  crowd from an irrational one are investigated, and different aggregation strategies are suggested based on different environments.

Part II, "Logical and Relational Learning, and Beyond", is concerned with two very important, well founded, and successful general paradigms for machine learning that are based on logic and relational analyses.

- **Marenglen Biba, Stefano Ferilli, and Floriana Esposito** ("Towards Multistrategic Statistical Relational Learning") discuss statistical relational learning, a growing field in machine learning that aims at the integration of logic-based learning approaches with probabilistic graphical models. Learning models in statistical relational learning consist in learning the structure (logical clauses in Markov logic networks) and the parameters (weights for each clause in Markov logic networks). Markov logic networks have been successfully applied to problems in relational and uncertain domains. So far the statistical relational learning models have mostly used the expectation-maximization (EM) for learning statistical parameters under missing values. In the paper, two frameworks for integrating abduction in the statistical relational learning models are proposed. The first approach integrates logical abduction with structure and parameter learning of Markov logic networks in a single step. During structure search, clause evaluation is performed by first trying to logically abduce missing values in the data and then by learning optimal pseudo-likelihood parameters using the completed data. The second approach integrates abduction with the structural EM by performing logical abductive inference in the E-step and then by trying to maximize parameters in the M-step.

- **Luc De Raedt** ("About Knowledge and Inference in Logical and Relational Learning") gives a gentle introduction to the use of knowledge, logic and inference in machine learning which can be regarded as a reinterpretation and revisiting of Ryszard Michalski's work ``A theory and methodology of inductive learning'' within the framework of logical and relational learning. At the same time some contemporary issues surrounding the integration of logical and probabilistic representations and types of reasoning are introduced.

- **Marta Fraňová and Yves Kodratoff** ("Two examples of computational creativity: ILP multiple predicate synthesis and the 'assets' in theorem proving") provide a precise illustration of what can be the idea of "computational creativity", that is, the whole set of the methods by which a computer may simulate creativity. The analysis is restricted to multiple predicate learning in inductive logic programming and to program synthesis from its formal specification. The authors show heuristics the goal of which is to provide the program with some kind of inventiveness. The basic tool for computational creativity is what is called an 'asset generator'. A detailed description of the authors' methodology for the generation of assets in program synthesis from its formal specification is given. In a conclusion a result is provided, which is a kind of challenge for the other theorem provers, namely how to 'invent' a form of the Ackerman function which is recursive with respect to the second variable instead of the first variable as the usual definitions are. In inductive logic programming multiple predicate synthesis, the assets have been provided by members of the inductive logic programming community, while their methodology tries to make explicit a way to discover these assets when they are needed.

- **Jan Rauch** ("Logical Aspects of the Measures of Interestingness of Association Rules") discusses the relations of the logical calculi of association rules and of the measures of interestingness of association rules. The logical calculi of association rules, 4ft-quantifiers, and known classes of association rules are first introduced. New 4ft-quantifiers and association rules are defined by the application of suitable thresholds to known measures of interestingness. It is proved that some of the new 4ft-quantifiers are related to known classes of association rules with important properties. It is shown that new interesting classes of association rules can be defined on the basis of other new 4ft-quantifiers, and several results concerning new classes are proved. Some open problems are mentioned.

Part III, "Text and Web Mining", is concerned with various problems and aspects of data mining and machine learning related to a great challenge we face nowadays that is related to the constantly growing role of the Internet and its related Web services which permeate all aspects of economy and human life. The papers in this part show how tools and techniques from broadly perceived machine learning and data/text mining can help the human being to fully utilize the power of these new services.

- **Katharina Morik and Michael Wurst** ("Clustering the Web 2.0") present two approaches to clustering in the scenario of Web 2.0 with a special concern of understandability in this new context. In contrast to the Semantic Web type approaches which advocate ontologies as a common semantics for homogeneous user groups, Web 2.0 aims at supporting heterogeneous user groups where users annotate and organize their content without a reference to a common schema so that the semantics is not made explicit. However, it can be extracted by using machine learning and hence the users are provided with new services.
- **Miroslav Kubat, Kanoksri Sarinnapakorn, and Sareewan Dendamrongvit** ("Induction in Multi-Label Text Classification Domains") describe an original technique for automated classification of text documents. It is assumed, first, that each training or testing example can be labeled with more than two classes at the same time which has serious consequences not only for the induction algorithms but also for how we evaluate the performance of the induced classifier. Second, the examples are usually described by very many attributes which makes induction from hundreds of thousands of training examples prohibitively expensive. Results of numerical experiments on a concrete text database are provided.
- **Boris Mirkin, Susana Nascimento, and Luís Moniz Pereira** ("Cluster-Lift Method for Mapping Research") present a method for representing research activities within a research organization by doubly generalizing them. The approach is founded on Michalski's idea of inferential concept interpretation for knowledge transmutation within a knowledge structure taken here to be a concept tree.. To be specific, the authors concentrate on the Computer Sciences area represented by the ACM Computing Classification System (ACM-CCS). Their cluster-lift method involves two generalization steps: one on the level of individual activities (clustering) and the other on the concept structure level (lifting). Clusters are extracted from the data on similarity between ACMCCS topics according to the working in the organization. Lifting leads to conceptual

generalization of the clusters in terms of "head subjects" on the upper levels of ACM-CCS accompanied by their gaps and offshoots. A real-world example of the representation is show.

- **Marzena Kryszkiewicz, Henryk Rybiński, and Katarzyna Cichoń** ("On Concise Representations of Frequent Patterns Admitting Negation") deal with the discovery of frequent patterns wchich is one of the most important issues in the data mining field. Though an extensive research has been carried out for discovering positive patterns, very little has been offered for discovering patterns with negation. One of the main difficulties concerning frequent patterns with negation is huge amount of discovered patterns as it exceeds the number of frequent positive patterns by orders of magnitude. The problem can be significantly alleviated by applying concise representations that use generalized disjunctive rules to reason about frequent patterns, both with and without negation. The authors examine three types of generalized disjunction free representations and derive the relationships between them. They also present two variants of algorithms for building such representations. The results obtained on a theoretical basis are verified experimentally.

Part IV, "Classification and Beyond", deals with many aspects, methods, tools and techniques related to broadly perceived classification which is a key issue in many areas, notably those related to the topics of the present volume.

- **Derek Sleeman, Andy Aiken, Laura Moss, John Kinsella, and Malcolm Sim** ["A system to detect inconsistencies between a domain expert's different perspectives on (classification) tasks"] discuss the range of knowledge acquisition, including machine learning, approaches used to develop knowledge bases for intelligent systems. Specifically, the paper focuses on developing techniques which enable an expert to detect inconsistencies in two (or more) perspectives which the expert might have on the same (classification) task. Further, the INSIGHT system is developed to provide a tool which supports domain experts exploring, and removing, the inconsistencies in their conceptualization of a task. The authors show a study of intensive care physicians reconciling two perspectives on their patients. The high level task which the physicians had set themselves was to classify, on a 5 point scale (A-E), the hourly reports produced by the Unit's patient management system. The two perspectives provided to INSIGHT were an annotated set of patient records where the expert had selected the appropriate class to describe that snapshot of the patient, and a set of rules which are able to classify the various time points on the same 5-point scale. Inconsistencies between these two perspectives are displayed as a confusion matrix; moreover INSIGHT then allows the expert to revise both the annotated datasets (correcting data errors, and/or changing the assigned classes) and the actual rule-set. The paper concludes by outlining some of the follow-up studies planned with both INSIGHT and this general approach.

- **Eduardo R. Gomes and Ryszard Kowalczyk** ("The Dynamics of Multiagent Q-learning in Commodity Market Resource Allocation") consider the commodity market (CM) economic model that offers a promising approach for the distributed resource allocation in large-scale distributed systems. The existing

CM-based mechanisms apply the economic equilibrium concepts, assuming that price-taking entities do not engage in strategic behaviour, and in this paper the above issue is addressed and the dynamics of strategic learning agents in a specific type of CM-based mechanism, called iterative price adjustment, is discussed. The scenario is considered in which agents use utility functions to describe preferences in the allocation and learn demand functions adapted to the market by reinforcement learning. The reward functions used during the learning process are based either on the individual utility of the agents, generating selfish learning agents, or the social welfare of the market, generating altruistic learning agents. The authors' experiments show that the market composed exclusively of selfish learning agents achieve results similar to the results obtained by the market composed of altruistic agents. Such an outcome is significant for a series of other domains where individual and social utility should be maximized but agents are not guaranteed to act cooperatively in order to achieve it or they do not want to reveal private preferences. This outcome is further analyzed, and an analysis of the agents' behaviour from the perspective of the dynamic process generated by the learning algorithm employed by them is also given. For this, a theoretical model of multiagent Q-learning with ε-greedy exploration is developed and applied in a simplified version of the addressed scenario.

- **Christian Borgelt** ("Simple Algorithms for Frequent Item Set Mining") introduces SaM, a split and merge algorithm for frequent item set mining. Its core advantages are its extremely simple data structure and processing scheme, which not only make it quite easy to implement, but also very convenient to execute on external storage, thus rendering it a highly useful method if the transaction database to mine cannot be loaded into main memory. Furthermore, the author's RElim algorithm is shown and different optimization options for both SaM and RElim are discussed. Finally, some numerical experiments comparing SaM and RElim with classical frequent item set mining algorithms (as, e.g., Apriori, Eclat and FP-growth) are given.

- **Michał Dramiński, Marcin Kierczak, Jacek Koronacki, and Jan Komorowski** ("Monte Carlo feature selection and interdependency discovery in supervised classification") consider applications of machine learning techniques in life sciences, Such applications force a paradigm shift in the way these techniques are used because rather than obtaining the best possible classifier, it is of interest which features contribute best to the classification of observations into distinct classes and what are the interdependencies between the features. A method for finding a cut-off between informative and non-informative features is given, followed by the development of a new methodology and an implementation of a procedure for determining interdependencies between informative features. The reliability of the approach rests on a multiple construction of tree classifiers. Essentially, each classifier is trained on a randomly chosen subset of the original data using only a randomly selected fraction of all of the observed features. This approach is conceptually simple yet computationally demanding. The method proposed is validated on a large and difficult task of modelling HIV-1 reverse transcriptase resistance to drugs which is a good example of the aforementioned paradigm shift.

- **Halina Kwaśnicka and Mariusz Paradowski** ("Machine Learning Methods in Automatic Image Annotation") are concerned with image analysis and more specifically automatic image annotation which grew from such research domains as image recognition and cross-lingual machine translation. Because of an increase in computational, data storage and data transfer capabilities of todays' computer technology, an automatic image annotation has become possible. Automatic image annotation methods, which have appeared during last several years, make a large use of many machine learning approaches, with clustering and classification methods as the most frequently applied techniques to annotate images. The chapter consists of three main parts. In the first, some general information concerning annotation methods is presented. In the second part, two original annotation methods are described. The last part presents experimental studies of the proposed methods.

Part V, "Neural Networks and Other Nature Inspired Approaches", deals with the development and applications of various nature inspired paradigms, approaches and techniques. Notably, diverse aspects related to neural networks, evolutionary computation, artificial immune systems, swarm heuristics, etc. are considered, showing their potentials and applicability.

- **Nikola Kasabov** ("Integrative Probabilistic Evolving Spiking Neural Networks Utilising Quantum Inspired Evolutionary Algorithm: A Computational Framework") considers integrative evolving connectionist systems (iECOS) that integrate principles from different levels of information processing in the brain, including cognitive, neuronal, genetic and quantum, in their dynamic interaction over time. A new framework of iECOS, called integrative probabilistic evolving spiking neural networks (ipSNN), utilizing a quantum inspired evolutionary optimization algorithm to optimize the probability parameters. Both spikes and input features in ipESNN are represented as quantum bits being in a superposition of two states (1 and 0) defined by a probability density function which allows for the state of an entire ipESNN at any time to be represented probabilistically in a quantum bit register and probabilistically optimized until convergence using quantum gate operators and a fitness function. The proposed ipESNN is a promising framework for both engineering applications and brain data modeling as it offers faster and more efficient feature selection and model optimization in a large dimensional space in addition to revealing new knowledge that is not possible to obtain using other models. As a further development of ipESNN, the neuro-genetic models – ipESNG, are indicated.
- **Boris Kryzhanovsky, Vladimir Kryzhanovsky, and Leonid Litinskii** ("Machine Learning in Vector Models of Neural Networks") present a review and some extensions of their works related to the theory of vector neural networks. The interconnection matrix is always constructed according to the generalized Hebbian rule which is well-known in machine learning area. The main principles and ideas are emphasized. Analytical calculations are based on the probabilistic approach. The obtained theoretical results are verified via computer simulations.

- **Hongbo Liu, Ajith Abraham, and Benxian Yue** ("Nature Inspired Multi-Swarm Heuristics for Multi-Knowledge Extraction") present a novel application of two nature inspired population-based computational optimization techniques, namely the Particle Swarm Optimization (PSO) and Genetic Algorithm (GA), for rough set reduction and multi-knowledge extraction. A Multi-Swarm Synergetic Optimization (MSSO) algorithm is presented for rough set reduction and multi-knowledge extraction. In the MSSO approach, different individuals encodes different reducts. The proposed approach discovers the best feature combinations in an efficient way to observe the change of positive region as the particles proceed throughout the search space. An attempt is made to prove that the multi-swarm synergetic optimization algorithm converges with a probability of 1 towards the global optimum. The proposed approach is shown to be very effective for multiple reduct problems and multi-knowledge extraction.
- **Tomasz Maszczyk, Marek Grochowski, and Włodzisław Duch** ("Discovering Data Structures using Meta-learning, Visualization and Constructive Neural Networks") discuss first several visualization methods which have been used to reveal hidden data structures, thus facilitating discovery of simplest but appropriate data transformations which can then be used to build constructive neural networks. This is an efficient approach to meta-learning, based on the search for simplest models in the space of all data transformations, as shown in the paper. It can be used to solve problems with complex inherent logical structure that are very difficult for traditional machine learning algorithms.
- **Vladimir Golovko, Sergei Bezobrazov, Pavel Kachurka, and Leanid Vaitsekhovich** ("Neural Network and Artificial Immune Systems for Malware and Network Intrusion Detection") consider neural networks and artificial immune systems as tools applicable to many problems in the area of anomaly detection and recognition. Since the existing solutions use mostly static approaches, which are based on the collection of viruses or intrusion signatures, detection and recognition of new viruses or attacks becomes a major problem. The authors discuss how to overcome this problem by integrating neural networks and artificial immune systems for virus and intrusion detection, as well as combining various kinds of neural networks in a modular neural system for intrusion detection.
- **Alexander O. Tarakanov** ("Immunocomputing for speaker recognition") proposes an approach to speaker recognition by intelligent signal processing based on mathematical models of immunocomputing,. The approach includes both low-level feature extraction and high-level ("intelligent") pattern recognition. The key model is the formal immune network including apoptosis (programmed cell death) and immunization both controlled by cytokines (messenger proteins). Such a formal immune network can be formed from audio signals using a discrete tree transform, singular value decomposition, and the proposed index of inseparability in comparison with the Renyi entropy. An application to the recognition of nine male speakers by their utterances of two Japanese vowels is shown, and the proposed approach outperforms main state of the art approaches of computational intelligence.

We are happy that we initiated and now are able to finalize this exceptional two-volume editorial project the scope and size of which is extraordinary. First, and most importantly, we have gathered astonishingly many eminent researchers and scholars from all parts of the world who have been actively working on a broadly perceived area of machine learning, data analysis, data mining, knowledge discovery, etc. They have contributed with great papers in which a synergistic and inspiring combination of a comprehensive state of the art material, deep technical analysis, novel applications, and much vision and new look on the past, present and future of the fields have been included. Second, the sheer size of this editorial project is exceptional, but the readers will obtain a rare view of what is the best, most original and promising in the areas. This all is certainly the best tribute that our research community can pay to Professor Ryszard Michalski. He had been inspiring all of us for so many years, and many of us owe him so much because his works, and contacts with him, have shaped our careers and maybe even life.

In an email to us, sent upon submitting her chapter to our editorial project, Professor Katharina Morik summarized it all in an unsurpassed way, albeit based on just one personal recollection: *"Ryszard would not want to have stories told and memories shared but just to see scientific work. Hence, I abstained from expressing my feelings [in the chapter submitted]. In 1986, I was just an unknown scientist in Berlin when inviting him to a workshop on knowledge representation and machine learning, admitting that I could not pay any royalty. To him, only the subject mattered. He came, and I learned a lot from him. He was kind of a godfather to machine learning those days, but he was open to the inductive logic programming approach which we were developing then. This gave me confidence that in science you don't need to have connections, be famous, or offer a lot of money: just work hard, discuss, and share the enthusiasm, that's it. Ryszard has never lost this attitude"*.

We wish to thank all the authors for their excellent contributions and an efficient collaboration in this huge and demanding editorial project. Special thanks are also due to Dr. Tom Ditzinger, Senior Editor, and Ms. Heather King, both from Engineering/Applied Sciences of Springer, who have provided much encouragement and support.

Warsaw   Jacek Koronacki
August 2009   Zbigniew W. Raś
                                                              Sławomir T. Wierzchoń
                                                              Janusz Kacprzyk

Table of Contents

Part IV: Classification and Beyond

Part V: Neural Networks and Other Nature Inspired Approaches

Part I
General Issues

Knowledge-Oriented and Distributed Unsupervised Learning for Concept Elicitation

Witold Pedrycz

Department of Electrical & Computer Engineering,
University of Alberta, Edmonton AB T6R 2G7 Canada
and
System Research Institute, Polish Academy of Sciences,
Warsaw, Poland
pedrycz@ee.ualberta.ca

Abstract. In this study, we discuss a new direction of unsupervised learning and concept formation in which both domain knowledge and experimental evidence (data) are considered together. This is a reflection of a certain paradigm which could be referred to as knowledge-oriented clustering or knowledge mining (as opposed to data mining). We offer the main concepts and in selected cases present algorithmic details. The distributed way of forming information granules which is realized at the level of individual locally available data gives rise to higher order information granules (type-2 fuzzy sets, in particular).

Keywords: concept elicitation, unsupervised learning, fuzzy clustering, collaborative processing, domain knowledge, type-2 fuzzy sets.

1 Introduction

Clustering and fuzzy clustering have been regarded as a synonym of structure discovery in data. The result, no matter what technique has been used, comes as a collection of information granules which serve as a quantification of concepts [5][6][12][serving as descriptors of the phenomenon behind the data. In essence, in pattern recognition and system modeling, abstraction, information granulation (and discretization as its particular example) and concepts are ultimate underpinnings of the area [3] [10][14][19][26][28].

We witness an interesting paradigm shift: clustering is no longer a data intensive pursuit whose findings are exclusively based upon processing numeric entities but to a larger extent embraces some domain knowledge which is present in any problem of practical relevance. Clustering is not a "blind" pursuit any longer but enhances its sophistication by dealing with a variety of knowledge hints whose usage augments the quality of findings and quite often make them more attractive to the end user. In a nutshell, the processes of concept formation, discovery of relationships, building associations and quantifying trends. This important leitmotiv of data processing has been succinctly put forward in [13] in the following manner " *knowledge mining, by*

J. Koronacki et al. (Eds.): Advances in Machine Learning II, SCI 263, pp. 3–21.
springerlink.com   © Springer-Verlag Berlin Heidelberg 2010

which we mean the derivation of high-level concepts and descriptions from data through symbolic reasoning involving both data and relevant background knowledge".

The idea of knowledge-based clustering [25] has been one of the streams of algorithmic pursuits carried out in the realm of fuzzy clustering which can be regarded as a realization of the general idea presented above.

To focus our investigations and make it more appealing from the perspective of algorithmic investigations and ensuing implementations, we use the Fuzzy C-Means (FCM) [4] that is regarded as a vehicle to illustrate the main ideas. It could be stressed, however, that the conceptual developments are by no means restricted to this category of unsupervised learning, cf. also [9][11].

The objective of the study is three-fold. First, we develop a concept of collaborative design of information granules [1][2][16][20] [27][29][30] where we stress a role of distributed processing which to a significant extent is implied by the nature of data which become distributed. Second, we construct a coherent framework of knowledge-based unsupervised learning showing that data and a variety of knowledge hints can be used in an orchestrated manner. Third, we demonstrate that information granules such as e.g., type-2 fuzzy sets are the constructs that emerge as a result of aggregation of locally constructed concepts that are within the realm of individual data. These objectives imply a structure of the study. In Section 2, we discuss main modes of collaboration and elaborate on the architecture of unsupervised learning. Along with these, we also provide all necessary preliminaries. Section 3 is focused on the algorithmic details of the single-level scheme of collaboration. The knowledge-based clustering is discussed in Section 4 while Section 5 is concerned with a certain category of knowledge hints which come under the name of so-called knowledge viewpoints. Section 6 brings an idea of justifiable granularity which allows us to form type-2 fuzzy sets or higher order information granules, in general.

2 The Modes of Collaboration and Architecture of Unsupervised Learning

In this section, we discuss some prerequisites, introduce required terminology and notation and then move on to the main categories of collaboration along with the underlying architectural considerations.

2.1 Preliminaries: Notation and Terminology

Distributed data where the distribution could be sometimes very visible in terms of the location of sources of data (such as for instance those encountered in networks of stores, banks, health care institutions and wireless sensor networks). We can distinguish two ways (modes) of organization of collaboration which imply the corresponding optimization scheme. It is instructive to start with graphic symbols to be used throughout the study, Figure 1.

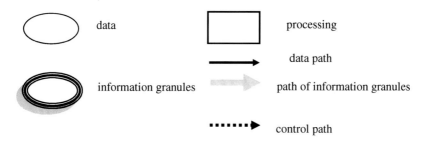

Fig. 1. Basic symbols used in the study

There are two three categories of symbols: (a) reservoirs of data (collections of experimental evidence) and information granules (forming an abstraction of numeric data), (b) algorithms (processing modules) which are used to transform numeric entities into information granules, and (c) communication linkages. We distinguish clearly between interaction (communication) realized at the level of data (e.g., numeric entities) and concepts (which are abstract objects) – information granules. Furthermore we include some control linkages whose role is to affect the parametric setting of the algorithms. All these graphic symbols will be beneficial in the visualization the essence of various architectures, especially when stressing the nature of collaboration and a form of the hierarchy of the emerging concepts.

The distributed nature of processing is of general interest and can be encountered in the context of multievel description of data, collaborative concept formation, and related security and privacy issues, cf. [7][8]18][23][24].

The FCM clustering scheme (which we have considered as an experimental framework) is concerned with a finite N-element set of n-dimensional patterns (data) $\mathbf{x}_1, \mathbf{x}_2, \ldots, \mathbf{x}_N \in \mathbf{R}^n$ N position in n-dimensional Euclidean space. The structure in this data is searched for by minimizing the well-known objective function, cf. [4].

$$Q = \sum_{i=1}^{c} \sum_{k=1}^{N} u_{ik}^{p} \parallel \mathbf{x}_k - \mathbf{v}_i \parallel^2 \tag{1}$$

with $\mathbf{v}_1, \mathbf{v}_2, \ldots, \mathbf{v}_c$ being a set of "c" prototypes. $\mathbf{v}_i$ are the prototypes and U stands for the c by N partition matrix. The parameter "p", p> 1, known as a fuzzification coefficient is used to control the shape of the obtained membership functions (information granules). The distance function used in the FCM is the standard Euclidean in which the corresponding features are normalized by including the corresponding variance, that is $\parallel \mathbf{x} - \mathbf{y} \parallel^2 = \sum_{j=1}^{n} \frac{(x_j - y_j)^2}{\sigma_j^2}$ where σ_j^2 is the variance of the j-th attribute (feature).We also note that once the optimization of Q has been completed, the membership degrees for any new **x** can be derived on a basis of the existing prototypes, that is

$$u_{rs} = \frac{1}{\sum_{j=1}^{c} \frac{\left\| \mathbf{x}_s - \mathbf{v}_r \right\|^2}{\left\| \mathbf{x}_s - \mathbf{v}_j \right\|^2}} \tag{2}$$

and

$$\mathbf{v}_r = \frac{\sum_{k=1}^{N} u_{rk}^p \mathbf{x}_k}{\sum_{k=1}^{N} u_{rk}^p} \tag{3}$$

r=1, 2, .. , c, s=1,2,…, N.

2.2 The Single-Level Architecture of Collaboration

In this topology, see Figure 2, each data set (data site) is available and processed locally. The communication is realized between all or selected nodes and is completed through the exchange of information granules that are concepts formed locally. The collaboration is intended to reconcile the differences and establish a coherent framework within which the findings are shared and actively exploited at the local level. The architecture exhibits a single layer as all nodes are located at the same level.

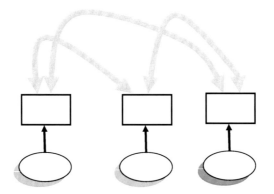

Fig. 2. A single level architecture of collaborative unsupervised learning

Depending on the nature of links, we can distinguish between the architectures of varying level of collaborative connectivity. The extreme case is the one where each node communicates with all remaining nodes in the structure. On the other end of the spectrum of the connectivity scale positioned are the structures which are very loosely connected in which the communication links are very sparse. There are a wealth of structures which from the perspective of connectivity are positioned somewhere in-between these extreme cases.

2.3 Hierarchical Architecture of Collaboration

The single-level architectures of collaboration can give rise to a vast number of variants of hierarchies whose some of the representative examples are included in Figure 3.

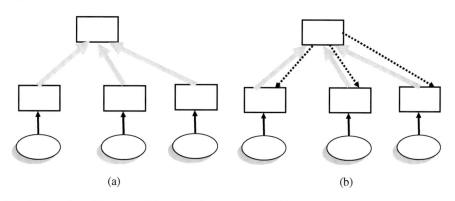

(a)   (b)

Fig. 3. Examples of single-level hierarchical structures of collaborative clustering; (a) one-directional (bottom-up), and (b) bidirectional with control paths to lower level of hierarchy

The generic two-level structure is illustrated in Figure 3 (a). Here the collaboration is realized at the higher level of the structure. The processing does not concern the data but information granules being produced at the lower level of the hierarchy. The processing at the consecutive levels is completed in serial fashion. This means that once the information granules have been formed at the lower level, they are processed at the next layer. The communication is one-directional. Whatever the results (information granules) are constructed there; they do not affect the processing at the lower level (which has been completed before processing at the higher level takes place). The structure illustrated in Figure 3(b) differs from the previous one by the existence of the control mechanism whose role is to affect the processing environment at the lower level depending upon the nature of the computing realized at the higher level. The changes made to the processing environment (processing modules) are intended to improve the quality and consistency of the overall structural (conceptual) findings. For instance, the number of concepts (clusters) at the lower level could be adjusted. Likewise, some parameters of the clustering could be modified as well (in the FCM, we can adjust the values of the fuzzification coefficient which modify the shape of the membership functions, modify the distance function, etc.).

The structures illustrated in Figure 4 attest to the architectural diversity of the hierarchies:

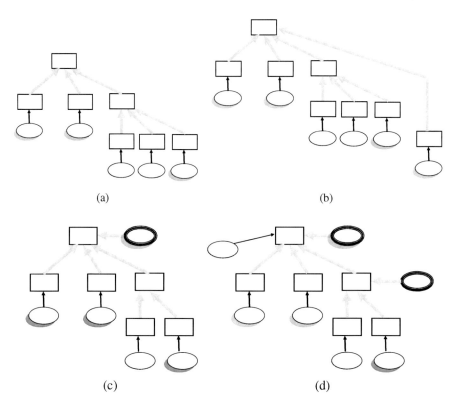

(a)   (b)

(c)   (d)

Fig. 4. Selected examples of hierarchies

In Figure 4(a), there are several levels of hierarchy so the results of processing propagate from one layer to the next one. In some cases, several layers could be by-passed where some findings are sent off several layers up; this topology is shown in Figure 4 (b). The structure visualized in Figure 4(c) exhibits an interesting effect of joint processing of information granules (coming from the lower level of the hierarchy) and numeric data provided from some other source. The way in which these data are processed is indicative of their relevance – they are regarded to be equally important as the information granules produced at the lower level of the hierarchy. We might envision another situation like the one portrayed in Figure 4 (c) where at the higher level of the hierarchy we encounter an auxiliary source of information granules (being reflective of the domain knowledge available there).

3 Algorithmic Developments for a Single-Level Scheme of Collaboration

In this section, we present a concise statement of the problem. Prior to that, however, some notation is worth fixing. To a significant extent, the notation is a standard one which we encounter in the literature. We use boldface to denote vectors in some n-dimensional space, say, $\mathbf{x}, \mathbf{y}, \mathbf{z} \in \mathbf{R}^n$. We consider a finite number of datasites (data sets), denoted here by D[1], D[2],..., D[P] which are composed of the patterns (data) represented in the same feature space (space of attributes). For the ii-th dataset D[ii], the resulting structural information is conveyed in the form of the partition matrix U[ii] with c[ii] clusters formed there. The standard objective function minimized by the FCM takes on the form

$$Q[ii] = \sum_{k=1}^{N} \sum_{i=1}^{c[ii]} u_{ik}^m[ii] \left\| \mathbf{x}_k - \mathbf{v}_i[ii] \right\|^2 \tag{4}$$

Along with the partition matrices, fuzzy clustering produces a collection of prototypes. For the ii-th datasite we have the prototypes $\mathbf{v}_1[ii], \mathbf{v}_2[ii],..., \mathbf{v}_c[ii][ii]$.

Schematically, we can portray this situation of collaboration in Figure 5 which underlines a fact that all communication and collaboration occurs at the level of information granules.

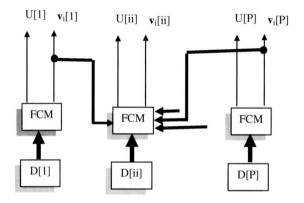

Fig. 5. Mechanisms of collaboration realized through communication of granular findings (partition matrices)

Having established the general setup, we move with the algorithmic development of the optimization process. In what follows, we concentrate on the optimization process of a collaborative formation of information granules by starting with an augmented objective function and deriving detailed formulas for the partition matrix and the prototypes. We assume that the granularity of findings at all datasites is the same, that is c[1] = c[2] =... = c[P] =c. This particular level of granularity of findings has to be agreed upon prior to any engagement in any collaborative activities. Two fundamental ways of communication the findings between the datasites are considered. They directly imply a certain format of the objective function to be considered. The communication of the structural findings is again realized in terms of the prototypes $\mathbf{v}_i[jj]$. The objective function to be minimized takes on the form

$$Q[ii] = \sum_{k=1}^{N}\sum_{i=1}^{c} u_{ik}^2[ii]\left\|\mathbf{x}_k - \mathbf{v}_i[ii]\right\|^2 + \beta\sum_{\substack{jj=1\\jj\neq ii}}^{P}\sum_{k=1}^{N}\sum_{i=1}^{c} u_{ik}[ii]^2 \parallel \mathbf{v}_i[ii] - \mathbf{v}_i[ii\mid jj]\parallel^2 \quad (5)$$

The minimization of (5) is realized with respect to the partition matrix U[ii] and the prototypes $\mathbf{v}_i$[ii]. The second term of the objective function is used to achieve agreement between the structures produced at the individual sites. Not reporting here the detailed calculations (which are somewhat tedious but not necessarily complicated) the resulting formulas governing the update of the partition matrix and the prototypes are the following

$$u_{rs}[ii] = \cfrac{1}{\displaystyle\sum_{j=1}^{c}\cfrac{\left\|\mathbf{x}_s - \mathbf{v}_r[ii]\right\|^2 + \beta\sum_{\substack{jj=1\\jj\neq ii}}^{P}\left\|\mathbf{v}_r[ii] - \mathbf{v}_r[jj]\right\|^2}{\left\|\mathbf{x}_s - \mathbf{v}_j[ii]\right\|^2 + \beta\sum_{\substack{jj=1\\jj\neq ii}}^{P}\left\|\mathbf{v}_j[ii] - \mathbf{v}_j[jj]\right\|^2}} \quad (6)$$

and

$$v_{rt}[ii] = \cfrac{\displaystyle\sum_{k=1}^{N} u_{rk}^2[ii]x_{kt} + \beta\sum_{\substack{jj=1\\jj\neq ii}}^{P}\sum_{k=1}^{N} u_{rk}^2[ii]v_{rt}[jj]}{\displaystyle\sum_{k=1}^{N} u_{rk}^2[ii]\bigl(1+\beta(P-1)\bigr)} \quad (7)$$

r=1, 2…c; t=1, 2,…,n[ii]

The detailed derivations presented so far are now embedded as the essential part of the organization of the overall collaboration process. There are two underlying processes which are carried out consecutively. We start with the fuzzy clustering procedures that are run independently at each datasite for a certain number iterations. The stopping criterion is the one that is typically used in the FCM algorithm, namely we monitor the changes in the values of the partition matrices obtained in the consecutive iterations and terminate the process when the Tchebyshev distance between the partition matrices does not exceed a certain predefined threshold ε; say $\max_{i,k} |u_{ik}(iter+1) - u_{ik}(iter)| < \varepsilon$ with $u_{ik}(iter)$ being the (i, k) th entry of the partition matrix produced at the iteration "iter". At this point, datasites exchange the findings by transferring partition matrices, as illustrated in Figure 5, and afterwards an iterative process which realizes the minimization of (6) or (7) takes place. Again when the convergence has been reported, the results (partition matrices) are exchanged (communicated) between the datasites and the iterative computing of the partition matrices and the prototypes resumes. The communication scheme between the agents when they present their findings could be organized in a number of different ways. In the scheme presented so far, we adhered to a straightforward scenario such that

(a) agents broadcast their findings to all participants of the collaborative process, (b) once the results of clustering have been obtained (and each agent might exercise its own termination criterion, say the maximal number of iterations, no significant changes in the values of the objective function, or others) it communicates its readiness to share findings to others. At this moment, the granular results could be communicated to the other agents. Once the agent has received them from all others, it initiates its own clustering computing (following the scheme presented in the previous sections).

4 Incorporation of Domain Knowledge in the Schemes of Unsupervised Learning: Selected Categories of Knowledge Hints

Available domain knowledge which can be effectively used in the navigation of the search for the structure in the data can be quite diversified. The clustering realized in the presence of domain knowledge comprises two fundamental steps, that is (a) formalization of domain knowledge, and (b) incorporation the formalized knowledge "hints" into the optimization framework of unsupervised learning; we can envision that this could be realized through a suitable extension of the objective function Q that is to be minimized. Irrespectively of the algorithmic realization, the essence of the augmented objective function is the same: the modified (expanded) Q becomes a result of a seamless integration of data and knowledge.

The format of knowledge hints that might arise can be quite diversified. Here we elaborate on some main categories and then show how they could be included in the objective function.

A collection of labeled patterns. There is a set of labeled data whose class membership is available. Usually such a set is quite small in comparison with the data to be handled in unsupervised learning. For instance, in handwritten character recognition, there could be thousands of characters to organize and only a small fraction of labeled patterns. Those could be patterns that were found difficult to classify (say poorly written) or they were identified to be interesting by a human expert given the nature of the problem at hand. The percentage of the labeled patterns is low because of the classification costs so labeling all data becomes highly impractical and very tedious. One can envision a similar scenario when classifying temporal signals in diagnostics problems as these signals cannot be labeled on a continuous basis and only its small fraction could be inspected by human and assign to different categories. These labeled data can serve as a collection of "anchor" points over which the structure can be developed. In essence, the objective function to be formed has to include the component to be optimized with respect to the partition matrix and the prototypes (which is the same as in the "standard" clustering with not supervision) and quantify the closeness of this structure with the labels of the labeled patterns. The original objective function can be augmented in an additive fashion as follows

$$Q = \sum_{i=1}^{c} \sum_{k=1}^{N} u_{ik}^{p} \parallel \mathbf{x}_k - \mathbf{v}_i \parallel^2 + \gamma \sum_{i=1}^{c} \sum_{k \in D} (u_{ik} - f_{ik})^2 \parallel \mathbf{x}_k - \mathbf{v}_i \parallel^2 \qquad (8)$$

where the second term in the above expression is used to express a level of consistency achieved between the structure formed in the data and the constraints provided through the labels of the labeled data. Information about class membership (labels) is captured in terms of f_{ik}. The set of labeled data is denoted by **D**. The positive weight γ is used to strike a sound balance between the component of supervised and unsupervised learning. Higher values of γ stress the increasing relevance of the labeled patterns when searching for the structure in the data. One could note that the labeled data indicate the minimal number of clusters to be looked for in the entire data set however the number of clusters could be higher than that. The solution to this optimization problem is presented in [21][22]; to be noted that the solution is obtained for p = 2 (other values of "p" give rise to more complex scheme which includes solving polynomial equations).

Proximity nature of supervision. The previous scenario implies that we have a detailed information about the labels of some patterns. In some cases this might not be feasible however we are given hints in the form of "closeness or proximity of some patterns. For instance, one may have a hint of the form "patterns a and b are very *close*" or "pattern a is *distinct* from pattern b" The nature of relationship which holds between some selected pairs of patterns can be represented in the form of proximity relationship, denoted as prox(a, b). The proximity relationship exhibits the appealing property of symmetry. It is also reflexive. The transitivity requirement (whose numeric quantification is not satisfied) is not needed here. The proximity assumes values in-between 0 and 1 where 1 stands for the highest level of proximity. Interestingly, the proximity can be easily determined on a basis of the partition matrix. Hence, for two patterns k_1 and k_2, we have

$$\text{prox}(k_1, k_2) = \sum_{i=1}^{c} \min(u_{ik1}, u_{ik2}) \qquad (9)$$

Where prox(k_1, k_2) denotes the proximity level for this particular pair of patterns. We immediately take advantage of this relationship in the formation of the augmented objective function which reads as follows

$$Q = \sum_{i=1}^{c} \sum_{k=1}^{N} u_{ik}^{p} \| \mathbf{x}_k - \mathbf{v}_i \|^2 + \gamma \sum_{i=1}^{c} \sum_{k1,k2 \in P} (\text{prox}(k_1, k_2) - \xi_{k1,k2})(\| \mathbf{x}_{k1} - \mathbf{v}_i \|^2 + \| \mathbf{x}_{k2} - \mathbf{v}_i \|^2)$$

$$(10)$$

The set **P** is a collection of pairs of patterns for which the proximity values have been provided that is

$$\mathbf{P} = \{(k,l) \mid \text{prox}(k,l) = \xi_{kl}\} \qquad (11)$$

Note that the second term of this expression quantifies an extent to which the proximity values formed on a basis of the partition matrix coincide with those coming in the form of the knowledge hints. The weight factor γ plays the same role as discussed with regard to (xx). It is worth stressing that the knowledge available here comes in a far more "relaxed" form than those present in the previous case where the provided patterns have been labeled in an explicit manner. The number of clusters is not specified at all. The well-known cases of so-called *must-link* and *should-not-link* hints are subsumed by the far more general type of hints presented here by the proximity values. In essence, the must-link constraint corresponds to the proximity value equal to 1 while the should-not-link constraint comes with the proximity value equal to 0. The detailed optimization schemes are discussed in [17][25].

High-level functional constraints. The domain knowledge can be quantified in the form of some functionals expressed in terms of the corresponding membership grades (entries of the partition matrix). For instance, we know that some pattern $\mathbf{x}_k$ is "difficult" to classify no matter how many clusters we consider. This tells us that it is very likely that all the membership grades could come close to $1/c$. The effect of such potential cluster assignment could be easily quantify by the values of entropy H computed for the membership grades positioned in the k-th column of the partition matrix $\mathbf{u}_k$. Recall that the entropy $H(\mathbf{u}_k)$ assumes its maximum when $u_{ik} = 1/c$ and zeroes when some membership grade is equal to 1, $H(\mathbf{u}_k) = 0$. The domain hints of the form "pattern $\mathbf{x}_k$ is difficult to assign" or "there is no hesitation to allocate $\mathbf{x}_k$ in the structure of data" can be quantified through the entropy values. The augmented objective function comes in the form

$$Q = \sum_{i=1}^{c} \sum_{k=1}^{N} u_{ik}^{p} \parallel \mathbf{x}_k - \mathbf{v}_i \parallel^2 + \gamma \sum_{i=1}^{c} \sum_{k \in \mathbf{K}} (H(\mathbf{u}_k) - \rho_k)^2 \parallel \mathbf{x}_{kl} - \mathbf{v}_i \parallel^2 \quad (12)$$

Here its second term of (12) weighted by γ articulates differences between the knowledge hints (entropy values for the selected patterns) and the entropy values computed for the partition matrix. The sum is taken over all data tagged by the knowledge hints that is those data for which the entropy values have been provided

$$\mathbf{K} = \{ k | H(\mathbf{u}_k) = \rho_k \} \quad (13)$$

One may remark here that these types of constraints imply the optimization scheme that is more demanding as the partition matrix is included in the objective function in a nonlinear fashion. They call for the use of techniques of evolutionary optimization.

The next category of knowledge hints, referred to as knowledge viewpoints is more comprehensive as not pertaining to selected patterns but rather expressed over all data.

5 Knowledge Viewpoints in Unsupervised Learning

The viewpoints capture some insights at data expressed by the user/designer interacting with data or more generally interested in a description of the underlying phenomenon behind the generation of the data. Formally, the viewpoints are represented as set of "p" r-dimensional vectors in $\mathbf{R}^r$ where $p \leq c$ and $r \leq n$.

By considering the viewpoints to be of the same relevance as the prototypes themselves, we include them on the list of the prototypes. From the formal standpoint, the viewpoints are conveniently captured by two matrices called here B and F. The first one is of Boolean character and comes in the form

$$b_{ij} = \begin{cases} 1, & \text{if the j-th feature of the i-th row of B is determined by the viewpoint} \\ 0, & \text{otherwise} \end{cases}$$

$$(14)$$

Thus the entry of B equal to 1 indicates that the (i,j) the position is determined by a certain viewpoint. The dimensionality of B is c×n. The second matrix (F) being of the same dimensionality as B and includes the specific numeric values of the available viewpoints. Figure 6 includes a number of illustrative examples in case n =2 and c = 4.

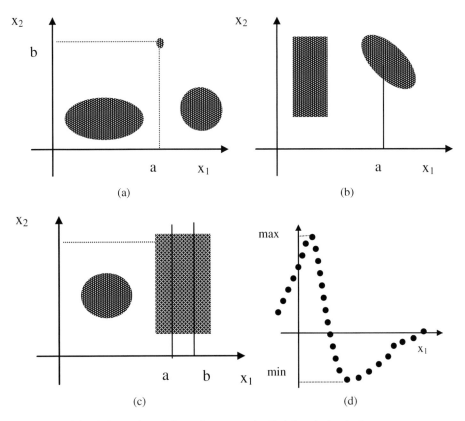

Fig. 6. Examples of viewpoints; see a detailed description in the text

In all examples included here the viewpoints are reflective of our domain knowledge about the problem in which the data were generated or convey our view at the problem in addition to the existing data. Let us elaborate on the examples illustrated in Figure 6:

In the first case shown there, Fig. 6 (a), this particular viewpoint stresses a relevance of a condensed cluster composed of a few patterns only (which otherwise could have been completely washed away given other clusters are far larger). The structure of viewpoints is captured by the following matrices (c=3)

$$B = \begin{bmatrix} 1 & 1 \\ 0 & 0 \\ 0 & 0 \end{bmatrix} \quad F = \begin{bmatrix} a & b \\ 0 & 0 \\ 0 & 0 \end{bmatrix}$$

In Figure 6(b) we encounter a single viewpoint which concerns only a single variable ($x1$) for which we form the viewpoint at $x1=a$ (in this way partially localizing the representative of the data). The matrices B and F come with the entries ($c=2$)

$$B = \begin{bmatrix} 0 & 0 \\ 1 & 0 \end{bmatrix} \quad F = \begin{bmatrix} 0 & 0 \\ a & 0 \end{bmatrix}$$

The viewpoints shown in Figure 6(c) emphasize that the elongated group of data should be looked at by eventual split along the x_1 coordinate. The matrices B and F are of the form ($c=3$)

$$B = \begin{bmatrix} 1 & 0 \\ 0 & 0 \\ 1 & 0 \end{bmatrix} \quad F = \begin{bmatrix} a & 0 \\ 0 & 0 \\ b & 0 \end{bmatrix}$$

The case shown in Figure 6(d) can be viewed as some temporal or spatial data (along x_1) with x_2 being the samples of amplitude. The intent of the viewpoints is to assure that the extreme points (max and min) are included in the structural findings provided by clustering. The viewpoints are described as follows ($c = 4$).

6 From Fuzzy Sets to Type-2 Fuzzy Sets

The discussed scenarios of collaborative clustering inherently invoke the concept of type-2 information granules (type-2 fuzzy sets) which manifest the diversity of the results produced at the individual data sites. In particular, the associated various points of view could result in different information granules and their diversified levels of specificity. Let us outline the essence of the processing in Figure 7.

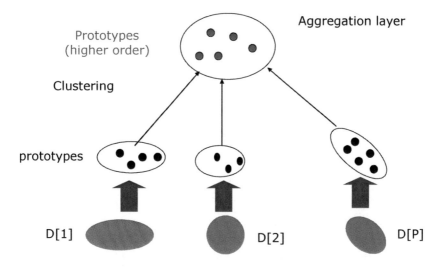

Fig. 7. A hierarchy of processing: from granulation of individual data D[1], D[2], ..., D[p] and fuzzy clustering realized in a two-phase mode

More specifically, we encounter a certain system (phenomenon) which is perceived from various perspectives (points of view) and on a basis of data D[1], D[2], …, D[P] and afterwards the structure is captured via a collection of information granules. Given the granulation supplied by the FCM algorithm, we encounter families of prototypes associated with the data. More formally, we have

Data D[1]: $\mathbf{v}_1[1]$, $\mathbf{v}_2[1]$,…, $\mathbf{v}_{c_1}[1]$ with the number of clusters equal to c_1

Data D[2]: $\mathbf{v}_1[2]$, $\mathbf{v}_2[2]$,… $\mathbf{v}_{c_c}[1]$ with the number of clusters equal to c_2

Data D[p]: $\mathbf{v}_1[1]$, $\mathbf{v}_2[p]$,… $\mathbf{v}_{c_p}[1]$ with the number of clusters equal to c_p

Note that the number of clusters (prototypes) could vary between data sets. The obtained prototypes are considered together and viewed as a more synthetic data set which is again clustered at the higher level producing generalized prototypes, say $\mathbf{z}_1$, $\mathbf{z}_2$, …,$\mathbf{z}_c$. The crucial phase is to associate each of these prototypes with the prototypes we started with at the lower processing level. The assignment mechanism exploits the maximum association between the given $\mathbf{z}_j$ and the prototype in each D[1], D[2[, …, D[p] which is linked with one of the prototypes to the highest extent. By doing this, we end up with the arrangement of the following form (mapping) which associates z_j with the family of the corresponding prototypes,

$$\mathbf{v}_{i_1}[1], \ \mathbf{v}_{i_2}[2],...., \mathbf{v}_{i_p}[P]$$

$$\lambda_{i_1} \ , \ \lambda_{i_2},....,\ \ \lambda_{i_p}$$

The essence of this mapping is illustrated in Figure 8.

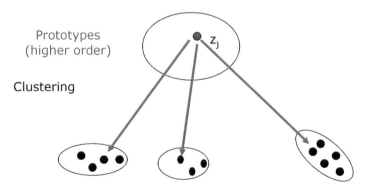

Fig. 8. The determination of correspondence between the prototypes

Given the degrees of membership of the prototypes formed at the higher level (when considering all prototypes obtained at the lower level), the calculations are governed by the expressions

$$\lambda_{ij}[ii] = \frac{1}{\sum_{k=1}^{c[ii]} \left(\frac{\parallel \mathbf{v}_i[ii] - \mathbf{z}_j \parallel}{\parallel \mathbf{v}_k[ii] - \mathbf{z}_j \parallel} \right)^2}$$

(15)

$$\lambda_{i_0}[ii] = \max_{i=1,2,\dots,c[ii]} \lambda_{ij}$$

Alluding to the components of the mapping, we can represent a collection of the prototypes associated with z_j in a form of a certain information granule. The development of this information granule is guided by the *principle of justifiable granularity*. The details of the concept are presented in the Appendix.

As the result of the above minimization, we obtain the granular (interval-valued) representation of the numeric data. Repeating the same procedure for all variables in case of the multidimensional case we end up with the prototypes in the form of hypercubes, H_1, H_2, ..., and H_c. Let us stress that the granular character of the prototypes is a direct consequence of the variability we have encountered because of the processing of several views at data. The distance of any numeric entity x from the hyperboxes can be determined by considering the following bounds, see Figure 9

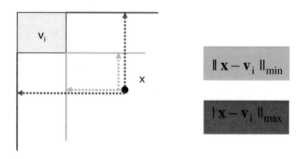

Fig. 9. The determination of the distance bounds between numeric datum and hyperbox information granules; the distances are computed for individual coordinates and here the bounds are determined

These two bounds of distances lead to the calculations of the membership functions. We obtain the following bounds on the membership grades
- lower membership

$$u_i^-(\mathbf{x}) = \frac{1}{\sum_{j=1}^{c} \left(\frac{\parallel \mathbf{x} - \mathbf{v}_i \parallel_{max}}{\parallel \mathbf{x} - \mathbf{v}_j \parallel_{min}} \right)^{\frac{2}{p-1}}}$$

(16)

- upper membership

$$u_i^+ (\mathbf{x}) = \cfrac{1}{\displaystyle\sum_{j=1}^{c} \left(\cfrac{\| \mathbf{x} - \mathbf{v}_i \|_{min}}{\| \mathbf{x} - \mathbf{v}_j \|_{max}} \right)^{\frac{2}{p-1}}} \tag{17}$$

In essence the resulting information granule becomes a fuzzy set of type-2; more specifically an interval type-2 fuzzy set. Examples of the membership functions of fuzzy sets of type-2 are shown in Figure 10. The range of the interval-valued membership grades depends upon the granularity of the prototypes which is visible in Figure 10(a) and 10(c). The different granularity of the prototypes results in quite asymmetric behavior of membership functions as illustrated in Figure 10(b).

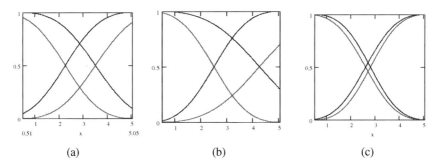

(a)   (b)   (c)

Fig. 10. Examples of interval valued fuzzy sets defined in **R** for several cases of prototypes: (a) V_1= [-0.5 0.5] and V_2 = [5.0 6.5], (b) V_1= [0.0 0.5] and V_2 = [5.0 8.05], (c) V_1= [0.2 0.5] and V_2 = [5.0 5.1]

7   Conclusions

Concept elicitation is one of the cornerstones of data analysis. The study presented here underlines several crucial methodological and algorithmic points that become more and more visible in the area. Indisputably, concepts come hand in hand with the formalism of information granules; irrespectively of their character (say, intervals or fuzzy sets). The granularity of the concepts relates to their generality. The distributed nature of concept formation and ways of aggregating them lead to the idea of higher order granular constructs. In this way, we show not only a compelling reason behind the emergence of constructs such as e.g., type-2 fuzzy sets but also offer a constructive way of their design – an issue that has not been tackled in the literature.

Acknowledgements

Support from the Natural Sciences and Engineering Research Council of Canada (NSERC) and Canada Research Chair (CRC) Program is gratefully acknowledged.

References

1. Bargiela, A., Pedrycz, W.: Recursive information granulation: aggregation and interpretation issues. IEEE Trans. Systems, Man and Cybernetics-B 33(1), 96–112 (2003)
2. Bargiela, A., Pedrycz, W.: Granular Computing: An Introduction. Kluwer Academic Publishers, Dordrecht (2003)
3. Bellman, R., Kalaba, R., Zadeh, L.: Abstraction and pattern classification. Journal of Mathematical Analysis and Applications 13(1), 1–7 (1966)
4. Bezdek, J.C.: Pattern Recognition with Fuzzy Objective Function Algorithms. Plenum Press, New York (1981)
5. Breiman, L., Friedman, J., Olshen, R., Stone, R.C.: Classification and Regression Trees. Wadsworth International Group, Belmont (1984)
6. Chau, M., Zeng, D., Chen, H., Huang, M., Hendriawan, D.: Design and evaluation of a multi-agent collaborative Web mining system. Decision Support Systems 35(1), 167–183 (2003)
7. Chen, Y., Yao, Y.Y.: A multiview approach for intelligent data analysis based on data operators. Information Sciences 178(1), 1–20 (2008)
8. Costa da Silva, J., Klusch, M.: Inference in distributed data clustering. Engineering Applications of Artificial Intelligence 19, 363–369 (2006)
9. Duda, R.O., Hart, P.E., Stork, D.G.: Pattern Classification, 2nd edn. J. Wiley, New York (2001)
10. Grzymala-Busse, J.W., Stefanowski, J.: Three discretization methods for rule induction. International Journal of Intelligent Systems 16, 29–38 (2001)
11. Hoppner, F., et al.: Fuzzy Cluster Analysis. J. Wiley, Chichester (1999)
12. Hunt, E., Marin, J., Stone, P.: Experiments in Induction. Academic Press, New York (1966)
13. Kaufman, K.A., Michalski, R.S.: From data mining to knowledge mining. Handbook of Statistics, vol. 24, pp. 47–75. Elsevier, Amsterdam (2005)
14. Michalski, R.S.: A theory and methodology of inductive learning. Artificial Intelligence 20, 111–161 (1983)
15. Michalski, R.S., Bratko, I., Kubat, M.: Machine Learning and Data Mining: Methods and Applications. J. Wiley, London (1998)
16. Moore, R.: Interval analysis. Prentice-Hall, Englewood Cliffs (1966)
17. Loia, V., Pedrycz, W., Senatore, S.: P-FCM: a proximity-based fuzzy clustering for user-centered web applications. Int. J. of Approximate Reasoning 34, 121–144 (2003)
18. Merugu, S., Ghosh, J.: A privacy-sensitive approach to distributed clustering. Pattern Recognition Letters 26, 399–410 (2005)
19. Mitra, S., Hayashi, Y.: Neuro-fuzzy rule generation: survey in soft computing framework. IEEE Transaction on Neural Networks 11(3), 748–768 (2000)
20. Pawlak, Z., Skowron, A.: Rough sets and Boolean reasoning. Information Sciences 177(1), 41–73 (2007)
21. Pedrycz, W., Waletzky, J.: Neural network front-ends in unsupervised learning. IEEE Trans. on Neural Networks 8, 390–401 (1997a)
22. Pedrycz, W., Waletzky, J.: Fuzzy clustering with partial supervision. IEEE Trans. on Systems, Man, and Cybernetics 5, 787–795 (1997b)
23. Pedrycz, W.: Distributed fuzzy systems modeling. IEEE Transactions on Systems, Man, and Cybernetics 25, 769–780 (1995)
24. Pedrycz, W.: Collaborative fuzzy clustering. Pattern Recognition Letters 23, 675–686 (2002)
25. Pedrycz, W.: Knowledge-Based Clustering: From Data to Information Granules. J. Wiley, Hoboken (2005)
26. Pedrycz, W., Gomide, F.: Fuzzy Systems Engineering. J. Wiley, Hoboken (2007)
27. Pedrycz, W., Park, B.J., Oh, S.K.: The design of granular classifiers: A study in the synergy of interval calculus and fuzzy sets in pattern recognition. Pattern Recognition 41(12), 3720–3735 (2008)

28. Quinlan, J.R.: Induction of decision trees. Machine Learning 1, 81–106 (1986)
29. Zadeh, L.A.: Towards a theory of fuzzy information granulation and its centrality in human reasoning and fuzzy logic. Fuzzy Sets and Systems 90, 111–117 (1997)
30. Zadeh, L.A.: Toward a generalized theory of uncertainty (GTU)-—an outline. Information Sciences 172, 1–40 (2005)

Appendix: The Principle of Justifiable Granularity

We briefly introduce the principle of justifiable granularity. Let us consider a finite family of pairs $\mathbf{X} = (x_i, \mu_i)$ where $x_i \in \mathbf{R}$ and μ_i denotes a degree of membership or more generally a weight (with the values coming from the unit interval) which quantifies an extent to which x_i is deemed essential in the context of $\mathbf{X}$. Furthermore there is a typical element in $\mathbf{X}$ with the weight (membership) equal to 1. Describe this pair as $(z_0, 1)$. The principle of justifiable granularity is to capture the variability (diversity) of numeric data $\mathbf{X}$ by a certain information granule in such a way that (a) the information granule "captures" elements of X to a significant extent, and (b) information granule is specific enough (so that its semantics becomes clearly articulated). While these requirements are intuitively appealing, there are still open questions as to the formalism of information granules to be used in the realization of this construct and the detailed construction of the granule itself (which calls for the formalization of the two requirements introduced above). One could easily anticipate a variety of approaches. In what follows, as one of the viable alternatives, we consider set-based formalism of information granulation.

Given the set of pairs (x_i, μ_i), see Figure A-1, we are interested in representing these membership values by spanning an interval $[x_., x_+]$ around x_0 so that it realizes an intuitively appealing procedure: increase *high* membership values to 1 and reduce to 0 *low* membership values. In this sense, we form an interval capturing the diversity residing within the pairs (x_i, μ_i). The formal rule behind the construction of this interval reads as follows

$$\text{if } z_i \in [z_-, z_+] \text{ then elevate to membership grades to } 1$$

$$\text{if } z_i \notin [z_-, z_+] \text{ then reduce membership grades to } 0$$

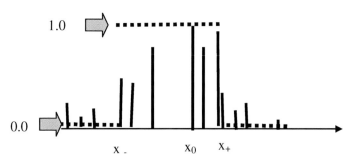

Fig. A-1. Computing the interval representation of numeric values through the principle of justifiable granularity by elevating and suppressing respective membership grades

The bounds of the interval [x., x₊] are subject to optimization with the criterion such that the total changes to the membership degrees (being equal either to 1-μ_i or μ_i) are made as small as possible. The values of x . and x₊ are determined in such a way so that they lead to the minimal value of the following performance index (which expresses cumulative changes made to the membership grades)

$$\text{Min}_{x_1, x_2 \in \mathbf{R}: x_1 \leq x_2} \left\{ \sum_{a_i \in [x_1, z_2]} (1 - \mu_i) + \sum_{a_i \notin [z_1, z_2]} \mu_i \right\} \tag{A.1}$$

Note that the interval [x., x₊] is not necessarily symmetric around z_0 and its location depends upon the distribution of x_is and their weight values.

We can consider another realization of the principle of the principle of justifiable granularity by taking into consideration not only the required changes made to the membership degrees (weights) but the location of the individual x_i's which leads us to the area–based criterion. Without any loss of generality, let us make some additional assumptions. Consider that $x_0=0$. the set of elements of **X** which are higher than 0 are ordered in an increasing manner, that is $x_1 < x_2 < x_n$. We compute the following sum (which represents areas of the corresponding regions).

The optimal positive threshold d in **R** is such for which the expression $S_2 + (d-S_1)$ attains minimum where the areas S_1 and S_2 are illustrated in Figure A-2.

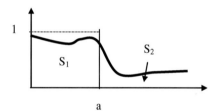

a

Fig. A-2. Optimization of threshold (a) with the use of the area criterion

The principle of justifiable granularity produces an information granule out of some numeric entities. Two cases are of particular interest here:

(a) if **X** includes numeric data (say, readings of some instrument) then the interval representation of **X** becomes a granular measurement – an aggregate of a collection of numeric measurements.

(b) if **X** consists of membership grades obtained for the same element of the universe of discourse, the obtained interval of membership degrees gives rise to the interval-valued fuzzy set. In other words, we can say that the diversity in the collection of fuzzy sets can be quantified in the form of a single type-2 fuzzy set (here interval-valued fuzzy set).

Toward Interactive Computations:
A Rough-Granular Approach

Andrzej Skowron and Marcin Szczuka

Institute of Mathematics, The University of Warsaw
Banacha 2, 02-097 Warsaw, Poland
`skowron@mimuw.edu.pl`

Abstract. We present an overview of Rough Granular Computing (RGC) approach to modeling complex systems and processes. We discuss the granular methodology in conjunction with paradigms originating in rough sets, such as approximation spaces. We attempt to show the methodology aimed at construction of complex concepts from raw data in hierarchical manner. We illustrate, how the inclusion of domain knowledge, relevant ontologies, and interactive consensus finding leads to more potent granular models for processes.

Keywords: granular computing, interactive computations, rough sets, vague concept approximation, interaction of granules.

1   Introduction

Information granulation plays an important role in the process of reasoning under uncertainty from the experimental data and domain knowledge, where constructions or computations are preformed on objects called granules. One of the key issues in such computations is related to the approximation of granules. The framework on which the approximations of granules are based in the rough set approach to granularity is called Rough Granular Computing (RGC).

Complex granules, essential to approximation of vague concepts are often learned from distributed environments using hierarchical approach. In such learning different kinds of granules may be produced. Examples include indiscernibility or similarity classes, patterns, rules, sets of rules, approximation spaces, classifiers, clusters, process models or agents. More compound granules are often obtained as a result of interaction between more elementary ones.

In this short paper we outline some of the approaches studied within the rough granular computing (RGC) framework and their applications in learning of complex granules and granule interactions. The elements of methodology developed on the basis of RGC were applied to real-life projects. Applications of techniques originating in RGC are related to, among others, unmanned area vehicle (UAV) control, robotics, prediction of risk patterns from medical and financial data, sunspot classification, and bioinformatics.

The paper starts with brief summary of fundamental notions and paradigms in RGC. Next, we discuss the approach to construction of complex concepts (granules) from more elementary ones using hierarchical learning. We then present the

J. Koronacki et al. (Eds.): Advances in Machine Learning II, SCI 263, pp. 23–42.
springerlink.com

possible use of concept ontologies in the process of constructing complex granules. We also comment on possibilities of using the described RGC approach to the problem of process mining. We conclude by pointing out that the issues in RGC that we have discussed are a particular manifestation of problems taht we try to address within the broader *Wisdom Technology* (WisTech) research direction.

2   Selected Issues in RGC

In this section, we discuss some basic issues in RGC. This approach is based on rough set methods as outlined in [32,33,35]. RGC is an approach aiming at constructive definition of computations over basic objects called *granules*. The goal is to make possible searching for solutions of problems specified with use of vague concepts. We consider granules being (constructive) definitions of sets used in assembling compound objects which are satisfying the given (possibly vague) specification to, at least, satisfactory degree. Granules are usually defined by means of granule systems [45,46]. In a granule system we distinguish basic building blocks called *elementary granules* and we introduce operations making it possible to build new granules from existing ones, either elementary or constructed in previous steps. Among various possible types of operations on granules, one can distinguish two special – *fusion* and *decomposition*. For more readings on RGC, the reader is referred to [31,36].

The elementary granules in a given granule system are obtained in the process called *granulation*. The granulation essentially divides the universe of items (objects) in discourse into a family of blobs (granules). Each granule contains the collection of objects that we deem indistinguishable in terms of constructed system. Granules may or may not intersect, depending of the general assumption regarding granule system. Granular representation can be viewed as an attempt to mimic the human way of achieving data compression and it plays a key role in implementing the divide-and-conquer strategy in human-like problem solving [56]. The RGC approach combines rough set methods with methods based on granular computing (GC) [2,36,56], borrowing also from other soft computing paradigms.

2.1   Synthesis of Complex Objects Satisfying Vague Specifications

One of the central issues related to granule systems is the definition of inclusion and closeness relations (measures) for granules. These measures should be defined for granules with varying complexity structure. For this purpose we can use the concept of rough inclusion borrowed from *rough mereology* [38] as a starting point.

In real-life applications, we often deal with problems where not only is the information about objects partial, but also, to make things even more difficult, the specification of problems is written in natural language. Inevitably, such specifications involve vague or/and imperfect concepts. In view of that, problems that we are trying to solve can be characterized as searching for complex objects satisfying a given specification to a satisfactory degree [38]. These complex objects should be synthesized from more elementary ones using available operations. This is directly corresponding to the idea of constructing compound granules from elementary ones. In the following section, we discuss the approach searching for relevant granules which rewrites this problem into optimization problem in GC.

2.2   Optimization in Discovery of Compound Granules

The problem considered in this section is the one of perception evaluation as a means of optimizing various tasks [15]. The solution to this problem hearkens back to early research on rough set theory and approximation.

The evaluation of perception we investigate here is at the level of approximation spaces. The quality of an approximation space relative to a given approximated set of objects is a function of the description length of an approximation of the set of objects and the approximation quality of this set. We intend to show how the approximation spaces translate to granular computing. In GC, the focus is on discovering granules satisfying selected criteria. These criteria are expected to be "optimal", in this manner taking the inspiration from the *minimal description length* (MDL) principle proposed by Jorma Rissanen in 1983.

First, we recall the definition of an approximation space from [44]. They are examples of specialized, parameterized relational structures. Tuning parameters makes it possible to search for relevant approximation spaces relative to given concepts. Such approximation spaces can be treated as granules used for concept approximation.

Definition 1. *A parameterized approximation space is a system* $AS_{\#,\$} = (U, I_{\#}, \nu_{\$})$, *where*

- U *is a non-empty set of objects,*
- $I_{\#} : U \to P(U)$ *is an uncertainty function, where* $P(U)$ *denotes the power set of* U,
- $\nu_{\$} : P(U) \times P(U) \to [0,1]$ *is a rough inclusion function,*

and $\#, \$$ *denote vectors of parameters (the indexes* $\#, \$$ *will be later omitted whet it does not lead to confusion).*

The uncertainty function defines for every object $x \in U$, a set of objects that are described similarly to x. The set $I(x)$ is called the neighborhood of x in the sense of approximation space (see, e.g., [33,44]).

The rough inclusion function $\nu_{\$} : P(U) \times P(U) \to [0,1]$ defines the degree of inclusion of X in Y, where $X, Y \subseteq U$.

In the simplest case it can be defined by (see, e.g., [33,44]):

$$\nu_{SRI}(X,Y) = \begin{cases} \frac{card(X \cap Y)}{card(X)}, & \text{if } X \neq \emptyset, \\ 1, & \text{if } X = \emptyset. \end{cases}$$

The lower and the upper approximations of subsets of U are defined as follows.

Definition 2. *For any approximation space* $AS_{\#,\$} = (U, I_\#, \nu_\$)$ *and any subset* $X \subseteq U$, *the lower and upper approximations are defined by*
$$LOW\left(AS_{\#,\$}, X\right) = \{x \in U : \nu_\$\left(I_\#\left(x\right), X\right) = 1\},$$
$$UPP\left(AS_{\#,\$}, X\right) = \{x \in U : \nu_\$\left(I_\#\left(x\right), X\right) > 0\}, \text{ respectively.}$$

The lower approximation of a set X with respect to the approximation space $AS_{\#,\$}$ is the set of all objects that can be classified with certainty as objects of X with respect to $AS_{\#,\$}$. The upper approximation of a set X with respect to the approximation space $AS_{\#,\$}$ is the set of all objects which can be possibly classified as objects of X with respect to $AS_{\#,\$}$.

Several known approaches to concept approximation can be presented using the language of approximation spaces (see, e.g., references in [44]). For more specific details regarding approximation spaces, the reader is referred to [6], [35], and [47].

The key to granular computing is the information granulation process that leads to the formation of information aggregates (with inherent patterns) from a set of available objects. A corresponding methodological and algorithmic issue is the formation of transparent (understandable) information granules, inasmuch as they should provide a clear and understandable description of patterns present in the sample of objects [2,36]. Such a fundamental property can be formalized by imposing a set of constraints that must be satisfied during the information granulation process. Usefulness of these constraints is measured by *quality* of the approximation space:

$$Quality_1 : Set\_AS \times P(U) \to [0,1],$$

where U is a non-empty set of objects and Set_AS is a set of possible approximation spaces with the universe U.

Example 1. If $UPP(AS, X)) \neq \emptyset$ for $AS \in Set\_AS$ and $X \subseteq U$ then

$$Quality_1(AS, X) = \nu_{SRI}(UPP(AS, X), LOW(AS, X)) = \frac{card(LOW(AS, X))}{card(UPP(AS, X))}.$$

The value $1 - Quality_1(AS, X)$ expresses the degree of completeness of our knowledge about X, given the approximation space AS.

Example 2. In applications, we frequently use yet another quality measure, analogous to the minimal length principle [40,49], which also takes into account description length of the approximation. Let us denote by $description(AS, X)$ the description length of approximation of X in AS. One way of calculating the description length of approximation is by taking the sum of description lengths of algorithms that are testing the membership for neighborhoods used in construction of the lower approximation, the upper approximation, and the boundary region of the set X. Then, the quality $Quality_2(AS, X)$ can be defined by

$$Quality_2(AS, X) = g(Quality_1(AS, X), description(AS, X)),$$

where g is a function used for fusion of $Quality_1(AS, X)$ with $description(AS, X)$ (their respective values). This function g shall be made relevant to the task. It can, for example, reflect weights given by experts relative to given criteria.

One can consider different optimization problems related to a given class Set_AS of approximation spaces. For example, for a given $X \subseteq U$ and a threshold $t \in [0, 1]$, one can search for an approximation space AS satisfying the constraint $Quality_2(AS, X) \geq t$.

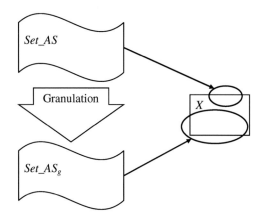

Fig. 1. Granulation of parameterized approximation space

Example 3. Another example involves searching for an approximation space which additionally satisfies the constraint $Cost(AS) < c$, where $Cost(AS)$ denotes the cost of approximation space AS (measured by the number of attributes used to define neighborhoods in AS) and c is a given threshold. In the this example we also consider costs of searching for relevant approximation spaces within a given family defined by a parameterized approximation space (see Figure 1). Since any parameterized approximation space $AS_{\#,\$} = (U, I_\#, \nu_\$)$ is a family of approximation spaces, the cost of searching in such a family for an approximation space relevant for a given approximation of concept (set) X, can be treated as a factor of the quality measure. Hence, the quality measure for approximation of X in $AS_{\#,\$}$ can be defined by

$$Quality_3(AS_{\#,\$}, X) = h(Quality_2(AS, X), Cost\_Search(AS_{\#,\$}, X)),$$

where AS is the result of searching in $AS_{\#,\$}$, $Cost\_Search(AS_{\#,\$}, X)$ is the cost of searching in $AS_{\#,\$}$ for AS, and h is a fusion function. For the purposes of this example we are assuming that the values of $Quality_2(AS, X)$ and $Cost\_Search(AS_{\#,\$}, X)$ are normalized to interval $[0, 1]$. Fusion operator h can be defined as a combination of $Quality_2(AS, X)$ and $Cost\_Search(AS_{\#,\$}, X)$ of the form

$$\lambda Quality_2(AS, X) + (1 - \lambda)Cost\_Search(AS_{\#,\$}, X),$$

where $0 \leq \lambda \leq 1$ is a weight expressing mutual importance of quality and cost in their fusion. Note, that we assumed that the fusion functions g, h in the definitions of quality are monotone w.r.t each argument.

Let $AS \in Set\_AS$ be an approximation space relevant for approximation of $X \subseteq U$, i.e., AS is the optimal (or semi-optimal), relative to measure $Quality_2$. By $Granulation(AS_{\#,\$})$ we denote a new parameterized approximation space obtained by granulation of $AS_{\#,\$}$. For example, $Granulation(AS_{\#,\$})$ can be obtained by reducing the number of attributes or alternating inclusion degrees (i.e., changing possible values of the inclusion function). Let AS' be an approximation space in $Granulation(AS_{\#,\$})$ obtained as a result of searching for optimal (semi-optimal) approximation space in $Granulation(AS_{\#,\$})$ for approximation of X.

Taking into account that parameterized approximation spaces are examples of parameterized granules, one can generalize the above example of parameterized approximation space granulation to the case of granulation of parameterized granule system.

2.3   Hierarchical Modeling of Granule Structures

Modeling relevant granules such as patterns, approximation spaces, clusters or classifiers starts from relational structures corresponding to their attributes. For any attribute (feature) a we consider a relational structure $\mathcal{R}_a = (V_a, \{r_i\}_{i \in I})$, where V_a is a set of values of the attribute a. Examples of such relational structures defined over the attribute-value set V_a are: $(V_a, =)$, $(V_a, \leq)$, where $\leq$ is a linear order on V_a, or $(V_a, \leq, +, \cdot, 0, 1)$, where $V_a = \mathbb{R}$ and $\mathbb{R}$ is the set of reals. By L_a we denote a set of formulas interpreted over $\mathcal{R}_a$ as subsets of V_a. It means that if $\alpha \in L_a$ then its semantics (an object corresponding to its meaning) $\|\alpha\|_{\mathcal{R}_a}$ is a subset of V_a. For example, one can consider an example of discretization of $\mathbb{R}$ by formulas $\alpha_1, \ldots, \alpha_k$ with interpretation over $\mathcal{R}_a = (\mathbb{R}, \leq, +, \cdot, 0, 1)$, where $\|\alpha_i\|_{\mathcal{R}_a}$ for $i = 1, \ldots, k$ create a partition of $\mathbb{R}$ into intervals.

If $\mathcal{A} = (U, A)$ is an information system and $a \in A$ then $\|\alpha\|_{\mathcal{R}_a}$ can be used to define semantics of α over $\mathcal{A}$ by assuming

$$\|\alpha\|_{\mathcal{A}} = \{x \in U : a(x) \in \|\alpha\|_{\mathcal{R}_a}\}.$$

Hence, any formula α can be treated as a new binary attribute of objects from U (see Figure 2).

If $\mathcal{A}^* = (U^*, A^*)$ is an extension of $\mathcal{A} = (U, A)$, i.e., $U \subseteq U^*$, $A^* = \{a^* : a \in A\}$, and $a^*(x) = a(x)$ for $x \in U$, then $\|\alpha\|_{\mathcal{A}} \subseteq \|\alpha\|_{\mathcal{A}}^*$.

In the next step of modeling, relational structures corresponding to attributes can be fused. Let us consider an illustrative example. We assume $\mathcal{R}_{a_i} = (V_{a_i}, r_{\mathcal{R}_{a_i}})$ are relational structures with binary relation $r_{\mathcal{R}_{a_i}}$ for $i = 1, \ldots, k$. Then, by $\mathcal{R}_{a_1} \times \ldots \times \mathcal{R}_{a_k}$ we denote their fusion defined by a relational structure over $(V_{a_1} \times \ldots V_{a_k})^2$ consisting of relation $r \subseteq (V_{a_1} \times \ldots V_{a_k})^2$ such that for any $(v_1, \ldots, v_k)$ and $(v'_1, \ldots, v'_k)$ drawn from $V_{a_1} \times \ldots \times V_{a_k}$ we have

$$(v_1, \ldots, v_k) r_{r_{\mathcal{R}_{a_1} \times \ldots \times \mathcal{R}_{a_k}}} (v'_1, \ldots, v'_k)$$

if and only if $v_i r_{\mathcal{R}_{a_i}} v'_i$ for $i = 1, \ldots, k$. One can extend this example by imposing some additional constraints. For example, if $V_{a_1} = \mathbb{R}$ then the constraints can be defined by a binary relation $r_\varepsilon \subseteq \mathbb{R}^2$ defined by $x \; r_\varepsilon \; y$ iff $|x - y| < \varepsilon$, where $\varepsilon \in (0, 1)$ is a threshold.

In the process of searching for (sub-)optimal approximation spaces, different strategies may be used. Let us consider an example of such strategy presented in [48]. In this example, $DT = (U, A, d)$ denotes a decision system (a given sample of data), where U is a set of objects, A is a set of attributes and d is a decision. We assume that for any object $x \in U$, only partial information equal to the A-signature of x (object signature, for short) is accessible, i.e., $Inf_A(x) = \{(a, a(x)) : a \in A\}$. Analogously, for any concept we are only given

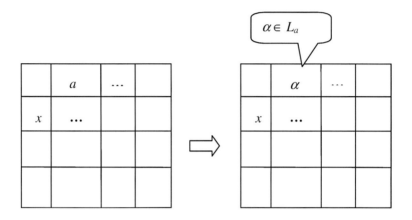

Fig. 2. New attribute defined by a formula α from L_a

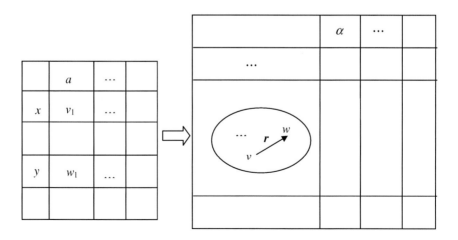

Fig. 3. Granulation to tolerance classes. r is a similarity (tolerance) relation defined over signatures of objects.

a partial information about this concept by means of a sample of objects, e.g., in the form of decision table. One can use object signatures as new objects in a new relational structure $\mathcal{R}$. In this relational structure $\mathcal{R}$ some relations between object signatures are also modeled, e.g., defined by the similarities of these object signatures (see Figure 3).

Discovery of relevant relations between object signatures is an important step in searching for relevant approximation spaces. In this way, a class of relational structures representing perception of objects and their parts is constructed. In the next step, we select a language $\mathcal{L}$ consisting of formulas expressing properties over the defined relational structures and we search for relevant formulas in $\mathcal{L}$. The semantics of formulas (e.g., with one free variable) from $\mathcal{L}$ are subsets of object signatures. Note, that each object signature defines a neighborhood of objects from a given sample (e.g., decision table DT) and another set on the whole universe of objects being an extension of U. Thus, each formula from $\mathcal{L}$ defines a family of sets of objects over the sample and also another family of sets over the universe of all objects. Such families can be used to define new neighborhoods for a new approximation space by, e.g., taking their unions. In the process of searching for relevant neighborhoods, we use information encoded in the available sample. More relevant neighborhoods make it possible to define more relevant approximation spaces (from the point of view of the optimization criterion). Following this scheme, the next level of granulation may be related to clusters of objects (relational structures) for a current level (see Figure 4).

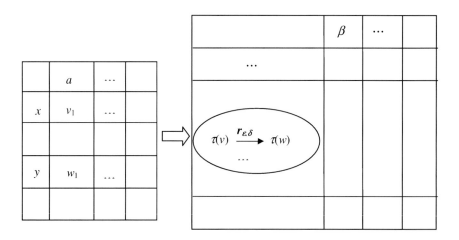

Fig. 4. Granulation of tolerance relational structures to clusters of such structures. $r_{\varepsilon,\delta}$ is a relation with parameters ε, δ on similarity (tolerance) classes.

In Figure 4 τ denotes a similarity (tolerance) relation on vectors of attribute values, $\tau(v) = \{u : v \tau u\}$, $\tau(v) \ r_{\varepsilon,\delta} \ \tau(w)$ iff $dist(\tau(v), \tau(w)) \in [\varepsilon - \delta, \varepsilon + \delta]$, and $dist(\tau(v), \tau(w)) = inf\{dist(v', w') : (v', w') \in \tau(v) \times \tau(w)\}$ where $dist$ is a distance function on vectors of attribute values.

One more example is illustrated in Figure 5, where the next level of hierarchical modeling is created by defining an information system in which objects are time windows and attributes are (time-related) properties of these windows.

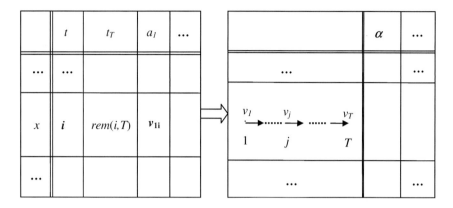

Fig. 5. Granulation of time points into time windows. T is the time window length, $v_j = (v_{1j}, \ldots, v_{Tj})$ for $j = 1, \ldots, T$, $rem(i, T)$ is the remainder from division of i by T, α is an attribute defined over time windows.

It is worth mentioning that quite often this searching process is even more sophisticated. For example, one can discover several relational structures (e.g., corresponding to different attributes) and formulas over such structures defining different families of neighborhoods from the original approximation space. As a next step, such families of neighborhoods can be merged into neighborhoods in a new, higher degree approximation space.

The proposed approach is making it possible to construct information systems (or decision tables) on a given level of hierarchical modeling from information systems from lower level(s) by using some constraints in joining objects from underlying information systems. In this way, structural objects can be modeled and their properties can be expressed in constructed information systems by selecting relevant attributes. These attributes are defined with use of a language that makes use of attributes of systems from the lower hierarchical level as well as relations used to define constraints. In some sense, the objects on the next level of hierarchical modeling are defined using the syntax from the lover level of the hierarchy. Domain knowledge is used to aid the discovery of relevant attributes (features) on each level of hierarchy. This domain knowledge can be provided, e.g., by concept ontology together with samples of objects illustrating concepts from this ontology. Such knowledge is making it feasible to search for relevant attributes (features) on different levels of hierarchical modeling (see Section 3). In Figure 6 we symbolically illustrate the transfer of knowledge in a particular application. It is a depiction of how the knowledge about outliers in handwritten digit recognition is transferred from expert to a software system. We call this process *knowledge elicitation*. Observe, that the explanations given by expert(s) are expressed using a subset of natural language limited by using concepts from provided ontology only. Concepts from higher levels of ontology are gradually approximated by the system from concepts on lower levels.

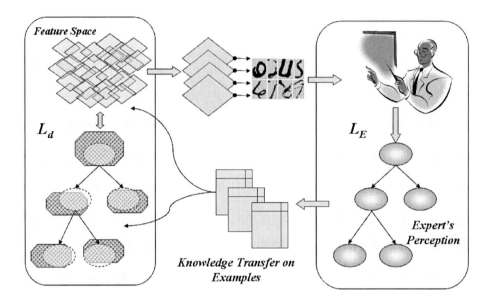

Fig. 6. Expert's knowledge elicitation

This kind of approach is typical for hierarchical modeling [3]. This is, in particular, the case when we search for a relevant approximation space for objects composed from parts for which some approximation spaces, relevant to components, have already been found. We find that hierarchical modeling is required for approximation of complex vague concepts, as in [27,37].

3   Ontology Approximation in RGC

Approximation of complex, possibly vague concepts requires a hierarchical modeling and approximation of more elementary concepts on subsequent levels in the hierarchy along with utilization of domain knowledge. Due to the complexity of these concepts and processes on top levels in the hierarchy one can not assume that fully automatic construction of their models, or the discovery of data patterns required to approximate their components, would be straightforward. We propose to include in this process the discovery of approximations of complex vague concepts, performed interactively with co-operation of domain experts. Such interaction allows for more precise control over the complexity of discovery process, therefore making it computationally more feasible. Thus, the proposed approach transforms a typical data mining system into an equivalent of experimental laboratory in which the software system, aided by human experts, attempts to discover: (i) approximation of complex vague concepts from data under some domain constraints, (ii) patterns relevant to user (researcher), e.g., required in the approximation of vague components of complex concepts.

The research direction aiming at interactive knowledge construction has been pursued by our team, in particular, toward the construction of classifiers for complex concepts (see, e.g., [3,4,7,8,9] and also [11,22,25,26,28]) aided by domain knowledge integration. Advances in recent years indicate a possible expansion of the research conducted so far into discovery of models for processes involving complex objects from temporal or spatio-temporal data.

The novelty of the proposed RGC approach for the discovery of approximations of complex concepts from data and domain knowledge lies in combining, on one side, a number of novel methods of granular computing developed using the rough set methods and other known approaches to the approximation of vague, complex concepts (see, e.g., [3,4,6,7,8,9], [18,22,25,26,28,32,33,35,36,55,56]) with, on the other side, the discovery of structures from data through an interactive collaboration with domain experts (see, e.g., [3,4,6,7,8,9],[18,22,25,26,28,36]). The developed methodology based on RGC was applied, to various extent, in real-life projects including: unmanned area vehicle control, robotics, prediction of risk patterns from temporal medical and financial data, sunspot classification, and bioinformatics. For technical details please refer to [3,4,6,7,8,9] and [18,22,25,26,28,36]).

4   Toward RGC for Process Mining

The rapid expansion of the Internet has resulted not only in the ever growing amount of data therein stored, but also in the burgeoning complexity of the concepts and phenomena pertaining to those data. This issue has been vividly compared in [13] to the advances in human mobility from the period of walking afoot to the era of jet travel. These essential changes in data have brought new challenges to the development of new data mining methods, especially that the treatment of these data increasingly involves complex processes that elude classic modeling paradigms. Types of datasets currently regarded "hot", like biomedical, financial or net user behavior data are just a few examples. Mining such temporal or complex data streams is on the agenda of many research centers and companies worldwide (see, e.g., [1,41]). In the data mining community, there is a rapidly growing interest in developing methods for process mining, e.g., for discovery of structures of temporal processes from observations (recorded data). Works on process mining, e.g., [10,20,52,54] have recently been undertaken by many renowned centers worldwide[1]. This research is also related to functional data analysis (cf. [39]), cognitive networks (cf. [30]), and dynamical system modeling in biology (cf. [12]).

---

[1] http://www.isle.org/~langley/,
   http://soc.web.cse.unsw.edu.au/bibliography/discovery/index.html

In [23,24] we outlined an approach to discovery of processes from data and domain knowledge which is based on RGC philosophy. In the following section 5 we discuss some issues related to granule interactions in process mining that draw on these previous publications.

5   Interaction of Granules in RGC

Interactions between granules are rudimentary for understanding the nature of interactive computations [14]. In the RGC framework it is possible to model interactive computations performed on granules of different complexity aiming at construction of approximations of complex vague concepts. Approximations of such concepts are capable of adaptive adjustment with the changes of underlying data and domain knowledge. Hence, the decision making algorithm based on the approximation of such vague concepts is also adaptively changing. This is somewhat contrary to a statement from [53]: "algorithms [...] are metaphorically dump and blind because they cannot adapt interactively while they compute".

Models of interaction can be discrete or continuous, static (i.e., restricted to input-output relations) or dynamic (i.e., defining computations representing interactions). An example of static interaction is the one between a given information system IS and a value vector v of attributes returning the message YES iff there exists in IS an object with the attribute-value vector v. Continuous model of dynamic interaction can be, e.g., defined by relevant differential equation which models a (part of) real-life dynamical system.

In this section, we discuss some examples of interactions of granules showing the richness and complexity of granule interactions which should be modeled in RGC. The first example is related to discovery of concurrent systems from information systems.

Back in 1992, Zdzisław Pawlak (cf. [34]) proposed to use data tables (information systems) as specifications of concurrent systems. In this approach any information system can be considered as a representation of a (traditional) concurrent system: attributes are interpreted as local processes of the concurrent system, values of attributes – as states of these local processes, and objects – as global states of the considered system. Several methods for synthesis of concurrent systems from data have been developed (see, e.g., [29,42,43,51]). These methods are based on the following steps. First, for a given information system S we generate its (formal) theory $Th(S)$ consisting of a set of selected rules over descriptors defined by this system. These rules describe the coexistence constraints of local states within global states specified by S. Next, we define a maximal extension $Ext(S)$ of S consisting of all objects having descriptions consistent with all rules in $Th(S)$. Finally, a Petri net with the set of reachable markings equal to $Ext(S)$ is generated. There have been also developed methods for synthesis of Petri nets from information systems based on decomposition of information systems into the so called components defined by reducts. This approach is making it possible to represent a given information system by a set of interacting local processes defined by some functional dependencies extracted from data. Interactions between local processes are represented by rules over descriptors extracted from data too. It is worth mentioning that the ability to produce from an information system a structure that is essentially (is similar to) a Petri net brings significant profits. Petri nets and similar structures have been studied for decades, and nowadays we have quite potent collection of tools that make use of these notions, at our disposal.

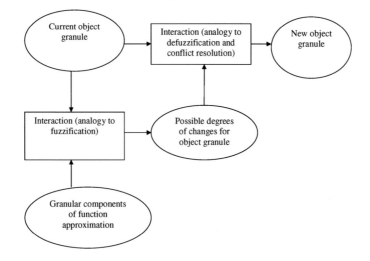

Fig. 7. Interaction of currently considered granule with the granule represented by approximation of function describing changes of states

Our second example is related to learning of state changes for agents interacting with dynamically changing environments. One possible approach can be analogous to modeling by differential equations. However, instead of assuming the definition of the functions describing these changes we propose to approximate these functions from experimental data using domain knowledge [23,24]. Once approximation of functions describing changes is done, we couple it with descriptions of indiscernibility (similarity) classes in which the current state is included in order to identify indiscernibility (similarity) classes for the next state(s). This requires some special interaction of granule representing uncertain information about the current state and the granule represented by approximation of functions describing changes between consecutive states. This interaction is illustrated in Figure 7. First, the granule of object is interacting with components of function approximation. This step is, in some sense, analogous to fuzzification in fuzzy control. In the case of rule based classifier, this step involves search for inclusion degrees of object granule and patterns represented by the left hand sides (antecendents) of rules. This may be perceived as matching membership degrees in fuzzy controller. Finally, the results of the interaction are fused to form a granule representing the next state. Again, this step is analogous to defuzzification in fuzzy controller. In the case of rule based classifier, this step is based on the conflict resolution strategy or voting strategy making it possible to select or construct the final decision granule in presence of possibly contradictory, partially matching rules.

We perceive the idea described above as very important direction for further research on methods for discovery of process trajectory approximation from data and domain knowledge.

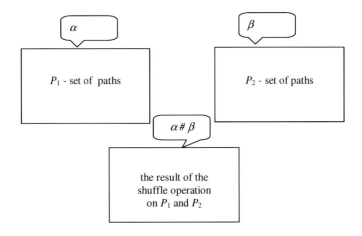

Fig. 8. Interaction of coexisting granules representing processes (sets of paths) P_1 and P_2 creates a new process represented by the result of shuffle operation on P_1 and P_2. α and β are satisfying the co-existing constraints.

The next example is an extension of the one from section 2.2, with the information system in which the objects are represented by time windows and attributes describe properties of these time windows (see also Figure 5). Assuming that some relations in time are preserved between time windows (such as *after*, *before*), one can construct (on the next level of modeling) an information system with objects representing paths ("trajectories") of time windows and attributes corresponding to properties of such paths. Please, observe that indiscernibility or similarity classes of such information system are sets of paths, identified by the vector of attribute values (or description of similarity class). Hence, each such class can be treated as the semantics of a process represented by the description of the corresponding indiscernibility (similarity) class. Such processes can further interact. Constraints for vector of attribute values describing the processes are making it possible to select processes which can, in a sense, co-exist. The process resulting from such co-existence may be described as the result(s) of *shuffling* operation, well known from formal language theory, performed on sets of paths treated as (syntactical) description of these two processes (see Figure 8). In this way, we construct objects (their descriptions) for new information system on the next level of hierarchical modeling. Attributes derived for such objects describe properties of newly constructed process(es). In this particular case, the interaction is only realized on the level of selecting processes by applying constraints on attribute value vectors.

More advanced interaction of processes may occur if we consider the situation when each path in a given process is represented by a vector of attribute values. Such a situation may occur when, for instance, paths from the lower level undergo clustering. Then, some additional constraints can be related to paths of the resulting process constructed from paths of interacting, lower-level processes. They may represent results of synchronization of two or more processes. For example, in any path of the process obtained as a result of interaction between two lower-level processes states with a certain distinguished property should separate (appear in-between) states with another specific property.

It should be noted that in practical approaches to modeling it is often necessary to use relevant names (labels) for the constructed processes, tantamount to their position and rôle in concept hierarchy (or corresponding ontology). To answer to this requirement one may use methods of inducing, e.g., Petri nets from examples of paths (see, e.g., [20]).

Another way of looking at modeling of interactions is by employing the agent-oriented framework. The depiction of agents' interactions with environment(s) is essentially based on observation, that each agent perceives only a partial (and possibly vague) information about environment. On the basis of the perceived information and its own state the agent derives (creates) some granules, with the goal of changing the state of environment to its favor. These granules are involved in interactions with the environment and granules originating in other agents. Using either competitive or cooperative strategies (coalitions of) agents involved in interactions form a resulting action which changes the environment(s) in a way that is in some accordance with components (agent-specific granules). The approaches that use elements of such interactive agent co-operation are nowadays popular in multiagent systems [19,50].

In the following, final example we describe an application of domain knowledge in modeling of interactions. We use sentences from (a limited subset of) the natural language coupled with so called *behavioral graphs* (cf. [5]) to define relationships (interactions) that occur between parts of a complex object. In this example we show such description for the task of recognizing whether at a given moment the observed road situation leads to imminent danger or not. The modeling of the system that ultimately is capable of recognizing the extremely compound concept of *dangerous situation* on the basis of low-level measurements, is indeed hierarchical. In the Figure 9 we present a behavioral graph for a single object-vehicle on a road. This behavioral graph appears in between the lowest level (sensor measurements) and the highest level (dangerous situation) in the hierarchy of concepts.

A composition of behavioral graphs, appearing on lower level in the hierarchy, can be used to represent behavior (and interaction) of a more compound part consisting of, e.g., two vehicles involved in the maneuver of overtaking (see Figure 10). Please note, that the concept of *overtaking* is built of components which at some point were also approximated from the lower level concepts. This is a case of, e.g., *changing lane* or *A passing B* (refer to Figure 10).

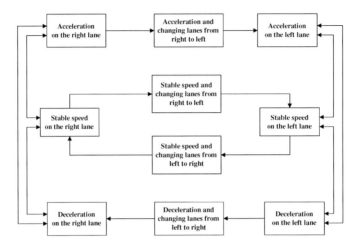

Fig. 9. A behavioral graph for a single object-vehicle

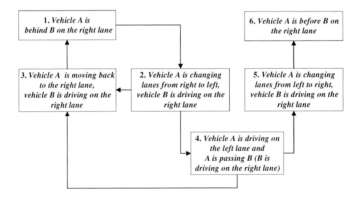

Fig. 10. A behavioral graph for the maneuver of overtaking

The identification of the behavioral pattern of a complex object on the basis of sensory data cannot go forward without (approximation of) ontology of concepts. It is this ontology that makes it possible to link the low level measurements (sensory concepts) with the high level description of behavioral patterns [3,4,6,7,9,18,36]. By means of this ontology we establish that – following our road example – in order to know what the *overtaking* is, one has to define a concept of *A passing B*, as well as link both *A* and *B* to an object-vehicle structure (see Figure 9).

6   Conclusions

We discussed some issues in modeling of interactive computations over granules. All these issues are closely related to research directions within the *Wisdom Technology* (WisTech) research programme, as outlined recently in [16,17,18].

Wisdom is commonly understood as the ability of *rightly judging* based on available knowledge and interactions. This common notion can be refined. By *wisdom*, we understand an adaptive ability to make judgments correctly to a satisfactory degree (in particular, making satisfactory decisions), having in mind real-life constraints. The intuitive nature of wisdom understood in this way can be metaphorically expressed by the so-called *wisdom equation* as shown in (1).

$$wisdom = adaptive\ judgment\ +\ knowledge\ +\ interaction \qquad (1)$$

Wisdom could be treated as a certain kind of knowledge. Especially, this type of knowledge is important at the highest level of hierarchy of meta-reasoning in intelligent agents.

WisTech is a collection of techniques aimed at further advancement of technologies that make it possible to acquire, represent, store, process, discover, communicate, and learn *wisdom* in designing and implementing intelligent systems. These techniques include approximate reasoning by agents (or agent teams) about vague concepts describing real-life, dynamically changing, and (usually) distributed systems in which these agents are operating. Such systems consist of several autonomous agents operating in highly unpredictable environments and interacting with each other. WisTech can be treated as the unifying successor of database technology, information management, and knowledge engineering techniques. In this sense, WisTech is the combination of the component technologies, as represented in equation (1), and offers an intuitive starting point for a variety of approaches to designing and implementing computational models for intelligent systems.

There are many ways to build computational models that are based on WisTech philosophy. In this paper we have outlined just one of them, which based on the Rough Granular Computing approach.

Acknowledgments. This research has been supported by the grant N N516 368334 from Ministry of Science and Higher Education of the Republic of Poland.

References

1. Aggarwal, C. (ed.): Data Streams: Models and Algorithms. Springer, Berlin (2007)
2. Bargiela, A., Pedrycz, W.: Granular Computing: An Introduction. Kluwer Academic Publishers, Dordrecht (2003)
3. Bazan, J., Peters, J.F., Skowron, A.: Behavioral pattern identification through rough set modelling. In: Ślęzak, D., Yao, J., Peters, J.F., Ziarko, W.P., Hu, X. (eds.) RSFDGrC 2005. LNCS (LNAI), vol. 3642, pp. 688–697. Springer, Heidelberg (2005)
4. Bazan, J., Skowron, A.: On-line elimination of non-relevant parts of complex objects in behavioral pattern identification. In: Pal, S.K., Bandyopadhyay, S., Biswas, S. (eds.) PReMI 2005. LNCS, vol. 3776, pp. 720–725. Springer, Heidelberg (2005)
5. Bazan, J.G., Peters, J.F., Skowron, A.: Behavioral pattern identification through rough set modelling. In: Ślęzak, D., Yao, J., Peters, J.F., Ziarko, W.P., Hu, X. (eds.) RSFDGrC 2005. LNCS (LNAI), vol. 3642, pp. 688–697. Springer, Heidelberg (2005)

6. Bazan, J., Skowron, A., Swiniarski, R.: Rough sets and vague concept approxima-
   tion: From sample approximation to adaptive learning. In: Peters, J.F., Skowron,
   A. (eds.) Transactions on Rough Sets V. LNCS, vol. 4100, pp. 39–62. Springer,
   Heidelberg (2006)

7. Bazan, J., Kruczek, P., Bazan-Socha, S., Skowron, A., Pietrzyk, J.J.: Risk pattern
   identification in the treatment of infants with respiratory failure through rough set
   modeling. In: Proceedings of IPMU 2006, Paris, France, July 2-7, pp. 2650–2657.
   Éditions E.D.K, Paris (2006)

8. Bazan, J., Kruczek, P., Bazan-Socha, S., Skowron, A., Pietrzyk, J.J.: Automatic
   planning of treatment of infants with respiratory failure through rough set modeling. In: Greco, S., Hata, Y., Hirano, S., Inuiguchi, M., Miyamoto, S., Nguyen, H.S.,
   Słowiński, R. (eds.) RSCTC 2006. LNCS (LNAI), vol. 4259, pp. 418–427. Springer,
   Heidelberg (2006)

9. Bazan, J.: Rough sets and granular computing in behavioral pattern identification
   and planning. In: Pedrycz, W. et al [36], pp. 777–799 (2008)

10. Borrett, S.R., Bridewell, W., Langely, P., Arrigo, K.R.: A method for representing
    and developing process models. Ecological Complexity 4(1-2), 1–12 (2007)

11. Doherty, P., Łukaszewicz, W., Skowron, A., Szałas, A.: Knowledge Representation
    Techniques: A Rough Set Approach. Studies in Fuzziness and Soft Computing,
    vol. 202. Springer, Heidelberg (2006)

12. Feng, J., Jost, J., Minping, Q. (eds.): Network: From Biology to Theory. Springer,
    Berlin (2007)

13. Friedman, J.H.: Data mining and statistics. What's the connection? Keynote Ad-
    dress. In: Proceedings of the 29th Symposium on the Interface: Computing Science
    and Statistics, Houston, Texas (May 1997)

14. Goldin, D., Smolka, S., Wegner, P.: Interactive Computation: The New Paradigm.
    Springer, Heidelberg (2006)

15. Jankowski, A., Peters, J., Skowron, A., Stepaniuk, J.: Optimization in discovery of
    compound granules. Fundamenta Informaticae 85(1-4), 249–265 (2008)

16. Jankowski, A., Skowron, A.: A WisTech paradigm for intelligent systems. In: Peters,
    J.F., Skowron, A., Düntsch, I., Grzymała-Busse, J.W., Orłowska, E., Polkowski, L.
    (eds.) Transactions on Rough Sets VI. LNCS, vol. 4374, pp. 94–132. Springer,
    Heidelberg (2007)

17. Jankowski, A., Skowron, A.: Logic for artificial intelligence: The Rasiowa - Pawlak
    school perspective. In: Ehrenfeucht, A., Marek, V., Srebrny, M. (eds.) Andrzej
    Mostowski and Foundational Studies, pp. 106–143. IOS Press, Amsterdam (2007)

18. Jankowski, A., Skowron, A.: Wisdom Granular Computing. In: Pedrycz, W. et al
    [36], pp. 229–345 (2008)

19. Luck, M., McBurney, P., Preist, C.: Agent Technology. Enabling Next Generation
    Computing: A Roadmap for Agent Based Computing. AgentLink (2003)

20. de Medeiros, A.K.A., Weijters, A.J.M.M., van der Aalst, W.M.P.: Genetic process
    mining: An experimental evaluation. Data Mining and Knowledge Discovery 14,
    245–304 (2007)

21. Mitchell, M.: Complex systems: Network thinking. Artificial Intelligence 170(18),
    1194–1212 (2006)

22. Nguyen, H.S., Bazan, J., Skowron, A., Nguyen, S.H.: Layered learning for con-
    cept synthesis. In: Peters, J.F., Skowron, A., Grzymała-Busse, J.W., Kostek, B.z.,
    Świniarski, R.W., Szczuka, M.S. (eds.) Transactions on Rough Sets I. LNCS,
    vol. 3100, pp. 187–208. Springer, Heidelberg (2004)

23. Nguyen, H.S., Skowron, A.: A rough granular computing in discovery of process models from data and domain knowledge. Journal of Chongqing University of Post and Telecommunications 20(3), 341–347 (2008)

24. Nguyen, H.S., Jankowski, A., Skowron, A., Stepaniuk, J., Szczuka, M.: Discovery of process models from data and domain knowledge: A rough-granular approach. In: Yao, J.T. (ed.) Novel Developments in Granular Computing: Applications for Advanced Human Reasoning and Soft Computation. IGI Global, Hershey (submitted, 2009)

25. Nguyen, T.T.: Eliciting domain knowledge in handwritten digit recognition. In: Pal, S.K., Bandyopadhyay, S., Biswas, S. (eds.) PReMI 2005. LNCS, vol. 3776, pp. 762–767. Springer, Heidelberg (2005)

26. Nguyen, T.T.: Outlier and exception analysis in rough sets and granular computing. In: Pedrycz, W. et al [36], pp. 823–834 (2008)

27. Nguyen, T.T., Skowron, A.: Rough-granular computing in human-centric information processing. In: Bargiela, A., Pedrycz, W. (eds.) Human-Centric Information Processing Through Granular Modelling. Studies in Computational Intelligence. Springer, Heidelberg (in press, 2009)

28. Nguyen, T.T., Willis, C.P., Paddon, D.J., Nguyen, S.H., Nguyen, H.S.: Learning Sunspot Classification. Fundamenta Informaticae 72(1-3), 295–309 (2006)

29. Pancerz, K., Suraj, Z.: Discovering concurrent models from data tables with the ROSECON. Fundamenta Informaticae 60(1-4), 251–268 (2004)

30. Papageorgiou, E.I., Stylios, C.D.: Fuzzy Cognitive Maps. In: Pedrycz, W., et al [36], pp. 755–774 (2008)

31. Pal, S.K., Polkowski, L., Skowron, A. (eds.): Rough-Neural Computing: Techniques for Computing with Words. Cognitive Technologies. Springer, Berlin (2004)

32. Pawlak, Z.: Rough sets. International Journal of Computer and Information Sciences 11, 341–356 (1982)

33. Pawlak, Z.: Rough Sets: Theoretical Aspects of Reasoning about Data. System Theory. In: Knowledge Engineering and Problem Solving, vol. 9. Kluwer Academic Publishers, Dordrecht (1991)

34. Pawlak, Z.: Concurrent versus sequential: the rough sets perspective. Bulletin of the EATCS 48, 178–190 (1992)

35. Pawlak, Z., Skowron, A.: Rudiments of rough sets. Information Sciences 177(1), 3–27 (2007); Rough sets: Some extensions. Information Sciences 177(1), 28–40; Rough sets and boolean reasoning. Information Sciences 177(1), 41–73

36. Pedrycz, W., Skowron, A., Kreinovich, V. (eds.): Handbook of Granular Computing. John Wiley & Sons, New York (2008)

37. Poggio, T., Smale, S.: The mathematics of learning: Dealing with data. Notices of the AMS 50(5), 537–544 (2003)

38. Polkowski, L., Skowron, A.: Rough mereology: A new paradigm for approximate reasoning. International Journal of Approximate Reasoning 51, 333–365 (1996)

39. Ramsay, J.O., Silverman, B.W.: Applied Functional Data Analysis. Springer, Berlin (2002)

40. Rissanen, J.: Minimum-description-length principle. In: Kotz, S., Johnson, N. (eds.) Encyclopedia of Statistical Sciences, pp. 523–527. John Wiley & Sons, New York (1985)

41. Roddick, J.F., Hornsby, K., Spiliopoulou, M.: An updated bibliography of temporal, spatial and spatio- temporal data mining research. In: Roddick, J., Hornsby, K.S. (eds.) TSDM 2000. LNCS (LNAI), vol. 2007, pp. 147–163. Springer, Heidelberg (2001)

42. Skowron, A., Suraj, Z.: Rough Sets and Concurrency. Bulletin of the Polish Academy of Sciences 41, 237–254 (1993)
43. Skowron, A., Suraj, Z.: Discovery of Concurrent Data Models from Experimental Tables: A Rough Set Approach. In: First International Conference on Knowledge Discovery and Data Mining, pp. 288–293. AAAI Press, Menlo Park (1995)
44. Skowron, A., Stepaniuk, J.: Tolerance approximation spaces. Fundamenta Informaticae 27, 245–253 (1996)
45. Skowron, A., Stepaniuk, J.: Information granules and rough-neural computing. In: Pal, S.K., Polkowski, L., Skowron, A. (eds.) Rough-Neural Computing: Techniques for Computing with Words. Cognitive Technologies, pp. 43–84. Springer, Heidelberg (2003)
46. Skowron, A., Stepaniuk, J.: Rough sets and granular computing: Toward rough-granular computing. In: Pedrycz, et al.[36], pp. 425–448 (2008)
47. Skowron, A., Stepaniuk, J., Peters, J., Swiniarski, R.: Calculi of approximation spaces. Fundamenta Informaticae 72(1-3), 363–378 (2006)
48. Skowron, A., Synak, P.: Complex patterns. Fundamenta Informaticae 60(1-4), 351–366 (2004)
49. Ślęzak, D.: Approximate entropy reducts. Fundamenta Informaticae 53(3-4), 365–387 (2002)
50. Sun, R. (ed.): Cognition and Multi-Agent Interaction. From Cognitive Modeling to Social Simulation. Cambridge University Press, New York (2006)
51. Suraj, Z.: Rough set methods for the synthesis and analysis of concurrent processes. In: Polkowski, L., Lin, T.Y., Tsumoto, S. (eds.) Rough Set Methods and Applications: New Developments in Knowledge Discovery in Information Systems. Studies in Fuzziness and Soft Computing, vol. 56, pp. 379–488. Springer-Verlag/Physica-Verlag, Heidelberg (2000)
52. Unnikrishnan, K.P., Ramakrishnan, N., Sastry, P.S., Uthurusamy, R.: Network Reconstruction from Dynamic Data. SIGKDD Explorations 8(2), 90–91 (2006)
53. Wegner, P.: Why interaction is more powerful than algorithms. Communications of the ACM 40(5), 80–91 (1997)
54. Wu, F.-X.: Inference of gene regulatory networks and its validation. Current Bioinformatics 2(2), 139–144 (2007)
55. Zadeh, L.A.: A new direction in AI-toward a computational theory of perceptions. AI Magazine 22(1), 73–84 (2001)
56. Zadeh, L.A.: Generalized theory of uncertainty (GTU)-principal concepts and ideas. Computational Statistics and Data Analysis 51, 15–46 (2006)

Data Privacy: From Technology to Economics

Stan Matwin[1,2] and Tomasz Szapiro[3]

[1] School of Information Technology and Engineering,
University of Ottawa, Ottawa, Canada
[2] Institute for Computer Science, Polish Academy of Sciences, Warsaw
[3] Division of Decision Analysis and Support, Warsaw School of Economics,
Warsaw, Poland

Abstract. In this paper, we attempt to relate two different approaches to data privacy – the technological approach, embodied in the current Privacy-preserving Data Mining work, and the economic regulations approach. Our main thesis is that none of these two approaches alone will be able to address the important, growing data privacy issues. We advocate a hybrid system, combining both approaches in a complementary manner. We present a view of privacy in the context of an accepted taxonomy of economic goods, stating the question: if privacy is exchanged and traded, then what kind of good is it? We also show that, viewed in the light of an established economic growth theory, involving privacy in the growth process leads to a steady state growth.

1 Introduction

Privacy-preserving Data Mining (PPDM) has produced a steady stream of advanced solutions addressing different aspects of privacy in the context of data mining. However, the PPDM field is still in its early days – for instance, there is no agreement about the meaning of privacy. The various proposed definitions are related to the data itself, and by and large ignore psychological and social aspects of privacy, see Spiekerman *et al.* (2001), Berendt and Teltzrow (2005) for some exceptions to the former). It has been observed by Clifton (2005) that interactions between people and organizations are an inherent aspect of privacy. Therefore, it seems interesting to turn to economics for inspiration and better understanding of privacy and for creative policies. Results of Varian (1996), Acquisiti (2004), Rasi (2004), as well as Kargupta *et al.* (2007) encourage us to take this view.

The Inaugural Workshop on the Economics of Information Security at the University of California-Berkeley in 2002 provided the first insight into the problems of economics of privacy as distinct from security and of the economics of digital rights management[1]. We continue this trend, and we argue in this paper that privacy preserving techniques, as we present them here, cannot alone resolve some of the consequences of loss of control of privacy. We introduce the view of privacy as an economic good and discuss impact of organization of market for privacy on economy. We show that purely administrative regulation of privacy market as well as the

---

[1] The State of Economics of Information Security, A Journal of Law and Policy in the Information Society, Volume 2, Number 2.

J. Koronacki et al. (Eds.): Advances in Machine Learning II, SCI 263, pp. 43–74.
springerlink.com   © Springer-Verlag Berlin Heidelberg 2010

regulation-free (privacy) market lead to the ineffectiveness, inefficiency, and waste: the level of goal achievement is lower than it could have been, given level of engaged resources. Our claim is that market for privacy requires a complex, hybrid solution. Such solution, however, may appear too expensive to be economically rational. Thus we ponder chances of emergence of economic growth (increase of production due to the increase of privacy) - as a potential source of funding for the regulatory changes.

Economic growth considered as a process was thoroughly investigated by economists (see e.g. Lucas (1998). The main goal of that research is focused on identification and explanation of differences in economic growth in different economic systems. That research identifies different factors influencing the growth, models them using mathematical techniques[2] and verifies the models on empirical data. The list of growth factors involves rate of investment, capital, level of productivity, innovations, and technological progress. More recently, human[3] and social capital[4] are investigated also as growth factors. The human capital is usually measured by length of life, numerical characteristics of educational systems (direct or complex, e.g. HDI - Human Development Index) etc., while the social indicators include criminality, participatory democracy and involvement in political life, see e.g. Putnam (2001).

Such a variety of growth factors calls for a framework capable to integrate different types of variables. This problem has been noticed in economics, which aspires to develop a general understanding of economic processes, and to go beyond simply providing intuitive explanations of specific situations. Such frameworks allow debating macroeconomics policies in a logical, reasoned manner. E.g. Robert Solow (1988) some thirty years after his seminal 1956 paper said "... *there was* [in the 1956 paper] *a paragraph that I am proud of: it made the point that growth theory provides a framework within which one can seriously discuss macroeconomic policies that not only achieve and maintain full employment but also make a deliberate choice between current consumption and current investment, and therefore between current consumption and future consumption...".*

We accept this view when adopting Solow's approach to investigate privacy as a factor of growth. In order to be able to do this, however, we have to position the concept of privacy (more exactly, data privacy as we define it below) with respect to such basic notions of economics as goods and markets. We therefore first describe privacy as an economic good and discuss possible regulations of the market for privacy.

Our main idea can be summarized as follows. As mentioned earlier, let us view privacy as a good in the economic sense (i.e. non-material entity consumed to satisfy human needs). We consume privacy by revealing information about ourselves to an interested party for a consideration (this consideration may be either monetary, or informational, e.g. a recommendations/advice). Abundant examples are provided by practices of e-commerce. They are generally consistent with the prevailing non-technical definition by Alan Westin (1967): privacy is the ability of individuals "...*to determine for themselves when, how, and to what extent information about them is communicated to others*".

---

[2] The most important classical growth models were presented by Ramsey (1928), Harrod (1939ab) and Domar (1946), and Robert Solow (1956). To get deeper insight, see Acemoglu (2009).
[3] See e.g. Becker (1964) and Becker and Murphy (1990).
[4] See e.g. Routledge and von Amsberg (2002).

Transactions involving privacy are price-free[5], in the sense that none of the parties involved in a privacy transaction – the person giving away her information and the organization acquiring it – uses the price of the information exchanged. In particular, the owner (provider, or donor) of privacy is in no position to valuate (price) her information. The acquirer is at least able to assess the value of aggregated information, obtained from many sellers. This makes the transaction asymmetric from informational point of view. For a privacy transaction to take place, the perceived benefit to the seller (provider of information) must exceed the value of privacy he is giving up. Likewise, the expected value of the transaction to the buyer (organization) must exceed the value of the consideration given.

The classic economics view includes two main kinds of goods (and two less frequent kinds as well, see below). *Private goods* are normally bought from some finite pool of goods, thereby decreasing this pool (subtractability). Access to goods is limited to those paying for them (excludability). Public goods are normally available to all, and do not diminish by the fact of being consumed by one person (e.g. art in a free public museum). Seeing privacy as a good raises an immediate taxonomical question: what kind of good is it?

It is known that free market of private goods sometimes proves to be not efficient, due to, among others, asymmetry of information, and requires some kind of intervention. This is also the case in price-free privacy transactions. Also market of public goods, not always efficient (e.g. due to market externalities), calls for interventions. Externalities often plague privacy transactions, as the received information is divulged to third parties. As observed by James Moor, "*Personal information about us is well greased and slides rapidly through computer systems around the world, around the clock*", see Moor (1997).

Privacy has characteristics of both a private good (one can refuse to "sell" it by not providing one's data) and public good (consumption by one consumer does not preclude other members of society from consuming them to some extent). See a more detailed discussion of privacy as an economic good in sec. 3.1.

Therefore, for at least two reasons, privacy market calls for regulatory intervention. While such intervention is usually understood as intervention of the state, it does not necessarily need to be one. For instance, it can take form of a technological framework, such as the PPDM. We claim that sound PPDM has to incorporate market-related views of exchange and communication.

Success of a systemic change, such as a regulatory intervention, depends on ability to formulate a social contract and agree to it. Regulatory interventions are not cost free, as they involve the cost of change and new current costs of functioning of a changed system. Thus agreement about such a contract strongly depends on ability to finance the change. There are two sources of such financing – firstly, it can be covered from the budget or regulatory institution and secondly, it can be covered from future increase of budget achieved as result of increased performance of the changed system. The first possibility requires restructuring of the budgeting and thus it destabilizes the current political balance. Achievement of agreement is therefore very

---

[5] Acquisti and Grossklags (2005) noticed that "*...First, individuals are willing to trade privacy for convenience or bargain the release of personal information in exchange for relatively small rewards. Second, individuals are seldom willing to adopt privacy protective technologies...*

difficult as it involves losses for some stakeholders. Thus we arrive at the second option – expectations of economic growth resulting from the change.

This argument leads to the question about influence of trading privacy as a good on the economy overall. We attempt to develop a simple growth model aimed at proving that - under simple assumptions, intuitively interpretable and based on a theoretical economic record– there exists a so called *steady state* of economy. Economy in a steady state leads to return at the same rate as invested assets. If one believes that regulatory changes lead to better allocation of resources, then he assumes increase of product. Increase of product results in increase of investment, which in turn results in the growth of the economy. As privacy is treated here as one of resources, we expect more effective allocation of privacy will also lead to increase of product and the growth of the economy.

In the presentation of this early phase of research we decided to consider a natural and intuitive redescription of Solow model. The basic Solow model is considered by the economists as an appropriate tool in analyzing of growth and differences among countries. We change the classical Solow model in two aspects. Firstly, we are not interested in further comparative analysis but in the existence of a steady state. Secondly, we neglect dynamics of labor in belief that it is relatively slow when compared to speed of stock (investment changes and technological changes), which put privacy issues in the front line of research.

The paper is organized as follows. Sec 2 provides a review of technological tools for data privacy protection, and develops a simple taxonomy of the existing approaches which allows us to summarize the wealth of research in this field and establish later a connection with the economic aspects of the results of data processing. Sec. 3 views privacy in an economic perspective by viewing privacy as n economic good. We discuss the positioning of privacy within the existing categories of economic goods. We present the concept of a steady state of market. In sec. 4 we discuss different types of data privacy (individual and aggregated) and their relationships. We argue the need for a hybrid, technological and regulatory vacy steady state in the growth model of economic market.

Finally, we want to observe that this volume, dedicated to the memory of Ryszard Michalski, seems to us be a highly appropriate venue for the work on privacy and data mining. Ryszard was very interested in economic and social implications of technologies, and in the hybrid solutions to difficult problems, e.g. in combining Cognitive Science and Computer Science approaches to finding interesting relationships in data. The concept we advocate here also promotes a hybrid solution – a mix of regulation, market mechanism and technology to preserve privacy rights.

2   Data Mining and Privacy Preserving Techniques

2.1   Basic Dimensions of Privacy Techniques

The data mining field has for the last ten years shown growing interest in the techniques that ensure data privacy in the data mining context. One may identify the source of this work in the statistical data disclosure work of the 1970's and 19080's, see Domingo-Ferrer (2002) for a review of this work. We review here these ten years of research, attempting to provide a taxonomy that can provide the main dimensions allowing the understanding and reviewing of this emerging research field.

Note that since we are not presenting a complete review of PPDM work, but just highlight and outline different approaches, we use just one or two representative papers for the different dimensions of our taxonomy, rather than providing a thorough literature review.

Privacy related techniques can be characterized by a) the kind of source data modification they perform, e.g. data perturbation, randomization, generalization and hiding; b) the machine learning algorithm that works on the data and how is it modified to meet the privacy requirements imposed on it; and c) whether the data is centralized or distributed among several parties, and – in the latter case – what is the distribution based on.

But even at a more basic level, it is useful to view privacy-related techniques along just two fundamental dimensions. The first dimension defines what are we actually protecting as private – is it the data itself, or the model (the results of data mining). As we show below, the latter knowledge can also lead to identifying and revealing information about individuals. The second dimension defines the protocol of the use of the data: is the data centralized and owned by a single owner, or is the data distributed among multiple parties? In the former case, the owner needs to protect the data from revealing information about individuals represented in the data when the data is being used to build a model by someone else. In the latter case, we assume that the parties have limited trust in each other, i.e. they are interested in the results of data mining performed on the union of the data of all the parties while not trusting the other parties with seeing their own data without first protecting it against disclosure of information about individuals to other parties.

In this article we structure our discussion of the current work on PPDM in terms of the taxonomy proposed in Table 1. which gives us the following bird's eye view of this research field:

Table 1. Classification taxonomy to systematize the current work in PPDM

	Centralized	**Distributed**
Data modification	Agrawal and Srikant (2000), Evfimievski et al., (2002).	Vaidya and Clifton (2002).
Algorithm/ result modification	Oliveira et al., (2004).	Jiang and Atzori, (2006).

2.2 Data Modification

This subfield emerged in 2000 with the seminal paper by Agrawal and Srikant (2000). They stated the problem as follows: given data in the standard attribute-value representation, how can an accurate decision tree be built so that, instead of using original attribute values x_i, the decision tree induction algorithm takes input values $x_i + r$, where r belongs to a certain distribution (Gaussian or uniform). This is a data perturbation technique: the original values are changed beyond recognition, while the distributional properties of the entire data set that decision tree induction uses remain the same, at least up to a small (empirically, less than 5%) degradation in accuracy. There is a clear trade-off between the privacy assured by this approach, and the quality of the model compared to the model obtained from the original data. This line of research has been continued in Evfimievski *et al.*, (2002), where the approach is extended to association rule mining. As a note of caution about these results, H. Kargupta et al. (2003) have shown in how the randomization approaches are sensitive to attack. They demonstrate how the noise that randomly perturbs the data can be viewed as a random matrix, and that the original data can be accurately estimated from the perturbed data using a spectral filter that exploits some theoretical properties of random matrices. It is therefore clear that, in spite of their technical sophistication, technological solutions alone will not fully address the privacy problem, and that one has to complement them with a regulatory component.

Other perturbation approaches targeting binary data involve changing (flipping) values of selected attributes with a given probability, see Zhan and Matwin (2004), Du and Zhan (2003), or replacing the original attribute with a value that is more general in some pre-agreed taxonomy, see Iyengar(2002). Generalization approaches often use the concept of k-anonymity: any instance in the database is indistinguishable from other $k-1$ instances (for every row in the database there are $k-1$ identical rows). Finding the least general k-anonymous generalization of a database (i.e. moving the least number of edges upward in a given taxonomy) is an optimization task, known to be *NP*-complete. There are heuristic solutions proposed for it, e.g. Iyengar (2002) uses a *genetic algorithm* for this task. Friedman *et al.* (2006) show how to build k-anonymity into the decision tree induction.

The simplest and most widely used privacy preservation technique is anonymization of data (also called de-identification). In the context of de-identification, it is useful to distinguish three types of attributes.

Explicit identifiers allow direct linking of an instance to a person, e.g. a cellular phone number or a driver's license number to its holder.

Quasi-identifiers, possibly combined with other attributes, may lead to other data sources capable of unique identification. For instance, Sweeney(2001) shows that the quasi-identifier triplet <date of birth, 5 digit postal code, gender>, combined with the voters' list (publicly available in the US) uniquely identifies 87% of the population of the country. As a convincing application of this observation, using quasi-identifiers, Sweeney was able to obtain health records of the governor of Massachusetts from a published dataset of health records of all state employees in which only explicit identifiers have been removed.

Finally, non-identifying attributes are those for which there is no known inference linking to an explicit identifier. Usually performed as part of data preparation, anonymization removes all explicit identifiers from the data.

While anonymization is by far the most common privacy-preserving technique used in practice, it is also the most fallible one. In August 2006, AOL published for the benefit of the Web mining research community 20 million search records (queries and URLs the members visited) from 658,000 of its members. AOL had performed what it believed was anonymization, in the sense that it removed the names of the members. However, based on the queries – which often contained information that would identify a small set of members or a unique person – it was easy in many cases to manually re-identify the AOL member using secondary public knowledge sources. An inquisitive New York Times journalist identified one member and interviewed her.

Sweeney (2001) is to be credited with sensitizing the privacy community to the fallacy of anonymization: *"Shockingly, there remains a common incorrect belief that if the data look anonymous, it is anonymous"*. Even if information is de-identified today, future data sources may make re-identification possible. As anonymization is very commonly used prior to model building from medical data, it is interesting that this type of data is prone to specific kinds of re-identification, and therefore anonymization of medical data should be done with particular skill and understanding of the data. Malin (2005) shows how the main four de-identification techniques used in anonymization of genomic data are prone to known, published attacks that can re-identify the data. Moreover, he points out that for quasi-identifiers there will never be certainty about de-identification, as new attributes and data sources that can lead to a linkage to explicitly identifying attributes are constantly being engineered as part of genetics research. Again, particularly for health data, technological solutions need to be complemented by regulations.

2.3 Algorithm Modification

Is it true that when the data is private, there will be no violation of privacy? The answer is no. In some circumstances, the model may reveal private information about individuals. Atzori *et al.* (2005) give an example of such situation for association rules: suppose the association rule $a_1 \wedge a_2 \wedge a_3 \Rightarrow a_4$ has *support sup* = 80, *confidence conf* = 98.7%. This rule is 80-anonymous, but considering that

$$sup(\{a_1, a_2, a_3\}) = \frac{sup(\{a_1, a_2, a_3, a_4\})}{conf} = \frac{80}{.0987} \approx 81.05$$

and given that the pattern $a_1 \wedge a_2 \wedge a_3 \wedge a_4$ holds for 80 individuals, and the pattern $a_1 \wedge a_2 \wedge a_3$ holds for 81 individuals, clearly the pattern $a_1 \wedge a_2 \wedge a_3 \wedge \neg a_4$ holds for just one person. Therefore the rule unexpectedly reveals private information about a specific person. Atzori *et al.* (2005) proposes to apply k-anonymity to patterns instead of data as in the previous section. The authors define inference channels as *itemsets* from which it is possible to infer other itemsets which are not k-anonymous as in the above example. They then show an efficient way to represent and compute inference channels which, once known, can be blocked from the output of an association rule finder. The inference channel problem is also discussed in Oliveira *et al.* (2004), where an itemset *"sanitization"* removes itemsets that lead to sensitive (non-*k*-anonymous) rules.

We will return to algorithmic modifications in sec. 4, where we show that it fills an important gap in capturing data privacy as a certain type of economic good.

2.4 Distributed Data

Most of the work mentioned above addresses the case of centralized data. The distributed situation, however, is often encountered and has important applications. Consider, e.g., several hospitals involved in a multi-site medical trial that want to mine the data describing the union of their patients. This increases the size of the population subject to data analysis, thereby increasing the scope and the importance of the trial. In general, if we abstractly represent the database as a table, there are two collaborative frameworks in which data is distributed. Horizontally partitioned data is distributed by rows (all parties have the same attributes, but different instances – as in the medical study example). Vertically partitioned data is distributed by columns (all parties have the same instances; some attributes belong to specific parties, and some, e.g. the class, are shared among all parties – as in the vehicle data analysis example).

To illustrate the importance o such data protocols, we can cite the following real-life scenario for a vertical partition: certain models of Ford passenger cars with Firestone tires from a specific factory had unusually high tire failures, resulting in numerous rollover accidents. Both manufacturers had their own data on the accidents, but did not want to disclose them to each other. Only mining all the data (the union of Ford and Firestone data) may have led to a solution of this problem.

Another real-life scenario illustrates the horizontal data partition: several research hospitals participate in a medical study. They all have patient data relevant to the study and obtaining collecting the same information about the patients (the same attributes). If they could mine the union of their data, the study results will carry substantively more weight as based on a much larger sample, than results obtained by each hospital on its own data. Hospitals, however, are prevented for the reasons of privacy regulations in place, from disclosing their data to each other.

Occupational Health and Safety (OHS) is another application where distributed data is shared in various ways among different parties. In this application, we first have a combined database describing people's jobs and their characteristics, together with people's medical examination and test results. This database will be ideally fed by different employees that a person has had over the years, as well as by different medical data providers. These organizations will not necessarily wish or be able to share their data. We have here a vertical partition of the data. In addition, it may be interesting, for the purpose of a research project, to combine the data (from different OHS databases) about people who have over the years worked for similar types of employers. Again, owners of these databases may not be authorized to share their data with other organizations. This is the horizontal dimension of the data partition.

An important branch of research on learning from distributed data while parties do not reveal their data to each other is based on results from Computer Security, specifically from cryptography and from the Secure Multiparty Computation (SMC). Particularly interesting is the case when there is no trusted external party – all the computation is distributed among parties that collectively hold the partitioned data. SMC has produced constructive results showing how any Boolean function can be computed from inputs belonging to different parties so that the parties never get to know input values that do not belong to them. These results are based on the idea of splitting a single data value between two parties into "shares" so that none of them knows the value but they can still do computation on the shares using a gate such as *exclusive or*, see Yao (1986). In particular, there is an SMC result known as Secure Sum: k parties have private values x_i and they want to compute $\Sigma_i x_i$ without disclosing their x_i to any other party. This result, and similar results for value comparison and other simple functions, are the building blocks of many privacy-preserving machine learning algorithms. On that basis, a number of standard *classifier* induction algorithms, in their horizontal and vertical partitioning versions, have been published, including decision tree (*ID3*) induction, *Naïve Bayes*, the *Apriori* association rule mining algorithm, Vaidya and Clifton (2002), Kantarcioglu and Clifton (1998), and many others.

We can observe that data privacy issues extend to the use of the learned *model*. For horizontal partitioning, each party can be given the model and apply it on the new data. For vertical partitioning, however, the situation is more difficult: the parties, all knowing the model, have to compute their part of the decision that the model delivers, and have to communicate with selected other parties after this is done. For instance, for decision trees, a node n applies its test, and contacts the party holding the attribute in the child c chosen by the test, giving c the test to perform. In this manner, a single party n only knows the result of its test (the corresponding attribute value) and the tests of its children (but not their outcomes). This is repeated recursively until the leaf node is reached and the decision is communicated to all parties.

A different approach involving cryptographic tools other than Yao's (200 circuits is based in the concept of homomorphic encryption , see Paillier (1999). Encryption e is homomorphic with respect to some operation $*$ in the message space if there is a corresponding operation $*'$ in the ciphertext space such that for any messages m_1, m_2, $e(m_1)*'e(m_2) = e(m_1 * m_2)$. The standard RSA encryption is homomorphic with $*'$ being logical multiplication and $*$ logical addition on sequences of bytes. To give a flavor of the use of homomorphic encryption, let us see in detail how this kind of encryption is used in computing the scalar product of two binary vectors.

Assume just two parties, Alice and Bob. They both have their private binary vectors $A_{1,...,N}$, $B_{1,...,N}$. In association rule mining, A_i and B_i represent A's and B's transactions projected on the set of items whose frequency is being computed. In our protocol, one of the parties is randomly chosen as a key generator. Assume Alice is selected as the key generator. Alice generates an encryption key (e) and a decryption key (d). She applies the encryption key to the sum of each value of A and a digital envelope R_i*X of A_i (i.e. $e(A_i i + R_i * X)$), where R_i is a random integer and X is an integer which is greater than N. She then sends $e(A_i + R_i *X)$s to Bob. Bob computes the multiplication

$$M = \prod_{j=1}^{N} [e(A_j + R_i *X) \times B_j]$$

when $B_j = 1$ (since when $B_j j = 0$, the result of multiplication doesn't contribute to the frequency count). Now,

$$M = e(A_1 + A_2 + \cdots + A_j + (R_1 + R_2 + \cdots + R_l) * X)$$

due to the property of homomorphic encryption. Bob sends the result of this multiplication to Alice, who computes

$$[d(e(A_1 + A_2 + \cdots + A_j + (R_1 + R_2 + \cdots + R_l) * X)]) \bmod X =$$

$$= (A_1 + A_2 + \cdots + A_l + (R_1 + R_2 + \cdots + R_j) * X) \bmod X$$

and obtains the scalar product. This scalar product is directly used in computing the frequency count of an itemset, where N is the number of items in the itemset, and A_i,

B_i are transactions of Alice and Bob projected on the itemset whose frequency is computed.

While more efficient than the SMC based approaches, homomorphic encryption methods are more prone to attack as their security is based on a weaker security concept, Paillier (1999), than Yao's approach. In general, cryptographic solutions have the advantage of protecting the source data while leaving it unchanged: unlike data modification methods, they have no negative impact on the quality of the learned model. However, they have a considerable cost in terms of complexity of the algorithms, computation cost of the cryptographic processes involved, and the communication cost for the transmission of partial computational results between the parties, Subramaniam *et al.* (2004). Their practical applicability on real-life sized datasets still needs to be demonstrated.

The discussion above focuses on protecting the data. In terms of our diagram, we have to address its right column. Here, methods have been proposed to address mainly the North-East entry of the diagram. Vaidya and Clifton (2002) propose a method to compute association rules in an environment where data is distributed. In particular, their method addresses the case of vertically-partitioned data, where different parties hold different attribute sets for the same instances. The problem is solved without the existence of a trusted third party, using Secure Multiparty Computation (SMC). Independently, we have obtained a different solution to this task using homomorphic encryption techniques (Zhan et al. 05). Moreover, Jiang and Atzori (2006) have obtained a solution for the model-protection case in a distributed setting (South-East quadrant in Table 1). Their work is based on a cryptographic technique and addresses the case of vertical partitioning of the data among parties. Again, it is related to the privacy in aggregated goods as discussed in Sec. 4.

2.5 Data Privacy and the Two-Tiered Data Representation

In 1992, [Bergadano et al. 92] introduced the idea of a two-tiered concept representation, and showed how concepts in such representation can be learned from examples. In a two-tiered representation, the first tier captures explicitly basic concept properties, and the second tier characterizes allowable concept's modifications and context dependency. While the goal of this representation was to allow to capture and represent different degrees of typicality of instances, we believe that the two-tiered approach to data representation is also applicable to data privacy.

In terms of the different kinds of attributes discussed in sec. 2.2 above, quasi-identifiers do not identify the instance on their own, but when combined can provide such identification. Non-identifiers do not identify the data in any known way. We believe that this can be naturally rendered in a two-tiered representation in the following way. The first tier will consist of the attributes which are the non-identifiers and it will be used to learn a concept description which can then be rendered public. The second tier could consist of the quasi-identifier attributes together with the second-tier inference rules that will specify how the second tier information can be added to the first tier, and by whom.

Let us see an example. Suppose a patient in an electronic health record is represented by her identifying number, data of birth, address, and relevant medical information: records of visits, tests, procedures, etc. Explicit identifiers and quasi-identifiers will be the identifying number, the date of birth, the address, etc. We will place this kind of information in the *second tier*, as we do not want it to be involved in the data mining process. It could later be used to infer knowledge specific to a given person, if the person authorizes access to this knowledge about her. We can, at the same time, use and exchange the data in the first tier to the extent to which this data is anonymous. To ensure anonymity, we could subject it to one of the techniques described in sec. 2.2.

3 Basic Economics Concepts Related to Data Privacy

In order to discuss privacy from an economics point of view, we need to introduce the basic economic concepts we will use in our perspective on privacy. This Section is devoted to the terminology and some axiomatic principles underlying economic analysis of privacy. Some of presented comments may seem simplified, but expanding this presentation, although possible and interesting in itself, would not serve the clarity of the argument.

3.1 Motivation of Agents, Goods and Their Economic Classification

Following Maslow (1943) we assume that human beings are motivated to act by their needs. Humans cannot exist if their needs in a hierarchy (from hunger to self-actualization) are not satisfied. The need for privacy is nowadays accepted universally. Satisfaction of higher-level needs follows satisfaction of basic needs.

Modern societies insert human rights in the middle of this hierarchy. The Declaration of Independence and Bill of Rights state that satisfaction of richness of human needs satisfaction is *unalienable right*[6] and demand rights' protection[7].

Economic literature focuses on means to satisfy basic needs. Value of satisfaction from consumption of a good, expressed numerically, is called its *utility*. Individuals (called *economic agents* or shortly *agents*) are utility maximizers (their decisions are driven by *utility maximization principle* or *homo oeconomicus rationality*). This means that given the choice among consumption of product of higher and lower utility, they will select the former. Utility is usually assumed to be non-decreasing[8] and concave, i.e. the rate of utility growth decreases[9]. Advanced needs are investigated in psychology, sociology and law more frequently then in economics.

Agents can take role of producers or consumers. Producers own assets transformed into goods. Consumers pay for the possibility of using the goods. Each individual can participate in different markets and in different roles.

---

[6] Declaration of Independence:... *We hold these truths to be self-evident, that all men are created equal, that they are endowed by their Creator with certain, that among these are Life, Liberty and the pursuit of Happiness.*

[7] Bill of Rights: ...*No State shall make or enforce any law which shall abridge the privileges or immunities of citizens of the United States; nor shall any State deprive any person of life, liberty, or property, without due process of law; nor deny to any person within its jurisdiction the equal protection of the laws.*

[8] This requirement reflects the assumption that agents are greedy – given continuity of available product they always prefer more then less.

[9] This requirement is interpreted as risk aversion of agents: increase of utility related to consumption of additional unit of a product of agent whose consumes at some level is smaller then in case of consumption at lower level.

In economics means which are consumed in order to satisfy human needs are called *goods*. Goods can be directly consumed or invested to produce consumable goods. Goods are exchangeable. Usually individuals acquire goods in exchange transactions. Goal of a transaction is to provide parties with a good to be consumed in order to satisfy needs.

There are *substantial goods* and *non-substantial goods*. Bread bought in a store is an example of a substantial good as it serves to satiate hunger. Goods which indirectly lead to satisfaction of needs are known as non-substantial goods. For instance, ownership rights purchased by a person can be used to achieve economic effect which will lead to satisfaction of a need. Thus copyright, license, concessions, patents, brands (logo) and know-how are examples of classes of non-substantial goods.

Let us consider the following two characteristics of goods: the possibility of *non-exclusive consumption* of a good and *depletability* (*subtractability*) of a good. It is the nature of a good that decides about of the consequence of exclusion of others. One can be excluded from eating pastries, because of a high prices, or because someone else has already eaten all the pastries. One cannot, however, be excluded from consuming the results of upgrading of the quality of drinking water in a city-wide system. The quantity of a good can either decrease as the effect of consumption (e.g. bread), or not (e.g. light from a street lamp). These features of goods can be considered jointly, i.e. a good can be non-subtractable and non-excludable: upgraded water available in tubes for everybody is non-excludable, and at the same time drinking it does not decrease its quality for the others (non-subtractability). This discussion leads us to the following matrix classification of goods.

Table 2. Basic features of good – excludability and subtractability – define four classes of economic goods

	Subtractable	**Non-subtractable**
Excludable	*Private Goods*	*Club Goods*
Non-Excludable	*Common Assets*	*Public Goods*

Private Goods when consumed exclude consumption by other consumers, and they are subtractable, e.g. bread. Consumption of *Public Good* (e.g. military security or fire protection) neither excludes their consumption by others, nor it decreases the availability of the good. *Club Goods* (e.g. a painting in a private museum), when consumed, do not become less available for the other consumers (non-subtractability), but some individuals (the ones who do not buy tickets) are not able to consume them (excludability). Finally *Common-pool Goods* (e.g.fish in the ocean) decrease availability when consumed but consumers cannot be excluded from consumption.

It is important that the nature of goods be decided independently of the organization of the market.

3.2 Market Ineffectiveness and Regulatory Interventions

Markets are characterized by organization of transactions. A markets is defined by the regulations and norms respected in transactions into which participants enter in that market. Different organizations of transactions determine different markets. In *free markets*, conditions of transactions (e.g. price) are decided by parties and no external intervention occurs. In central planned economics, external intervention (e.g. official price lists) was a rule.

There are two views on market functioning: the first is related to *goal orientation,* and the second - to performance. Evaluation of the functioning of the market from the goal orientation perspective is called effectiveness. Evaluation of market from the performance perspective is called efficiency.

The goal-oriented evaluation of a market is related to the goals of transactions. If a market enables achievement of transactions' goals, then it is *effective*. The existence of homelessness, extreme poverty, hunger and diseases show that needs of populations are difficult to satisfy universally. While the right to satisfy needs is universal, real economies do not provide goods at a satisfactory levels. This shows that markets nowadays are not effective institutions - they fail to satisfy the needs.

The performance-oriented evaluation of a market measures the extent of goal achievement. This implies that the goal is not satisfied universally, but only to a certain extent. As a first approximation we can say that a market works in an optimal way if a higher level of performance is not attainable.

Every market regulation attempts to organize exchanges in such way that resulting allocation of goods is possibly profitable. This observation leads to the concept of *efficiency* (and a related concept of *optimality*).

The allocation A of goods is said to be more *efficient* than the allocation B if degree of satisfaction of population is higher in the case of A[10]. If – given allocation A – there is no allocation B which is more efficient, then the allocation is said to be *Pareto-optimal* (in sequel it will be called shortly *optimal*). Let us observe that many optimal allocations may exist.

Thus a minimal requirement for market is to be *efficient,* i.e. to exclude inefficient (non-optimal) allocations.

Real markets for public and private goods are not effective. Moreover, markets for private goods are not efficient – they admit non-optimal allocations.

Non-optimality occurs e.g. in situations when one buys the over-priced good as a result of lack of knowledge about the availability of this product at a lower price. Lack of information leads to additional cost without the increase of satisfaction. Situation when parties have different information about conditions of transactions is called *information asymmetry.*

If likelihood of punishment can be neglected, then people (utility maximizers) are tempted to violate regulation (e.g. non-customers use customer parking). In these situations we say that they face the so called *moral hazard.*

Information asymmetry and moral hazard lead to externalities and adverse selection – two mechanisms responsible for market inefficiency. To explain this, let us continue the parking example. If a free store parking is frequently used by nonconsumer, then cost of this service is transferred to other customers (it is reflected in the margins and paid for by the customers). Such transfer of cost (and sometimes of profit) to an external party is called an *externality.*

---

[10] Let us notice that in reality ineffective allocations although socially undesirable can occur.

Unjustified usage of customer parking may result in difficulties in parking and loss of customers and decrease of profit of the store. The store may react to this situation by increasing the margins and thus may accelerate further loss of customers. This leads to lower profit and may encourage the store to further increase its margins and will cause further losses of customers. This mechanism is called *adverse selection.*

Thus moral hazard and information asymmetry are mechanisms responsible for appearance of externalities and adverse selection in the market and may result in market inefficiency[11] (Fig. 1.)

Impacts of externalities and information asymmetry can be decreased by *interventions* in the market[12]. These interventions can take form of formal regulations.

Ineffectiveness and inefficiency of free market make people turn to state (*governmental*) control of market organization (state-controlled economy). Historical experience and theoretical arguments show that state controlled economy is also ineffective and inefficient[13].

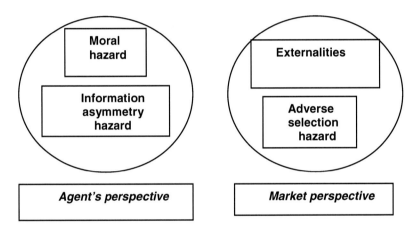

Fig. 1. Sources of market inefficiency have general nature – do not depend of nature of good

Are there intermediate levels of market control / intervention which may lead to optimal states? Literature suggests two interesting solutions. One is called here *technological* intervention. Technological developments enable introduction of mechanism which exclude free-riding. Use of password-protected WiFi at airports is an example of technological intervention which excludes externality.

---

[11] Influence of information asymmetry and externalities on decision making process was studied by Acquisti and Grossklags (2005).

[12] Arrow (1963) shows in medical services area that in occurrence of non-optimal states evolutionally non-market regulations are introduced.

[13] Governmental institutions can be coerced by decision makers, and they can ignore the needs of population. Furthermore, state regulation requires non-market, administrative driven evaluations of product in economy and control of performance of employees. This raises the question of credibility of information processed in economic decisions. State control does not remove temptations to transfer burdens and privileges between state and agents – therefore does not exclude non-optimal allocations. See also Kornai (1980).

Another possible solution is and intervention based on social contract, called in sequel *contractual* intervention.

Contractual solution is an attempt to improve optimality of a non-excludable good through change of its basic characteristics – making it exclusive. Basic research is good example of this situation. It is socially agreed that results of basic research are public – nobody can exclude any person from the use of the Maxwell equation. It is also accepted that basic research is funded by the state[14]. Therefore basic research suffers all the pains of a state controlled enterprise. A social contract could turn basic research into an excludable good: this can be accomplished by means of an appropriate legal solution. Allowing patenting of research results obtained in public universities changes a public good (research) into a private one. Such contractual change improves efficiency of research. However, making research a private good is also not satisficing. Free market funding of basic research is difficult due to moral hazard and adverse selection. These mechanisms, together with short-term profit business orientation, may negatively the development of science. Higher profitability of application-oriented research creates temptation to give up basic research (an externality, resulting from free availability of results of basic research) and turn to applied science. With time, decreased investment in basic research decreases the rate of scientific progress and finally results in decrease in application and decrease of profits (adverse selection). Thus solely contractual regulations are not sufficient.

Let us now see a hybrid solution. Let us consider an fictitious state where by law there are only two public TV channels. The first channel offers information and the second - entertainment. There is a public regulator, with a budget to satisfy population needs in both areas. The budget is created from taxes and relatively low individual monthly payments. In this situation both channels offer public good – watching TV by a person does not exclude others from consumption and there is no subtraction. We may expect inefficiency, and in fact indeed there is a moral hazard, a temptation for free riding – consumers withholding their monthly payments. Cost of administrative procedures to recover from free riding makes the risk of punishment negligibly low. Decrease of income from monthly payments results in decrease of budget, which in turn results in the decrease of quality of programming. Lowering quality of TV programming further weakens public's readiness to pay for public TV. Adverse selection forces the public regulator to allocate additional public funds to public TV, in order to preserve TV quality. This is obviously an inefficient allocation, since public TV is supported by the taxes of people who do not own TV sets, without an increase of their utility.

Contractual regulation in this context could be achieved through legal rule which would allow private TV to broadcast entertainment programming and commercials as well as collect payments for these services. Technological part of the solution consists of the encoding of the private TV broadcast, so that only the paying customers can receive it (why this technological solution is this generally allowed for private television, while not for public television?)

The mission of public television is to guarantee satisfaction of people's needs (see the Section 2.1). Exclusion from access to information programming – the basic mandate of public media - may result in information asymmetry and may lead to serious

---

[14] Basic research is so expensive that collective founding is needed.

political consequences (in an extreme case – totalitarian abuse of authority). Thus exclusion from informational services is not possible. However, exclusion from consumption of entertainment is possible and therefore allowing private producers in this market is acceptable. Let us observe that contractual regulation of entertainment based on market of privately broadcasted, coded programs leads to decrease of externality of the public, fee-funded TV as described above, to a necessary.

Let us point out that in this case – a hybrid solution involving contractual regulation, and therefore the social contract – costs of change are split between the private and public sector. The former invested in change based on future profit predictions. These predictions are based on the knowledge of the business and the knowledge of the market. The issue of the cost of the public broadcaster requires a more complex cost-benefit analysis. Expected public benefits are difficult to measure, are usually not cost effective, and require new sources of funding (e.g. economic growth, which results from the change of the economic regulations.)

As shown in the above example, regulatory interventions are not free - they require funding. It is therefore important to check whether introduction of changes will not dismantle the economy. In the next chapter we briefly review the concept of an economic growth in which investment leads to growth of the product and thus creates resources to fund the change. This approach will be continued in sec. 5 in the specific context of privacy.

3.3   Growth in Economy with a Privacy Market

We consider an economic system described using the concept of its *state* at the time *t*. The state of economy is described using set of n attributes $\{x_i\}_{i=1,2,...,n}$. Change of attributes in time (called *evolution*) is described as a function which is defined on a time interval[15]). When attributes can be described as *input* or *output attributes*, they then describe economic processes. Input attributes are usually *resources*, output attributes are *products*. The process is described as mapping from the input space to the output space.

It is worth to remember that from purely theoretical point of view, in the rough analysis it is sufficient to compress the concept of resource to its most general category. In this Section resources are capital technology and labor. Capital includes hardware, land, and finance. *Technology* involves technical solutions, organization, knowledge and information processing. *Labor* involves not only resources of available work (measured e.g. in hours) but also human capital – education, entrepreneurship and health. The economic assumption is that they multiplicatively influence the production – there is no production if one of them is eliminated. Therefore production will be described as a function defined as product of simple expressions depending on measures of resources. Given constant capital, we see that the product of technology and labor describes effectiveness of labor, so it is called *effective labor*. Let us observe that resources play also a dual role – if they can be consumed then they are also *products*.

The dynamics of the system is described by system of axioms which can be derived from the economic interpretation. This system of axioms can be represented in a mathematical form, usually as system of differential equations.

---

[15] The functions are always assumed to be sufficiently regular to guarantee correctness of mathematical operations performed. Also the time interval is not specifically denoted.

We say that the economic system remains in a *steady state* if its evolution is characterised by equal increases of capital and production[16]. This means that increase of capital (i.e. investment) results in increase of production (and there are resources to fund regulatory changes and constant costs of regulation). The additional capital can be invested in resources used for further production, or for regulatory changes.

In this section we present shortly basic economic concepts used to analyse economic growth dynamics. In next section these concepts will be extended to involve data privacy goods.

Let us denote by t a time moment, by K, L, A – input attributes interpreted as capital, labour, and technology, respectively. Y denotes the output attribute describing product. Consequently production is described using the *production function F,* Y=sF(K,A,L) and AL is *effective labor.*

We assume the following set of axioms describing interdependence of variable (economic mechanisms):

- Production is split into consumption and savings (which is invested);
- Invested (increase of stock) part is directly proportional to production;
- Increase of labor at each moment is directly proportional to population size[17];
- Increase of technology at each moment is directly proportional to technology level;

If we consider evolutions e(t)=[K(t),A(t)L(t)], then these axioms can be described using the following system of differential equations[18].

$$\frac{dK}{dt} = sF(K(t),A(t)L(t)), \quad \frac{dL}{dt} = nL(t), \quad \frac{dA}{dt} = gA(t)$$

The condition for steady state is

$$\frac{dK}{dt} = \frac{dY}{dt}$$

The evolution of economy can be determined if production function is known. E.g. for Cobb-Douglas function[19] $F(K,L)=Y(t)=K(t)^{\alpha}(A(t)L(t))^{1-\alpha}$, one gets the following formula for evolution starting from $(K_0=0, A_0 L_0)$:

$$\frac{dK}{dt} = sF(K(t),A(t)L(t)) = sF(t,K(t),A_0 L_0 e^{(n+g)t}).$$

---

[16] The dynamics seem mysterious if rates are not equal, since there is difficulty with explanation of reasons for outputs smaller or greater then inputs.

[17] Labor and population are equated here – as we assume that a constant part of population is in the workforce, so that any change in population results in an equivalent change of the workforce, and therefore quantities of population and labor can be treated as equivalent.

[18] The derivative dK/dt – change of stock is explained as investment.

[19] The Cobb-Douglas functional form of production functions is widely used to represent the relationship of an output to inputs. It was proposed by Knut Wicksell (1851-1926), and tested against statistical evidence by Charles Cobb and Paul Douglas in 1928. It has important her feature of *constant returns to scale*, i.e. I describes situations that, if assets L and K are each increased by rate r, then the product Y increases by r.

Hence

$$K(t) = \sqrt[1-\alpha]{\frac{s}{n+g}} \cdot (A_0 L_0) \cdot e^{(n+g)t}, \ L(t) = L_0 e^{nt}, \ A(t) = A_0 e^{gt}.$$

To investigate the existence of steady states it is convenient to introduce auxiliary functions $y(t) = \frac{Y(t)}{A(t)L(t)}$ (production resulted from "unit of effective work AL") and $k(t) = \frac{K(t)}{A(t)L(t)}$ (stock resulted for "unit of effective work AL"). Simple computation proves that $y(t) = F(K,1) = f(k)$, where $f(k)$ is simplified notation for $F(K,1)$.

Direct computation proves that under assumed set of axioms, there exists the evolution k^* of a system in steady state and it is a solution of the following equation for $f(k) = F(K,1)$:

$$\frac{dk^*}{dt} = sf(k^*) - k^*(g+n).$$

Let us observe that if $k > k^*$, than $\frac{dk}{dt} < 0$, which means that system evolution leads to values k' smaller then k until k' is not smaller than k^*. If $k' < k^*$, then $\frac{dk}{dt} > 0$, which means that system evolution leads to values k'' bigger than k' until k'' is not bigger k^*. Then the state attribute starts decreasing again.

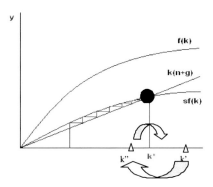

Fig. 2. Arrows show convergence to the steady state

Thus the equation for evolution in steady state describes in fact the *adjustment mechanism* which is responsible for convergence of economy to the steady state trajectory, see Figure 2.

4 Data Privacy Goods from an Economic Viewpoint

4.1 Data Goods

Data influence satisfaction of human needs. According to Maslow (1943), the most evident human need for information occurs at cognitive level of his hierarchy – we need information to know, to understand, and explore. Satisfaction of cognitive needs conditions self-actualization. We accept the view according to which provision of information which serves to satisfy human needs is to be treated as human right. Thus data are goods in economic sense. As a data serve to satisfy human need, they are called *data goods*. There are *non-substantial goods*. Let us consider two types of Personal Data Goods (PDG) – *individual personal data goods* (IPDG) and *aggregated personal data goods* (APDG).

Data about my age and about my research achievements are my personal goods – my IPDGs. They allow me to satisfy my need for self-actualization. Through a contractual agreement, my dean can used my IPDG, merge it with others' IPDGs, determine my performance index an, e.g. rank me within a group of peers. In this manner, she creates my APDG.

According to Alan Westin's standard definition privacy is the ability of individuals "*...to determine for themselves when, how, and to what extent information about them is communicated to other*" (Westin 1967). From a purist economics point of view privacy can also be treated as need[20]. Everybody experienced violation of privacy and makes an efforts to protect themselves from a similar experiences in future. In our discussion, the ability to control IPDG (APDG) will be called individual (aggregated) personal *data privacy, hence denoted IPDP (APDP)*. The need for privacy implies that means which serve to satisfy the privacy needs can be treated as goods.

Table 3. Classification of data goods

	Individual	**Aggregated**
Personal Data Good (PDG)	IPDG	APDG
Personal Data Privacy (PDP)	IPDP	APDP

They are discussed in detail in four subsequent sections devoted to IPDG, IPDP, APDG, and APDP respectively[21].

4.1.1 Individual Personal Data Good

Law is to protect property - including *personal goods* of every person. There is no formal definition of personal good as it would depend on different values respected differently by different people. Legal regulations usually list here: health, freedom, honor, dignity, freedom of conscience but also name, image, confidentiality of correspondence, inviolability of private property, creativity in the arts and sciences.

Personal goods are culturally rooted and related to values shared by a given society. They are difficult to valuate, but possessing them can influence the economic situation of their owner.

Personal goods are related to every person. An individual can produce individual personal data goods, e.g. movie star can attempt to improve her media image image.

---

[20] Probably privacy is not the basic need, see e.g. Moor (1997) for further discussion of this issue.

[21] It has been pointed out that data that is not considered private today (e.g. someone's single hair) may be universally considered private with time. E.g. in ten years it may become cheap and easy to obtain a person's genetic make-up, including their propensity to specific diseases, from a person's single hair.

However, IPDG cannot be sold or inherited. Usually an individual is well-informed on ownership of her IPDGs.

For example, non-regulated personal data transaction occurs when data good seller (IPDG) (a buyer of e.g. apples in a store) gives hers telephone number to data buyer/acquirer (seller of apples). In this case there is no price – no money is involved.

Let us comment on this price free transaction. Why would a buyer provide a seller with his personal information? This behavior is rooted in Lucas's (1972) theory of rational expectations which is based on assumption that agents' decisions reflect expectations concerning future states of the economy as well as past and currents experiences[22]. Thus investment is driven by their expected profitability. Privacy donation in form of disclosure of telephone number can be viewed therefore as investment which will be returned later. This donation does not exclude the possibility to disclose the telephone number to other parties.

IPDGs are therefore Club Goods: they are excludable and non-subtractable.

4.1.2 Individual Personal Data Privacy

Processing of IPDGs is subject of scrupulous protection. This protection is strictly defined especially in the area of so called sensitive personal data (e.g. ethnic or race origin, political opinions, religion, health, genetic code, personal habits, sentences).

The perception of personal data violations is extremely subjective, culture dependent[23]. Our perception of privacy can be very different from the reader's one. Still, there are effective legal tools to cope with this problem.

Violation of personal goods can be accepted only with permission by owner[24] of personal goods or by law. Otherwise he who harms owner of a personal good is obliged by law to give up her actions, and if it is too late, then she has to compensate the losses to the harmed person. As an example of a privacy transaction, let us consider situation when one makes available her private flat for a film production. In this case the expected profitability usually results in the pricing of the loss of privacy (i.e. in our view, the sale of privacy). The price reflects the value of privacy for the owner of the flat.

Individuals can produce privacy protecting access to his IPD (e.g. resigning from providing data on websites). Individual can also consume privacy (e.g. consciously controlling his availability for telephone calls).

Contrary to IPDG, the IPDP can be sold, e.g. a movie star can sell rights to her image to a cosmetics firm for a price (profit). Cosmetic firm then consumes the star image for advertising purposes.

---

[22] This approach extends framework of theory of adapted expectations, according to which only information on past and current states of economy are processed to build expectations used in decision analyses.

[23] This view is widely documented in literature of the subject.

[24] This is the issue of the so called *opt-in* and *opt-out* approach to seeking a person's permission to use their data. The opt-out approach will use the data unless the person whose data is used explicitly *disallows* such use, e.g. when filling out an on-line form and submitting personal data when registering for a service. The opt-in approach will authorize the use of that data only when a person explicitly *allows* such use (most likely, at the time of data collection). Due to the differing legal frameworks, the opt-out approach prevails in North America, while the opt-in approach is compulsory in the European Union. Both options accompany the situation of giving away one's privacy. Opting in gives more, and opting out – less control to the provider of data, who gives away some amount of his privacy.

An individual is usually well-informed on ownership of her IPDP. IPDP is a private good: it is excludable and subtractable.

4.1.3 Aggregated Personal Data Good

Innovation in data processing and the increase of scale of collection, storage and processing of massive amounts of individuals' data results in producing new data describing individuals – e.g. consumer profiles. Machine Learning algorithms enable finding patterns in personal data and lead to new knowledge about an individual.

As an example one can consider a worker whose data are is included in the database representing individuals' records in an Occupational Health and Safety system. Processing of workers population data allows creating knowledge about their health risk. The information on health risk has obviously value for a person considering a given job, who could change her decision about taking this job (mechanism similar to use of weather forecast). But this information also has value for employers, insurance companies etc. Therefore this knowledge can be useful for the satisfaction of needs of both the worker and the employer (decreasing uncertainty related to their decisions). Aggregated data may also e.g. pharmaceutical companies in designing their research and product development efforts.

There are three groups of actors in a typical good production and exchange scenario. The first one is a population of individuals whose records are collected, stored in Data Bases (DBs) and processed according to requests. The second group consists of subjects who own DBs - - we will call them Data Bases Administrating Institutions (Data Custodian). The third group are market customers – usually corporations or state institutions.

Herself, an individual cannot produce APDG since she cannot operate on DB records. However, an individual can consume APDG (e.g. finding information of about her health risk).

Also Data Custodian cannot produce APDG without individuals' participation since DB requires the creation and use of individual records - not possible without individual's participation. Data Custodian can produce privacy (e.g. hiding information of risk related to individual's health)[25]. However, for Data Custodian, the APDG she produces is usually non-personal good but common asset (excludable and non-subtractable). We can observe here that such actions of the data custodian correspond the the algorithmic modification kind of PPDM, discussed in sec. 2.3.

Market customers organizations cannot produce APDG either, but they can buy APDG (e.g. market surveys, or Occupational Health and Safety risk profiles). As we see, APDG is a common asset, but its nature is not stable. In particular, firms can resell APDG, which becomes then a private good. In particular if they resell APDG which changes the characteristics of APDG from a common asset a to public good.

Let us notice that right to ownership of APDG is *distributed*. This right is in a sense spread in a population and shared by this population and the Data Custodian.

Distribution of ownership is responsible or the fact that individual is not always informed on ownership of her IPDG.

Let as summarize. The knowledge obtained from APDG processing has three important features. It cannot be obtained by individual herself since she cannot operate

---

[25] The custodian could achieve this by protecting the individual data within APDG. This has not yet become the norm, despite the availability of technical tools that provide this kind of protection (Sec. 2).

on DB records. It can be used to satisfy human needs (i.e. is a data good). It can play role of personal data good, or private good or public good depending on parties who exchange the good.

It seems that moral hazard and information asymmetry are unavoidable features of any market of IPDG. This dismantles any hope for the efficiency of such markets.

For different actors APDG has different characteristics as a good. For individuals it a club good: excludable and (non)subtractable. For Data Custodian it is a Common Asset; for companies it is a Common Asset as well.

4.1.4 Aggregated Personal Data Privacy

If an individual is well-informed about her APDG, then she can consume her APDP. As a personal good any APDG (a profile of a personal data of a group of individuals) creates the issue of APDG control, i.e. creates APDP. Most PPDM tools used to protect IPDP fail for APDP. Data processing created a new threat for the privacy of individuals whom the data represent.

We see that an individual faces difficulties with control of his APDG because she may not be informed on its existence (e.g. in case if Data Custodian has not informed individual data providers on results of data aggregation and inference - before or after data processing)[26].

The threat for APDP is increased by the distributed ownership. APDP requires – at least theoretically - control of a population and of the Data Custodian.

If an individual is well-informed about her IPDG, then she can consume her APDP.

Again it seems that moral hazard and information asymmetry are unavoidable features of any market of IPDP, and that efficiency of such markets is impossible.

APDP is a private goods: excludable and subtractable.

4.2 Data Market Ineffectiveness and Regulatory Interventions

Data Market is defined by regulations and norms respected in economic transactions dealing with data exchanges. Its main characteristics were discussed above and are listed in Table 4.

Let us make two important remarks. Firstly, every data good can be transformed to public good through publication. Secondly any privacy is a private good.

From sec. 3.1 we conclude that markets for personal data goods and privacy cannot be effective thus regulation is needed. Let us return to description of transactions and consider types of data market organization.

We assume that each individual has some knowledge of his personal data (certainly individual, and occasionally- aggregated) and ability to control their use – data privacy. This knowledge represents a for him a value which we call *data utility* (specifically: *personal good utility* and *privacy utility*).

The efficiency (Pareto-optimality) concept applies the allocation of goods (see sec. 3.2) to *personal data good*, which characterizes states of the data market. Again, there can exist many optimal allocations.

Since data privacy – individual or aggregated - is a private good, thus we may expect existence of a moral hazard and information asymmetry leading to externalities

---

[26] There is no privacy for other actors: Data Custodian and market customers do not treat the APDG of an individual as personal good.

and adverse selection and to non-efficient states. As for any other good, in situation of information asymmetry, data/privacy provider may accept a lower than possible price.

The temptation to undervalue one's own privacy jeopardizes the privacy of one's family, leading to moral hazard. As the result, the family may be exposed to hardship (e.g. unwelcome media interest), or – alternatively - enjoy extra profits, which proves that externalities can occur. Again, as in the general case, leads to non- optimality.

Table 4. Data Market Characteristics. Table entries contain different characteristics of personal and aggregated data in their top part, and the classification of these kinds of data in terms of the economics goods in their bottom part (italics).

	Individual Data	**Aggregated Data**
Personal Good	Culturally rooted and related to values shared by societies. Difficult to price Influence owner's economic situation. Related to every person. Individual can produce IPDG. IPDG cannot be sold or inherited. Individual is well-informed about ownership of her IPDGs.	Actors: population, Data Custodian, market customers. Herself, individual cannot produce APDG. Individual can consume APDG. Herself, Data Custodian cannot produce APDG. Data Custodian can produce (profit from it and own it) privacy (APDP) as common asset (excludable and non-subtractable). Market customers cannot produce APDG. Market customers can buy APDG as common asset. If APDG is sold to public media it becomes a public good. Ownership of APDG is distributed. Individual is not always informed about ownership of IPDG. For individual - club goods: excludable, (non)subtractable
	Club goods: excludable, non-subtractable	*For Data Custodian and firms - Common Asset*
Privacy	Individual can produce privacy. Individual can consume privacy. IPDP can be sold. An individual is well-informed on ownership of her IPDP.	Individual cannot produce APDP. Individual can buy APDP. Difficult control of her APDG.
	Private goods: excludable, subtractable	*Private goods: excludable, subtractable*

As far as APDP is concerned, let us consider two groups of exchanges. In the first group there are exchanges between individuals and the Data Custodian, in the second - between the Data Custodian and market customers (firms or state institutions). In the first group exchanges are usually price-free. In the second group transactions are priced. Exchange in the first group can be treated as a specific economic transaction. There is a flow of the investment good (personal data) to the Data Custodian. Data processing is usually financed and generating profit or at least cost refund. In exchange, the owner of personal data may have access to the APDG good (the result of processing of his personal data). When he connects his personal data to the this APDG, he creates new personal knowledge - a new good in the form of IPDG. The personal data owner can then control the privacy of this good. In the second group we have a typical market situation. Participants of the market are Data Custodian and firms. Data Custodian offers aggregated or anonymous data. Both sides attempt to price the value of privacy. It is interesting that this is precisely a market valuation of privacy, since on the one hand - without individuals giving up their privacy firms would not get the information, and on the other – without expectation of profit firms would not accept the price they need to pay the individuals for their IPDG.

Let us return to example of Occupational Health and Safety records. We have two groups of actors: individuals, and organizations. Both groups face temptation to abuse good faith of others - moral hazard.

Firstly, the individual experiences moral hazard to hide information about his health in order to improve his position in the labor market or as insurance company customer[27]. This will cause adverse selection (decrease quality of labor) and will result in privacy market inefficiency.

Secondly, employers or insurers are tempted to apply discrimination practices based on the acquired APDG-based knowledge. From economic point of view they are tempted to externalize - to decrease cost of labor.

Conclusion is straightforward: *free, unregulated data market is to suffer non-optimality*. Free market has to face regulations. And such regulations are continuously and gradually introduced since half of a century. Nevertheless, a partly regulated economy which uses only administrative mechanisms to control privacy data market appears not effective. Time and again we learn about serious leakages of sensitive data, despite the regulations in place. Again, technological solutions, from the standard data encryption to PPDM representations presented earlier, exist to alleviate (but not fully solve) this problem.

The US data privacy market was thoroughly investigated in a paper by Laudon (1996)[28]. While some of the context information should be updated, e.g. due to the raise of internet, his analysis of roots, foundations and mechanisms of data markets remains right, and his conclusions seems even stronger a decade later.

Let us reword Laudon's original statements: data privacy market exists, it is ineffective and unfair and its ineffectiveness can be removed only through economically driven control of privacy transactions. To this end he proposes a specific market structure and explains its organization of transactions in the form of a National Information Market (NIM). Laudon's Market is organized as a stock exchange. Problems

---

[27] See also Varian (2004).

[28] As we have shown in sec. 4.2, free data privacy markets cannot work.

of asymmetry and externalities are to a large extent- .resolved through regulation This solution adds some new problems similar to the ones already known to financial markets regulations[29].

Let us add, however, that another serious difficulty with NIM is related to the expected cost and complexity of this solution.

Economic reasoning shows that all free data markets may be not effective – effectiveness calls for regulatory intervention. Loudon has shown that existing data privacy markets are partly regulated in a purely administrative sense and that they are ineffective.

Laudon's proposition of the NIM seems premature although it attempts to expand administrative regulation with intervention of an economic regulation enabling free trade of personal information. Without addressing the specificity of data privacy information, Laudon addresses issues of moral hazard and asymmetry of information attempting to reduce their undesired impact. He foresees problems of societal nature, neglects issues of non-optimality for different classes of goods, and does not take into account solutions provided by the IT technology (such as PPDM).

Acquisti (2002) reveals similar doubts about the existence of satisfactory purely economic regulation to privacy trade - he says "...*It is clear that the economic incentives have failed to generate alone workable solutions: it seems like privacy is more difficult to sell than to protect...*"

Our claim is that including technological (data processing and algorithmic) solutions some of these problems can be overcame. Still, it is not inexpensive. Therefore funding of this change has to be reflected in growth of economy. This issue is discussed in the next Section.

4.3 Growth in Economy with Privacy Market

As previously, we consider an economic system described by *states* at the time t. We consider n attributes $\{x_i\}_{i=1,2,...,N} = \pi_I$ .The attribute $x_{N-1}=\pi_A$ is called the *individual data privacy, and* $x_N=\pi$ is called the *aggregated data privacy*. We assume that changes of labor are slow and small enough to consider them constant and negligible in analysis of results of fast data processing. In our view, privacy is produced by individuals as they take decisions controlling the use or non-iuse of their data, therefore the labor involved is the decision process that leads to the control actions. .Again, we consider evolutions of input and output attributes, which now describe economic processes involving privacy. The dynamics of the system is again described by system of axioms derived from economic interpretation and presented in a mathematical form. We say that *the economic system is in a privacy steady state*, if unit of growth of privacy results from unit of growth of an input attribute[30]) We assume the following set of axioms describing interdependence of variable (economic mechanisms):

---

[29] Laudon presents also a sound and honest analysis of problems related with his proposed NIM. A long list of difficulties shows that to set up a NIM one has win a political and regulatory struggle to introduce fundamental modifications of the social, legal and economic frameworks. Social and legal issues are in implementing NIM are related to the problem of social approval to sell a basic right (privacy), the problem of inequality (the rich and powerful would benefit from NIM? to a greater extent than the poor), radical changes in the American property law. Laudon also points out the problem of costs related to NIM (business transaction and administrative costs) as well as the issue of the necessary trade regulation.

[30] Other cases will be considered elsewhere.

- Privacy plays dual role: individual data privacy it is an input asset, while aggregated data privacy is an output asset (it can be reinvested or consumed).
- Individual data privacy can be produced using, stock products, and technology.
- Aggregated data privacy requires individual data privacy to be produced.
- Privacy is therefore split into individual and aggregated parts (only APDP is invested and the result of this investment is an increase of the capital and decrease of individual data privacy.
- Increase of growth in aggregated privacy is directly proportional to level of individual privacy;
- Increase of labor at each moment is directly proportional to population size;
- Increase of technology at each moment is directly proportional to technology level;

Let us denote by t a time moment, by K and A– input attributes interpreted as capital and technology, while π_I (π_A) is the attribute describing individual (aggregated) data privacy. Let as consider production function $Y=sG(K,A,\pi_I,\pi_A)$ and aggregated data privacy production function P_A, $P_A=\sigma H(K,A,\pi_I)$.

The product $J= KA\pi_I\pi_A$ is called *effective information resource*. We assume that the function $G(K,A,\pi_I,\pi_A)=F(K,J)$ and that F is homogeneous in both variables (i.e. $F(cK,cJ)=c\ F(K,J)$).

If we consider evolutions $\varepsilon(t)=[K(t),A(t),\pi_I,\pi_A]$, then these axioms can be described using the following system of differential equations[31].

$$\frac{dK(t)}{dt} = sF(K(t),A(t)\ \pi_I(t)\ \pi_A(t))=sF(K,J) \qquad (A_1)$$

$$\frac{dA(t)}{dt} = gA(t), \qquad (A_2)$$

$$\frac{d\pi_I(t)}{dt} = a\pi_I(t) \qquad (A_3)$$

$$\frac{d\pi_A(t)}{dt} = b\pi_A(t) \qquad (A_4)$$

We moreover assumed also that the function

$$G(K,A,\pi_I,\pi_A)=F(K,J)=K^\alpha J^{1-\alpha}, \text{ where } 0<\alpha<1, \qquad (*)$$

Theorem 1. Let $\tau\in\mathcal{C}^1$ and the superposition $\tau{\circ}G$ is of the form:

$$\tau{\circ}G = sG(K(t),A(t)L(t)H(t))$$

and it satisfies the Lipschitz condition. Then, given the initial condition $K(t_0)=K_0$, $A(t_0)=A_0$, $\pi_I(t_0) = \pi_I^0$ and $\pi_A(t_0) = \pi_A^0$, there exists a neighborhood of t_0, where the mapping $\tau(t)=A(t)\pi_I(t)\pi_A(t)$ has unique solution[32], which we input to A_1:

---

[31] The derivative dK/dt – change in stock - corresponds to investment (according to the assumption it is equal to savings).

[32] Explicitly, $A(t)= A_0e^{gt}$, $\pi_I(t) = \pi_I^0 e^{at}$, and $\pi_A(t) = \pi_A^0 e^{bt}$, thus

$\tau(t)=A(t)I(t)A(t)=\left(g+a+b\right)\pi_A^0 e^{gt}e^{at}e^{bt} = \left(g+a+b\right)\pi_A^0 e^{(g+a+b)t}$

$$\frac{dK}{dt} = sF(K(t), A(t)\pi_1(t)\pi_{AI}(t)) = sF(K(t), (g + a + b)A_0\pi_A^0\pi_A^0 e^{(g+a+b)t}))$$

Theorem 2. A. The solution of the equation A_1 i.e. the trajectory $\tau:[T_0, T_1] \rightarrow \mathbf{R}^3$, $\tau(t) = (K(t), A(t), \pi_1(t) \pi_A(t))$, of economy $\mathcal{G}_+$, with production process represented by the Cobb-Douglas production

$$Y(t) = K(t)^\alpha (A(t)\pi_1(t)\pi_1(t))^{1-\alpha}$$

and the initial state $\tau(t_0) = (K_0, A_0, \pi_A^0, \pi_A^0)$, is of the form:

$$A(t) = A_0 \cdot e^{gt}, \quad \pi_1(t) = \pi_1^0 e^{at}, \quad \pi_A(t) = \pi_A^0 e^{bt},$$

$$K(t) = \sqrt[1-\alpha]{\frac{s}{g + \pi_A^0 + \pi_A^0}}(A_0\pi_A^0\pi_A^0) \cdot e^{(g+a+b)t}$$

B. For the economy described with production process described by the Cobb-Douglasa function, there is a steady growth trajectory. The condition: s=g+a+b then holds.

Let us denote $\delta = a+b$ $k(t) = \dfrac{K(t)}{J(t)}$ and $y(t) = \dfrac{Y(t)}{J(t)}$. Moreover

Lemma 1. Let $f(x) = F(x, 1)$, then

A. $y = \dfrac{Y}{J} = \dfrac{F(K, J)}{J} = F(\dfrac{K}{J}, 1) = F(k, 1) = f(k)$

B. In steady state the following equation holds

$$\frac{dk}{dt} = sf(k) - k(g+a+b) \cdot$$

The interpretation was given in Figure 2. It explains adjustment dynamics of market of privacy, as in sec. 2.3.

5 Discussion

In order to create the *language* in which we can express this result about the economics of privacy, we found it useful to view privacy as an economic good. A contribution of this paper is the discussion of privacy in the framework of the standard quadrant of the economic classification of goods. This taxonomy of goods introduces private, public, and club goods, and common assets. We show how privacy, individual and aggregate, fits in this quadrant. This has interesting implications both economically, as well as for the technical work on privacy preserving methods (PPDM).

Economics has recognized extensions to Solow's 1956 model with human and social capital. In this paper we propose to extend it even further, by including privacy as yet another factor. We show that such an extension works, in the sense that the resulting growth in privacy, viewed as one of the resources, leads to a steady state.

We also argue that it is interesting to connect this discussion with the technology of privacy, i.e. PPDM. We have made the first step in this direction. We want to observe that recent work in PPDM on protecting of privacy of *models* (as opposed to protection of privacy of data, as most of PPDM algorithms and representations do), when seen from the point of views of privacy as a good, fills an important gap, as it addresses producing privacy at the *aggregate* level (APDG in our notation).

Finally, having reviewed the economic and technological aspects of trading and protecting privacy, we show that neither technology nor economic regulation existing today seem to be able to resolve issues related to fundamental challenges of markets for privacy.

Acknowledgements

S. Matwin's research is supported by the Natural Sciences and Engineering Research Council of Canada, and the Ontario Centres of Excellence. This work has been done during the first author's stay at the Artificial Intelligence Research Institute (IIIA) the Consejo Superior de Investigaciones Científicas (CSIC), Barcelona, Spain, as well as during the second author's research visit at the University of Ottawa. T. Szapiro's research is supported by the Foundation for Polish Science. Both authors thank Morvarid Sehatkar for her help with an earlier version of this paper.

References

Acemoglu, D.: Introduction to Modern Economic Growth. Princeton University Press, Princeton (2009)

Acquisti, A.: Privacy in Electronic Commerce and the Economics of Immediate Gratification. In: Proceedings of ACM Electronic Commerce Conference (EC 2004), pp. 21–29. ACM Press, New York (2004)

Acquisti, A., Grossklags, J.: IEEE Security & Privacy, 24–30 (January/February 2005)

Agrawal, R., Srikant, R.: Privacy-preserving data mining. ACM SIGMOD Record, 439–450 (2000)

Arrow, K.J.: Uncertainty and the welfare economics of medical care. American Economic Review 53, s.941–s.973 (1963)

Becker, G.S.: Human capital: a theoretical and empirical analysis. Princeton University Press, Princeton (1964)

Becker, G.S., Murphy, K.M.: Human capital, fertility, and economic growth. Journal of Political Economy (1990)

Berendt, B., Teltzrow, M.: Addressing Users' Privacy Concerns for Improving Personalization Quality: Towards an Integration of User Studies and Algorithm Evaluation. In: Mobasher, B., Anand, S.S. (eds.) ITWP 2003. LNCS (LNAI), vol. 3169, pp. 69–88. Springer, Heidelberg (2005)

Bergadano, F., Matwin, S., Michalski, R., Zhang, J.: Learning Two-Tiered Descriptions of Flexible Concepts: The POSEIDON System. Machine Learning 8(1), 5–43

Clifton, C.W.: What is Privacy? Critical Steps for Privacy-Preserving Data Mining. In: Workshop on Privacy and Security Aspects of Data Mining (2005)

Domar, E.: Capital expansion, rate of growth and employment. Econometrica (April 1946)

Domingo-Ferrer, J. (ed.): Inference Control in Statistical Databases. LNCS, vol. 2316. Springer, Heidelberg (2002)

Du, W., Zhan, Z.: Using randomized response techniques for privacy-preserving data mining. In: Proceedings of The 9th ACM SIGKDD International Conference on Knowledge Discovery and Data Mining, p. 510 (2003)

Evfimievski, A., Srikant, R., Agrawal, R., Gehrke, J.: Privacy preserving mining of association rules. In: Proceedings of the eighth ACM SIGKDD international conference on Knowledge discovery and data mining, pp. 217–228 (2002)

Friedman, A., Schuster, A., Wolff, R.: k-Anonymous Decision Tree Induction. In: Fürnkranz, J., Scheffer, T., Spiliopoulou, M. (eds.) PKDD 2006. LNCS (LNAI), vol. 4213, pp. 151–162. Springer, Heidelberg (2006)

Harrod, R.F.: An essay in dynamic theory. The Economic Journal 49, 193 (1939a)

Harrod, R.F.: Second essay in dynamic theory. The Economic Journal 70, 278 (1939b)

Iyengar, V.S.: Transforming data to satisfy privacy constraints. In: Proceedings of the eighth ACM SIGKDD international conference on Knowledge discovery and data mining, pp. 279–288 (2002)

Jiang, W., Atzori, M.: Secure Distributed k-Anonymous Pattern Mining. In: Proceedings of the Sixth International Conference on Data Mining. IEEE Computer Society, Los Alamitos (2006)

Kargupta, H., Das, K., Liu, K.: Multi-party, Privacy-Preserving Distributed Data Mining Using a Game Theoretic Framework. In: Kok, J.N., Koronacki, J., Lopez de Mantaras, R., Matwin, S., Mladenič, D., Skowron, A. (eds.) PKDD 2007. LNCS (LNAI), vol. 4702, pp. 523–531. Springer, Heidelberg (2007)

Kantarcioglu, M., Clifton, C.: Privacy-preserving distributed mining of association rules on horizontally partitioned data. IEEE Transactions on Knowledge and Data Engineering 16, 1026–1037 (2004)

Kornai, J.: Economics of Shortage. North-Holland, Amsterdam (1980)

Laudon, K.: Markets and Privacy. Communications of the ACM 39(9) (September 1996)

Lucas, R.E.: Expectations and the neutrality of Money. Journal of Economic Theory 4, 103–124 (1972)

Lucas, R.E.: On the mechanics of economic development. Journal of Monetary Economics 22(1) (1998)

Malin, B.A.: An Evaluation of the Current State of Genomic Data Privacy Protection Technology and a Roadmap for the Future. Journal of the American Medical Informatics Association 12, 28 (2005)

Maslow, A.H.: A Theory of Human Motivation. Psychological Review 50, 370–396 (1943)

Moor, J.: Towards a Theory of Privacy in the Information Age. Computers and Society 27, 27–32 (1997)

Oliveira, S.R.M., Zaïane, O.R., Saygin, Y.: Secure Association Rule Sharing. In: Dai, H., Srikant, R., Zhang, C. (eds.) PAKDD 2004. LNCS (LNAI), vol. 3056, pp. 74–85. Springer, Heidelberg (2004)

Paillier, P.: Public-key cryptosystems based on composite degree residuosity classes. In: Stern, J. (ed.) EUROCRYPT 1999. LNCS, vol. 1592, p. 223. Springer, Heidelberg (1999)

Putnam, R.: Social Capital - Measurement and Consequences. In: Helliwell, J.F. (ed.) The Contribution of Human and Social Capital to Sustained Economic Growth and Well-Being; International Symposium Report, Quebec (2001)

Ramsey, F.: The mathematical theory of saving. Economic Journal 38 (1928)

Rasi, G.: Privacy as Quality in Modern Economy. In: The 26th International Conference on Privacy and Personal Data Protection, Wroclaw, Poland (2004)

Routledge, B.R., von Amsberg, J.: Social Capital and Growth. Journal of Monetary Economics 50(1), 167–193 (2002)

Solow, R.: Growth theory and after. The American Economic Review 78(3) (June 1988)

Solow, R.: A contribution to the theory of economic growth. Quarterly Journal of Economics 70 (1956)

Spiekermann, S., Grossklags, J., Berendt, B.: E-privacy in 2nd generation E-Commerce: privacy preferences versus actual behavior. In: Proceedings of the ACM Conference on Electronic Commerce (EC 2001), Tampa, FL, October 14-17 (2001)

Subramaniam, H., Wright, R.N., Yang, Z.: Experimental analysis of privacy-preserving statistics computation. In: Proc. of the VLDB Worshop on Secure Data Management, pp. 55–66 (2004)

Sweeney, L.: Computational Disclosure Control: A Primer on Data Privacy Protection, Massachusetts Institute of Technology, Dept. of Electrical Engineering and Computer Science (2001)

Vaidya, J., Clifton, C.: Privacy Preserving Association Rule Mining in Vertically Partitioned Data. In: Conference on Knowledge Discovery in Data, pp. 639–644 (2002)

Varian, H.: Economic Aspects of Personal Privacy. U.S. Dept. of Commerce, Privacy and Self-Regulation in the Information Age (1996)

Varian, H.: System Reliability and Free Riding. In: Camp, L.J., Lewis, S. (eds.) Economics of Information Security. Advances in Information Security, vol. 12, pp. 1–15. Kluwer Academic Publishers, Dordrecht (2004)

Westin, A.: Privacy and Freedom. Atheneum, NY (1967)

Yao, A.: How to Generate and Exchange Secrets. In: 27th FOCS (1986)

Zhan, Z., Matwin, S.: Privacy-prteserving Data Mining in Electronic Surveys. In: ICEB 2004, pp. 1179–1185 (2004)

Zhan, J., Matwin, S., Chang, L.: Privacy-Preserving Collaborative Association Rule Mining. In: DBSec, pp. 153–165 (2005)

Appendix – Proofs

The proof of Theorem 1. The existence of solution is immediate consequence of existence theorem which is well known in differential equation theory, por. Dubnicki et al. (1996).

The proof Theorem 2. A. The formula is derived form the direct calculation:

$$\frac{dK(t)}{dt} = sK^{\alpha}(A_0\pi_A^0\pi_A^0)^{1-\alpha} =$$

$$= sK^{\alpha}(A_0e^{gt}\pi_A^0e^{at}\pi_A^0e^{bt})^{1-\alpha} = sK^{\alpha}(A_0\pi_A^0\pi_A^0)^{1-\alpha}e^{(g+a+b)(1-\alpha)t}$$

Thus $\dfrac{1}{K(t)^{\alpha}} \cdot \dfrac{dK(t)}{dt} = s(A_0\pi_A^0\pi_A^0)^{1-\alpha}e^{(g+a+b)(1-\alpha)t}$

After integration we get to:

$$\int_{t_0}^{t}\frac{1}{K^{\alpha}(\tau)} \cdot \frac{dK(\tau)}{d\tau}d\tau = \int_{t_0}^{t}s(A_0\pi_A^0\pi_A^0)^{1-\alpha}e^{(g+a+b)(1-\alpha)\tau}d\tau$$

$$\frac{1}{1-\alpha}K(t)^{1-\alpha} = s(A_0\pi_A^0\pi_A^0)^{1-\alpha} \cdot \frac{e^{(g+a+b)(1-\alpha)t}}{(g+a+b)(1-\alpha)}$$

$$K(t)^{1-\alpha} = s(A_0\pi_A^0\pi_A^0)^{1-\alpha} \cdot \frac{e^{(g+a+b)(1-\alpha)\tau}}{(g+a+b)},$$

$$K(t) = \sqrt[1-\alpha]{\frac{s}{g+a+b}} \cdot s(A_0\pi_A^0\pi_A^0) \frac{e^{(g+a+b)\tau}}{(g+a+b)}. \qquad \square$$

The proof of Theorem 2. B. Let us check the steady state condition.

$$Y(t) = K^{\alpha}(A(t)\pi_A^0(t)\pi_A^0(t))^{1-\alpha} =$$

$$\left(\sqrt[1-\alpha]{\frac{s}{g+a+b}} \cdot (A_0\pi_A^0\pi_A^0) \cdot e^{(g+a+b)t}\right)^{\alpha} (A_0\pi_A^0\pi_A^0 e^{(g+a+b)t})^{1-\alpha} =$$

$$= \left(\sqrt[1-\alpha]{\frac{s}{g+a+b}} \cdot (A_0\pi_A^0\pi_A^0) \cdot e^{(g+a+b)t}\right)^{\alpha} (A_0\pi_A^0\pi_A^0)^{1-\alpha} e^{(g+a+b)(1-\alpha)t}$$

$$= \left(\sqrt[1-\alpha]{\frac{s}{g+a+b}} \cdot\right)^{\alpha} (A_0\pi_A^0\pi_A^0)e^{(g+a+b)}$$

Thus $\dfrac{dY(t)}{dt} = \left(\sqrt[1-\alpha]{\dfrac{s}{g+a+b}} \cdot\right)^{\alpha} A_0\pi_A^0\pi_A^0(g+a+b)e^{(g+a+b)t}$

$$\frac{dK(t)}{dt} = \sqrt[1-\alpha]{\frac{s}{g+\pi_A^0+\pi_A^0}} A_0\pi_A^0\pi_A^0(g+a+b)e^{(g+a+b)t}$$

Thus the condition $\dfrac{dY(t)}{dt} = \dfrac{dK(t)}{dt}$ implies $\left(\sqrt[1-\alpha]{\dfrac{s}{g+a+b}} \cdot\right)^{\alpha-1} = 1$, hence s=g+a+b.

Savings cannot exceed the production so 0<s<1, thus 0<g+a+b<1.   □

The proof of Lemma 1. A – immediate. To prove B let us compute LHS.

$$\frac{dk}{dt} = \frac{d}{dt}\left(\frac{K}{J}\right) = \frac{\dfrac{dK}{dt}J - K\dfrac{dJ}{dt}}{J^2} = \frac{\dfrac{dK}{dt}A\pi_I\pi_A - K\dfrac{d(A\pi_I\pi_A)}{dt}}{(A\pi_I\pi_A)^2} =$$

$$= \frac{\dfrac{dK}{dt}}{A\pi_I\pi_A} - \frac{K\dfrac{d(A\pi_I\pi_A)}{dt}}{(A\pi_I\pi_A)^2 \cdot dt}$$

$$= \frac{1}{A\pi_I\pi_A}\frac{dK}{dt} - \frac{K}{(A\pi_I\pi_A)^2}\left(\frac{dA}{dt}\pi_I\pi_A + A\frac{d\pi_I}{dt}\pi_A + A\pi_I\frac{d\pi_A}{dt}\right)$$

$$= \frac{1}{A\pi_I\pi_A}\frac{dK}{dt} - \frac{K}{(A\pi_I\pi_A)^2}\frac{dA}{dt}\pi_I\pi_A + \frac{K}{(A\pi_I\pi_A)^2}A\frac{d\pi_I}{dt}\pi_A + \frac{K}{(A\pi_I\pi_A)^2}A\pi_I\frac{d\pi_A}{dt} =$$

$$= \frac{1}{J}sY - \frac{K}{J}g - \frac{K}{J}a - \frac{K}{J}b =$$

$$= s\frac{Y}{J} - \frac{K}{J}(g+a+b) = sy - I(g+a+b) = sf(k) - I(g+a+b)\cdot$$

We see that $\dfrac{dk}{dt} = sf(k) - k(g+a+b)$, hence

$$\frac{dk}{dt} \geq 0 \Leftrightarrow sf(k) - k(g+a+b) \geq 0 \Leftrightarrow sf(k) \geq k(g+a+b),$$

$$\frac{dk}{dt} \leq 0 \Leftrightarrow sf(k) \leq k(g+a+b)$$

If k* (the trajectory in steady state) solves the above equation (the graphs of functions sf(k)+hk and k(δ+n+g) intersect), then any perturbation of k* has to converge to k*. If k> k* (k> k*), then k decreases (increases).

Adapting to Human Gamers Using Coevolution

Phillipa M. Avery[1] and Zbigniew Michalewicz[2]

[1] Department of Computer Science, University of Adelaide, South Australia
pippa@cs.adelaide.edu.au
[2] School of Computer Science, University of Adelaide, South Australia;
also at the Institute of Computer Science,
Polish Academy of Sciences, Warsaw, Poland,
and Polish-Japanese Institute of Information Technology, Warsaw, Poland
zbyszek@cs.adelaide.edu.au

Abstract. No matter how good a computer player is, given enough time human players may learn to adapt to the strategy used, and routinely defeat the computer player. A challenging task is to mimic this human ability to adapt, and create a computer player that can adapt to its opposition's strategy. By having an adaptive strategy for a computer player, the challenge it provides is ongoing. Additionally, a computer player that adapts specifically to an individual human provides a more personal and tailored game play experience. To address this need we have investigated the creation of such a computer player. By creating a computer player that changes its strategy with influence from the human strategy, we have shown that the holy grail of gaming – an individually tailored gaming experience, is indeed possible. We designed the computer player for the game of TEMPO, a zero sum military planning game. The player was created through a process that reverse engineers the human strategy and uses it to coevolve the computer player.

1 Introduction

A common method of representing a computer player is by a static strategy for game-play. The representation of the strategy could be a neural network, a set of *IF – THEN* rules (fuzzy or not), decision trees or many other means. Regardless of the representation, the use of a static strategy results in a computer player that becomes obsolete once the human player adapts to it. When a computer player ceases to challenge the human player, it is no longer fun to play against. Thus, a game-play experience that is unique depending on the circumstance, and the human game strategy used, has been of significant importance in recent years. Games such as Fable (Lionhead Studios) and Star Wars: Knights of the Old Republic (BioWare), which change the game scenario dependant on the choices the player makes (to be a good, or evil character) have been very successful. The 'choose your own adventure' style has proven to be a lucrative venture, and adaptive adversaries are key to continued success. It seems that the ability for a computer player to adapt to an individual human player is the holy grail of game-play.

Additionally, the ability to provide an adaptive computer player that tailors itself to the human player has a lot of potential in the training industry. It is now fairly well accepted that playing computer games is beneficial for teaching purposes [5]. The ability to provide a fun and challenging way of learning has clear benefits. The problem lies in

J. Koronacki et al. (Eds.): Advances in Machine Learning II, SCI 263, pp. 75–100.
springerlink.com   © Springer-Verlag Berlin Heidelberg 2010

finding a way to provide the individualistic training for each student. Currently this lies in providing a level rating (e.g. easy, normal and hard) that the player can choose. This system is inadequate however, as the classification of student levels into (normally) three levels of expertise has obvious disadvantages. Instead, by adapting to the individual human, the game play is not standardized for a wide category of expertise. As identified by Charles et. al. [7], in relation to adaptation in games:

> Adaptation as such is strongly connected to learning and we may use it to learn about a player in order to respond to the way they are playing, for example by adjusting a computer opponent's strategy so as to present a more appropriate challenge level.

By providing an experience that is tailored at a specific human player, the player can experience a game that simultaneously gains in difficulty as their experience and skill in the game increases.

The adaptation to the human has a number of advantages. Firstly, it grows in strength with the human. When the human player first plays the game, the strategy creation for the game is fairly weak. The human player then starts to increase in strength as understanding of the tactics increase, and more complex strategies are developed. The same can be said for the coevolutionary algorithm. The initial generations of a coevolutionary computer player create very basic and not overly intelligent rule development. Then, as each game is played, it has a chance to encompass and counter the opposition's strategy of game-play and create better rules. We wanted to use this potential of the coevolutionary system to create a computer player that coevolves to adapt to a human player relative to the human's capability. By including the human strategy into the coevolutionary system, the system can evolve strategies that have adapted to the human strategies. As the human gains more experience in playing the game, so does the computer player.

To achieve the goal of creating an adaptive computer player, we needed to create a way of including the human strategy in a form the coevolutionary system could understand and use. This essentially involved creating a method of reverse engineering the human strategy from the outcome of the game-play. We decided to create an iterative system that would record the data (the human choices and the game environment) from each game played, and use that data to create a model of the human. The model would then be in a form that could be included in the coevolutionary process. To create the model for our coevolutionary system, it would need to be of the same fuzzy logic rule base structure as the other individuals. The human model is then added as a supplementary individual to the coevolutionary populations, so that the coevolution can evolve with, and against the model.

By reverse engineering the human's strategy into a set of rules and adding the model created to the coevolutionary system, we are able to influence the coevolutionary process. At the beginning of the human's learning progress, he or she will not have any experience in playing the game, and are likely to have minimal strategy development. This stage is probably the most difficult to reverse engineer, as the human is more random in strategic choices. Consider the idea of beginner's luck in games such as poker, expert players can have trouble reading beginners, as the beginner makes seemingly random choices due to inexperience. However, even the very general (and probably not effective) rules reverse engineered at this stage have a chance of affecting the co-evolution. It is unlikely that the human rules will be considered the best individual in the population for elitism, but they still have a chance of effecting the next generation through selection and crossover.

As the human starts to gain experience in the game, he or she develops 'winning' strategies. It is likely that the human will then repeat the same or similar strategies over concurrent games if the strategy continues to work. It is here that the adaptive coevolutionary system really comes into play. As the humans form a clearer pattern for reverse engineering, the rules being modelled and added to the coevolutionary system have a greater impact. Now the coevolution can act to directly overcome the human strategy, and create new strategies that are a greater challenge to the human player. This allows us to develop a system that grows and improves along with the human the computer player is competing against. A tailored system that provides the best challenge for the individual human.

This chapter discusses the issues involved with creating a computer player using co-evolution that can adapt to humans, and the methods we used to create this system. We begin with a brief background discussion on the game of TEMPO and related work. We then discuss the mechanism used to create our adaptive computer player. After implementing this system, we ran a user study to observe effectiveness of the system, and the way human players interact with the computer player. The results of this user study are provided, along with discussion and analysis. We conclude the chapter by discussing the findings, and the areas of research that have been identified for future work.

2   The Game of TEMPO

The use of strategic thinking is not limited to game playing, and many of the strategies used by players can be carried into real world situations. This can be seen in a number business and defence organizations, where the organization is essentially competing against rivals for a strategic dominance in their field of expertise. This real world situation can be directly compared to a zero-sum game of strategy, where there are two or more competitors, and only one can win.

In the area of defence, this game playing can occur when countries engage in espionage and weapon research and manufacturing. The ultimate purpose of this is to maintain sufficient utilities to win a war against rival countries if the need arises, but not spend excessively. The maintenance of only *sufficient* utilities is important, as it implies enough utilities to win a war, without neglecting other budgeting issues. To achieve this fine line, the personnel who perform the resource allocation need to know how to think strategically. The resource allocation is made difficult due to influences such as the political motivations of current (and future) governments, the changing field of the technologies used, and of course the opposing countries with their own changing environments [13].

The US Department of Defence (DoD) realized the difficulty involved in the task, and attempted to give their personnel an advantage through the creation of a management system known as the Planning, Programming, and Budgeting System (PPBS). The PPBS put into place a framework for the decision making of defence budgeting, and incorporated a set way of planning for current and future objectives [1]. As part of the set-up of this new system, a major training program was initiated to enable the personnel to use the complex new system. The game of TEMPO was created by H. Hatry, F. Jackson and P. Leer of General Electric's TEMPO think tank as part of this initiative [2], and was used by the DoD in the training of their system personnel [13]. The game enabled the personnel to practice the strategies they would use in the creation of the resource allocation, and subsequent yearly DoD budget. Since its creation in the early 1960s, the TEMPO game has been used to teach resource allocation to over 20,000 students.

The original game of TEMPO was a paper game where opponents were pitted against each other, and their decisions recorded for review by trainers. The efficiency of this was not optimal, as the time taken for game-play with other students limited the functionality. Steps were taken to automate the game with a computer player for the opposition. This allowed the students to play the game on their own time, and the results were automatically recorded and available for the trainers. The computer player system provided a greater learning environment for the students, and was successfully used for training the DoD personnel. The creation of the player opened up a new area of research into AI for resource allocation games. The next section describes player objectives for the TEMPO game.

2.1   Game Objectives

TEMPO is a zero sum game played between two opposing parties by allocating resources in a cold war style simulation. The goal of the game is to acquire more *offensive utilities* than the opposition before war breaks out. The decision making process requires allocating the yearly budget on the following:

1. Operating existing forces.
2. Acquiring additional forces.
3. Intelligence and counter intelligence.
4. Research and development.

The forces of the game are comprised of weapons that are grouped into four unit types: Offensive A, Offensive B, Defensive A and Defensive B (OA, OB, DA and DB respectively). Each of these units has its own weapons, such as OA1, OA2 and so on. Each weapon has its own attributes (discussed further in the next section), with its own power capability given as *utils*. It is the utils of the weapons that are currently operated for the year that give a player his or her score.

The purchase of intelligence is also provided to give insight into the opponent's tactics. Counter intelligence is used to prevent the opposition gaining this insight. Lastly, investment in research and development is available to provide for future weaponry. The use of research and development in the game allows budget allocation to provide for better weapons in future years. It was excluded in the computer player developed for the DoD however, and was not included in this research. We would like to include it in future research, but analysis on its implementation is required.

The resource allocation involved in the game is conceptually simple; determine what force category is needed and allocate accordingly. The reality is however very different, as the combinations of allocation plans can be high due to the amount of areas to allocate to. This complexity is then magnified by the changing environment that occurs yearly, such as the increase in the chance of war breaking out, and the addition of new weaponry.

The complexity is representative of a number of real world situations in the corporate and defence world alike, where resource allocation can be a very complicated and difficult task to manage. To understand *how* to make an allocation, a person must have a good understanding of the strategies and mechanisms used in the process. Only through personal achievement and practice can they truly understand the value of various strategies. This is where TEMPO is a great mechanism to practice the techniques needed to develop a well thought out and balanced real-world allocation. The process could also be easily translated into a business training environment instead of a military one.

Now that we have introduced the game, we delve further into its mechanisms. The next section gives a detailed description of the game used in this research, and the rules of play.

2.2 How to Play

The goal of playing TEMPO is to obtain more offensive utilities than your opposition in a cold war scenario when the end of the game is reached, and war breaks out. To do this, a player must allocate a budget on a yearly basis to operate and acquire weapons, and purchase intelligence. Each year the environment changes, and more choices become available. This section gives details on exactly how the game works, and what information a player is given.

Player's Environment				Previous Year's Data					
		Budget		Player			Enemy		
Year	Pwar	Avail	Left	Type	Offensive	Defensive	Type	Offensive	Defensive
2	12.00%	11530	11530	A	1518	0	A	200	0
				B	200	325	B	0	465

Current Year Allocation:									
Weapon	MaxAcq	AcCost	Inventory	OptCost	Utils	Opted	Bought	ToOpt	ToBuy
OA1	15	75	0	150	120	30	0	0	0
OB1	25	50	0	30	20	30	0	0	0
DA1	25	40	0	20	15	25	25	0	0
OB1	25	100	0	60	50	20	20	0	0
OA2	35	75	0	35	60	0	0	0	0
DA3	25	100	0	50	200	0	0	0	0

Fig. 1. Example screen of a year's game-play

Figure 1 shows an example year of game-play excluding the intelligence component. The player's environment section shows the yearly information necessary to make decisions. This includes the current year of game-play (*year*), the percentage chance of war occurring at the end of the current year (*pwar*) and the given budget for the year (*budget*). The budget is represented as the amount given (available) for the year, and the amount left after spending. Each player in the game starts with the same environmental values, but has slightly different values for consecutive years, as each value is increased by a limited random amount. At the end of each year, the average of the *pwar* values for both players (represented in range [0..1]) is compared against a randomly generated number from the same range, and if the generated number is less than the *pwar* value, war breaks out and the game is over.

Using the amount given in the *budget*, the player can purchase weapons from the current year's available weapon list. Each year new weapons may become available, possibly with better attributes then previous weapons. The attributes for each weapon are:

1. *MaxAcq* – the maximum acquisition number for the weapon each year.
2. *AcqCost* – the cost to acquire (buy) a single unit of the weapon.
3. *Inventory* – the amount of weapons given to the player in inventory for the year (these are then available for operation).
4. *OptCost* – the cost to operate the weapon for the current year.
5. *Utils* – the power value for the weapon.

Each weapon can be in one of two stages during the game years. These stages are *acquisition* and *operation*. When beginning the game, a player has initial units in their inventory. To obtain additional weapons, they must be acquired. You can acquire up to the MaxAcq number of weapons during a year, and each one bought will cost the indicated amount in AcqCost. The weapons acquired in the current year of game-play, will be available to operate the next year. Operating a weapon activates the weapon for the year. The available weapons to operate are any weapons in inventory, any weapons bought the previous year, and any previously operated weapons. If a weapon is not operated in the current year it is lost for future use. If a weapon is operated, it is then 'used' for that year, and the utils for the weapon are added to the player's total utils.

At the end of each year, the total weapon utils for each category/type are summed. For example, if a player purchases units of OA1 with total utils of 300, and OA2 with total utils of 100, then the total OA for the year is 400 utils. Offensive weapons of a particular type are countered by defensive weapons of the same type, and vice versa. For example, if Player A has 400 OA utils at the end of a year, and Player B has 100 utils of DA, then the result at the end of the year would be 300 utils of OA left for Player A. Extra defensive utils however, are wasted budget. For example, if Player B had 200 OA utils in the same scenario, and Player A had 300 DA utils, then Player B would have 0 OA utils left, and Player A would have wasted the cost for the extra 100 DA utils. This example is extended and shown in table 1 for clarity.

Table 1. Example TEMPO net offensive util scoring

Player A		Player B	
	Type A		
$OA_{playerA}(OA1 + OA2 + ... OAn)$	400	$OA_{playerB}(OA1 + OA2 + ... OAn)$	200
$DA_{playerA}(DA1 + DA2 + ... DAn)$	300	$DA_{playerB}(DA1 + DA2 + ... DAn)$	100
Net Offensive A $(OA_{playerA} - DA_{playerB})$	300	Net Offensive A $(OA_{playerB} - DA_{playerA})$	0
	Type B		
$OB_{playerA}(OB1 + OB2 + ... OBnn)$	200	$OB_{playerB}(OB1 + OB2 + ... OBn)$	600
$DB_{playerA}(DB1 + DB2 + ... DBn)$	300	$DB_{playerB}(DB1 + DB2 + ... DAn)$	100
Net Offensive B $(OB_{playerA} - DB_{playerB})$	100	Net Offensive B $(OB_{playerB} - DB_{playerA})$	300
Total Net Offensive Utils	**400**	**Total Net Offensive Utls**	**300**

If war had broken out in the year represented in table 1, Player A would have won the game by 100 utils (Player A total net offensive utils - Player B total net offensive utils). Correspondingly, Player B would have lost by -100 utils. This example shows how a player cannot win the game simply by maximizing offensive weapons, as the other player can cancel these with defensive ones. Additionally, a sliding scale 'diminishing returns' function is applied when currently operated utils in any one force type (e.g. OA1) produces more than 2000 utils. The adjustments applied are shown in table 2, with figure 2 depicting the diminishing returns distribution.

The amount of total net utils for OA, OB, DA and DB from the previous year are displayed to the player (on the top right of the screen in figure 1). If the player purchases intelligence, then they are also given the opposition results from the previous year (which are displayed to the right of the player's results), although these may be skewed somewhat through the opposition purchasing counter intelligence.

Table 2. Util adjustments to reflect diminishing returns

Gross utils (GU)	Adjusted utils
1 - 2000	Same as Gross utils
2001 - 3000	2000 + .9 × (GU-2000)
3001 - 4000	2900 + .8 × (GU-3000)
4001 - 5000	3700 + .7 × (GU-4000)
5001 - 6000	4400 + .6 × (GU-5000)
6001 - 7000	5000 + .5 × (GU-6000)
7001 - 8000	5500 + .4 × (GU-7000)
8001 - 9000	5900 + .3 × (GU-8000)
9001 - 10,000	6200 + .2 × (GU-9000)
10,001 - 11,000	6400 + .1 × (GU-10,000)
11,000 - ∞	6500

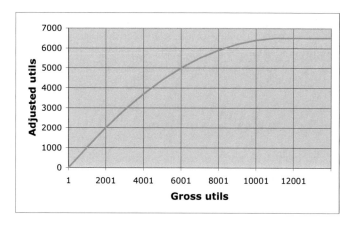

Fig. 2. Diminishing return distribution for util adjustments

Each year of game-play gives an increase in the *pwar* and *budget*, but it also gives new weapons with their own unique attributes. This allows the player to more choice as time goes on, but also increases the complexity of the choices.

The intelligence component of the game involves allocating part of the budget to purchase intelligence into the opposition's allocations. The intelligence is broken into two parts: the intelligence into the opposition's results (*INTEL*), and the counter intelligence used to stop the opposition from seeing your results (*CI*). When a player purchases INTEL, the opposition results for the previous year are given to the player. If CI is bought, the player is only told if opposition utils in a particular category *exist* or not. The original version of TEMPO included a boolean decision for both INTEL and CI. The INTEL was broken into offensive and defensive INTEL, with a set price that you either purchase it at or not. The CI was also a set price with the same boolean choice. For various reasons, this was then changed and INTEL and CI were broken down into the different types of Offensive and Defensive Intelligence A and B (OIA, OIB, DIA, DIB). The boolean mechanism of purchase was also changed and replaced with a maximum cost for each type. The player could then decide the degree of INTEL/CI to be purchased. The percentage of INTEL/CI purchased then effects the quality of the results obtained back.

There are various common tactics for TEMPO that human players learn through game-play. After witnessing a number of games being played by humans, there emerge some common strategies that can be beneficial general tactics. These include such things as using the *pwar* variable to determine how you concentrate your allocation, for example if it is low, you might choose to focus your attention on building up your operational offensive weapons. Another common tactic is to focus your allocations on the weaponry that gives you value for money, that is the weapons that have the highest amount of utils for the cost used to purchase/operate them. There are many other tactics in addition to the ones mentioned, however as yet there is no magical strategy that will win against any opponent.

The dynamic environment of TEMPO, combined with the increasing complexity caused by the number of weapons available can make the allocation decision process difficult. This is amplified by the uncertainty of what the opponent is doing at the same time. It is only once you have committed your allocation for the year that you can find out what purchases the opposition has made (if you chose to purchase intelligence), and even then the information may be corrupted due to the opposition's purchase of CI. It can be a difficult game for human players to master, and the creation of a computer player is a challenging task.

3   Related Work on Adapting to Humans

While there has been much research on evolving computer players to beat human players, there has not been much research on evolving computer players that are designed to be *challenging* for humans. This means creating computer players that are specifically tailored to give the human player a challenge, not just trying to find the optimal way to beat them every time [8, 22].

One area of research in strategic planning is that of predicting an opponent's strategy and modifying your strategy accordingly. This is a large area of investigation, which ranges from opponent modelling [9] to human behaviour recognition fields [12]. Carberry performed a review on one of the more important techniques of plan recognition in [6]. The plan recognition technique uses inference to determine an opponent's plan and goal through observed actions. This is achieved through chaining actions to goals reachable through these actions. For example, on observation a person at the store purchases eggs, milk and flour. From this it might be reasonable to infer that they plan to bake something. They then add some Maple Syrup and ice cream to their purchase. It now might be reasonable to infer that they are specifically making pancakes. As humans we do this all the time, but this can be difficult for computers to achieve, especially as the search space of possibilities expands.

The individuals developed using the TEMPO coevolutionary system create a static rule base that human players can adapt to beat over time. Our long term goal is to create a system where the individuals are evolving and adapting to beat a particular human player during real-time game-play. To find a way to adapt to a human player, we investigated two fields of research: ways to extract rules representing a human's strategy from the output of their game-play, and ways that the system can use these rules to adapt.

The most relevant work is by Louis et. al. with their Case-Injected Genetic Algorithm (CIGAR) research [14, 15, 16, 17, 18]. The CIGAR system uses a database (the case base) of problems mapped to solutions (cases) to prompt the GA to come up with better and more human-like solutions. The basic CIGAR system is a GA that starts with an empty case base and a randomly initialized GA. Once the GA is run, the best individuals (represented as cases) are saved in the case base. When another similar problem comes along, instead of starting with a randomly initialized GA, the case base from the previous problem is used to inject a percentage of the population with previously favourable individuals, and so on. The individuals to select from the case base are chosen by similarity to the current best individual in the GA, using a hamming distance metric. Interestingly, it was noted that individuals that were too advanced for the current population would actually hamper the evolution. The selected individuals then replace the worst individuals in the GA.

The work applying CIGAR to games used a strike force real time strategy (RTS) game, where the computer player has to allocate its resources to a set of aircraft platforms (the blue team). The platforms then attack the human players forces (the red team), which are represented by buildings that the aircraft can target, and defensive installations that can attack the computer players air force. The CIGAR system for this works in the same manner as described above. In this case it also includes cases learnt from humans playing the blue team in previous installations, by storing the human moves in the case base. Thus, the case base consists of human derived cases, and cases discovered by the GA from playing against human and computer players. The cases are then chosen using the same similarity metric, with the possibility of a human case being chosen – and the evolution learning from human game-play.

This approach differs from ours in a number of ways. Firstly the TEMPO system uses a coevolutionary mechanism, not a GA. This changes the way it can use the memory, and adapt to human players. The coevolutionary system can be run without any human interaction, and is constantly changing and adapting. There is no search for an overall optimum, just a way to beat the current opposition. There are also differences in the way individuals are represented. The TEMPO coevolutionary system uses individuals representing a strategy in fuzzy logic. To use human knowledge in the TEMPO system, the human strategy must also be represented as a fuzzy rule base, which can be difficult.

Additionally, the TEMPO game is not a RTS game – it is a turn-based game where each player makes his or her decisions simultaneously. Each year of game-play, the environment of the game changes and becomes more complex, and the knowledge on what the opposition is doing is minimal and can be misleading (if counter intelligence has been used). The players being developed in the TEMPO system are developed to encompass all this and develop generalised strategies for game-play.

Finally there is the way the entire adaptive system works. Our system to creates a player that is tailored to an individual human. This means that the human rules have to be obtained from, and added to, the currently evolving system. This differs from the Louis et. al. research, where the human knowledge was obtained from humans playing past games.

Other relevant research by Ponsen et. al. [19, 20, 21] extends the dynamic scripting method developed by Spronck et. al. [23, 24, 25] for RTS games. The dynamic scripting method is used to change the strategy rules (the script) for an opponent during game-play. Rules that perform well in a particular dynamic situation are given a higher weighting, and are then more likely to be selected. The rules themselves are manually designed for the specific game being implemented (similar to the way most current AI for games is done).

Ponsen extends the dynamic scripting to RTS by changing the script during successive stages of the game as more resources become available. In addition, an offline evolutionary algorithm was applied that attempted to create scripts to counter well-known optimized tactics (for the game of WARGUS). Static players were used to measure against and the fitness was adjusted for losses and wins against the static player. Aha et. al. [3] addressed some of the disadvantages of using a static player. Aha et. al. used case-base reasoning to select scripts for game-play against random opponents chosen from eight different opponent scripts. The evolutionary algorithm developed by Ponsen et. al. was used to develop scripts for each of the eight opponent scripts, and the case-based system (CAT) then chose which tactic to use (from possibly different scripts) at each stage of the game.

The research using dynamic scripting has many similarities to the research presented in this paper. However, once again the research does not use coevolution, and the mechanisms for the research are very different. The overall goal is slightly different as well, as the main goal for the TEMPO system is for training purposes.

Lastly, the idea of coevolving against human players was investigated by Funes et. al. [10, 11]. In this work, the game of Tron was used to evolve against human players on the internet. The game-play involved the human playing against a strategy developed using Genetic Programming (GP) coding. The evolution process contains two separately evolving populations. The foreground population contains a population of 100 individuals that play against human players for evaluation. Computer players are randomly selected from the population for game-play. In each generation the worst 10 individuals are replaced by the top 10 from the background population. The individuals are evaluated by playing a set number of games (5 for the 90 'veteran' players, and 10 for the 10 'new' players) against human players, the results of all games played against a human are then used to calculate the fitness. The second population (the background population) is used to evolve strategies through self-play, training a population of 1000 individuals against a training set consisting of the best individual from the foreground population, and the top 24 individuals from the previous background population generation.

The research using Tron showed how coevolution with humans could be effectively used. However, the learning of the foreground population was very slow (with at least 550 games needed each generation). We are using coevolution to adapt real-time (after every game played) to a single human player, using a method of human opponent modelling to coevolve with. The method of our coevolution is also different, without a direct evolution against the human player, and with two populations of individuals performing self-play to learn strategies. The previously discussed differences in game and individual representation also apply. Also, our testing involved a controlled test environment, which online learning cannot readily achieve.

4   Coevolving with Humans

It is commonly known that playing a static coevolved player against the same human repeatedly, allows the human to determine a counter-strategy that is dominant over the static computer player. The first time the human plays against the strategy however, it is unknown and could possibly be difficult to beat. The question then arises as to whether the need for adaptation is indeed necessary. Could we not just continue the coevolutionary process, and pick different individuals to play against the human each time? There could well be enough difference in strategy represented through the coevolutionary process itself to provide a challenge for each new game played.

While there has not been much recorded research on this topic, intuitively it would seem that a similar occurrence to the static scenario would eventuate. This reasoning is based on observation through our previous research, where the coevolution reached a plateau. This plateau is visible through the baseline measurement technique used, where the results show very little change. Even though it is constantly changing and undulating, the evolution does not tend to make great leaps in development. Thus, even though the human player would be playing a different player each time, the strategies being developed by the player are similar in strength to previous ones, and the human player could learn to overcome them. Additionally, the tailored system allows the players to directly counter the human strategy making, and hence provide a greater individual challenge. In teaching terms this is a great advantage.

There are a number of ways that the human models could be used in the coevolutionary process. Originally we thought of having a separate human population, similar to the memory population in previous research. However, the extra processing time required for selecting and evaluating a separate population was deemed excessive for the purpose. Additionally, if evaluation were the only influence the human model had on the populations, poorer human strategies would have little to no effect when they are constantly beaten. Including the human models into the populations has a direct effect on the individuals being created (through selection and crossover).

To include the human in the process, we needed to find a way to coevolve against the human model and the other randomly evolving players. This allows the system to create players that are still finding randomly evolving strategies that can take into account, and counter, the human player's actions. By including the human model in the coevolutionary process, when a new 'best' individual is chosen from the system to play against a human player, this individual has been able to incorporate and counter the stronger elements of the human strategies. Now the individual can provide a new challenge for the human.

5   Representing Human Strategies

Representing (modelling) humans is a research field in itself and can be very difficult to do. To minimize this issue, we chose to very loosely reverse engineer the human choices as a model of the strategies used. The model used would also need to be of the same format as the coevolutionary individuals in the TEMPO system. Doing this allows the human model to be directly inserted into the coevolutionary system for the process described above.

The reverse engineering of the human was as follows. When a human plays a game against the computer player, the data of the game is recorded. This data includes the choices the human made, and the environmental data for each game year. The data is then used in an evolutionary system to find individuals that model the human by mimicking the human choices. To evolve the human model, individuals represented in the same way as the coevolved individuals are randomly initialized. Each individual then plays the exact same game as the human, against the same computer player as the human played. The individual is evaluated as the difference in outcome and allocated resources to what the human achieved. The closer the individual comes to the same results as the human, the better the fitness.

Additionally we added changeable constant weightings to the evaluation. The weightings were applied to the differences in the outcome, weapons bought and intelligence/counter intelligence bought between the individuals and the human value. By adding weights, we are able to sway the evaluation importance of each of the parameters for the purpose of creating more realistic rules. For example, it might be that getting a closer outcome (total offensive utils at the end of the game) to the human was more important than creating rules that allocated the resources in the same manner as the human, and vice versa. Hence the evaluation function has the following evolutionary variables:

1. *human_NetUtils* – the total net offensive utils for the human player at the end of their game.
2. *individual_i_NetUtils* – the total net offensive utils the individual scored (when playing the same game).
3. *Years* – the total number of years the game played for before war broke out.
4. *human_IntelChoice*, *human_OptChoice*, and *human_BgtChoice* – the data arrays of human allocations made for intelligence, weapons operated and weapons bought respectively for the year.

5. *individual$_i$_IntelChoice*, *individual$_i$_OptChoice*, and *individual$_i$_BgtChoice* – the allocations the individual made for the year corresponding to the human allocations.
6. *NetUtilsWeight*, *IntelWeight* and *WeapWeight* – the constant weights applied to the different evaluation areas as described.

The described evaluation function *eval* is implemented as:

$$\begin{aligned}
eval(individual_i) = \; & abs(human\_NetUtils - individual_i\_NetUtils)\,NetUtilsWeight + \\
& \sum_{t=1}^{Years}((abs(human\_IntelChoice_t - individual_i\_intelChoice_t)\,IntelWeight) + \\
& (abs(human\_OptChoice_t - individual_i\_OptChoice_t)\,WeapWeight) + \\
& (abs(human\_BgtChoice_t - individual_i\_boughtChoice_t)\,WeapWeight)),
\end{aligned}$$

We experimented with different values for the constants, with different preference weights for the resources and outcome. For the final process, the NetUtils constant was assigned the highest preference with a weight of five, followed by the Weapon constant with a weight of three. The INTEL/CI constant was given a weighting of one.

We implemented the human reverse engineering mechanism using an evolutionary algorithm, with a single population. The variation operators used were two-point crossover and mutation, where chosen genes were replaced with a random value. Crossover was run on the population first, followed by mutation. To avoid premature convergence on a suboptimal solution, we also forced the individuals to have unique genotype.

We experimented with selection operators, and used ranked selection with elitism in the final process. After much experimentation and manual changing of the parameters to determine a good result, the final evolutionary parameters chosen were as follows. The process ran for 150 generations, with a population of 100 individuals. An elitism ratio of 5% was used, with a 50% crossover ratio, and a 10% mutation ratio. No rule penalty was applied to minimize the rules used in the rule bases, as this seemed to occur naturally.

The rules evolved using this method give a rough estimation of possible strategies the human used. It by no means represents the human strategy exactly, which is in many ways a good thing. Our task is not to try and create an optimized computer player against a human player, but to create a computer player that is *challenging* for the human player. Even if it is only evolving against a rough estimate of the human player, for a single game-play situation, the evolutionary process is still given the opportunity to counter the human strategies.

6   The Human Adaptive Coevolutionary Process

To incorporate all the ideas described above, we needed to develop an entire system for game-play. We named the system the Human Adaptive Coevolutionary Process (HACP). This section describes the coevolutionary system used to create the computer player, followed by an overview of how HACP is composed.

6.1   The Coevolutionary System

HACP uses a coevolutionary algorithm comprised of two populations of individuals, and a memory population used to evaluate against. This section describes the coevolutionary mechanism used.

6.1.1 Individual Representation

The representation of the fuzzy logic system used a structure that can be seen in figure 3. There are $m = w + q$ rules (where w is the maximum number of weapon rules, and q is the maximum number of intelligence rules). Each rule is built from the following (figure 3 expands rule 3): U_3 is a Boolean defining if the rule is used, B_{i3} is a boolean defining if input i is used, MF_{i3} is the triangular membership function used for the input i, and Y_3 is the output in range [0..1] for Rule$_3$.

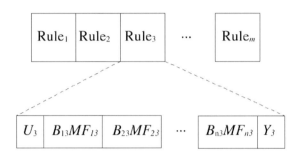

Fig. 3. Structure of a chromosome

To implement the triangular membership functions, we used a representation of the Takagi and Sugeno fuzzy system. We created a single triangular membership function for an input dimension, corresponding to one of the linguistic variables used. Each input type has its own set of membership functions. The membership functions either represent the different types of input (e.g. category is either Offensive – 0, or Defensive – 1), or a range measurement (very low – 0, low – 1, medium – 2, high – 3, very high – 4).

We also needed to represent the linguistic variables. Given the total minimum and maximum of a particular input, and the number of membership functions needed, the linguistic variable would create the required number of membership function objects for the input type. Each membership function has a minimum, centre and maximum value. A diagram showing a typical membership function is shown in figure 4. The membership function is represented as an isosceles triangle with a y value of 1.0. The bottom edge is of length $maximum - minimum$, and the centre value is used to divide the triangle into the left and right right-angle triangles. When passed an input value ($u \varepsilon U$), the slope-intercept equation: $y = mu + b$ is used to calculate the membership degree. In our implementation $m = 1/(centre - minmax)$ where $minmax$ is either minimum or maximum depending on which half of the triangle u is in. We then use $b = -minmax/(centre - minmax)$.

The fuzzy controller takes each input value for the year (e.g. budget, weapon category etc.), and matches the value against the fuzzy *IF* part of the rules, getting the membership degree for each input that triggered a fuzzy rule. The fuzzy *AND* product rule was used to sum all the membership values for all the used inputs in the rule. The weighted average of all the rules was then taken as

$$y(\mathbf{u}) = \frac{\sum_{l=1}^{m} w^l y^l}{\sum_{l=1}^{m} w^l}, \tag{1}$$

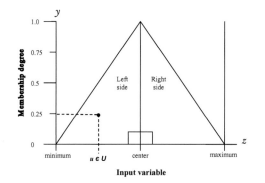

Fig. 4. Membership function example

where **u** is the input vector for the rule base, m is the number of rules in the rule base, y^l is the crisp output value of rule l, and w^l is the product of the membership degrees for all triggered input values. We define

$$w^l = \prod_{i=1}^{n} \mu\, var_i^l\, u_i\,, \tag{2}$$

where n is the number of inputs, and $\mu\, var_i^l\, u_i$ is the membership degree (μ) of input u_i in the corresponding linguistic variable var_i, for the rule l.

The above process is run for each of the different weapons available, with some common input values for all weapons (such as pwar, budget etc.), and some weapon specific values (such as category, type, utils etc.). The total $y(\mathbf{u})$ value obtained from all the triggered rules for each weapon with its specific **u** vector is then used to allocate a percentage of the budget (the *ratio*) for the specified weapon. The allocation is performed by normalising all the weapon $y(\mathbf{u})$ values with

$$ratio_i = \frac{y(\mathbf{u})_i}{\sum_{j=1}^{m} y(\mathbf{u})_j}\,,$$

where i is the weapon index currently being allocated a ratio, and m is the number of weapons available for purchase at the time.

The intelligence rule base goes through the same process.

6.1.2 The Coevolutionary Algorithm Implemented

Our coevolutionary system is based on two competing populations to evolve individuals represented by fuzzy logic rule bases. The outline of our algorithm is shown in figure 5.

To initialize the populations, each individual creates a gene array for the weapon and intelligence rule bases, and randomly assigns integer values from the range [0..99]. These values are then mapped from genotype to the appropriate phenotype, depending on the allele requirements, as part of the evaluation phase. For example, if the gene locus requires a phenotype to a linguistic variable, then the minimum phenotype value is the

procedure The Coevolutionary Algorithm
begin
    $t \leftarrow 0$
    initialize $P_1(t)$
    initialize $P_2(t)$
    while(**not** termination-condition) **do**
    begin
        evaluate $P_1(t)$ against $P_2(t)$
        evaluate $P_2(t)$ against $P_1(t)$
        select $P_{1\_elites}(t)$ from $P_1(t)$
        select $P_{2\_elites}(t)$ from $P_2(t)$
        select $P_{1\_intermediate}(t)$ from $P_1(t)$
        select $P_{2\_intermediate}(t)$ from $P_2(t)$
        alter $P_{1\_intermediate}(t)$
        alter $P_{2\_intermediate}(t)$
        select $P_1(t+1)$ from $P_{1\_elites}(t) \cup P_{1\_intermediate}(t)$
        select $P_2(t+1)$ from $P_{2\_elites}(t) \cup P_{2\_intermediate}(t)$
        $t \leftarrow t+1$
    end
end

Fig. 5. The coevolutionary algorithm implemented

first membership function number, and the maximum is the total number of membership functions for the variable (e.g. $0 - 4$).

After the initialization, the individuals are evaluated against the other population. The basic system applies the following evaluation technique (with modifications introduced later). We iterate over each population, with each individual played against r randomly chosen individuals from the opposition population. For our experiments $r = 20$. A single game-play involves a complete game, through to the final year when war breaks out and a total net utils is allocated for the game. In each game the two players distribute their budget as determined by the rule base. At the end of the game, the total net utils for each player is determined and the fitness evaluation variables updated accordingly. These variables keep track of the number of games played by the individual, the total net utils for all games played, and the won and loss count for the games (a draw counts as a loss for this purpose). Each individual plays a minimum n games, but with a possible (but improbable) maximum $n + (O_p n)$ times, where O_p is the opposition population number. The maximum is due to the random selection for game-play by the opposition's evaluation round. The evaluation function consists of a number of evaluation variables, defined as:

- *wonRatio* – the total number of wins divided by the total games played.
- *totalNetUtils* – the sum of all total net utils (both positive and negative) for all games played.
- *gamesPlayed* – the total number of games played whilst evaluating the individual.

- *weapRulePenalty* and *intelRulePenalty* – the constant parameters used to determine the weight of the rule penalty for the rule bases (weapon and intelligence respectively).
- *usedWeaponRuleNum* and *usedIntelRuleNum* – the number of rules that were marked as used for each rule base.
- *usedWeapInputNum* and *usedIntelInputNum* – the number of input variables used for each used rule, in each rule base.
- *weapInputNum* and *intelInputNum* – the total possible usable inputs for each rule base.
- *netUtilsPenalty* – assigned the highest net utils scored from all games played.
- *lossCounter* – counts all games lost or drawn from the games played.

The evaluation function *eval* is then calculated as follows:

$$eval(individual_i) = wonRatio + (10e^{-6} ((totalNetUtils/gamesPlayed) -$$
$$weapRulePenalty (usedWeapRuleNum + usedWeapInputNum/weapInputNum) -$$
$$intelRulePenalty (usedIntelRuleNum + usedIntelInputNum/intelInputNum) +$$
$$10e^{-6} (netUtilsPenalty \times lossCounter / gamesPlayed)))$$

Once all the individuals have been evaluated, the populations are sorted by the individual's fitness score (which maximises the evaluation function). After sorting, the elite individuals from each population are collected as $P_{1\_elites}(t)$ and $P_{2\_elites}(t)$ respectively. The number of elites saved is determined by the evolutionary parameter *elitismRatio*, which was set to 10% for most experiments. The intermediate populations $P_{1\_intermediate}(t)$ and $P_{2\_intermediate}(t)$ are then selected from $P_1(t)$ and $P_2(t)$ through tournament selection ($k = 2$). The $P_{intermediate}(t)$ populations are of size $Pn - (Pn \times elitismRatio)$, where Pn is the size of the corresponding $P(t)$ population. We conducted experiments with rank selection, but found that tournament selection worked better.

The intermediate populations are then altered, with either mutation or crossover applied. The evolutionary parameter *xoverRatio* defines the chance of crossover occurring instead of mutation. If the mutation operator is used, the chosen parent has mutation applied at a rate defined in the evolutionary parameter *mutationRatio*. If mutation is applied to a gene, there is a 10% chance of a big mutation being applied where the gene is randomly reassigned a value. Otherwise a small mutation of plus or minus 1 is applied (with boundary checking put in place). If the crossover operator is chosen, a randomly selected two-point crossover is applied. The variation operators were carried over from the Johnson et. al. research, but the size of the individuals make at least two-point crossover necessary. Possible future work could investigate k-point crossover.

Once the alterations have been applied to the intermediate populations, the survivor populations ($P_1(t + 1)$ and $P_2(t + 1)$) are created. These are created as $P_1(t + 1) = P_{1\_elites}(t) \cup P_{1\_intermediate}(t)$ and $P_2(t + 1) = P_{2\_elites}(t) \cup P_{2\_intermediate}(t)$.

In addition to this, a memory mechanism was added mimicking the short and long term memory of humans (STM and LTM respectively). The memory consisted of an extra population populated with the best individual from each population from each generation. Individuals were then selected from the memory for evaluation against the other two populations. The selection was performed using a mechanism to replicate

STM and LTM. To implement the STM function, the top ten individuals in the memory were identified, and an additional amount of games played against them. The LTM was then applied by playing another set of games against the entire memory. So where previously the fitness was created by playing r games, now the games against the memory is split into r_1 games against the opposition, plus r_{2S} games played against the STM, and r_{2L} games against the LTM. The memory individuals selected for the LTM gameplay are selected with linear time based probability, so the more recent individuals have a higher chance of being selected. The STM is also selected from with the same linear time based probability, but as the size is a static 10 individuals, as the entire population grows, the probability becomes more uniform. Our used parameters included a window of the top ten individuals (the STM), and played against these individuals for $r_{2S} = 10$ games. This was followed by $r_{2L} = 10$ games against the LTM (the entire memory). The games played against the opposition remained the same ($r_1 = 20$).

6.2 The HACP System

HACP incorporates the described coevolutionary system with a graphical user interface (GUI) to play against human players. This was then combined with the human reverse engineering (modelling) system described in section 5. The process flow can be seen in figure 6.

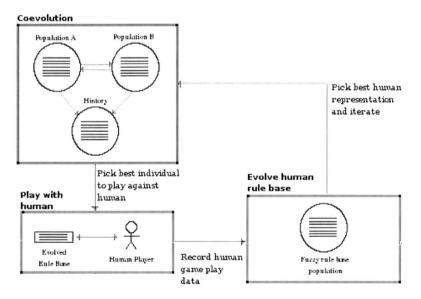

Fig. 6. The human adaptive coevolutionary process

The process consists of the following steps:

1. The process begins by coevolving two populations against each other and the memory population, using the STM and LTM as described previously.

2. After a set number of generations, the best individual from the currently winning population is chosen to play against the human.
3. The data from this game is then recorded and passed to the evolutionary human modelling system.
4. The modelling system then evolves a rule base that mimics the actions the human made. This system runs for a set number of generations before the best individual is selected.
5. The best individual is then placed into the coevolutionary system, replacing the worst individuals from each population, and the whole process iterates again from the beginning.

To ensure the human model affects the coevolutionary populations, the population size was cut down to 15 individuals. Thus, even if the human model has a weaker fitness than the individuals in the population, it still has an affect through selection. The first time the coevolution is run, it runs for 300 generations. This is enough to develop a beginner level player that buys some form of weaponry. The consecutive iterations of coevolution then only have 100 generations to coevolve a new player, which allows reasonable time (generally < 1 minute) in between games with the human.

The number of sample games played against the opposition and memory for evaluation was also cut down to decrease the time taken. For this process, the individual is evaluated by playing $r_1 = 5$ games against the opposition, $r_{2S} = 5$ against the Short Term Memory, and $r_{2L} = 5$ against the Long Term Memory. Crossover was applied at rate of 30%, and if crossover was not applied to the individual, mutation was applied for each gene at a rate of 30% with a 10% chance of a large mutation.

The final component of the process is the human modelling system. This is added using the evolutionary process described in section 5. The human modelling system added on average an extra minute to the entire process.

The GUI was coded in Java and communicated to the C++ code through http sockets. Figure 7 shows a screen shot of the GUI mid game.

The environment section shows the current year, the chance of war breaking out at the end of the current year (the *pwar*), the budget for the year, and the amount of the budget left as the user allocates to weapons and INTEL/CI. The previous year's data section shows the total OA, OB, DA and DB utils left over from the previous year (once the opposition's corresponding utils have been subtracted) for both the player, and the opposition. The opposition utils are only shown if INTEL has been purchased, otherwise "UNKNOWN" is displayed. If the opposition has purchased CI, then the value shown in the enemy's previous year data may be incorrect to some degree. The previous year's INTEL section displays the amount of INTEL bought by the player for the previous year. See [4] for more details on the INTEL/CI implementation used.

The Intelligence Allocation table and Counterintelligence Allocation table both have three columns. Each row in the table represents a different type of INTEL/CI category, with the associated cost. The last column in each table is the user entry field, where the user can enter the budget amount allocated to the category. The cost of INTEL/CI is deliberately low to encourage purchase.

The Weapons table displays all the weapons available for the year. Each row represents a different weapon with the corresponding attributes for the weapon. The first

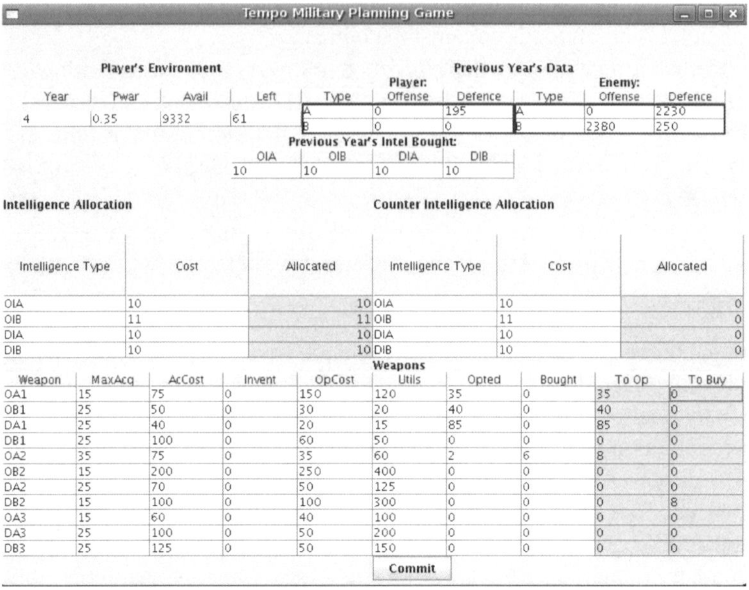

Tempo Military Planning Game

Year	Pwar	Avail	Left	Type	Player: Offense	Defence	Type	Enemy: Offense	Defence
	Player's Environment					Previous Year's Data			
4	0.35	9332	61	A	0	195	A	0	2230
				B	0	0	B	2380	250

Previous Year's Intel Bought:

OIA	OIB	DIA	DIB
10	10	10	10

Intelligence Allocation | **Counter Intelligence Allocation**

Intelligence Type	Cost	Allocated	Intelligence Type	Cost	Allocated
OIA	10	10	OIA	10	0
OIB	11	11	OIB	11	0
DIA	10	10	DIA	10	0
DIB	10	10	DIB	10	0

Weapons

Weapon	MaxAcq	AcCost	Invent	OpCost	Utils	Opted	Bought	To Op	To Buy
OA1	15	75	0	150	120	35	0	35	0
OB1	25	50	0	30	20	40	0	40	0
DA1	25	40	0	20	15	85	0	85	0
DB1	25	100	0	60	50	0	0	0	0
OA2	35	75	0	35	60	2	6	8	0
OB2	15	200	0	250	400	0	0	0	0
DA2	25	70	0	50	125	0	0	0	0
DB2	15	100	0	100	300	0	0	0	8
OA3	15	60	0	40	100	0	0	0	0
DA3	25	100	0	50	200	0	0	0	0
DB3	25	125	0	50	150	0	0	0	0

Commit

Fig. 7. Screenshot of the TEMPO GUI

column gives the name of the weapon, consisting of the category (Offensive/Defensive), type (A/B) and number (1,2,3). The second column shows maximum number of weapons that can be acquired (bought) each year. The third column is the cost to purchase a single unit of the weapon. The fourth column represents the inventory for the weapon – the number of weapons given to the player for 'free' at the beginning of the game. The fifth column gives the cost to operate (use) a unit of the weapon for a year. The sixth column gives the number of utils the weapon has (the power ability of the weapon). The seventh and eighth columns show the weapons that have been opted and bought (respectively) in the previous year. Finally, the ninth and tenth columns are the user input columns to allocate the budget to opt previously bought and opted (or available in inventory) weapons, and purchase new weapons for the coming year.

Once a player has made their allocations, the commit button is pressed and play either continues into the next year, or war breaks out and the game results are displayed.

7   User Study

Using the HACP system, we ran a user study to test the effectiveness with humans. The purpose of the user study was to obtain users with no experience of TEMPO, and use the HACP system as a way of training these users. We also wanted to test the effectiveness of using a system that adapts to a human, as opposed to a static player or a coevolving player with no knowledge of the human strategy. The users represented strategic game players of varying demographics.

To achieve this experiment, we created an application with three consecutive stages. Each stage would involve the human playing 4 consecutive games against a computer player, with the results for each game recorded. We chose 4 games due to time constraints, but ideally more games would be beneficial. The first stage involved running the user against the same static computer player for all 4 games. The static player was previously evolved and consisted of 19 weapon rules, 8 intelligence rules and 8 counter intelligence rules, with the two most active rules as:

1. IF Category IS Offensive and Type IS B
   THEN Evaluation IS medium
2. IF OperationCost IS Very Low
   THEN Evaluation IS high

The intelligence and counter intelligence rules were rarely activated.

After playing the 4 games against this static player, the human was then informed that the next stage was about to start. Stage 2 consisted of the human playing 4 games against the coevolutionary system. The system was first run for 300 generations (with the same evolutionary parameters as described in section 6), and the individuals for both populations in the final generation were saved. This was the starting point for all the human players, and the best individual from this coevolutionary run was chosen as the starting individual to play against the human for Stage 2. Once the first game was completed, the coevolutionary system was then restarted from where it left off, and another 100 generations were run. The best individual from the best population (according to fitness) was then chosen to play against the human, and this process was then iterated for the rest of the games. While the first coevolved player in Stage 2 was the same for each test subject, consequent players were different due to the coevolution.

Stage 3 was conducted in the same manner as Stage 2. However, in this stage, the HACP system was used, and the additional steps of finding a human model and including it in the coevolutionary system were applied.

7.1   User Study Results

The results for Stage 1 are depicted in table 3. The table shows the results for each human player for each game played against the static computer player. The score is the total net utils for the human player at the end of the game. Positive results show a win, while negative ones are a loss. The number of years the game played for is also recorded for comparison purposes. The final column in the table shows the total number of wins each player had in the stage. The last row in the table gives the average score and years played for each game, and the average games won for Stage 1.

As expected, the results show that most of the players had an overall loss for this round. The loss does however decrease over the progression of the stage, depicting player learning. From analysing the results and questioning the players, we found that the average games played before the players felt confident in their game-play were 4–5 games. We also note that there were some players who developed good strategies from Stage 1, and performed well against the computer player throughout the stages.

The Stage 2 results are depicted in table 4, in the same format as the Stage 1 results. By this stage, most of the players are confident in the game-play and begin to develop

Table 3. User Study Stage 1 Results

	Stage 1								
	Game 1		Game 2		Game 3		Game 4		
Player No.	Score	Years	Score	Years	Score	Years	Score	Years	Wins
1	-135	4	655	8	0	2	0	3	1
2	-1920	7	1165	2	2480	8	2160	6	3
3	0	3	-4275	8	-2750	4	-50	2	0
4	460	4	-7100	7	2370	8	425	3	3
5	-6661	9	3230	3	-6440	7	-6012	8	1
6	-4340	7	1660	5	0	7	-2000	6	1
7	0	7	-620	5	-1640	5	0	7	0
8	2465	8	2000	5	2610	10	1400	6	4
9	0	7	-1090	5	-99	2	1165	2	1
10	0	8	2713	3	-1290	6	-1550	4	1
11	-1190	6	-3890	7	-1820	7	2730	6	1
12	2260	5	-4820	5	-4640	5	-4940	5	1
Average	**-1029.18**	**6.36**	**-504.73**	**5.27**	**-598.09**	**6.00**	**-157.45**	**4.82**	**1.45**

Table 4. User Study Stage 2 Results

	Stage 2								
	Game 1		Game 2		Game 3		Game 4		
Player No.	Score	Years	Score	Years	Score	Years	Score	Years	Wins
1	3860	5	-540	4	-495	4	1300	3	2
2	8940	7	4412	8	6589	9	-3004	6	3
3	-2518	7	1200	2	2955	7	-2599	8	2
4	-40	3	6100	7	5039	7	1092	5	3
5	-1155	4	-3358	5	659	6	2940	3	2
6	0	8	3260	6	700	4	0	9	2
7	-1440	3	3070	5	-840	3	0	7	1
8	1065	5	2910	5	615	5	2965	4	4
9	5188	9	1210	7	1278	5	3840	2	4
10	-1236	4	4800	8	289	4	4640	2	3
11	2965	6	4070	5	2604	7	0	3	3
12	-4620	5	-7656	7	-9316	8	-9186	5	0
Average	**1420.82**	**5.55**	**2466.73**	**5.64**	**1763.00**	**5.55**	**1015.82**	**4.73**	**2.64**

some strategies to win. There is a dramatic increase in the average scores for the entire stage, although there is still a slight learning curve in some participants. To note, the first game played against is actually a simpler (but different) player than the static one played in Stage 1, so the number of negative scores tends to show players are still learning the game. By the end of this stage we only have a single player that has not

Table 5. User Study Stage 3 Results

	Game 1		Game 2		Game 3		Game 4		
Player No.	Score	Years	Score	Years	Score	Years	Score	Years	Wins
1	2785	6	-649	10	-2600	4	0	5	1
2	3960	6	-490	6	-3290	6	2780	6	2
3	-4601	5	-3085	6	730	4	-1899	9	1
4	3765	6	442	9	582	7	2926	8	4
5	47	3	428	3	316	3	-342	4	3
6	-1440	3	3140	4	1640	6	-2090	4	2
7	230	4	1500	6	2148	2	5750	7	4
8	2455	6	3695	7	1019	8	400	3	4
9	4705	8	-586	7	1500	1	110	6	3
10	4995	7	445	7	1500	1	5630	8	4
11	3335	6	2450	6	3210	4	1740	5	4
12	-6418	6	-4894	7	-1180	3	-5050	6	0
Average	**1839.64**	**5.45**	**662.73**	**6.45**	**614.09**	**4.18**	**1364.09**	**5.91**	**2.91**

won a game, with the majority of players winning at least 2 games. By the end of this stage, players are beginning to play around with strategies, or have found a 'winning' strategy they continue to use.

The Stage 3 results are shown in table 4, and follow the same format as before. The first game has the same player as Stage 2, Game 1 (the same coevolutionary starting point), but this time round the majority of players win. At Stage 3, Game 2, the first round of the HACP process has been performed, with the coevolution taking into account the human game-play. The Game 2 results for this stage show an overall success with the HACP system. A large amount of the players who won in Stage 2, Game 2 either lost the game or had the amount of utils won by greatly decreased. The 3rd game in Stage 3 shows this same trend happening with some players, while others succeed with a new winning strategy. The same thing is seen in Stage 3, Game 4.

The other notable thing from Stage 3 is with player 12, who was the only player that had difficulty learning the game. The results from this stage show that even though the losses were continuing, they were not by as much. It appears that even in this case the game-play was slowly helping. Whether this is due to a longer learning curve or the HACP system however is still questionable.

8   Conclusions and Future Work

The total wins for Stage 3 were slightly higher than Stage 2, which was expected due to the player learning curve continuing through into Stage 2. However, the results clearly show that there was not a large jump in results between Stage 2 and 3, and that the system did in fact continue to challenge and teach people. This conclusion was reinforced through the informal verbal feedback process at the end of the user study. The majority of the users stated that they found Stage 3 more challenging than Stage 2 when asked if they noticed any difference between the two stages. There were also some users who specifically stated that strategies they had developed and used in Stage 2 needed re-evaluation in Stage 3 when the computer player overcame their strategy. The human players were then forced to think of different strategies.

There were also players who were able to dominate the game from Stage 1. This is likely due to the choices made to cater for average users. For example, to allow for an easier starting point, we only evolved the starting computer player for 300 generations. With a population of 15 individuals this only allows creation of a very simplistic player. Additionally, the short number of games played meant that the HACP system in Stage 3 did not have much time to evolve against more complex human strategies. The human modelling system also struggled more to create these strategies. Even with these deficiencies however, the human players with stronger strategies noted that they thought Stage 3 was beginning to increase in difficulty. Future research should test impact of additional games on the adaptation.

While the HACP system does seem to have considerable benefit for the majority of users, the extreme ends of the learning curve do not seem to benefit as much. One way to address this problem is to adapt the evolutionary parameters to the capability of the human player. The idea here is to adaptively restrict or encourage the coevolutionary process to match the proficiency of the individual human player.

If a human player is doing particularly poorly and losing every game, he or she soon feels discouraged and stops playing. Hampering the success of the coevolutionary process can address this issue, be it through a reduction in generations or population size, or applying an additional weight to the evaluation function. The weight could change depending on how much the individual won by against the human player, with the result of allowing 'lesser' individuals to obtain a higher fitness. The human players who continually win against the computer player also need a mechanism to make the play more interesting. The same concept could be applied to these human players, but in reverse. When a computer player loses to a human, the generations and/or population size could adaptively *increase*. The evaluation weighting could be applied to progressively increase the loss penalty for successive losses. There are also many other possibilities that could be applied.

The other area we would like to improve is the development of the human model. Each time a new game is played, a new model is reverse engineered with no reference to previous games. This process wastes a vast amount of data that could be used to refine the model, as usually humans build strategies from previous game-play. One mechanism we have thought of applying is to use the results of the previous year to influence the fitness of the current models being evolved. If a model has some similar tactics to the previous year, then it is rewarded. This concept also has inadequacies however, as it only forms a single link to the previous year, and does not take into account long term strategy.

Overall, we had a great deal of success with the implementation of the HACP system, with a number of users stating that they had fun trying to beat the computer player. Being able to learn new and better strategies on their own time allows students individual training that caters for their own needs. The HACP system provides a good mechanism for applying such training. The HACP system also has benefits for creation of computer players in many commercial games (such as turn-based strategy games), where each turn the strategy could be updated to incorporate the human player's strategy. The applications are many, and this research is just a beginning. We have shown that it can work, and making it work in other games is an exciting challenge.

Acknowledgements. The authors would like to thank all the participants for the user study in this research. Their time and effort were very much appreciated. We would also like to acknowledge MNiSW grant number N516 384734.

References

1. The Planning, Programming, and Budgeting System (PPBS) directive. Tech. rep., USA Department of Defense (1984)
2. TEMPO military planning game – explanation and rules for players. Tech. rep. (2003)
3. Aha, D., Molineaux, M., Ponsen, M.: Learning to win: Case-based plan selection in a real-time strategy game. In: Muñoz-Ávila, H., Ricci, F. (eds.) ICCBR 2005. LNCS (LNAI), vol. 3620, pp. 5–20. Springer, Heidelberg (2005)
4. Avery, P., Greenwood, G., Michalewicz, Z.: Coevolving strategic intelligence. In: IEEE Proceedings for Congress on Evolutionary Computation, Hong Kong, China (2008)
5. Becker, K.: Teaching with games: the minesweeper and asteroids experience. Journal of Computing Sciences in Colleges 17, 23–33 (2001)
6. Carberry, S.: Techniques for plan recognition. In: User Modeling and User-Adapted Interaction, vol. 11, pp. 31–48. Kluwer Academic Publishers, Dordrecht (2001)
7. Charles, D., McNeill, M., McAlister, M., Black, M., Moore, A., Stringer, K., Kucklich, J., Kerr, A.: Player-centered game design: Player modeling and adaptive digital games. In: DiGRA 2005 Conference: Changing Views – Worlds in Play (2005)
8. Davis, I.L.: Strategies for strategy game AI. In: AAAI 1999 Spring Symposia, pp. 24–27 (1999)
9. Epstein, S.L.: Game playing: The next moves. In: Proceedings of the Sixteenth National Conference on Artificial Intelligence, pp. 987–993. AAAI Press, Menlo Park (1999)
10. Funes, P., Pollack, J.: Measuring progress in coevolutionary competition. In: Animals to Animats 6: Proceedings of the Sixth International Conference on Simulation of Adaptive Behavior, pp. 450–459. MIT Press, Cambridge (2000)
11. Funes, P., Sklar, E., Juillé, H., Pollack, J.: Animal-animat coevolution: Using the animal population as fitness function. In: Animals to Animats 5: Proceedings of the Fifth International Conference on Simulation of Adaptive Behavior, pp. 525–533. MIT Press, Cambridge (1998)
12. Giordano, J.C., Reynolds Jr., P.F., Brogan, D.C.: Exploring the constraints of human behavior representation. In: Proceedings of the 2004 Winter Simulation Conference, pp. 912–920 (2004)
13. Johnson, R.W., Michalewicz, Z., Melich, M.E., Schmidt, M.: Coevolutionary tempo game. In: Congress on Evolutionary Computation, Portland, Oregon, vol. 2, pp. 1610–1617 (2004)
14. Louis, S.J., McDonnell, J.: Learning with case-injected genetic algorithms. IEEE Transactions on Evolutionary Computation 8, 316–328 (2004)
15. Louis, S.J., McDonnell, J.: Playing to train: Case injected genetic algorithms for strategic computer gaming. In: GECCO (2004)
16. Louis, S.J., Miles, C.: Combining case-based memory with genetic algorithm search for competent game AI. In: ICCBR Workshops, pp. 193–205 (2005)

17. Louis, S.J., Miles, C.: Playing to learn: Case-injected genetic algorithms for learning to play computer games. IEEE Transactions on Evolutionary Computation 9, 669–681 (2005)
18. Miles, C., Louis, S., Cole, N., McDonnell, J.: Learning to play like a human: case injected genetic algorithms for strategic computer gaming. In: Congress on Evolutionary Computation, vol. 2, pp. 1441–1448 (2004)
19. Ponsen, M., Muoz-Avila, H., Spronck, P., Aha, D.: Automatically generating game tactics via evolutionary learning. AI Magazine 27, 75–84 (2006)
20. Ponsen, M., Spronck, P.: Improving adaptive game AI with evolutionary learning. In: Computer Games: Artificial Intelligence, Design and Education, pp. 389–396 (2004)
21. Ponsen, M., Spronck, P., Muoz-Avila, H., Aha, D.: Knowledge acquisition for adaptive game AI. Science of Computer Programming 67, 59–75 (2007)
22. Scott, B.: AI game programming wisdom. In: Rabin, S. (ed.) The Illusion of Intelligence, pp. 16–20. Charles River Media (2002)
23. Spronck, P.: Adaptive game AI. Ph.D. thesis, Maastricht University (2005)
24. Spronck, P., Ponsen, M., Sprinkhuizen-Kuyper, I., Postma, E.: Adaptive game AI with dynamic scripting. Machine Learning 63, 217–248 (2006)
25. Spronck, P., Sprinkhuizen-Kuyper, I., Postma, E.: Online adaptation of computer game opponent AI. In: Proceedings of the 15th Belgium-Netherlands Conference on Artificial Intelligence, vol. 2003, pp. 291–298 (2003)

Wisdom of Crowds in the Prisoner's Dilemma Context

Mirsad Hadzikadic and Min Sun

College of Computing and Informatics, UNC Charlotte
9201 University City Blvd, Charlotte, NC 28223
`mirsad@uncc.edu, msun@uncc.edu`

Abstract. This paper provides a new way of making decisions– using the wisdom of crowds (collective wisdom) to handle continuous decision making problems, especially in a complex and rapidly changing world. By simulating the Prisoner's Dilemma as a Complex Adaptive System, the key criteria that separate a wise crowd from an irrational one are investigated, and different aggregation strategies are suggested based on different environments.

1 Introduction and Background

Decision-making has been the subject of research from several perspectives. From a cognitive perspective, the decision making process is regarded as a continuous process characterized by the interaction with the environment. From a normative perspective, the analysis of individual decisions is concerned with the logic of decision making and rationality and the invariant choice it leads to [2]. Generally speaking, decision making is the process of selecting one course of action from several alternative actions. It involves using what you know (or can learn) to get what you want [3]. The decision making techniques used in everyday life include 1) flipping a coin, 2) asking friends or experts, 3) evaluating the disadvantages and advantages, 4) performing the cost-benefit analysis, etc. Since decision-making involves expertise, information, experience, emotions, relationships, and goals, many computer-based Decision Support Systems are promoted to help people make decisions in complicated situations for either individual or business purposes. Although knowledge-based decision support systems have been widely used, managers sometimes feel disappointed with their performance because of: 1) difficulties in collecting useful information in a specific field; 2) the cost of setting up and updating knowledge databases; 3) inherent inadequacies in dealing with complex and rapidly changing environments; and 4) difficulties in determining the proper decision-making model/strategy, especially for problems in social sciences or economics which involve numerous human interactions and uncertain personal feelings. With these concerns in mind, a new concept for making decisions is introduced-- the (modified) wisdom of crowds.

The idea of using the wisdom of crowds for decision-making is originally introduced by J. Surowiecki [1]. In the book, he argues that under certain circumstances the performance of a crowd is often better than that of any single member of the group. This idea appears to be appropriate for explaining the behavior of financial markets as

J. Koronacki et al. (Eds.): Advances in Machine Learning II, SCI 263, pp. 101–118.

expressed by the Nobel-winning economist William Sharpe [5]. Similarly, the concept of "wise crowds" might be useful to decision makers to solve complex problems. For example, collective voting has already been successfully used by some search engines, including Google [6]. Even though there are many case studies and anecdotes which demonstrate the importance of collective wisdom, there are also authors supporting the opposite conclusion, some of them cited in the text "Extraordinary Popular Delusions and the Madness of Crowds," by Charles MacKay [7]. Since Surowiecki dealt with these concerns in his book, we will not repeat those arguments here.

In the following sections, Surowiecki's theory is extended to address a continuous decision-making problem, one that deals with a complex and rapidly changing world with interactions. The key criteria separating the wise crowd from the irrational one are investigated using a computer-based simulation. Also, the ability to learn is added to make both individuals and crowds "smarter" over time. Finally, a relationship between the size of crowds and their performance (aggregation strategies) is suggested for varying environments.

2   Methodology

In this paper, a simulation using the framework of Complex Adaptive Systems is designed and implemented to demonstrate the wisdom of crowds in the context of the Prisoner's Dilemma problem. The Prisoner's Dilemma is a type of non-zero-sum game developed in the game theory. The basic idea is for two suspects who committed a crime to decide whether to "cooperate" with each other or to "defect." Cooperating is the best outcome for both--they go free as there is no proof that they committed the crime. However, as they do not trust each other, they are enticed to defect from the agreement and confess the crime, thus getting a lighter sentence than their partner in crime. Of course, the worst-case scenario is if they both defect, thus securing a lengthy prison sentence for both.

In order to establish a crowd, we extended the two-player game into a many-players situation. Our Prisoner's Dilemma game involves hundreds of players (crowd) playing against each other pair-wise, which allows for exploration of various aggregation strategies. The details describing the Prisoner's Dilemma in this context are introduced in Section 3.

The Complex Adaptive System (CAS) framework represents a dynamic network of agents (representing cells, species, individuals, firms, nations, etc.) acting in parallel, while constantly reacting to what the other agents are doing [8, 9]. A system is considered complex if it is agent-based and exhibits non-linear behavior, feedback loops, self-organization, and emergence [10]. Such a system is considered adaptive if it has the capacity to change and learn from experience.

The control in a CAS is distributed. Any coherent behavior of the system has to arise from the competition and cooperation among the agents (constituent parts) themselves. The overall behavior of the system is a result of the decisions made by individual agents in each cycle [8]. The system often exhibits the property of self-organization. Self-organization is a process in which the internal organization of the system increases in complexity without being guided or managed by an outside source. Self-organizing systems frequently demonstrate emergent properties [9].

Examples of CAS include the stock market, social insect and ant colonies, the biosphere and the ecosystem, the brain, the immune system, and any human social group-based endeavor [11-13]. Hence, it is natural to describe the Prisoner's Dilemma as a complex adaptive system in order to reveal spontaneous reactions among individual players, as well as the wisdom hidden inside the group as a whole.

3   Wisdom of Crowds in the Context of Prisoners' Dilemma

3.1   Theories of Wisdom of Crowds

A "crowd" in Surowiecki's book [1] is any group of people who can act collectively to make decisions and solve problems. The wisdom of crowds theory simply suggests that a collective can solve a problem better than most of the members in the group can by acting alone. As MacKay [7] pointed out, not all crowds (groups) are wise. One needs to look no further than the stock market and its many examples of fads and market bubbles. Consequently, efforts have been made to understand under what circumstances the wisdom of crowds may take effect. Surowiecki suggests the following key criteria to separate wise crowds from irrational ones [1]:

- *Diversity of opinion* - Each person should have private information even if it's just an eccentric interpretation of the known facts.
- *Independence* - People's opinions aren't determined by the opinions of those around them.
- *Decentralization* - People are able to specialize and draw on local knowledge.
- *Aggregation* - Some mechanism exists for turning private judgments into a collective decision.

Three distinct problems have been specified in which crowds may be smarter than individuals [1]. The first is a *needle-in-the haystack* problem where some people in the crowd may know the answer while many, if not most, do not. The second is a *state estimation* problem, where some person may get lucky to hit the precise answer (while not being aware of their "accuracy" in advance), but the group does not. Finally, there is a *prediction* problem, where the answer has yet to be revealed [14, 15]. For the prediction problem, the unrevealed answer can be either fixed (e.g., the prediction of the next Oscar winner does not change the answer itself) or it can be "fluid" (e.g., the return on your next investment where your action might affect the answer).

The well-known example for the "needle in the haystack" problem is the show "Who Wants to Be a Millionaire." In this show, a contestant is asked a series of multiple-choice questions leading to the ultimate prize of $1M. The contestant can choose from one of three options for getting help to answer the question she does not know: (1) eliminate two of the four possible answers, (2) call a predetermined "expert" for counsel, or (3) poll the studio audience. The results show that the experts provide the correct answer a respectable two-thirds of the time, while the audience – a group of folks with nothing better to do on a weekday afternoon – return the correct answer over 90 percent of the time.   The success of polling lies in the fact that, assuming randomness in the answers provided, even a small percentage of the people in the crowd who know the correct answer can add a noticeable advantage to that answer, which helps it stand out using the majority rule.

The "stated estimation" problem normally defines the "guess a quantity number" situations. An interesting characteristic of this type of problem is that although one or several of the crowd members may come close to predicting the correct value/quantity of the target variable, none of them know it for sure when they offer the guess. The well-known example is the "Francis Galton's surprise." The crowd at a county fair was asked to guess the weight of an ox that was exhibited at the fair. The person with the most accurate answer was promised a prize. Everyone tried his or her best to provide the right answer, while maintaining the secrecy of the guess. The participants included some experts (e.g., butchers) and many non-experts. It was obvious that the experts stood a better chance of wining the prize than the non-experts. However, since the target number was a continuous/real number, the non-experts still had a small chance of hitting the most precise number by luck and win the competition. To his surprise, Galton discovered that the average of all the responses was, in fact, closer to the ox's true butchered weight than the individual estimates of most crowd members, including those made by the cattle experts.

Let's look closer into this stated estimation problem. The collective error can be described as [14]:

$$Collective\ error = Average\ individual\ error - Prediction\ diversity$$

The average individual error combines the squared errors of all of the participants, while the prediction diversity combines the squared difference between the individuals and the average guess. This equation tells us [14]:

1. The crowd's aggregate prediction is always better than those of most individuals in it, regardless of whether the crowd has a normal or skewed distribution of answers. Sometimes it can even be better than the best individual, given enough diversity in the right direction.
2. We can reduce the collective error by either increasing the accuracy or increasing the diversity of the crowds.

Other types of problems have been grouped into the third category: the prediction problems. An interesting story is told in Surowiecki's book [1], regarding a submarine lost at sea. The task was to locate the submarine with a very limited knowledge of when and under what weather conditions the submarine went down. A group of specialists with a wide range of expertise was asked to offer their best independent/ individual guesses regarding the various scenarios for submarine's trajectory in the last moments. Although no one knew exactly what happened, by building a composite picture of the projected submarines movements a remarkably accurate guess was formed and the submarine was found. In this case, even though no single individual in the group knew any of the exact answers, the group as a whole produced them all. This story suggests that even if the crowd is not aware of how much useful information each individual has, the appropriate aggregation of partially available information can provide the best answer.

In order to test this idea, we designed and implemented a simulation that can aggregate information from a "crowd" in the context of the Prisoner's Dilemma problem. Furthermore, this simulation is used to explore the effectiveness of the wisdom of crowds when the right answer is not fixed, but rather a continuous decision-making is called for. The following sections provide the details of the simulation.

3.2 Wisdom of Crowds in the Prisoner's Dilemma Game

Since it was first raised by Merrill Flood and Melvin Dresher in the 1950's [16], a lot of research has been done on the Prisoner's Dilemma (PD) problem, especially after Robert Axelrod introduced the concept of the iterated prisoner's dilemma in his book "The Evolution of Cooperation" [17]. The PD is a typical type of non-zero-sum game explored in the game theory, based on the well-known expression of PD, the Canonical PD payoff matrix [17], which shows the non-zero net results for the players.

		Player B	
		Cooperate	Defect
Player A	Cooperate	3,3	0,5
	Defect	5,0	1,1

Some of the best-known strategies for solving this game are listed below [17,25]:

- *Tit-For-Tat* -- Repeat opponent's last choice
- *Tit-For-Two-Tats* – Similar to Tit-For-Tat, except that the opponent must make the same choice twice in a row before it is reciprocated
- *Grudger* -- Co-operate until the opponent defects. Then, always defect (unforgiving)
- *Pavlov* - Repeat the last choice if it led to a good outcome
- *Adaptive* - Start with the set of pre-selected choices (c, c, c, c, c, c, d, d, d, d, d) , then after the initial 11 moves, select actions which give the best average score; re-calculated after each move.

Finding the strategy to gain the highest number of points is the ultimate problem for the Iterated Prisoner's Dilemma game. Every year, the IPD tournament [18] is held to evaluate strategies from different competitors. Also, the genetic algorithms have been widely used [19, 20] to discover the best strategy. Currently, memory- and outcome-based strategies such as Tit-For-Tat [21] and Pavlov [21] are regarded as the most effective ones [22-25].

Extending the "two-player" game to the "many players" context brings about the situation where hundreds of players (a crowd) play together/against each other. With no central control, players begin to play "cooperate" or "defect" based on their own strategies. After each round, points are added up for each player. Consequently, a potential "smart" crowd is formed. This decentralization of strategies for playing is interpreted as a set of diverse opinions held in the crowd. Then, a simple polling of playing strategies serves as the aggregation method for understanding the vote/wisdom of the crowd.

As opposed to the needle-in-the-haystack problem and stated estimation problem, Prisoner's Dilemma states a different type of problem -- dynamic prediction problem [14]. The term "dynamic" is used because the outcome is influenced not only by each play, but also by each player's history of previous predictions. The introduction of this "dynamic" process helps us evaluate performance of various strategies in different crowds over time, which is similar to the decision-making process or cognitive behavior of agents in the real life.

Although more complicated, the participating crowd in the context of Prisoner's Dilemma satisfies the four key criteria to get a smart crowd:

a. Diversity and Decentralization

Page [14] divides diversity into four frameworks:

- Perspective: ways of representing situations and problems
- Interpretations: ways of categorizing or partitioning perspectives
- Heuristics: ways of generating solutions to problem
- Predictive Models: ways of inferring causes and effects.

Definition for decentralization is the dispersion or distribution of functions and powers, specifically the delegation of power from a central authority to regional and local authorities. As one of the key criteria forming a smart crowd, decentralization emphasizes that people are able to specialize and draw on local knowledge [14].

In the Prisoner's Dilemma setting, each agent is given a memory and a strategy. The memory serves to record and accumulate new knowledge, which represents diversity in two ways: the agent's game history with a certain player, and the accumulation of local knowledge. The agent's strategy is the ability to choose either to cooperate or to defect based on the information stored in the memory. The strategy also represents diversity in two ways: diversity in the ways of generating solutions to the problem, and diversity in the ability to draw conclusions from the local knowledge, since the agent does so without the preset upper-level/centralized guidance. This diversity and decentralization among the agents are guaranteed through the combination of interpretations and heuristic frameworks described above, as well as through the process of dispersed decision-making.

b. Independence

The Prisoner's Dilemma as played in our system (the Iterated Prisoner's Dilemma) allows communication between, and learning from other, agents. This aspect is fundamentally different from Surowiecki's approach. However, we still provide a "controller" for ensuring agent independence in the system, which enables us to experiment with both independence-securing and learning-enabling environments.

c. Aggregation

Aggregation means combining outputs/solutions from different entities into higher-level entities. In the Prisoner's Dilemma game, aggregation assumes deriving a group-level solution by combining the individual members' contributions (or solutions), regardless of whether these contributions are duplicate, contradictory, or incomplete. The most commonly used methods for this type of aggregation are sampling, polling, and voting.

3.3 Implementation

In order to design a CAS for the Prisoner's Dilemma game, first we need to create: 1) individual "player-agents" who can "cooperate" or "defect" when playing the game based on their own strategy, and 2) special "aggregator-agents" who represent the wisdom of crowds by acting as aggregators of various groups within the crowd of agents. These aggregators also participate in the game, but they have a different decision-making procedure. Since agents play against each other repeatedly without a central control (via random selection), we assign each agent a memory that is used to store information (knowledge) about their previous "matches." The player-agents initially "receive" a randomly allocated strategy that they use to select their actions, based on the information they have. The aggregator-agents are given the ability to make their decisions upon consulting with their "advisory group," formed from the set of player-agents selected by each aggregator-agent.

The question now becomes: what kind of strategies should be available to the agents? One way to approach this problem is to understand how humans perceive and approach problems. This is obviously related to human personality factors. Raymond Cattelle's suggests that there are 16 personality factors [26] that influence human perception of and approach to problems. To keep things manageable in this project, we selected three personality factors to describe the way people perceive problems: dominance, vigilance, and openness to change.

a. Dominance

Agents that are less dominant are: deferential, cooperative, adverse to conflict, submissive, humble, obedient, easily led, docile, and accommodating. An agent that is perceived as dominant is characterized as: forceful, assertive, aggressive, competitive, stubborn, and bossy.

b. Vigilance

Agents low in vigilance indicate behavior that is: trusting, unsuspecting, accepting, unconditional, and easy-going. A highly vigilant agent is characterized as suspicious, skeptical, distrustful, and oppositional.

c. Openness to change

Not-so-open-to-change agents are defined as: traditional, attached to the familiar, conservative, and respectful of traditional ideas. Highly open agents are defined as: analytical, critical, freethinking, and flexible.

In the Prisoner's Dilemma simulation, the action of each agent includes methods for perceiving and solving problems. The methods for perceiving problems can be described by considering the questions described in Figure 1:

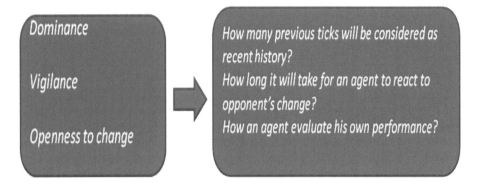

Fig. 1. Personality vs. Action Mode

a.   How many previous ticks will be considered as a recent history?
Agents with a "conservative personality" prefer consulting a longer history; otherwise they are open to change and only care about the most recent history.

b.   How long will it take for an agent to react to an opponent's change in behavior?
Agents with a "vigilant personality" are more suspicious of negative behavior. They are also easier to make hostile. Otherwise, they are less sensitive to betrayal.

c.   How does an agent evaluate his own performance?
Agents with a "domineering personality" are more aggressive and competitive, thinking of their opponents relative to their own gain or loss. Otherwise, they care only about their own absolute gain.
   The methods for solving problems can be described with the following rules[4]:

a. Repeat the opponent's last action
b. Assume an action opposite to the opponent's last action
c. Co-operate
d. Defect
e. Repeat own last action
f. Assume an action opposite to your own last action.

Consequently, in the system each player-agent is described using a chromosome-like structure: [4]

Agent Number	Basic Strategy	Limitation	Reaction1	Reaction2

Where:

- *Agent Number* identifies each player.
- *Basic Strategy* indicates the strategy an agent chooses to guide its behavior.
- *Limitation* modifies the Basic Strategy as described below. Taken together, these two numbers define the judgment of the situation the agent is facing.
- *Reaction1* defines the behavior of the agent if the situation described by Basic Strategy + Limitation applies in the current case/match.
- *Reaction2* defines the behavior of the agent if the situation described by Basic Strategy + Limitation does not apply in the current case/match.

There are five basic strategies:

0. The agent does not care what happened before.
1. The agent takes into consideration the total number of times the opponent cooperated or defected in the past.
2. The agent takes into consideration whether during the previous X number of matches/time (X defined by Limitation) the opponent cooperated or defected (X times in a row).
3. The agent takes into consideration the average number of points it got previously by cooperating/defecting when playing against the same opponent.
4. The agent takes into consideration whether the number of points it got from the last play is less than three points.

Reaction1 and Reaction2 can assume one of the following values:

0. Repeat opponent's last action
1. Assume an action opposite to opponent's last action
2. Co-operate
3. Defect
4. Repeat own last action
5. Assume an action opposite to its own last action

For example, Competitor 001 shown below simply repeats the opponent's last action. This is a typical Tit-for-Tat.

001	0	0	0	0

Competitor 101 repeats the opponent's last action if its' opponent cooperated the last two times/matches; otherwise it chooses an action opposite to its own last action.

101	2	2	0	5

As can be seen from the above, our agents do not simply "cooperate" or "defect." They choose to "repeat" or "reverse" an action performed earlier either by their opponents or by themselves. This may be more similar to the way people behave in real life. This process also aggregates redundant strategies often present in evolutionary algorithms.

Another parameter, "forgiveness," could be added to the chromosome to represent the random or predefined chance to cooperate (when defecting for a long time) or to defect (to test the opponent after cooperating for a long time). Using "forgiveness" the chromosome could represent even a greater variety of strategies.

Also, the parameter called "history-weight" is added to the chromosome to represent the different attitudes that agents could have regarding their own history. They may choose to regard every match in their entire history equally, or they may adjust how much emphasis they want to put on either their earlier matches or their most recent ones.

Aggregator-agents represent special participants (competitors) in the game. On each turn, aggregator-agents choose to cooperate or to defect according to the opinions from their chosen player-agent group. Unlike the regular player-agents, aggregator-agents have no strategy that can give them guidance regarding cooperation or defection; their only strategy is to decide (a) which player-agent group they want to listen to, and (b) the manner in which they plan to aggregate the group's advice.

Each Aggregator-agent is described using a chromosome-like structure:

Agent Number	Selection Strategy	Select_Number	Aggregation Strategy

Where:

- *Agent Number* identifies each aggregator-agent.
- *Selection Strategy* indicates the strategy used to select a player-agent group.
- *Select_Number* indicates how many player-agents are chosen to form the group; it can be any number between 1 and the total number of player-agents.
- *Aggregation Strategy* indicates the strategy used for aggregation.

There are 4 selection strategies:

1. The agent chooses the top Select_Number player-agents ranked by points.
2. The agent chooses the bottom Select_Number player-agents ranked by points.
3. The agent chooses the top N and bottom (Select_Number–N) player-agents ranked by points.
4. The agent chooses Select_Number player-agents randomly.

There are 2 aggregation strategies:

1. The agent chooses the majority opinion
2. The agent chooses the minority opinion.

As shown in Figure 2, all agents are scattered randomly in the display area (90*90 grid in the NetLogo environment) with player-agents represented by red dots and aggregator-agents represented by yellow person-shaped images. A set of basic strategies are assigned randomly to each agent. Agents move randomly in the display area (the speed of agents can be changed via the control panel). If two agents happen to be in the same neighborhood (8-neighbor grid) a meeting is initiated. Agents play a match based on the strategy they follow and the information they have about each other. After each play, the points are added and the agents move on to the next match [4].

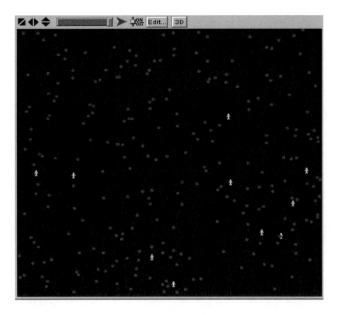

Fig. 2. Application

4 Experiments

We have conducted numerous experiments to demonstrate the power of the (expanded) wisdom of crowd concept. This section provides a summary of the four major groups of experiments.

4.1 Experiment 1: Player-Agents' Performance in Fixed Crowds

In this experiment, we focus on the player-agents' performance in fixed crowds, where "fixed" denotes the same group of players playing the whole time, and no evolution or learning takes place. These settings satisfy all four of Surowiecki's criteria.

The performance of the player-agents is summarized in Figure 3. The jagged blue line shows the highest average-score (the winner's score). For each player-agent, the average-score is calculated as the total number of points gained from all the plays, divided by the total number of plays. The purple line shows the average of player-agent average-scores. It is computed as the sum of average-scores divided by the number of all player-agents, thus outlining the average performance of the whole society of agents. Finally, the black line shows the basic strategy, denoted by its numerical representation, chosen by the player-agent winner [4].

The chart in Figure 3 shows a smooth line for the best performance with a score slightly above 3, while the average performance records a score of slightly below 3. In this fixed society, the best performer is a greedy player who takes advantage of the

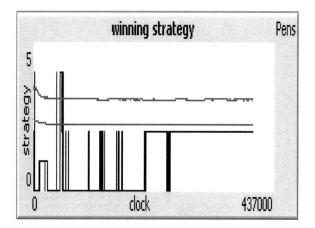

Fig. 3. Player-agents' performance in fixed crowds

"naïve" cooperating players by defecting all the time. The best recommendation for the aggregator-agent, therefore, is simply to listen to the best player-agent; i.e., always defect [4].

The reason that this society remains stable is because neither the player-agents nor the whole crowd has a goal (fitness function), which in the real life rarely happens. So we introduce such a goal in the later experiments.

4.2   Experiment 2: Player-Agents' Performance in Evolutionary Crowds

In this experiment, we focus on the player-agents' performance in evolutionary crowds, where "evolutionary" means every certain number of steps/plays some of the players are replaced with preferred (higher scoring) player-agents. The whole society/crowds try to reach the goal--obtaining higher scores by eliminating the less competitive player-agents. These setting also satisfy all four of Surowiecki's criteria.

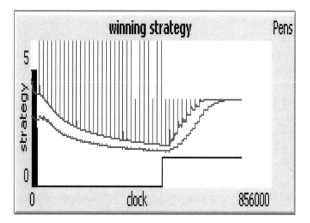

Fig. 4. Player-agents' performance in evolutionary crowds

The chart in Figure 4 shows more volatility as the crowd is changing over time. The player-agents with the highest score gradually replace the lowest scoring agents. In the beginning, evolution shows preference for the greedy players and eliminates the naïve ones. This causes the score of the best performers to decrease and the average score to increase, thus making the crowd "smarter." Later on, after retaining too many greedy players and no naïve ones, the greedy ones die out and are replaced by those who are "smart" enough to both cooperate and defect according to a specific situation. The crowd thus ends up with the score of 3, which suggests that the final outcome/strategy is to "cooperate," and the best decision recommended for the aggregator-agent is simply to cooperate [4].

4.3   Experiment 3: Player-Agents' and Aggregator-Agents' Performance with the Learning Ability in Evolutionary Crowds

Adding the learning ability to the player-agents enables them to learn individually and to improve their decisions.   Although this violates one of Surowiecki's criteria –independence – it is crucial for success in real life. Experiments show that by keeping enough diversity of opinion in the crowd, the aggregate wisdom of the crowd can still perform better than most individual members, even better than the best individual. Consequently, in this experiment we focus on the player-agents' performance with learning ability in evolutionary crowds.

Each experiment was run ten times with a different random seed.   During each round, 250 player-agents were placed on the grid.   After player-agents had had a chance to play against and learn from each other for a certain learning period, the aggregator-agents with different strategies were introduced into this game. Aggre1, 5, 9.., 250 represent the aggregator-agents with different aggregation strategies. For example, an aggregator-agent whose strategy is to consult player-agents with the highest scores may choose to follow the advice of the group of player-agents having the current highest score, and we call it aggre1. This is likely a wise strategy for the aggregator-agents. Similar strategies include the best players, median players, and average players.

The charts in Figure 5 show the performance of player-agents and aggregator-agents, after certain duration of learning, using ten different seeds (formation of crowds).

By introducing the ability to learn, the performance of player-agents and aggregator-agents show increasing volatility reflected in their scores for different seeds (crowds). When no learning happens, the performance of player-agents and aggregator-agent keeps relatively stable no matter what seed (formation of crowds) is used. Although the performance line for best-players is always on the top of the chart, depicting their superiority, we observe that the lines for Aggre19 and Aggre29 are close to the one for the best players (best_people), which suggests that the best way to make the decision by using the wisdom of the crowd in this situation is to listen to the top 10% performers in the crowd, so that the aggregators' performance will be similar (yet more stable) to the performance of the best individual in the crowd (though slightly lower). The best individual might change at each tick, while the performance of the aggregators remains high all the time.

As we introduced learning, more volatility occurred and the best player is not necessarily the all-time winner.   In the chart in Figure 5, which shows the situation after learning for

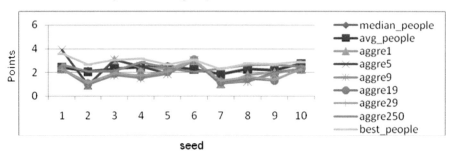

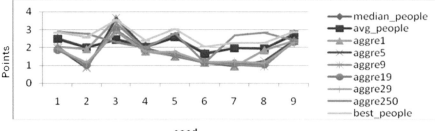

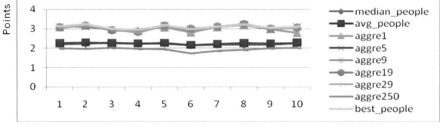

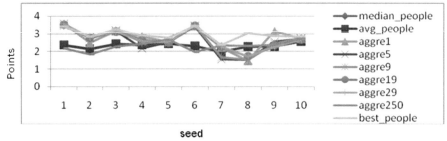

Fig. 5. Performance of Player-agent and Aggregator-agent in different seeds varying in duration of learning

150,000 ticks, the aggregator player performs better than the best player six times out of ten. This suggests that more than half the time making a decision using the wisdom of the crowd is even better than using the advice of the best individual in the crowd.

4.4 Experiment 4: Player-Agents' and Aggregator-Agents' Performance Varying with the Size of Crowds

The size of crowds, which is related to the diversity of opinions in the crowd, is another factor of agents' performance. In this final experiment, we focused on the player-agents'

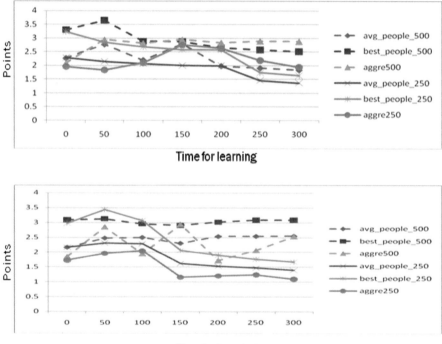

Fig. 6. Performance of Player-agent and Aggregator-agent varying in size of crowds

and aggregator-agents' performance while varying the size of the crowds. Two sets of experiments were run using different random seeds: 250 player-agents and 500 player-agents (Figure 6).

In Figure 6, avg_people_250 and best_people_250 represent the average player-agent and the player-agent with the current high score in a crowd of 250 player-agents, while aggre250 represents the aggregator-agent who chooses the strategy to listen to all 250 player-agents in the crowd. Similar explanation holds for avg_people_500, best_people_500 and aggre500.

The charts in Figure 6 show that despite the choice of different random seeds, the increased size of the crowd (which increases the diversity of opinion) results in a better performance for both player-agents and aggregator-agents. The aggregator-agent using the wisdom of the crowd performs better, most of the time, than the best player-agents in those crowds.

5   Lessons Learned and Future Work

In this paper, we extend the concept of wisdom of crowds to a continuous decision making problem – The Prisoner's Dilemma. A simulation using the concept of Complex Adaptive Systems is built to demonstrate the wisdom of crowds, thus testing Surowiecki's four criteria to form a smart crowd. However, it is hard to imagine a continuous decision-making example from the real world where members of the crowd are truly independent from each other. Therefore, by partially violating the independence criteria, we added the learning ability to the agents. Our experiments show that this addition makes both individual players and the aggregate-players smarter, while still guaranteeing the diversity of opinion. Furthermore, these experiments show that in a crowd where the "membership" can be defined dynamically, and where members of the crowd can communicate with each other and learn from each other, the wisdom of crowds approach is superior to the best performing members of the crowd.

Future work will focus on:

1.   Characterizing the structure of crowds more precisely, using variables such as size, density, and diversity.
2.   Identifying the behavior of the crowds with different agent settings: heuristic problem solving, differing behavior patterns, degrees of social influence, and varying speed of learning speed.
3.   Quantifying and qualifying the characteristics of aggregators.

6   Summary

The research on the wisdom of crowds reported in this paper provides us with a new way of decision-making. Unlike the widely used knowledge-based decision strategies which rely on collecting and analyzing specific knowledge for specific problems, the method proposed in this paper helps practitioners to better handle social science or economics problems that involve numerous human interactions, uncertain personal feelings, and dynamic changes.

By simulating the Prisoner's Dilemma game in a Complex Adaptive System, we investigated the key criteria that separate wise crowds from the irrational ones. We suggested different aggregation strategies for different environments. The further research on the collective wisdom will provide deeper insights along many of the dimensions only touched upon in this paper.

References

[1] Surowiecki, J.: The Wisdom of Crowds: Why the Many Are Smarter Than the Few and How Collective Wisdom Shapes Business, Economies, Societies and Nations (2004) ISBN 0-385-50386-5

[2] Kahneman, D., Tversky, A.: Choice, Values, Frames. The Cambridge University Press, Cambridge (2000)

[3] Walker, Katey, D.: Improving Decision-Making Skills. Manhattan, KS: Kansas State University Cooperative Extension Service, MF-873 (1987)

[4] Hadzikadic, M., Sun, M.: A CAS for Finding the Best Strategy for Prisoner's Dilemma. In: ICCS 2007, Boston (2007)

[5] Ferliel, A.: Inteview with William Sharpe. Investment Adviser (2004)

[6] Austin, D.: How Google Finds Your Needle in the Web's Haystack. American Mathematical Society Feature Column (2006)

[7] MacKay, C.: Extraordinary Popular Delusion and the Madness of Crowds (1841) ISBN 1853263494, 9781853263491

[8] Mitchell Waldrop, M.: Complexity: The Emerging Science at the Edge of Order and Chaos (1992)

[9] http://en.wikipedia.org/wiki/Complex_adaptive_system

[10] http://www.cscs.umich.edu/old/complexity.html

[11] Johnson, S.: Emergence: The connected lives of ants, brains, cities and software (2001) ISBN-10: 068486875X

[12] Kelso, J.A.S.: Dynamic patterns: The self-organization of brain and behavior (1995) ISBN-10: 0262611317

[13] Miller, J.H., Page, S.E.: Complex Adaptive Systems: An Introduction to Computational Models of Social Life (2007) ISBN-10: 0691127026

[14] Page, S.: The Difference: How the Power of Diversity Creates Better Groups, Firms, Schools, and Societies (2007) ISBN13: 978-0-691-12838-2

[15] Mauboussin, M.: Explaining the Wisdom of crowds. Legg Mason Capital Management (2007)

[16] Flood, M.M.: Some Experimental Games, Research Memorandum, RM-789, RAND Corporation (1958)

[17] Axelrod, R.: The Evolution of Cooperation. Basic Books, New York (1984)

[18] http://www.prisoners-dilemma.com/results/cec04/ ipd_cec04_full_run.html

[19] Axelrod, R.: Evolving New Strategies. Genetic Algorithm and Simulated Annealing (1987) ISBN-10: 0934613443

[20] Golbeck, J.: Evolving Strategies for the Prisoner's Dilemma. In: Advances in Intelligent Systems, Fuzzy System, and Evolutionary Computation (2002)

[21] http://www.iterated-prisoners-dilemma.net/
     prisoners-dilemma-strategies.shtml
[22] Fogel, D.: Evolving Behaviors in the Iterated Prisoner's Dilemma. Evolutionary
     Computation 1(1) (1993)
[23] Darwen, P., Yao, X.: On Evolving Robust Strategies for Iterated Prisoner's Dilemma. In:
     AI 1993 and AI 1994 Workshops on Evolutionary Computations, Melbourne, Australia.
     LNCS (LNAI). Springer, Heidelberg (1994)
[24] Kraines, D., Kraines, V.: Learning to Cooperate with Pavlov – an adaptive strategy for the
     Iterated Prisoner's Dilemma with Noise. Theory and Decision (1993)
[25] Kraines, D., Kraines, V.: Evolution of Learning among Pavlov Strategies in a Competitive
     Environment with Noise. Journal of Conflict Resolution (1995)
[26] Cattell, R.B.: The Scree Test for the Number of Factors. Multivariate Behavioral Research
     (1966)

Part II
Logical and Relational Learning, and Beyond

Towards Multistrategic Statistical Relational Learning

Marenglen Biba, Stefano Ferilli, and Floriana Esposito

Department of Computer Science, University of Bari, Via E. Orabona 4, 70124, Italy
biba@di.uniba.it, ferilli@di.uniba.it,
esposito@di.uniba.it

Abstract. Statistical Relational Learning (SRL) is a growing field in Machine Learning that aims at the integration of logic-based learning approaches with probabilistic graphical models. Markov Logic Networks (MLNs) are one of the state-of-the-art SRL models that combine first-order logic and Markov networks (MNs) by attaching weights to first-order formulas and viewing these as templates for features of MNs. Learning models in SRL consists in learning the structure (logical clauses in MLNs) and the parameters (weights for each clause in MLNs). Structure learning of MLNs is performed by maximizing a likelihood function (or a function thereof) over relational databases and MLNs have been successfully applied to problems in relational and uncertain domains. However, most complex domains are characterized by incomplete data. Until now SRL models have mostly used Expectation-Maximization (EM) for learning statistical parameters under missing values. Multistrategic learning in the relational setting has been a successful approach to dealing with complex problems where multiple inference mechanisms can help solve different subproblems. Abduction is an inference strategy that has been proven useful for completing missing values in observations. In this paper we propose two frameworks for integrating abduction in SRL models. The first tightly integrates logical abduction with structure and parameter learning of MLNs in a single step. During structure search guided by conditional likelihood, clause evaluation is performed by first trying to logically abduce missing values in the data and then by learning optimal pseudo-likelihood parameters using the completed data. The second approach integrates abduction with Structural EM of [17] by performing logical abductive inference in the E-step and then by trying to maximize parameters in the M-step.

1 Introduction

Traditionally, Machine Learning research has fallen into two separate subfields: one that has focused on logical representations, and one on statistical ones. Logical Machine Learning approaches based on logic programming, description logics, classical planning, rule induction, etc, tend to emphasize handling complexity. Statistical Machine Learning approaches like Bayesian networks, hidden Markov models, statistical parsing, neural networks, etc, tend to emphasize handling uncertainty. However, learning systems must be able to handle both for real-world applications. The first attempts to integrate logic and probability were made in Artifical Intelligence and date back to the works in [2,22,46]. Later, several authors began using logic programs to compactly specify Bayesian networks, an approach known as knowledge-based model construction [68].

J. Koronacki et al. (Eds.): Advances in Machine Learning II, SCI 263, pp. 121–142.
springerlink.com   © Springer-Verlag Berlin Heidelberg 2010

A central problem in Machine Learning has always been learning in rich representations that enable to deal with structure and relations. Much progress has been achieved in the relational learning field or differently known as Inductive Logic Programming [33]. On the other hand, successful statistical machine learning models with their roots in statistics and pattern recognition, have made possible to deal with noisy and uncertain domains in a robust manner. Powerful models such as Probabilistic Graphical Models [48] and related algorithms have the power to handle uncertainty but lack the capability of dealing with structured domains.

In Machine Learning, recently, in the burgeoning field of Statistical Relational Learning (SRL) [20] or Probabilistic Inductive Logic Programming [11], several approaches for combining logic and probability have been proposed. A growing amount of work has been dedicated to integrating subsets of first-order logic with probabilistic graphical models, to extend logic programs with a probabilistic semantics or integrate other formalisms with probability. Some of the logic-based approaches are: Knowledge-based Model Contruction [68], Bayesian Logic Programs [28], Stochastic Logic Programs [9,41], Probabilistic Horn Abduction [51], Queries for Probabilistic Knowledge Bases [44], PRISM [60], CLP(BN) [59]. Other approaches include frame-based systems such as Probabilistic Relational Models [43] or PRMs extensions defined in [47], description logics based approaches such as those in [8] and P-CLASSIC of [30], database query langauges [67], [54], etc.

All these approaches combine probabilistic graphical models with subsets of first-order logic (e.g., Horn Clauses). One of the state-of-the-art SRL approches is Markov logic [58], a powerful representation that has finite first-order logic and probabilistic graphical models as special cases. It extends first-order logic by attaching weights to formulas providing the full expressiveness of graphical models and first-order logic in finite domains and remaining well defined in many infinite domains [58,65]. Weighted formulas are viewed as templates for constructing Markov Networks (MNs) and in the infinite-weight limit, Markov logic reduces to standard first-order logic. In Markov logic it is avoided the assumption of i.i.d. (independent and identically distributed) data made by most statistical learners by using the power of first-order logic to compactly represent dependencies among objects and relations. In this paper we will focus on this SRL model.

The representation power and the robustness of SRL models to deal with uncertainty does not solve all the problems present in complex domains. Dealing with unknown or partially observed data is an important problem in Machine Learning. Most SRL models face this problem only from the parameter setting point of view by following similar approaches developed in the statistical machine learning field. The most used approach is Expectation-Maximization (EM) [13]. On the other side, in relational learning different approaches have been proposed that integrate multiple inference mechanisms in inductive learning to deal with incomplete data [16,26].

Multistrategic approaches to Machine Learning [38] aim at combining different inference strategies in order to take advantage of each of these during learning. One of these inference mechanisms is abduction. In the general inference schema, the fundamental equation $BK \cup T \models O$ involves a language L, a background knowledge BK and a theory T, that contains concept definitions accounting for some observations O.

Specifically, O stands for the extensional representation of concepts, while T is an intensional description, expressed in L, that explains such concepts together with BK. Deduction traces forward the equation, deriving O given T and BK, and hence it is a truth-preserving inference. Conversely, tracing the equation backward yields two falsity-preserving inferences (meaning that if O is false, then the hypothesis cannot be true): induction, when T is to be hypothesized given O and BK, or abduction, when BK is to be hypothesized given O and T (i.e., plausible/likely causes of given observations). Most approaches to relational learning rely on inductive mechanisms to fine-tune T in order to achieve the learning goal, but problems might arise due to the partial relevance of the available evidence O. Abduction could be exploited to overcome such a limitation by bridging the observations relevance gap. Indeed, it is able to capture default reasoning [57], a well-known form of reasoning to deal with incomplete information [27,50]. Thus, making these inference strategies work together would allow to take advantage of the benefits that each of them can bring. A step in this direction was proposed in [26], where the authors show how to learn with incomplete background data about the training examples by exploiting the hypothetical reasoning of abduction. Another approach is that in [16] where it was proposed a framework for the integration of abductive and inductive learning in an incremental ILP system.

In this paper we propose two frameworks that integrate logical abduction in an SRL model based on Markov logic. The novelty of the proposed approaches stands in the tight integration of structure and parameter learning of an SRL model in a single step inside which a logical abductive proof procedure and a statistical parameter estimation method are exploited. The first framework integrates logical abduction with structure and parameter learning of MLNs in a single step. During structure search guided by conditional likelihood, structure evaluation is performed by first trying to logically abduce missing values in the data and then by learning optimal pseudo-likelihood parameters using the completed data. The second approach integrates abduction with Structural EM of [17] by performing logical abductive inference in the E-step and then by trying to maximize parameters in the M-step.

2   Markov Networks and Markov Logic Networks

A Markov network (also known as Markov random field) is a model for the joint distribution of a set of variables $X = (X_1, X_2, \ldots, X_n) \in \chi$ [12]. It is composed of an undirected graph G and a set of potential functions. The graph has a node for each variable, and the model has a potential function ϕ_k for each clique in the graph. A potential function is a non-negative real-valued function of the state of the corresponding clique. The joint distribution represented by a Markov network is given by:

$$P(X = x) = \frac{1}{Z} \prod_k \phi_k(x_{\{k\}}) \tag{1}$$

where $x_{\{k\}}$ is the state of the kth clique (i.e., the state of the variables that appear in that clique). Z, known as the partition function, is given by:

$$Z = \sum_{x \in \chi} \prod_k \phi_k(x_{\{k\}}) \tag{2}$$

Markov networks are often conveniently represented as log-linear models, with each clique potential replaced by an exponentiated weighted sum of features of the state, leading to:

$$P(X = x) = \frac{1}{Z} \exp(\sum_{j} w_j f_j(x)) \tag{3}$$

A feature may be any real-valued function of the state. We will focus on binary features, $f_j \in \{0, 1\}$. In the most direct translation from the potential-function form, there is one feature corresponding to each possible state x_k of each clique, with its weight being $\log(\phi(x_{\{k\}}))$. This representation is exponential in the size of the cliques. However a much smaller number of features (e.g., logical functions of the state of the clique) can be specified, allowing for a more compact representation than the potential-function form, particularly when large cliques are present. MLNs take advantage of this.

A first-order KB can be seen as a set of hard constraints on the set of possible worlds: if a world violates even one formula, it has zero probability. The basic idea in Markov logic is to soften these constraints: when a world violates one formula in the KB it is less probable, but not impossible. The fewer formulas a world violates, the more probable it is. Each formula has an associated weight that reflects how strong a constraint it is: the higher the weight, the greater the difference in log probability between a world that satisfies the formula and one that does not, other things being equal.

A Markov logic network [58] L is a set of pairs $(F_i; w_i)$, where F_i is a formula in first-order logic and w_i is a real number. Together with a finite set of constants $C = \{c_1, c_2, \ldots, c_p\}$ it defines a Markov network $M_{L;C}$ as follows:

1. $M_{L;C}$ contains one binary node for each possible grounding of each predicate appearing in L. The value of the node is 1 if the ground predicate is true, and 0 otherwise.

2. $M_{L;C}$ contains one feature for each possible grounding of each formula F_i in L. The value of this feature is 1 if the ground formula is true, and 0 otherwise. The weight of the feature is the w_i associated with F_i in L. Thus there is an edge between two nodes of $M_{L;C}$ iff the corresponding ground predicates appear together in at least one grounding of one formula in L. An MLN can be viewed as a template for constructing Markov networks. The probability distribution over possible worlds x specified by the ground Markov network $M_{L;C}$ is given by

$$P(X = x) = \frac{1}{Z} \exp(\sum_{i=1}^{F} w_i n_i(x)) = \frac{1}{Z} \prod_{i} \phi_i(x_i)^{n_i(x)} \tag{4}$$

where F is the number of formulas in the MLN and $n_i(x)$ is the number of true groundings of F_i in x. As formula weights increase, an MLN increasingly resembles a purely logical KB, becoming equivalent to one in the limit of all infinite weights.

In this paper we focus on MLNs whose formulas are function-free clauses and assume domain closure (it has been proven that no expressiveness is lost), ensuring that the Markov networks generated are finite. In this case, the groundings of a formula are formed simply by replacing its variables with constants in all possible ways.

3 Structure and Parameter Learning of MLNs

3.1 Generative Structure Learning of MLNs

One of the approaches for learning MN weights is iterative scaling [12]. However, maximizing the likelihood (or posterior) using a quasi-Newton optimization method like L-BFGS has recently been found to be much faster [62]. Regarding structure learning, the authors in [12] induce conjunctive features by starting with a set of atomic features (the original variables), conjoining each current feature with each atomic feature, adding to the network the conjunction that most increases likelihood, and repeating. The work in [37] extends this to the case of conditional random fields, which are Markov networks trained to maximize the conditional likelihood of a set of outputs given a set of inputs.

The first attempt to learn MLNs was that in [58], where the authors used CLAUDIEN [10] to learn the clauses of MLNs and then learned the weights by maximizing pseudo-likelihood. In [29] another method was proposed that combines ideas from ILP and feature induction of Markov networks. This algorithm, that performs a beam or shortest first search in the space of clauses guided by a weighted pseudo-log-likelihood (WPLL) measure [3], outperformed that of [58]. Recently, in [39] a bottom-up approach was proposed in order to reduce the search space. This algorithm uses a propositional Markov network learning method to construct template networks that guide the construction of candidate clauses. In this way, it generates fewer candidates for evaluation. In [5], the authors proposed an algorithm based on the iterated local search metaheuristic and showed that using parallel computation, it is possible to improve over the previous algorithms. For every candidate structure, in all these algorithms, the parameters that optimize the WPLL are set through L-BFGS that approximates the second-derivative of the WPLL by keeping a running finite-sized window of previous first-derivatives.

3.2 Discriminative Structure and Parameter Learning of MLNs

Learning MLNs in a discriminative fashion has produced for predictive tasks much better results than generative approaches as the results in [64] show. In this work the voted-perceptron algorithm was generalized to arbitrary MLNs by replacing the Viterbi algorithm with a weighted satisfiability solver. The new algorithm is essentially gradient descent with an MPE approximation to the expected sufficient statistics (true clause counts) and these can vary widely between clauses, causing the learning problem to be highly ill-conditioned, and making gradient descent very slow. In [36] a preconditioned scaled conjugate gradient approach is shown to outperform the algorithm in [64] in terms of learning time and prediction accuracy. This algorithm is based on the scaled conjugate gradient method and very good results are obtained with a simple approach: per-weight learning weights, with the weight's learning rate being the global one divided by the corresponding clause's empirical number of true groundings.

However, for both these algorithms the structure is supposed to be given by an expert or learned previously and they focus only on the parameter learning task. This can lead to suboptimal results if the clauses given by an expert do not capture the essential dependencies in the domain in order to improve classification accuracy. On the other side, since to the best of our knowledge, no attempt has been made to learn the structure of

MLNs discriminatively, the clauses learned by generative structure learning algorithms tend to optimize the joint distribution of all the variables and applying discriminative weight learning after the structure has been learned generatively may lead to suboptimal results since the initial goal of the learned structure was not to discriminate query predicates.

Recently different attempts have been proposed for discriminative structure learning of MLNs. In [24] MLNs were restricted to non recursive definite clauses and the ILP system ALEPH [66] was used to generate a large number of potentially good candidates that are then scored using exact inference methods. In [4] the authors proposed another approach, they set parameters by maximizing likelihood and choose structures by conditional likelihood. Inference for each canidate clause is performed using the lazy version of the MC-SAT algorithm [53]. The authors propose some simple heuristics to make the problem tractable and show improvements in terms of predictive accuracy over generative structure learning approaches and discriminative weight learning algorithms.

4   First-Order Logic and Inductive Logic Programming

Relational learning is mostly related to first-order logic or more restricted formalisms. A first-order knowledge base (KB) is a set of sentences or formulas in first-order logic (FOL) [19]. Formulas in FOL are constructed using four types of symbols: constants, variables, functions, and predicates. Constant symbols represent objects in the domain of interest. Variable symbols range over the objects in the domain. Function symbols represent mappings from tuples of objects to objects. Predicate symbols represent relations among objects in the domain or attributes of objects. A term is any expression representing an object in the domain. It can be a constant, a variable, or a function applied to a tuple of terms. An atomic formula or atom is a predicate symbol applied to a tuple of terms. A ground term is a term containing no variables. A ground atom or ground predicate is an atomic formula all of whose arguments are ground terms. Formulas are recursively constructed from atomic formulas using logical connectives and quantifiers. A positive literal is an atomic formula; a negative literal is a negated atomic formula. A KB in clausal form is a conjunction of clauses, a clause being a disjunction of literals. A definite clause is a clause with exactly one positive literal (the head, with the negative literals constituting the body). A possible world or Herbrand interpretation assigns a truth value to each possible ground predicate.

Because of the computational complexity, KBs are generally constructed using a restricted subset of FOL where inference and learning is more tractable. The most widely-used restriction is to Horn clauses, which are clauses containing at most one positive literal. In other words, a Horn clause is an implication with all positive antecedents, and only one (positive) literal in the consequent. A program in the Prolog language is a set of Horn clauses. Prolog programs can be learned from examples (often relational databases) by searching for Horn clauses that hold in the data. The field of inductive logic programming (ILP) [33] deals exactly with this problem. The main task in ILP is finding an hypothesis H (a logic program, i.e. a definite clause program) from a set of positive and negative examples P and N. In particular, it is required that the hypothesis H covers all positive examples in P and none of the negative examples in N. The representation language for representing the examples together with the *covers* relation determines the ILP setting [56].

Learning from entailment is probably the most popular ILP setting and many well-known ILP systems such as FOIL [55], Progol [42] or ALEPH [66] follow this setting. In this setting examples are definite clauses and an example e is covered by an hypothesis H, w.r.t the background theory B if and only if $B \cup H \models e$. Most ILP systems in this setting require ground facts as examples. They typically proceed following a separate-and-conquer rule-learning approach [18]. This means that in the outer loop they repeatedly search for a rule covering many positive examples and none of the negatives (set-covering approach [40]). In the inner loop ILP systems generally perform a general-to-specific heuristic search using refinement operators [45,63] based on θ-subsumption [49]. These operators perform the steps in the search-space, by making small modifications to a hypothesis. From a logical perspective, these refinement operators typically realize elementary generalization and specialization steps (usually under θ-subsumption). More sophisticated systems like Progol or ALEPH employ a search bias to reduce the search space of hypothesis.

In the ILP setting of learning from interpretations, examples are Herbrand interpretations and an examle e is covered by an hypothesis H, w.r.t the background theory B, if and only if e is a model of $B \cup H$. A possible world is described through sets of true ground facts which are the Herbrand interpretations. Learning from interpretations is generally easier and computationally more tractable than learning from entailment [56]. This is due to the fact that interpretations carry much more information than the examples in learning from entailment. In learning from entailment, examples consist of a single fact, while in intereptations all the facts that hold in the example are known. The approach followed by ILP systems learning from interpretations is similar to those that learn from entailment. The most important difference stands in the generality relationship. In learning from entailment an hypothesis H_1 is more general than H_2 if and only if $H_1 \models H_2$, while in learning from interpretations when $H_2 \models H_1$. A hypothesis H_1 is more general than a hypothesis H_2 if all examples covered by H_2 are also covered by H_1. ILP systems that learn from interpretations are also well suited for learning from positive examples only [10].

5   Abduction

In this section we present Abductive Logic Programing and how an abductive proof procedure can be integrated in an Inductice Logic Programming approach for incremental theory revision.

5.1   Abuctive Logic Programming

Abductive Logic Programming (ALP) [15,31] is an extension of Logic Programming aimed at supporting abductive reasoning with theories (logic programs) that incompletely describe their problem domain. In ALP this incomplete knowledge is captured by an abductive theory, defined as a triple $(T, \mathcal{A}, \mathcal{I})$ where T is a (hierarchical) logic program, $\mathcal{A}$ is a set of abducible predicates, and $\mathcal{I}$ is a set of integrity constraints represented as program clauses.

Algorithm 1. Abductive Refutation Algorithm

---

abduce$(T, G, \Delta, \mathscr{AD}, \mathscr{I})$

{**input:** T: theory, G: Datalog goal (set of literals), Δ: initial abductive assumptions, $\mathscr{AD}$: the set of abducibles and default literals, $\mathscr{I}$: the integrity constraints;
output: Δ' final abductive assumptions;}
$\Delta' = \Delta$;
while $G \neq \emptyset$ **do**
  $L :=$ Select a literal from G;
  if $L \notin \mathscr{AD}$ **then**
    /* (**A1**) */ $G :=$ Resolvent of some clause of T with G on L;
  else if $L \in \Delta'$ **then**
    /* (**A2**) */ $G := G \setminus L$;
  else if $\overline{L}_J \notin \Delta$ and $\exists \Delta_C = consistency(T, L, \Delta' \cup \{L\}, \mathscr{AD}, \mathscr{I})$ **then**
    /* (**A3**) */ $G := G \setminus L$; $\Delta' := \Delta_C$;
  end if
end while

Algorithm 2. Consistency Derivation Algorithm

---

consistency$(T, L, \Delta, \mathscr{AD}, \mathscr{I})$

{**input:** T: theory, $L \in \mathscr{AD}$: a literal, Δ: initial abductive assumptions,
$\mathscr{AD}$: the set of abducibles and default literals, $\mathscr{I}$: the integrity constraints;
output: Δ' final abductive assumptions;}
$\Delta' := \Delta$;
$C := \bigcup$ of goals of the form : $-L_1, L_2, \ldots, L_n$ obtained by resolving the abducibles or default literal L with the integrity constraints $\mathscr{I}$ with no such goal been empty;
while $C \neq \emptyset$ **do**
  $B :=$ Select a goal from C; $M :=$ Select a literal from B;
  if $M \notin \mathscr{AD}$ **then**
    $H :=$ Resolvent of some clause of T with B on M;
    $C := \{C \setminus B\} \cup H$;
  else if $M \in \mathscr{AD}$ and $M \in \Delta'$ **then**
    /* (**F1**) */ $H := B \setminus M$; $C := \{C \setminus B\} \cup H$;
  else if $M \in \mathscr{AD}$ and $\overline{M} \in \Delta'$ **then**
    /* (**F2**) */ $C := C \setminus B$;
  else if $M \in \mathscr{AD}$ and $(M \notin \Delta', \overline{M} \notin \Delta')$ **then**
    /* (**F3**) */
    if $\exists \Delta_A = abduce(T, \overline{M}, \Delta', \mathscr{AD}, \mathscr{I})$ **then**
      $C := C \setminus B$; $\Delta' := \Delta_A$
    end if
  end if
end while

An abductive procedure can be exploited to deal with the problem of incompleteness by finding explanations that make hypotheses (abductive assumptions) on the state of the world, possibly involving new abducible concepts. The procedure is generally goal-driven by the observations that it tries to explain. Preliminarily, the top-level goal undergoes a transformation process that converts it into sub-goals. This provides a simple and unique modality for dealing with non-monotonic reasoning. Algorithm 1 sketches the classical abductive proof procedure proposed in [25]. After a literal is selected, if it is not abducible or a default one (A1), the procedure continues with a resolution step with clauses from T. Otherwise, if the fact has been already assumed abductively (and consistently) as true in previous steps (A2) it can be dropped (a case of successful proof). Otherwise (A3), a new fact may be assumed as true, provided that it is consistent with the current integrity constraints $\mathscr{I}$, which is verified by the consistency-check subroutine reported in Algorithm 2. The various branches in the consistency-check subroutine are similar to derivations except that, when dealing with an abducible or a default literal, if it has already been abduced (F1) then it is simply dropped (i.e. consistency is trivially proved); otherwise, if its complement has already been abduced or can be abduced (F2), the entire goal is dropped. In the last if-branch (F3), whenever the literal to be tested is an abducible or default one, but neither it nor its complement have been already abduced, the abductive procedure is called, in order to try hypothesizing it by abduction. Thus, the two procedures may call each other both when a new abductive assumption requires further consistency checks against the constraints and vice-versa.

Representing theories as *hierarchical* logic programs allows to maintain the Least Herbrand Models semantics, coping with negation by means of NAF [7] rule. Indeed, since the language of definite clauses with integrity constraints has been proven to subsume NAF [14], integrity constraints can be simulated using NAF as well. The advantage of adopting this semantics resides in the fact that $T \models P_1$, $T \models P_2$, ..., $T \models P_n$ implies that $T \models P_1 \wedge P_2 \wedge \cdots \wedge P_n$. Hence, positive/negative examples can be tested separately for completeness/consistency.

5.2   Integrating Abductive Inference in Inductive Learning

Algorithm 3 sketches the integration of an incremental inductive learning framework with an abductive proof procedure as proposed in [16]. Here, M represents the set of all positive and negative processed examples, E is the example currently examined, T is the theory generated so far according to M, $AbdT$ is the abduction theory, D is the set of facts hypothesized by the abductive derivation when successfully applied to a goal in theory T. *Generalize* and *Specialize* are the inductive operators used by the system to refine an incorrect theory. When a new observation is available, the abductive proof procedure is started, parameterized on the current theory, the example and the current set of past abductive assumptions. If the procedure succeeds, the resulting set of assumptions, that were necessary to correctly classify the observation, is added to the example description, otherwise the usual refinement procedure (generalization/specialization) is performed.

Several aspects of the strategy adopted in Algorithm 3 can be useful for our purposes of learning the structure of an SRL model. First, it can be useful to apply the abductive derivation on examples that are not correctly classified (i.e., generate an omission/commission error) by the current theory using deduction only. The system checks whether the example can be correctly explained by hypothesizing new facts by means of the abductive procedure reported in Algorithm 1. Indeed, if successful, such an application provides abduced facts that can be useful for extending the available knowledge of the world. The incremental strategy exploits this feature to complete the observations

---

Algorithm 3. Theory Revision extending an incremental inductive learning framework with an abductive proof procedure

---

Revise $(T; E; M; AbdT)$;
{**input:** T: theory, E: example, M: historical memory, $AbdT$: Abductive Theory;
output: T revised theory;}
$D \leftarrow E$
if $(Abductions = \text{Abduct}(T,E,D,AbdT))$ succeeds **then**
    Add to D the abduced literals $Abductions$; $M \leftarrow M \cup \{E \cup D\}$;
else $M \leftarrow M \cup E$
    if E is a positive example **then** Generalize(T,E,M);
    else Specialize(T,E,M);

---

in such a way that the corresponding examples are either covered (if positive) or ruled out (if negative) by the already generated theory, in order to avoid performing a revision of the theory whenever possible (only in case of failure the refinement operators are applied to modify/revise the theory). Abduction is thus exploited. The abductive proof procedure can be set in such a way that the set of abduced literals for each observation is minimal, which ensures that abducibles are used only when really needed, or maximal, which allows to make all possible consistent assumptions that can potentially provide new knowledge about the world. In [16] the minimal option is adopted in order to have a conservative behaviour, while here the maximal one could be more suitable to gain more information about the likelihood of the candidate theories. This is the approach that we follow here. Furthermore, in [16] the abduced information is attached directly to the observation that generated it, in order to keep observations independent from each other. However, this implies that the "completed" examples obtained this way must be available to subsequent abductions, so that the hypothesized facts can be considered in order to preserve consistency among the whole set of abduced facts. In our case, examples are to be exploited altogether, so there is no need to keep abductions attached to the corresponding observations, but a single initial goal including the conjunction of all available examples can be considered, which provides a unique set of abduced facts that explain the whole set of examples, are consistent among each other and can be exploited for the likelihood computation. Lastly, on the inductive side, another thing that can be borrowed is the exploitation of refinement operators that can modify a theory so that it can account for a new example on which it previously generated an omission/commission error. In our case, these operators can be exploited for guiding the move from a theory to one of its refinements, instead of randomly trying to apply all possible refinements.

6   Single Step Structure Learning with Abduction

In this section we describe how structure learning of MLNs in a single step can be combined with the procedure for logical abduction presented in the previous section. The algorithms we propose here are built upon the ideas that we presented in [4]. The parameters are set through maximum pseudo-log-likelihood (WPLL), and the structures are scored through conditional likelihood. The only difference regards the use of logical abduction to complete unknown values in the data during structure search and before computing the WPLL score for each structure.

The first difference between the full framework of MLNs proposed by [58] and the framework that we propose here is that in order to use ALP during structure search we need to restrict the clauses of our model MLN to Horn clauses. Most of relational learning is performed under this expressiveness power and the successes of ILP have shown that for many problems Horn logic is sufficient to deal with structured domains. Thus the structure learning algorithms that we propose here are an extension of those proposed in [4,5] in that here we perform logical abduction in the structure learning process and the language we follow here is based on Horn logic instead of full FOL. The second difference is that the algorithms proposed in [58], try to apply all possible refinements, while here we use ILP refinement operators to properly explore the search space.

6.1 Pseudo-likelihood

MLN weights can be learned by maximizing the likelihood of a relational database. Like in ILP, a closed-world assumption [19] is made, thus all ground atoms not in the database are assumed false. If there are n possible ground atoms, then we can represent a database as a vector $x = (x_1, ..., x_i..., x_n)$ and x_i is the truth value of the ith ground atom, $x_i = 1$ if the atom appears in the database, otherwise $x_i = 0$. Standard methods can be used to learn MLN weights following Equation 4. If the jth formula has $n_j(x)$ true groundings, by Equation 4 we get the derivative of the log-likelihood with respect to its weights by:

$$\frac{\partial}{\partial w_j} logP_w(X = x) = n_j(x) - \sum_{x'} P_w(X = x')n_j(x')$$ (5)

where x' are databases and $P_w(X = x')$ is $P(X = x')$ computed using the current weight vector $w = (w_1, ..., w_j)$. Thus, the jth component of the gradient is the difference between the number of true groundings of the jth formula in the data and its expectation according to the model. Counting the number of true groundings of a first-order formula, unfortunately, is a #P-complete problem.

The problem with Equation 5 is that not only the first component is intractable, but also computing the expected number of true groundings is also intractable, requiring inference over the model. Further, efficient optimization methods also require computing the log-likelihood itself (Equation 4), and thus the partition function Z. This can be done approximately using a Monte Carlo maximum likelihood estimator (MC-MLE) [21]. However, the authors in [58] found in their experiments that the Gibbs sampling used to compute the MC-MLEs and gradients did not converge in reasonable time, and using the samples from the unconverged chains yielded poor results.

In many other fields such as spatial statistics, social network modeling and language processing, a more efficient alternative has been followed. This is optimizing pseudo-likelihood [3] instead of likelihood. If x is a possible world (a database or truth assignment) and x_l is the lth ground atom's truth value, the pseudo-likelihood of x is given by the following equation (we follow the same notation as the authors in [58]:

$$P_w^*(X = x) = \prod_{l=1}^{n} P_w(X_l = x_l | MB_x(X_l))$$ (6)

where $MB_x(X_l)$ is the state of the Markov blanket of X_l in the data. (i.e., the truth values of the ground atoms it appears in some ground formula with). From Equation 4 we have:

$$P(X_l = x_l | MB_x(X_l)) = \frac{\exp(\sum_{i=1}^{F} w_i n_i(x))}{\exp(\sum_{i=1}^{F} w_i n_i(x_{[X_l=0]})) + \exp(\sum_{i=1}^{F} w_i n_i(x_{[X_l=1]}))} \quad (7)$$

Or we can take the gradient of pseudo-log-likelihood:

$$\frac{\partial}{\partial w_i} log P_w^*(X = x) = \sum_{l=1}^{n} [n_i(x) - P_w(X_l = 0 | MB_x(X_l)) n_i(x_{[X_l=0]}) - \quad (8)$$
$$P_w(X_l = 1 | MB_x(X_l)) n_i(x_{[X_l=1]})]$$

where $n_i(x_{[X_l=1]})$ is the number of true groundings of the ith formula when $X_l = 1$ and the remaining data do not change and similarly for $n_i(x_{[X_l=0]})$. To compute the expressions 7 or 8, we do not need to perform inference over the model. The optimal weights for pseudo-log-likelihood can be found using the limited-memory BFGS algorithm [34].

When computing $n_i(x_{[X_l=1]})$ and $n_i(x_{[X_l=0]})$, the usually followed approach is closed world assumption [19], i.e., all ground atoms not in the database are assumed false. Using logical abduction we can pontentially infer the truth value of these atoms and thus when we compute these counts we could have more accurate values that reflect the current data. Since the optimization of the weights by L-BFGS is performed on the estimates of the counts $n_i(x_{[X_l=1]})$ and $n_i(x_{[X_l=0]})$, an improved accuracy on these counts would also result in a more accurate parameter learning task. Thus the use of logical abduction is motivated by the fact that parameter estimation in satistical relational learning can benefit from completed data through logical procedures. To the best of our knowledge, this is the first approach to integrate a pure logical procedure for abductive inference with a statistical parameter estimation algorithm.

6.2   Structure Learning with Abduction

Structure learning can start from an empty network or from an existing KB. Algorithm iteratively generates refinements of the current structure and scores them by conditional likelihood. These refinements are generated using normal ILP refinement operators. Every neighbor of the current structure is obtained by a small generalization/specialization of a randomly chosen clause in the structure. Algorithm 5 performs Iterated Local Search [23,35] for the best model that fits the data. It starts by randomly choosing a clause CL_C in the current MLN structure. Then it performs a greedy local search to efficiently reach a local optimum MLN_S. At this point, a restart method is applied by randomly choosing a clause CL'_C from the clauses of MLN_S. Then again, a greedy local search is applied to MLN_S to reach another local optimum MLN'_S. The *accept* function decides whether the search must continue from the previous local optimum MLN_S or from the last local optimum MLN'_S. The *accept* function always accepts the best solution found so far.

Algorithm 4. MLNs Structure Learning with Abduction

Input: P:set of predicates, MLN:Markov Logic Network, RDB:Relational Database
CLS = All clauses in MLN;
LearnWPLLWeights(MLN,RDB);
BestScore = f(MLN,RDB);
BestModel = MLN;
repeat
    CurrentModel = FindBestModel(P,MLN,BestScore,CLS,RDB);
    if f(CurrentModel) $\geq f$(BestModel) **then**
        BestModel = CurrentModel;
        BestScore = f(MLN,RDB);
    end if
until BestScore does not improve for two consecutive steps
return BestModel;
f = CLL (conditional log-likelihood)

Algorithm 5. FindBestModel

Input: P:set of predicates, MLN:Markov Logic Network, BestScore: current best score, CLS:
List of clauses, RDB:Relational Database)
CL_C = Random Pick a clause in CLS;
$MLN_S = LocalSearch_{II}(CL_C,MLN,BestScore)$;
BestModel = MLN_S;
repeat
    CL'_C = *Random Pick a clause in* (MLN_S);
    $MLN'_S = LocalSearch_{II}(CL'_C,MLN_S,BestScore)$;
    if f(BestModel,RDB) $\geq f$(MLN'$_S$,RDB) **then**
        BestModel = MLN'$_S$;
        BestScore = f(MLN'$_S$,RDB)
    end if
    MLN_S = accept(MLN_S,MLN'$_S$);
until two consecutive steps have not produced improvement
Return BestModel
f = CLL (conditional log-likelihood)

For every candidate structure, the parameters that optimize the WPLL are set through L-BFGS. As pointed out in [29] a potentially serious problem that arises when evaluating candidate clauses using WPLL is that the optimal (maximum WPLL) weights need to be computed for each candidate. Since this involves numerical optimization, and needs to be done millions of times, it could easily make the algorithm too slow. In [12,37] the problem is addressed by assuming that the weights of previous features do not change when testing a new one. Surprisingly, the authors in [29] found this to be unnecessary if the very simple approach of initializing L-BFGS with the current weights (and zero weight for a new clause) is used. Although in principle all weights could change as the result of introducing or modifying a clause, in practice this is very rare. Second-order, quadratic-convergence methods like L-BFGS are known to be very fast if started near the optimum [62]. This is what happened in [29]: L-BFGS typically

---

Algorithm 6. LocalSearch$_{II}$

Input: (CL$_C$: clause chosen for refinement, MLN$_C$: current model, BestScore: current best score)
wp: walk probability, the probability of performing an improvement step or a
random step
repeat
  NBHD = Neighborhood of MLN$_C$ constructed using ILP refinement operators on the clause CL$_C$;
  for Each Candidate Structure MLN in NBHD **do**
    if MLN satisfies ILP coverage threshold **then**
      PerformLogicalAbduction(MLN,RDB);
      if all atoms have known truth values **then**
        LearnWPLLWeights(MLN,RDB);
      else
        LearnWPLLWeightswithEM(MLN,RDB);
      end if
    end if
  end for
  for Each structure scored MLN **do**
    score = f(MLN,RDB)
    if score $\geq$ BestScore **then**
      BestScore = score;
      MLN$_S$ = MLN
    end if
  end for
until two consecutive steps do not produce improvement
Return MLN$_S$;

---

converges in just a few iterations, sometimes one. We use the same approach for setting the parameters that optimize the WPLL.

In Algorithm 6, we generate NBHD, the neighborhood of MLN$_C$, by using ILP refinement operators. All structures in NBHD differ from MLN$_C$ by only one clause which is a generalization or specialization of the clause CL$_C$. Two modifications can be applied here with respect to the traditional setting. First of all, the structure refinement is not carried out randomly, but can be guided by the examples themselves, since they were purposely provided by an expert. Hence, each example that is not correctly classified by the current theory can be exploited to perform a generalization (if positive) or specialization (if negative) according to classical ILP operators. Application of such an operator will provide one or more (depending on the operator and on the generalization model adopted) alternative refinements of the original structure, each of which consists in a new structure obtained by refining a single clause in the original structure. Moreover, pruning criteria can be set in order to avoid working on refinements that are not regarded as promising or acceptable. For instance, one could require that each candidate structure fulfils a minimum coverage threshold in the logical sense, i.e., that the accuracy from the ILP point of view (how many positive examples are covered and how many negatives are not) is greater than a given minimum. We believe this heuristic can help exclude candidates that have a very low logical accuracy. Although there is a mismatch between the coverage criterion used by most ILP systems and the likelihood (or a function thereof) used by most statistical learners, a logical theory that does not explain any example from a logical interpretation would be less useful, and would contradict the idea that examples are purposely labelled by an expert and hence deserve some level of trust. Therefore, we decided to pose a threshold on the accuracy of candidate structures and learn weights only for those candidates that satisfy this threshold.

After the coverage check, we perform logical abduction using the theory of each structure and the examples in RDB. When the abductive proof procedure has potentially completed missing values in RBD, we check whether all the data have been completed. If this is the case then we can learn optimal WPLL weights without EM, otherwise we use EM. For very incomplete data, it is probable that the abductive proof procedure will not complete all the missing data. However, the partial completing of the data will potentially help the weight learning procedure to learn more accurate weights compared to the case when more data is missing.

After setting weights with WPLL, in order to score each MLN structure in terms of conditional likelihood (CLL), we need to perform inference over the network. A very fast algorithm for inference in MLNs is MC-SAT [52]. Since probabilistic inference methods like MCMC or belief propagation tend to give poor results when deterministic or near-deterministic dependencies are present, and logical ones like satisfiability testing are inapplicable to probabilistic dependencies, MC-SAT combines ideas from both MCMC and satisfiability to handle probabilistic, deterministic and near-deterministic dependencies that are typical of statistical relational learning. MC-SAT was shown to greatly outperform Gibbs sampling and simulated tempering in two real-world datasets regarding entity resolution and collective classification. MC-SAT produces probability outputs for every grounding of the query predicate on the test fold and these values can be used to compute the average CLL over all the groundings. In order to make the execution of MC-SAT tractable for every candidate structure, we follow the same heuristic that were proposed in [4], i.e., we score through MC-SAT only those candidate structures that show an improvement in WPLL, we use the lazy version of MC-SAT that is known as Lazy-MC-SAT [53] which reduces memory and time by orders of magnitude compared to MC-SAT, we pose a memory and time limit on the inference process thorugh Lazy-MC-SAT. As the experiments showed in [4], these heuristics proved quite successful in two real world domains. We denote this framework as Structure Learning with Abduction (SLA).

7   Structural EM with Abduction

In the presence of missing values a procedure normally used is Expectation- Maximization (EM) [13]. In this section we describe the EM algorithm and the Structural-EM algorithm that was first proposed in [17] to learn the structure of Bayesian Networks. Then we sketch a framework for integrating logical abduction in the Structural-EM algorithm and discuss the benefits that the statistical learning setting can have from logical abduction.

7.1   Expectation-Maximization and Structural EM

In the presence of missing values maximum likelihood parameter estimation is a numerical optimization problem, and all known algorithms involve nonlinear, iterative optimization and multiple calls to an inference algorithm. The most widely used algorithm for parameter estimation under hidden variables is Expectation-Maximization [13]. This algorithm proceeds in two steps, in the Expectation (E)-step it is computed the expectation of the previous model and the observed data and in the Maximization (M) step, the expected score is maximized. Thus, if we denote the previous model MLN_k and the parameters of the model $\lambda_{k,l}$ in the l step, then in the $l + 1$ the algorithm performs two steps:

E-Step: Computes the expectation of the log-likelihood given the old model $(MLN_k, \lambda_{k,l})$ and the observed data D, i.e.,

$$Q(MLN_k, \lambda | MLN_k, \lambda_{k,l}) = E[logP(D|MLN_k, \lambda)|MLN_k, \lambda_{k,l}].$$

where D denotes the completion of the data. The current model MLN_k, $\lambda_{k,l}$ and the observed data D give the conditional distribution and E denotes the expectation over it. The function Q is called the *expected score*.

M-Step: Maximize the expected score $Q(MLN_k, \lambda | MLN_k, \lambda_{k,l})$ w.r.t. λ, i.e., $\lambda_{k,l+1} = argmax_\lambda Q(MLN_k, \lambda | MLN_k, \lambda_{k,l})$.

Algorithm 4 can be instantiated using the EM. The problem, however, is the huge computational costs. To evaluate a single neighbor, the EM has to run for a reasonable number of iterations in order to get reliable ML estimates of λ_k . Each EM iteration requires a full inference on all data cases. In total, the running time per a neighbor evaluation is at least O(#EM iterations * size of data) which is intractable even for very simple problems. The idea of Structural EM [17] is to perform structure search inside the EM procedure. Algorithm 7 takes the current model $(MLN_k, \lambda_{k,l})$ and runs the EM algorithm for a while to get reasonably completed data. It then fixes the completed data cases and used them to compute the ML parameters λ_k of each neighbor MLN_k. The neighbor $(MLN_{k+1}, \lambda_{k+1})$ with the best improvement of the score is chosen for the next iteration.

7.2   Integrating Logical Abduction in Structural EM

Algorithm 8 shows how the abductive proof procedure can be plugged in the Structual EM algorithm. The logical abduction process is performed inside the E-step, in order to complete the available data. After the abductive process is completed, the EM approach fixes the current model $(MLN_k, \lambda_{k,l})$ and computes maximum pseudo-likelihood parameters of the neighbors of $(MLN_k, \lambda_{k,l})$. The neighbors are constructed using the ILP refinement. After weights have been set for each neighbor, the average CLL for each structure is then computed based on these weights using MC-SAT. The best model is the one that maximizes CLL. We call this framework Structural EM with Logical Abduction (SEMLA).

Algorithm 7. Structural EM

Input: (Current model: MLN_k, $\lambda_{k,l}$, RDB:relational data)
Perform random assignment of $\lambda_{0,0}$
repeat
  for k = 0, 1, 2, . . .
  repeat
    for l = 0, 1, 2, . . .
    $\lambda_{k,l+1} = argmax Q(MLN_k, \lambda | MLN_k, \lambda_{k,l})$
  until convergence is reached or $l = l_{max}$
  Find a model $MLN_{k+1} \in$ neighbors(MLN_k) that maximizes
  $max_\lambda Q(MLN_{k+1}, \lambda | MLN_k, \lambda_{k,l})$
  Set $\lambda_{k+1,0} = argmax Q(MLN_{k+1}, \lambda | MLN_k, \lambda_{k,l})$
until convergence is reached
neighbors(MLN_k) is computed using the ILP refinement operators.

Algorithm 8. Structural EM with abduction

Input: (Current model: MLN_k, $\lambda_{k,l}$, RDB:relational data)
Perform random assignment of $\lambda_{0,0}$
repeat
  for k = 0, 1, 2, . . .
  CL_C = Random Pick a clause in MLN_k;
  NBHD = Neighborhood of MLN_k constructed using ILP refinement operators on the clause
  CL_C;
  repeat
    for l = 0, 1, 2, . . .
    PerformLogicalAbduction(MLN_k,RDB);
    $\lambda_{k,l+1} = argmax W PLL Q(MLN_k, \lambda | MLN_k, \lambda_{k,l})$
  until convergence is reached or $l = l_{max}$
  Find a model $MLN_{k+1} \in$ NBHD that maximizes
  $max W PLL_\lambda Q(MLN_{k+1}, \lambda | MLN_k, \lambda_{k,l})$
  score each structure with CLL using MC-SAT
  Set $\lambda_{k+1,0} = argmax CLL Q(MLN_{k+1}, \lambda | MLN_k, \lambda_{k,l})$
until convergence is reached

The difference with the framework proposed in the previous section is that the abductive proof procedure in SEMLA is executed on the current model trying to complete the data based on the current theory. While in SLA the logical abductive process is performed on each of the neighbors of the current model exploiting a different theory which is obtained by refinement operators from the current theory. Another difference of SLA and SEMLA is that in SEMLA the E-step is performed for the current model with the current parameters, while in SLA the E-step is performed on each candidate structure separately with a different set of parameters. Finally, the M-step for SEMLA is performed on all the neighbors of the current model using the estimates on the current model and trying to maximize the likelihood of each neighbor, while in SLA the M-step uses the independent estimates on each of the candidate structures to maximize its likelihood.

From a computational complexity point of view, we expect SLA to be more expensive since logical abduction and the entire E-step is performed for each of the neighbors of the current model, while for SEMLA, both logical abduction and the E-step are performed only once for the current model and then used for all the neighbors. However, since the abduced atoms change with the available theory, in SLA the logical abduction process would produce abducibles according to the logical theory of each neighbor, thus the abduced truth values are directly related to each neighbor. In SEMLA this is not the case, for each neighbor the abduced truth values with the current model are used and these values may not be directly related with the theory of the neighbor.

8   Related Work

To the best of our knowledge this is the first proposal that tightly integrates in a task of structure learning algorithm, a logical approach to abduction with a statistical procedure for parameter learning. Previous related work has considered mostly statistical abduction as the principal form of inference and has considered logic simply as a representation formalism. One of the first approaches similar to ours is that of [51] where *probabilistic Horn abduction* was proposed. In this approach, a program contains non-probabilistic definite clauses and probabilistic disjoint declarations which are of the form $h_1 : p_1,...,h_n : p_n$ and an abducible atom h_i is considered true with probability p_i. This work focuses on the representation language issue trying to propose a simple language for integrating logic and probability and the authors does not deal with the structure learning problem. Moreover, it does not integrate any form of logic-based abductive proof procedure with statistical learning. Another approach is that in [61], where a logic-based framework is proposed and *statistical abduction* is introduced for representing and learning probabilistic knowledge. The abductive inference is made possible through the definiton of a probability distribution over abducibles. This makes possible to identify the best hypothesis as the most likely hypothesis and likelihood is maximized through statistical learning. The difference with our proposal is that the approach of [61] is purely statistical and the role of logic is purely sintactic, i.e., there is no pure logic-based proof procedure as in our two proposed frameworks. Moreover, the authors in [61] do not learn the structure of the model as we do here. Instead, they hand code the clauses of the model and only learn the statistical parameters of the model through an EM based algorithm. Finally, our SEMLA framework modifies the EM algorithm in a way that structure search can be performed inside the EM algorithm together with logical abduction. Finally, a similar approach is that proposed in [1,6] where the authors proposed Abductive Stochastic Logic Programs which is a framework that supports abduction in SLPs [41] to provide a probability distribution over abductive hypothesis based on a *possible world* semantics. Again the main difference with our proposed frameworks is that the approaches in [1,6] suppose to have an already learned structure in order to learn the parameters for the SLP. When the parameters of the SLP have been learned, this probabilistic program is used to define a probability distribution over the abducibles using a stochastic SLD derivation. The labelled hypothesised abducibles are chosen to maximize the likelihood. Therefore, since the structure of the model is first learned by ILP using "coverage" as guiding function, and all the following process involves only parameter learning, our proposals are different since we learn the structure by directly optimizing a likelihood based function. For an SRL this has proven to be the best way to learn a model as the results of [29] show, where ILP based approaches were outperformed by likelihood-guided approaches for the task of learning the structure of an SRL model. Moreover, we perform abduction during structure selection, while the approach of [1,6] uses a two step approach, first learns the structure with ILP (then the parameters) and then performs abduction. This two step approach has been shown in [32] to be inferiror in terms of accuracy compared to the single-step structure learning approach that we follow here.

9 Conclusion and Future Work

Statistical Relational Learning (SRL) is a growing field in Machine Learning that aims at the integration of logic-based learning approaches with probabilistic graphical models. Markov Logic Networks are one of the state-of-the-art SRL models that combine first-order logic and Markov networks (MNs) by attaching weights to first-order formulas and viewing these as templates for features of MNs. Learning models in SRL consists in learning the structure (logical clauses in MLNs) and the parameters (weights for each clause in MLNs). Structure learning of MLNs is performed by maximizing a likelihood function (or a function thereof) over relational databases and MLNs have been successfully applied to problems in relational and uncertain domains. However, most complex domains are characterized by incomplete data. Until now SRL models have mostly used Expectation-Maximization for learning statistical parameters under missing values. Multistrategic learning in the relational setting has been a successful approach to dealing with complex problems where multiple inference mechanisms can help solve different subproblems. Abduction is an inference strategy that has been proven useful for completing missing values in observations. In this paper we propose two frameworks for integrating abduction in an SRL model based on MLNs. The first tightly integrates logical abduction with structure and parameter learning of MLNs in a single step. During structure search guided by conditional likelihood, clause evaluation is performed by first trying to logically abduce missing values in the data and then by learning optimal parameters using the completed data. The second approach integrates abduction with Structural EM of [17] by performing logical abductive inference in the E-step and then by trying to maximize parameters in the M-step.

We intend to experimentally evaluate the proposed frameworks on complex relational domains with missing data. In order to evaluate the advantages of our approach, we intend to compare the accuracy performance against a pure statistical learner that uses EM to deal with missing values, a pure logical approach such as an ILP system and finally against another SRL approach that does not follow our approach to structure learning.

References

1. Arvanitis, A., Muggleton, S.H., Chen, J., Watanabe, H.: Abduction with stochastic logic programs based on a possible worlds semantics. Short Paper Proceedings of the 16th International Conference on Inductive Logic Programming, University of Corunna (2006)

2. Bacchus, F.: Representing and Reasoning with Probabilistic Knowledge. MIT Press, Cambridge (1990)
3. Besag, J.: Statistical analysis of non-lattice data. Statistician 24, 179–195 (1975)
4. Biba, M., Ferilli, S., Esposito, F.: Discriminative structure learning of markov logic networks. In: Železný, F., Lavrač, N. (eds.) ILP 2008. LNCS (LNAI), vol. 5194, pp. 59–76. Springer, Heidelberg (2008)
5. Biba, M., Ferilli, S., Esposito, F.: Structure learning of markov logic networks through iterated local search. In: Proceedings of 18th European Conference on Artificial Intelligence (ECAI). Frontiers in Artificial Intelligence and Applications, vol. 178, pp. 361–365 (2008)
6. Chen, J., Muggleton, S., Santos, J.: Abductive stochastic logic programs for metabolic network inhibition learning. In: Proceedings of Workshop Mining and Learning with Graphs, MLG 2007 (2007)
7. Clark, K.: Negation as failure. In: Gallaire, H., Minker, J. (eds.) Logic and databases, pp. 293–322. Plenum Press, New York (1978)
8. Cumby, C., Roth, D.: Feature extraction languages for propositionalized relational learning. In: Proceedings of the IJCAI 2003 Workshop on Learning Statistical Models from Relational Data, Acapulco, Mexico, IJCAII, pp. 24–31 (2003)
9. Cussens, J.: Parameter estimation in stochastic logic programs. Machine Learning 44(3), 245–271 (2001)
10. De Raedt, L., Dehaspe, L.: Clausal discovery. Machine Learning 26, 99–146 (1997)
11. De Raedt, L., Frasconi, P., Kersting, K., Muggleton, S. (eds.): Probabilistic Inductive Logic Programming - Theory and Applications. Springer, Heidelberg (2008)
12. Della Pietra, S., Della Pietra, V., Laferty, J.: Inducing features of random fields. IEEE Transactions on Pattern Analysis and Machine Intelligence 19, 380–392 (1997)
13. Dempster, A.P., Laird, N.M., Rubin, D.B.: Maximum likelihood from incomplete data via the em algorithm. Journal of the Royal Statistical Society, Series B 39, 1–38 (1977)
14. Eshghi, K., Kowalski, R.: Abduction compared to negation by failure. In: Levi, G., Martelli, M. (eds.) Proceedings of the 6th international conference on logic programming, pp. 234–255. The MIT Press, Cambridge (1989)
15. Esposito, F., Lamma, E., Malerba, P., Mello, D., Milano, M., Riguzzi, F., Semeraro, G.: Learning abductive logic programs. In: Proceedings of the ECAI 1996 workshop on abductive and inductive reasoning, Budapest, pp. 23–30 (1996)
16. Esposito, F., Semeraro, G., Fanizzi, N., Ferilli, S.: Multistrategy theory revision: induction and abduction in inthelex. Machine Learning 38(1-2), 133–156 (2000)
17. Friedman, N.: Learning belief networks in the presence of missing values and hidden variables. In: Fourteenth Inter. Conf. on Machine Learning, ICML 1997 (1997)
18. Furnkranz, J.: Separate-and-conquer rule learning. Artificial Intelligence Review 13(1), 3–54 (1999)
19. Genesereth, M.R., Nilsson, N.J.: Logical foundations of artificial intelligence. Morgan Kaufmann, San Mateo (1987)
20. Getoor, L., Taskar, B.: Introduction to Statistical Relational Learning. MIT, Cambridge (2007)
21. Geyer, C.J., Thompson, E.A.: Constrained monte carlo maximum likelihood for dependent data. Journal of the Royal Statistical Society, Series B 54, 657–699 (1992)
22. Halpern, J.: An analysis of first-order logics of probability. Artificial Intelligence 46, 311–350 (1990)
23. Hoos, H.H., Stutzle, T.: Stochastic Local Search: Foundations and Applications. Morgan Kaufmann, San Francisco (2005)
24. Huynh, T.N., Mooney, R.J.: Discriminative structure and parameter learning for markov logic networks. In: Proc. of the 25th International Conference on Machine Learning, ICML (2008)

25. Kakas, A., Mancarella, P.: On the relation of truth maintenance and abduction. In: Proc. 1st Pacific Rim International Conference on Artificial Intelligence (1990)
26. Kakas, A., Riguzzi, F.: Learning with abduction. New Generation Computing 18(3), 243–294 (2000)
27. Kakas, M., Kowalski, R., Toni, F.: Abductive logic programming. J. Logic. Comput., 718–770 (1993)
28. Kersting, K., De Raedt, L.: Towards combining inductive logic programming with bayesian networks. In: Rouveirol, C., Sebag, M. (eds.) ILP 2001. LNCS (LNAI), vol. 2157, pp. 118–131. Springer, Heidelberg (2001)
29. Kok, S., Domingos, P.: Learning the structure of markov logic networks. In: Proc. 22nd Int'l Conf. on Machine Learning, pp. 441–448 (2005)
30. Koller, D., Levy, A., Pfeffer, A.: P-classic: A tractable probabilistic description logic. In: Proc. of NCAI 1997, pp. 360–397 (1997)
31. Lamma, E., Mello, P., Milano, M., Riguzzi, F., Esposito, F., Ferilli, S., Semeraro, G.: Co-operation of abduction and induction in logic programming. In: Abductive and inductive reasoning: essays on their relation and integration. Kluwer, Dordrecht (2000)
32. Landwehr, N., Kersting, K., De Raedt, L.: Integrating naive bayes and foil. Journal of Machine Learning Research, 481–507 (2007)
33. Lavrac, N., Dzeroski, S.: Inductive Logic Programming: Techniques and applications. UK, Ellis Horwood, Chichester (1994)
34. Liu, D.C., Nocedal, J.: On the limited memory bfgs method for large scale optimization. Mathematical Programming 45, 503–528 (1989)
35. Loureno, H.R., Martin, O., Stutzle, T.: Iterated local search. In: Glover, F., Kochenberger, G. (eds.) Handbook of Metaheuristics, pp. 321–353. Kluwer Academic Publishers, Norwell (2002)
36. Lowd, D., Domingos, P.: Efficient weight learning for markov logic networks. In: Kok, J.N., Koronacki, J., Lopez de Mantaras, R., Matwin, S., Mladenič, D., Skowron, A. (eds.) PKDD 2007. LNCS (LNAI), vol. 4702, pp. 200–211. Springer, Heidelberg (2007)
37. McCallum, A.: Efficiently inducing features of conditional random fields. In: Proc. UAI 2003, pp. 403–410 (2003)
38. Michalski, R.S.: Inferential theory of learning. developing foundations for multistrategy learning. In: Michalski, R.S., Tecuci, G. (eds.) Machine Learning. A Multistrategy Approach, vol. IV, pp. 3–61. Morgan Kaufmann, San Francisco
39. Mihalkova, L., Mooney, R.J.: Bottom-up learning of markov logic network structure. In: Proc. 24th Int'l Conf. on Machine Learning, pp. 625–632 (2007)
40. Mitchell, T.M.: Machine Learning. The McGraw-Hill Companies, Inc., New York (1997)
41. Muggleton, S.: Stochastic logic programs. In: De Raedt, L. (ed.) Advances in inductive logic programming. IOS Press, Amsterdam (1996)
42. Muggleton, S.H.: Inverse entailment and progol. New Generation Computing Journal, 245–286 (1995)
43. Koller, D., Friedman, N., Getoor, L., Pfeffer, A.: Learning probabilistic relational models. In: Proc. 16th Int'l Joint Conf. on AI (IJCAI), pp. 1300–1307. Morgan Kaufmann, San Francisco (1999)
44. Ngo, L., Haddawy, P.: Answering queries from context-sensitive probabilistic knowledge bases. Theoretical Computer Science 171, 147–177 (1997)
45. Nienhuys-Cheng, S.-H., de Wolf, R.: Foundations of Inductive Logic Programming. Springer, Heidelberg (1997)
46. Nilsson, N.: Probabilistic logic. Artificial Intelligence 28, 71–87 (1986)
47. Pasula, H., Russell, S.: Approximate inference for first-order probabilistic languages. In: Proceedings of the Seventeenth International Joint Conference on Artificial Intelligence, pp. 741–748. Morgan Kaufmann, Seattle (2001)

48. Pearl, J.: Probabilistic reasoning in intelligent systems: Networks of plausible inference. Morgan Kaufmann, San Francisco (1988)
49. Plotkin, G.D.: A note on inductive generalization. Machine Intelligence 5, 153–163 (1970)
50. Poole, D.: A logical framework for default reasoning. Artif. Intell. 36, 27–47 (1988)
51. Poole, D.: Probabilistic horn abduction and bayesian networks. Artificial Intelligence 64, 81–129 (1993)
52. Poon, H., Domingos, P.: Sound and efficient inference with probabilistic and deterministic dependencies. In: Proc. 21st Nat'l Conf. on AI (AAAI), pp. 458–463. AAAI Press, Menlo Park (2006)
53. Poon, H., Domingos, P., Sumner, M.: A general method for reducing the complexity of relational inference and its application to mcmc. In: Proc. 23rd Nat'l Conf. on Artificial Intelligence. AAAI Press, Chicago (2008)
54. Popescul, A., Ungar, L.H.: Structural logistic regression for link analysis. In: Proceedings of the Second International Workshop on Multi-Relational Data Mining, pp. 92–106. ACM Press, Washington (2003)
55. Quinlan, J.R.: Learning logical definitions from relations. Machine Learning 5, 239–266 (1990)
56. De Raedt, L.: Logical settings for concept-learning. Artificial Intelligence 95(1), 197–201 (1997)
57. Reiter, R.: A logic for default reasoning. J. Artif. Intell. (13), 81–132 (1980)
58. Richardson, M., Domingos, P.: Markov logic networks. Machine Learning 62, 107–236 (2006)
59. Santos Costa, V., Page, D., Qazi, M., Cussens, J.: Clp(bn): Constraint logic programming for probabilistic knowledge. In: Proceedings of the Nineteenth Conference on Uncertainty in Artificial Intelligence, pp. 517–524. Morgan Kaufmann, Acapulco (2003)
60. Sato, T., Kameya, Y.: Prism: A symbolic-statistical modeling language. In: Proceedings of the Fifteenth International Joint Conference on Artificial Intelligence, pp. 1330–1335. Morgan Kaufmann, Nagoya (1997)
61. Sato, T., Kameya, Y.: A viterbi-like algorithm and em learning for statistical abduction. In: Proceedings of UAI 2000 Workshop on Fusion of Domain Knowledge with Data for Decision Support (2000)
62. Sha, F., Pereira, F.: Shallow parsing with conditional random fields. In: Proc. HLT-NAACL 2003, pp. 134–141 (2003)
63. Shapiro, E.: Algorithmic Program Debugging. MIT Press, Cambridge (1983)
64. Singla, P., Domingos, P.: Discriminative training of markov logic networks. In: Proc. 20th Nat'l Conf. on AI (AAAI), pp. 868–873. AAAI Press, Menlo Park (2005)
65. Singla, P., Domingos, P.: Markov logic in infinite domains. In: Proc. 23rd UAI, pp. 368–375. AUAI Press (2007)
66. Srinivasan, A.: The Aleph Manual,
    `http://www.comlab.ox.ac.uk/oucl/~esearch/areas/machlearn/Aleph/`
67. Taskar, B., Abbeel, P., Koller, D.: Discriminative probabilistic models for relational data. In: Proceedings of the Eighteenth Conference on Uncertainty in Artificial Intelligence, pp. 485–492. Morgan Kaufmann, Edmonton (2002)
68. Wellman, J.S., Breese, M., Goldman, R.P.: From knowledge bases to decision models. Knowledge Engineering Review 7 (1992)

About Knowledge and Inference in Logical and Relational Learning

Luc De Raedt

Department of Computer Science, Katholieke Universiteit Leuven,
Celestijnenlaan 200A, BE-3001 Heverlee, Belgium

Abstract. A gentle introduction to the use of knowledge, logic and inference in machine learning is given. It can be regarded as a reinterpretation and revisiting of Ryzard Michalski's *"A theory and methodology of inductive learning"* within the framework of logical and relational learning. At the same time some contemporary issues surrounding the integration of logical and probabilistic representations and reasoning are introduced.

1 Introduction

In his seminal paper *"A theory and methodology of inductive learning"* Ryszard Michalski [1983] introduced a logic for inductive learning and reasoning and showed how it could be used for learning structured concept descriptions from examples and background knowledge. Many of the ideas, concepts and techniques contained in this paper have influenced the field of machine learning, and are now, almost 30 years later, still actual. This includes the use of logic as a representation language for machine learning, the emphasis on producing understandable and interpretable descriptions, the role of knowledge, the view of induction as the inverse of deduction, dealing with structured and relational data, etc.

These issues are now being studied within the field of logical and relational learning [De Raedt, 2008]. This is the subfield of machine learning and artificial intelligence that is concerned with learning in expressive logical or relational representations. It is the union of inductive logic programming [Muggleton and De Raedt, 1994], (statistical) relational learning [Getoor and Taskar, 2007] and multi-relational data mining [Džeroski and Lavrač, 2001] and constitutes a general class of techniques and methodology for learning from structured data (such as graphs, networks, relational databases) and background knowledge.

The present paper provides a gentle introduction to theoretical and methodological aspects of the use of logic and inference in machine learning. While doing so, it focuses on those views that now constitute the foundations of logical and relational learning. In addition, it touches upon a popular topic of research: the

J. Koronacki et al. (Eds.): Advances in Machine Learning II, SCI 263, pp. 143–153.
springerlink.com

integration of logical and probabilistic reasoning in a machine learning context [Getoor and Taskar, 2007, De Raedt et al., 2008].

This paper is organized as follows: in Section 2, the problem of inductive learning is introduced using an example from structure-activity-relation prediction; in Section 3, various types of reasoning and logical inference are presented; Section 4 then touches upon the integration of probabilistic and logical reasoning for machine learning, and finally, Section 5 concludes.

2   Inductive Learning

Throughout this paper, we focus on the problem of inductive learning, which starts from a set of observations, a background theory and aims at inducing a definition of an underlying concept. Consider, for instance, the problem sketched in Figure 1. It contains five molecules, three of which are mutagenic and two which are not. The molecules correspond to observations or examples, and one usually distinguishes two types of examples, the positive (here: the mutagenic) from the negatives ones. The problem of (discriminative) inductive learning is then to find a hypothesis that allows one to distinguish the examples that belong to different classes. One hypothesis that can be used for this purpose is shown in grey on the figure. The task sketched here is inspired on the work of [Srinivasan et al., 1996] and is known under the name of structure-activity-relationship (SAR) prediction, which is an important step in the drug-discovery process. The example illustrates several important issues. First, the result of the induction process is a hypothesis (sometimes called a pattern or in molecular applications a structural alert) in the form of a graph, and it can be readily interpreted by human experts. Secondly, in this task, there is chemical *background knowledge* that should ideally be made available to the inductive learner. Background knowledge in this context can take the form of certain types of ring-structures and functional groups that have a chemical meaning. By focusing on and using such background knowledge the learner can often generate more meaningful patterns and may even be able to find such patterns faster.

In order to automate inductive learning, one must employ a representation language for examples, hypotheses and background knowledge. One possible and actually quite popular representation in data mining and machine learning is that of graphs [Washio and Motoda, 2003]. Indeed, molecules can be represented by their 2D graph structure (as in Figure 1). Hypotheses and functional groups then correspond to subgraphs. While using graph-based representations is fine for molecular datasets, it is less clear how to use them in many other types of applications, such as natural language and robotics. Therefore, we shall, as Ryszard Michalski, employ logical representations throughout the rest of this paper. More specifically, we shall employ concepts from logic programming (and logical and relational learning) to represent data, knowledge and hypotheses.

To illustrate how this works, consider the graphical structure of an example molecule. It can be represented by means of the following tuples, called *facts*:

logmutag(f1, 0.64) ←
lumo(f1, −1.785) ←
logp(f1, 1.01) ←
atom(f1, f1$_1$, c, 21, 0.187) ←
atom(f1, f1$_2$, c, 21, −0.143) ←
atom(f1, f1$_3$, c, 21, −0.143) ←
atom(f1, f1$_4$, c, 21, −0.013) ←
atom(f1, f1$_5$, o, 52, −0.043) ←
. . .

bond(f1, f1$_1$, f1$_2$, 7) ←
bond(f1, f1$_2$, f1$_3$, 7) ←
bond(f1, f1$_3$, f1$_4$, 7) ←
bond(f1, f1$_4$, f1$_5$, 7) ←
bond(f1, f1$_8$, f1$_9$, 2) ←
bond(f1, f1$_8$, f1$_{10}$, 2) ←
bond(f1, f1$_1$, f1$_{11}$, 1) ←
bond(f1, f1$_{11}$, f1$_{12}$, 2) ←
bond(f1, f1$_{11}$, f1$_{13}$, 1) ←

In this encoding, each entity is given a name and the relationships among the entities are captured. For instance, in the above example, the compound is named f1 and its atoms f1$_1$, f1$_2$, ... Furthermore, the relation atom/5 of arity 5 states properties of the atoms: the molecule they occur in (e.g., f1), the element (e.g., c denoting a carbon) and the type (e.g., 21) as well as the charge (e.g., 0.187). The relationships amongst the atoms are then captured by the relation bond/3, which represents the bindings amongst the atoms. Finally, there are also overall properties or attributes of the molecule, such as their *logp* and *lumo* values. Further properties of the compounds could be mentioned, such as the functional groups or ring structures they contain:

ring_size_5(f1, [f1$_5$, f1$_1$, f1$_2$, f1$_3$, f1$_4$]) ←
hetero_aromatic_5_ring(f1, [f1$_5$, f1$_1$, f1$_2$, f1$_3$, f1$_4$]) ←
. . .

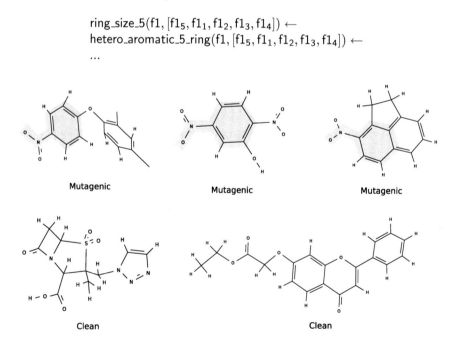

Mutagenic   Mutagenic   Mutagenic

Clean   Clean

Fig. 1. A structure-activity-relationship prediction example. Figure courtesy of Siegfried Nijssen.

The first tuple states that there is a ring of size 5 in the compound f1 that involves the atoms $f1_5, f1_1, f1_2, f1_3$ and $f1_4$ in molecule f1; the second one states that this is a heteroaromatic ring.

Using this representation it is possible to describe a possible hypothesis (or pattern) using a rule (also called clause)

$$\text{mutagenic}(M) \leftarrow \text{ring\_size\_5}(M, R), \text{element}(A1, R), \text{bond}(M, A1, A2, 2)$$

which actually can be directly transformed into English

Molecule M is mutagenic IF it contains a ring of size 5 called R and atoms A1 and A2 that are connected by a double (2) bond such that A1 also belongs to the ring R.

This shows that logical rules are readily interpretable and yield patterns that can be understood by human experts.

To decide whether a hypothesis would classify an example as positive, we need a notion of coverage. To illustrate this concept, reconsider the above example and let us assume that the example states that f1 is mutagenic, that is, mutagenic(f1). Then the example can be regarded as the rule:

$$\text{mutagenic}(f1) \leftarrow$$
$$\text{atom}(f1, f1_1, c), \ldots, \text{atom}(f1, f1_n, c),$$
$$\text{bond}(f1, f1_1, f1_2, 2), \ldots, \text{bond}(f1, f1_1, f1_3, 1),$$
$$\text{ring\_size\_5}(f1, [f1_5, f1_1, f1_2, f1_3, f1_4]), \ldots$$

and the hypothesis

$$\text{mutagenic}(M) \leftarrow \text{ring\_size\_5}(M, R), \text{atom}(M, M1, c)$$

then covers the example because the conditions in the rule are satisfied by the example when setting $M = f1$, $R = [f1_5, f1_1, f1_2, f1_3, f1_4]$ and $M1 = f1_1$. In terms of logic, the example e is a logical consequence of the rule h, which we shall write as $h \models e$. This notion of coverage forms the basis for the theory of inductive reasoning that we revisit in the next section.

The above illustration is based on the exposition of [De Raedt, 2008] of the well-known mutagenicity application of relational learning due to [Srinivasan et al., 1996], where the structural alert was discovered using the inductive logic programming system PROGOL [Muggleton, 1995] and the logical representations shown above. The importance of this type of application is clear when considering that the results were published in the scientific literature in the application domain [King and Srinivasan, 1996], that they were obtained using a general-purpose machine learning algorithm and were transparent to the experts in the domain.

3   Logical Inference

The coverage relation is one of the central concepts for inductive reasoning [Mitchell, 1982, Michalski, 1983] as it forms the basis for reasoning about hypotheses and their relationships. Especially important in this context is the notion of generality. One pattern is *more general* than another one if all examples that are covered by the latter pattern are also covered by the former pattern.

For instance, the rule

$$\mathsf{mutagenic}(\mathsf{M}) \leftarrow \mathsf{ring\_size\_5}(\mathsf{M}, \mathsf{R}), \mathsf{element}(\mathsf{A1}, \mathsf{R}), \mathsf{bond}(\mathsf{M}, \mathsf{A1}, \mathsf{A2}, 2)$$

is more general than the rule

$$\begin{aligned}\mathsf{mutagenic}(\mathsf{M}) \leftarrow \\ \mathsf{ring\_size\_5}(\mathsf{M}, \mathsf{R}), \mathsf{element}(\mathsf{A1}, \mathsf{R}), \\ \mathsf{bond}(\mathsf{M}, \mathsf{A1}, \mathsf{A2}, 2), \mathsf{atom}(\mathsf{M}, \mathsf{A2}, \mathsf{o}, 52, \mathsf{C})\end{aligned}$$

The former rule is more general (or, equivalently, the latter one is more specific) because the latter one requires also that the atom connected to the ring of size 5 be an oxygen of atom-type 52. Therefore, all molecules satisfying the latter rule will also satisfy the former one.

The generality relation is useful for inductive learning, because it can be used 1) to prune the search space, and 2) to guide the search towards the more promising parts of the space. The generality relation is employed by the large majority of logical and relational learning systems, which often search the space in a *general-to-specific* fashion. This type of system starts from the most general rule (the unconditional rule, which states that *all* molecules are active in our running example), and then repeatedly specializes it using a so-called refinement operator, which specializes a given hypotheses.

Using logical description languages for learning provides us not only with a very expressive and understandable representation, but also with an excellent theoretical foundation for the field. This becomes clear when looking at the generality relation. It turns out that the generality relation coincides with logical entailment. Indeed, the above examples of the generality relation clearly show that the more general rule logically entails the more specific one. So, the more specific rule s is a logical consequence of the more general one g, or, formulated differently, the more general rule logically entails the more specific one, that is, $g \models s$. Consider the simpler example: $\mathsf{flies}(\mathsf{X}) \leftarrow \mathsf{bird}(\mathsf{X})$ (if X is a bird, then X flies), which logically entails and which is clearly more general than the rule $\mathsf{flies}(\mathsf{X}) \leftarrow \mathsf{bird}(\mathsf{X}), \mathsf{normal}(\mathsf{X})$ (only normal birds fly). This property of the generalization relation provides us with an excellent formal basis for studying inference operators for learning. Indeed, because one rule is more general than another if the former entails the latter, deduction is closely related to specialization as deductive operators can be used as specialization operators. At the same time, one can obtain generalization (or inductive inference) operators by inverting deductive inference operators.

This can be illustrated using traditional deductive inference rules, which start from a set of formulae and derive a formula that is entailed by the original set. For instance, consider the *resolution* inference rule for propositional clauses:

$$\frac{h \leftarrow g, a_1, \ldots, a_n \text{ and } g \leftarrow b_1, \ldots, b_m}{h \leftarrow b_1, \ldots, b_m, a_1, \ldots, a_n} \tag{1}$$

This inference rule starts from the two rules above the line and derives the so-called *resolvent* below the line. This rule can be used to infer, for instance,

$$\text{flies} \leftarrow \text{blackbird, normal}$$

from

$$\text{flies} \leftarrow \text{bird, normal}$$
$$\text{bird} \leftarrow \text{blackbird}$$

An alternative deductive inference rule adds a condition to a rule:

$$\frac{h \leftarrow a_1, \ldots, a_n}{h \leftarrow a, a_1, \ldots, a_n} \tag{2}$$

This rule can be used to infer that

$$\text{flies} \leftarrow \text{blackbird}$$

is more general than

$$\text{flies} \leftarrow \text{blackbird, normal}$$

In general, a deductive inference rule can be written as

$$\frac{g}{s} \tag{3}$$

If s can be inferred from g and the operator is *sound*, then $g \models s$. Thus applying a deductive inference rule realizes specialization, and hence, deductive inference rules can be used as specialization operators. A *specialization operator* maps a hypothesis onto a set of its specializations. Because specialization is the inverse of generalization, *generalization operators* — which map a hypothesis onto a set of its generalizations — can be obtained by inverting deductive inference rules. The inverse of a deductive inference rule written in the format of Eq. 3 works from bottom to top, that is from s to g. Such an inverted deductive inference rule is called an *inductive* inference rule. This leads to the view of induction as the inverse of deduction; cf. also [Michalski, 1983, Jevons, 1874]. This view is operational as it implies that each deductive inference rule can be inverted into an inductive one, and that each inference rule provides an alternative framework for generalization.

An example generalization operator is obtained by inverting *adding condition* rule in Eq. 5. It corresponds to well-known *dropping condition* rule; cf. [Michalski, 1983]. It is also possible to invert the resolution principle of Eq. 4; cf. [Muggleton, 1987]. For instance, the resolution rule can be rewritten as

$$\frac{h \leftarrow g, a_1, \ldots, a_n \text{ and } g \leftarrow b_1, \ldots, b_m}{h \leftarrow b_1, \ldots, b_m, a_1, \ldots, a_n \text{ and } h \leftarrow g, a_1, \ldots, a_n} \tag{4}$$

and then it is possible to apply it in the inverse, that is, inductive direction. Using this inverted resolution rule one can induce

$$\text{bird} \leftarrow \text{blackbird}$$

from

$$\text{flies} \leftarrow \text{blackbird}, \text{normal}$$
$$\text{flies} \leftarrow \text{bird}, \text{normal}$$

In addition to inductive and deductive inference, there is also *abductive* inference. It can be obtained as a special case of the inverse resolution rule just mentioned. Indeed, consider the rule

$$\frac{h \leftarrow g \text{ and } g}{h \leftarrow g \text{ and } h} \tag{5}$$

It can be used to find possible explanations for phenomena. For instance, given the rule

$$\text{flies} \leftarrow \text{bird}, \text{normal}$$

and the observation flies the abductive inference rule can infer that it may be a bird that is normal. Although we have just shown that abductive inference can be considered a special case of inductive inference, abductive reasoning is typically used in a different context. While inductive reasoning aims at generalizing specific observations into general laws, abductive reasoning seeks explanations for a specific observation. The differences between abductive and inductive reasoning have resulted in quite some discussion in the scientific literature; cf. [Flach and Kakas, 2000].

Before deploying inference rules, it is necessary to determine their properties. Two desirable properties are *soundness* and *completeness*. These properties are based on the repeated application of inference rules in a proof procedure. Therefore, we write $g \vdash_r s$ when there exists a sequence of hypotheses $h_1, \cdots, h_n$ such that

$$\frac{g}{h_1}, \frac{h_1}{h_2}, \cdots, \frac{h_n}{s} \text{ using } r \tag{6}$$

A proof procedure with a set of inference rules r is then *sound* whenever $g \vdash_r s$ implies $g \models s$, and *complete* whenever $g \models s$ implies $g \vdash_r s$. In practice, soundness is always enforced while completeness is an ideal that is not always achievable in deduction. Fortunately, it is not always required in a machine learning setting. When working with an incomplete proof procedure, one should realize that the generality relation $\vdash_r$ is weaker than the logical one $\models$.

The formula $g \models s$ can now be studied under various assumptions. These assumptions are concerned with the class of hypotheses under consideration and the operator $\vdash_r$ chosen to implement the semantic notion $\models$. The hypotheses can be single rules, sets of rules (that is, clausal theories), or full first-order and even higher-order theories. Deductive operators that have been studied include θ-subsumption [Plotkin, 1970] (and its variants such as *OI*-subsumption[Esposito et al., 1996]) among single clauses, implication among single clauses and resolution among clausal theories. Each of these deductive notions results in a different framework for specialization and generalization. Due to its relative simplicity, θ-subsumption is by far the most popular framework. It is used by the vast majority of contemporary logical and relational learning systems. Rather than presenting the underlying structure in full detail, we illustrate the basics on the propositional subset of clausal logic. When performing operations on clauses, it is often convenient to represent the clauses by the sets of literals they contain. For instance, the clause flies $\leftarrow$ bird, normal can be represented as {flies, ¬bird, ¬normal}.

Using this notion for two propositional clauses c_1 and c_2,

$$c_1 \text{ subsumes } c_2 \text{ if and only if } c_1 \subseteq c_2 \qquad (7)$$

Thus, the clause flies $\leftarrow$ bird, normal subsumes the clause flies $\leftarrow$ bird, normal, pigeon. Observe that propositional subsumption is *sound*, which means that whenever c_1 subsumes c_2, it is the case that $c_1 \models c_2$, and *complete*, which means that whenever $c_1 \models c_2$, c_1 also subsumes c_2.

This model of inductive reasoning can now be extended to account for background knowledge. Background knowledge typically takes the form of a set of clauses B, which is then used by the covers relation. When learning in the presence of background knowledge B an example e is covered by a hypothesis h if and only if $B \cup h \models e$. This notion of coverage is employed in most of the work on inductive logic programming. In the flies example, one might employ the following two clauses for defining the bird predicate:

$$\text{bird} \leftarrow \text{blackbird} \qquad\qquad \text{bird} \leftarrow \text{ostrich}$$

Using these clauses as background theory B, the example flies $\leftarrow$ blackbird, normal is covered by the hypothesis flies $\leftarrow$ bird, normal.

The incorporation of background knowledge in the induction process has resulted in frameworks for generality *relative* to a background theory. More formally, a hypothesis g is more general than a hypothesis s relative to the background theory B if and only if $B \cup g \models s$. The only already seen inference rules that deal with multiple clauses are those based on (inverse) resolution. The other frameworks can be extended to cope with this generality relation following the logical theory of generalization. Various frameworks have been developed along these lines. Some of the most important ones are relative subsumption [Plotkin, 1971] and generalized subsumption [Buntine, 1988], which extend θ-subsumption towards the use of background knowledge. More details on frameworks for generalization and inductive and abductive reasoning can be found in [De Raedt, 2008, Nienhuys-Cheng and de Wolf, 1997].

4   Probabilistic Reasoning

Because the world is inherently uncertain, logic alone does not suffice for many application areas. It also explains why a lot of recent research in artificial intelligence and machine learning is concerned with combining logical and relational representations with probabilistic ones [Getoor and Taskar, 2007, De Raedt et al., 2008]. This line of research has resulted in a large number of

Table 1. Conditional probability tables for the carrier example

$P(\text{carrier}(\mathsf{X}) = true)$	carrier(X)	$P(\text{suffers}(\mathsf{X}) = true)$
0.6	$true$	0.7
	$false$	0.01

different probabilistic logical and relational representations. Rather than providing a detailed account of these representations, I provide some illustrative examples; the first one along the lines of [Sato and Kameya, 1997, Poole, 1993] is based on the idea of associating probabilities to facts or switches. Consider the following adaptation of the flies example:

$$\text{flies}(\mathsf{X}) \leftarrow \text{bird}(\mathsf{X}), \text{normal}(\mathsf{X})$$
$$\text{bird}(\mathsf{X}) \leftarrow \text{ostrich}(\mathsf{X})$$
$$\text{bird}(\mathsf{X}) \leftarrow \text{blackbird}(\mathsf{X})$$
$$\text{normal}(\mathsf{X}) \leftarrow \text{blackbird}(\mathsf{X})$$
$$0.8 :: \text{blackbird}(\mathsf{X}) \vee 0.2 :: \text{ostrich}(\mathsf{X}) \leftarrow$$

All formulae, except the last one, are typical rules. The last statement is a so-called probabilistic switch. It states that any X is either a blackbird (with probability 0.8) or an ostrich (with probability 0.2). If there are multiple switches, they are assumed to be independent of one another. A hypothesis containing switches can be regarded as a probability distribution over a set of hypotheses, where each hypothesis in the set contains one fact (or choice) for each switch and the probability of that hypothesis is given by the product of the probabilities of all the facts it contains. So, in the flies example, the set contains two hypotheses, one with probability 0.8 containing the fact blackbird(X) and the other one containing ostrich(X) with probability 0.2. Both hypotheses contain all non-probabilistic rules.

It should be stressed that there exist alternative ways of integrating logic and probability theory. For instance, several models, such as [Kersting and De Raedt, 2007, Costa et al., 2003, Getoor et al., 2001], integrate Bayesian networks [Pearl, 1988] with logic. These models represent joint probability distributions over relational states (sets of facts) in a compact manner. They employ conditional probability distributions for this purpose. For instance,

$$\text{suffers}(\mathsf{Person}) \mid \text{carrier}(\mathsf{Person})$$

states that the probability that a person suffers from a disease probabilistically depends on whether or not she is a carrier of that disease. One typically also needs to specify the prior probability of a person being a carrier, which would be captured by an expression of the form

$$\text{carrier}(\mathsf{Person})$$

Both statements would have associated probability tables, such as those in Table 1, for specifying the probability values.

The combination of probabilistic and logical representations affects both inference and learning. So is the coverage relation no longer a 1-0 criterion, but rather a probabilistic one. For instance, the probability that flies(tweety) is now 0.8, that is, the sum of the probabilities of the hypotheses in which the flies(tweety) is entailed. Furthermore, also the inference rules and types can now take into account the probabilistic information. For instance, it is well-known that abductive reasoning can be guided by the probabilistic information [Poole, 1993]. As an illustration, if one observes flies(tweety) there is a single abductive explanation for this fact, that is, blackbird(tweety) with probability 0.8. If on the other hand, one observes bird(tweety), there are two alternative explanations: blackbird(tweety) with probability 0.8 and ostrich(tweety) with probability 0.2, which shows that the probabilities of the explanations can be used to select the most likely one. Similar observations can be made for the carrier example.

Secondly, learning becomes more complex. When working with probabilistic representations, one now distinguishes the learning of the structure, that is, the rules, from the estimation of the parameters. Parameter estimation is an easier task, which explains why it has – so far – received the most attention.

To summarize, the introduction of probabilities into logical representations puts knowledge representation, inference and learning into a new perspective. Although there exists a consensus in the artificial intelligence community that (various types of) logic are well-suited for knowledge representation purposes and graphical models are well-suited for both reasoning about uncertainty and learning, we still lack a general theory and methodology for combining probability with logic and learning. The multitude of different alternative probabilistic logics that exist today and the difficulties to relate them shows that such a generally accepted framework is urgently needed. More details on these issues can be found in [Getoor and Taskar, 2007, De Raedt et al., 2008, De Raedt, 2008].

5   Conclusions

This paper has provided a gentle introduction to the use of logic, knowledge and inference in machine learning. As such it has revisited and reinterpreted some of the issues of Ryszard Michalski raised in his theory and methodology of inductive learning in terms of logical and relational learning. It has also discussed an extra dimension, the integration of probabilistic and logical representations and reasoning with principles of learning, which forms the subject of a lot of recent research. The reader interested in finding out more about these topics may want to consult [De Raedt, 2008].

References

Buntine, W.: Generalized subsumption and its application to induction and redundancy. Artificial Intelligence 36, 375–399 (1988)

Santos Costa, V., Page, D., Qazi, M., Cussens, J.: CLP(BN): Constraint logic programming for probabilistic knowledge. In: Meek, C., Kjærulff, U. (eds.) Proceedings of the 19th Conference on Uncertainty in Artificial Intelligence, pp. 517–524. Morgan Kaufmann, San Francisco (2003)

De Raedt, L.: Logical and Relational Learning. Springer, Heidelberg (2008)

De Raedt, L., Frasconi, P., Kersting, K., Muggleton, S.H. (eds.): Probabilistic Inductive Logic Programming. LNCS (LNAI), vol. 4911. Springer, Heidelberg (2008)

Džeroski, S., Lavrač, N. (eds.): Relational Data Mining. Springer, Heidelberg (2001)

Esposito, F., Laterza, A., Malerba, D., Semeraro, G.: Refinement of Datalog programs. In: Proceedings of the MLnet Familiarization Workshop on Data Mining with Inductive Logic Programing, pp. 73–94 (1996)

Flach, P., Kakas, A.C. (eds.): Abduction and Induction: Essays on Their Relation and Integration. Kluwer Academic Press, Dordrecht (2000)

Getoor, L., Taskar, B. (eds.): An Introduction to Statistical Relational Learning. MIT Press, Cambridge (2007)

Getoor, L., Friedman, N., Koller, D., Pfeffer, A.: Learning probabilistic relational models. In: Džeroski, S., Lavrač, N. (eds.) Relational Data Mining, pp. 307–335. Springer, Heidelberg (2001)

Jevons, W.S.: The Principles of Science: a Treatise on Logic and Scientific Method. Macmillan, Basingstoke (1874)

Kersting, K., De Raedt, L.: Bayesian logic programming: theory and tool. In: Getoor, L., Taskar, B. (eds.) An Introduction to Statistical Relational Learning. MIT Press, Cambridge (2007)

King, R.D., Srinivasan, A.: Prediction of rodent carcinogenicity bioassays from molecular structure using inductive logic programming. Environmental Health Perspectives 104(5), 1031–1040 (1996)

Michalski, R.S.: A theory and methodology of inductive learning. Artificial Intelligence 20(2), 111–161 (1983)

Mitchell, T.M.: Generalization as search. Artificial Intelligence 18, 203–226 (1982)

Muggleton, S.: Duce, an oracle based approach to constructive induction. In: Proceedings of the 10th International Joint Conference on Artificial Intelligence, pp. 287–292. Morgan Kaufmann, San Francisco (1987)

Muggleton, S.: Inverse entailment and Progol. New Generation Computing 13 (1995)

Muggleton, S., De Raedt, L.: Inductive logic programming: Theory and methods. Journal of Logic Programming 19/20, 629–679 (1994)

Nienhuys-Cheng, S.-H., de Wolf, R.: Foundations of Inductive Logic Programming. Springer, Heidelberg (1997)

Pearl, J.: Probabilistic Reasoning in Intelligent Systems: Networks of Plausible Inference. Morgan Kaufmann, San Francisco (1988)

Plotkin, G.D.: A note on inductive generalization. In: Machine Intelligence, vol. 5, pp. 153–163. Edinburgh University Press (1970)

Plotkin, G.D.: A further note on inductive generalization. In: Machine Intelligence, vol. 6, pp. 101–124. Edinburgh University Press (1971)

Poole, D.: Probabilistic Horn abduction and Bayesian networks. Artificial Intelligence 64, 81–129 (1993)

Sato, T., Kameya, Y.: Prism: A symbolic-statistical modeling language. In: Proceedings of the 15th International Joint Conference on Artificial Intelligence, pp. 1330–1339. Morgan Kaufmann, San Francisco (1997)

Srinivasan, A., Muggleton, S., Sternberg, M.J.E., King, R.D.: Theories for mutagenicity: A study in first-order and feature-based induction. Artificial Intelligence 85(1/2), 277–299 (1996)

Washio, T., Motoda, H.: State of the art of graph-based data mining. SIGKDD Explorations 5(1), 59–68 (2003)

Two Examples of Computational Creativity: ILP Multiple Predicate Synthesis and the 'Assets' in Theorem Proving

Marta Fraňová and Yves Kodratoff

Equipe Inférence et Apprentissage
Laboratoire de Recherche en Informatique
CNRS & Université Paris Sud
Bat. 490, 91405 Orsay, France
`mf@lri.fr, yk@lri.fr`

Abstract. We provide a precise illustration of what can be the idea of "computational creativity", that is, the whole set of the methods by which a computer may simulate creativity. This paper is restricted to multiple predicate learning in Inductive Logic Programming (ILP) and to Program Synthesis from its Formal Specification (PSFS). These two subfields of Computer Science deal with problems where creativity is of primary importance. They had to add to their basic formalisms, ILP and Beth's tableaux (for PSFS), sets of heuristics enabling the program to solve the problem of Multiple Predicate synthesis. Some of these are what is usually called a heuristic, that is, a method by which the execution speed of the program is supposed to be boosted, at the price of a lost of some solutions. This paper, inversely, shows heuristics the goal of which is to provide the program with some kind of inventiveness. The basic tool for computational creativity is what we call an 'asset generator' a specification of which is given in section 4, followed by a detailed description of our methodology for the generation of assets in PSFS. Since it may seem that our 'asset generation methodology' for PSFS relies essentially on making explicit the logician's good sense while performing a recursion constructive proof, and as an example of its efficiency, we provide in conclusion a result, a kind of challenge for the other theorem provers, namely: 'invent' a form of the Ackerman function which is recursive with respect to the second variable instead of the first variable as the usual definitions are.

In ILP multiple predicate synthesis, the assets have been provided by members of the ILP community, while our methodology tries to make explicit a way to discover these assets when they are needed.

1 Introduction, Motivations, and What Is a "Recursive Problem"?

The goal of this paper is presenting non trivial examples of a methodology of creativity. We could also say in a bit flaunting way that we aim at clarifying some essentials for a methodology of creativity, that is, to push forward the field of "computational creativity," a topic we discussed already in Franova et al. (1993) This research domain starts with the works of Newell and Simon (1972) and it has been quite well

J. Koronacki et al. (Eds.): Advances in Machine Learning II, SCI 263, pp. 155–173.
springerlink.com   © Springer-Verlag Berlin Heidelberg 2010

defined by Boden (1999). Our paper, however, does not aim at an analysis of the state of the art in computational creativity. It rather aims at showing how the specialists in Inductive Logic Programming (ILP) and in Programs Synthesis from their Formal Specifications (PSFS) have been dealing with the problem of the synthesis of multiple, mutually dependent, predicates. The problems met are of a recursive kind and they demand some creative computing of which we shall give several examples.

One of the hardest problems that ILP tackled is the one of the synthesis from examples of predicates that are mutually dependent. The base papers due to de Raedt and al. (1993a and b), and de Raedt and Lavrac (1996) provide an analysis of the problems raised by this question. They opened the way to a research stream illustrated by a score of works. Besides the authors cited in the body of this paper, one can also consult Martin and Vrain (1995), Zhang and Numao (1997), Fogel and Zaverucha (1998). The problem the ILP community worked upon is precisely what we call: a "recursive problem". We should not confuse *recursive programming*, which is writing in a recursive way an already known solution for a problem that might be recursive, and *solving recursive problems*, the solution of which is still unknown. The two techniques ask for very different creative methods. As an illustration, consider the following example. This is a (somewhat simplified) problem of recognizing a dot as being the one ending a sentence. This example asks for no special knowledge, and it shows the same difficulties as the two examples central to this paper. In order to make the difference between a 'sentence ending dot' and a dot 'part of or end of an acronym', we need to combine two sources of information. The first is that a dot ending a sentence (except in case of a typing error, of which we do not speak here) is always stuck to a string of letters forming a 'well-known' word. The second is that it is always followed by an empty space (except in case of a section ending, of which we do not speak here) and followed by a word starting with a capital letter, this word normally starts with a lower case letter. As long as we do not know any formal specification neither for a 'well-known word' nor for a 'word normally starting with a lower case letter', the problem is not to write a program, be it recursive or not, on this topic. The problem is finding a formal specification of these two informal specifications, possibly dependent on the application domain. In other words, we have to solve the 'recursive problem': sentence ending dot / unexpected upper case. Heitz's PhD thesis (2008) proposes an iterative solution to this recursive problem.

An ILP example is the one of the automatic synthesis of the mutually dependents predicates *odd* and *even* as it is done by the system ATRE of Malerba and al. (1998). This definition may seem unexpected but if we try to "impose recursion" to the synthesis system and if we give precedence to the relations that use once the successor function, it becomes the one to be expected:

even(X) :- zero(X)
odd(X) :- succ(Y,X), even(Y)
even(X) :- succ(Y,X), odd(Y).

This simple example shows that ATRE has been creative as compared to the definitions we learn at school : even(X) :- zero(X) ; odd(X) :- succ(0,X) ; odd(X) :- succ(Y,X), succ(Z,Y), odd(Z) ; even(X) :- succ(Y,X), succ(Z,Y), even(Z).

Before going inside our methodology for generating assets in PSFS, we will show what are the PSFS equivalents to some of the heuristics developed by ILP. After all, a formal specification can be seen as an infinite sequence of positive and negative examples that enables formally proving the equivalence between the specification and the synthesized program. The so-called 'classical approach of ILP (De Raedt and L. Lavrac, 1996) is very similar except that it 'only' relies on a finite sequence of positive and negative examples and provides and informal proof of equivalence: the synthesized program 'covers' the positive and does not cover the negative examples. Looking for multiple predicates makes the task of ILP much more difficult than synthesizing one predicate alone. This difficulty is even greater than could be expected as shown by De Raedt and al. (1993a and b). Since the task includes synthesizing several predicates in parallel, the order in which they are synthesized is central and very hard to control.

We gather these heuristics in three groups: the generalization/particularization methods, increasing the amount of background knowledge used and the discovery of new knowledge (the 'assets') relative to the domain. We can imagine that 'real' creativity belongs to the discovery of assets. One of our aims is to show that computational creativity merges the three approaches.

2 Generalization/Particularization Methods

As far as ILP starts from instantiated examples, it is obvious that the programs or predicates it will synthesize are generalizations of these instances. We will not insist on this point which has been much worked upon in the ILP literature, except to point out that two very opposite heuristics are used. Firstly, the possible clauses are built by generalization combined with subsequent specialization. Secondly, once some clauses are synthesized, it is necessary to choose which to keep by defining the notions of cover and of measure of interest.

The most classical method, which received scores of variations, is the one in FOIL (Quinlan, 1990), namely the entropy gain brought by each synthesized clause. The system is shown T_0 examples of which T^+_0 are positive and T^-_0 are negative. The entropy of the set of examples is

$$E_0 = -[((T^+_0/T_0) \log_2 (T^+_0/T_0)) + ((T^-_0/T_0) \log_2 (T^-_0/T_0))].$$

When clause C is synthesized, it is possible to check that it is in agreement with T^+_C positive examples and T^-_C negative examples. Let $T_C = T^+_C + T^-_C$, the entropy associated to C is

$$E_C = - T_C/T_0 [((T^+_C/T_C) \log_2 (T^+_C/T_C)) + ((T^-_C/T_C) \log_2 (T^-_C/T_C))].$$

The difference between these two values measures the gain in entropy due to clause C. This basic formula is one of the best-known to control the moves of the learning programs in the space of the possible hypotheses. In reality, it belongs partially to our topic because each inductive system, depending on the data it processes, can use a specific heuristic to find its way in the hypothesis space. For example, Kijsirikul and al. (1991) have developed a measure of interest that combines entropy gain and a syntactic distance between the final goal and the changes due to C. The logical equivalent of this heuristic is quite trivial: in PSFS, the lemmas and theorems are proved in the order they appear during the proof. Recursion proofs use another trivial heuristic: prove first the base case, then the general case.

When only one predicate is synthesized at a time, p, the consistency of each hypothesis, namely adding one clause to the knowledge base, is checked on the positive and negative examples. When several predicates are synthesized together, it is possible to go on using this process for each predicate p_i of the multiple hypothesis. It is clear that logics ask for checking the consistency of all the p_i in this hypothesis, without making any difference among them. However, this type of intertwining of all the proofs obviously increases too much the computation time. This is why, in order to solve a real life problem of image understanding, Esposito and al. (1998, 2000) had to introduce what they call their strategy of 'separate-and-parallel-conquer search'. This strategy relies on the classical 'divide and conquer' strategy. However, several 'conquer' are performed in parallel.

More precisely, the classical 'divide and conquer' strategy conquers (= learns) one isolated clause, which is added to the knowledge base. It then divides (= rejects) the examples covered by this clause. The process starts again on the examples that are left. The strategy used by Esposito and al. starts with an over-generalization covering one single example. The specializations enabling to avoid covering negative examples are checked in parallel. Each specialization use a gain heuristic in order to choose the specialization that will cover the less possible negative examples. All specializations covering none of the negative example are possible candidates to become the clause that will be synthesized.

3 Increasing Domain Knowledge

We must at first make a clear difference between domain knowledge (also called 'background knowledge') and the invention of assets. Obviously knowledge is always an asset but domain knowledge covers everything known about the problem. A simple example of it is the knowledge that recursive relations may exist, a very general type of knowledge. As a less simple example, which we will use in the following, is a 'classical' property of recursion (Peter 1967): Suppose we are studying a partially false relation $R(x, z)$ where x is the input variable and z the output variable. This suggests that we need to build a new recursive predicate $P(x)$, as we shall see later. When applying the induction hypothesis, it may happen that we can find a solution provided the condition $G(x)$ holds. In this case, the new predicate $P(x)$ is defined by $P(x)$:- $G(x), P(pred(x))$. On the contrary, asset invention is the discovery of still unknown structures amongst the data, and these structures are useful for finding a solution of the problem at hand.

We shall not give here the huge amount of knowledge implied by the very existence of ILP. The fact of being able to think that programs may be able to synthesize other programs supposes also a mass of background knowledge. This theme has been treated for ILP in general by the creators of FOCL (see Pazzani and Kibler 1992), and for multiple predicate synthesis by Giordana et al. (1993). More generally, all automatic learning inductive systems use what is called the "learning bias" that limit the size of the space of the clauses that can be learned. There are many kinds of such bias, such as the syntactical form of the clauses that can be learned. A quite evolved example of this case is found in a paper of Baroglio and Botta (1995) where the authors introduce clause templates for the predicates to synthesize. They give an example relative to the binary clauses p(x, y). They associate to these binary clauses a template such as p'(x, y) :- p(x, z), diff(z, y) when they need to use the predicate 'different' during the synthesis. Similarly, Brazdil and Jorge (1994) introduce 'sketches' that fit well intuition but that provide also links among the variables that are an important part of the solution. For example, when they want to synthesize from examples the predicate 'reverse', they use the 'sketch':

reverse (A, Z) :- $P1 (A, B, C), reverse (C, D), $P2 (D, B, Z). This says to the synthesis system that it must look for a recursive definition, the number of variables in the functionals $P1 and $P2, and that $P1 contains the input variable of 'reverse', and $P2 its output variable: a quite detailed 'background knowledge' that should be perhaps be rather called an educated guess.

4   Discovering New Knowledge about the Domain (The Assets)

4.1   Including in Background Knowledge the Synthesized Clauses

This is normal practice in ILP where the newly learned clauses are added to the knowledge base of the clauses already known. The classical action associated to this choice is to delete the examples covered by the just added clause. The process will then converge since the synthesis will stop when all the positive examples are covered. It is also possible, inversely, to add to the existing learning examples ('ground facts') the facts that can be deduced from the synthesized clauses. This is what Jorge and Brazdil (1996) call "iterative bootstrap induction." This obviously complexifies the program since new facts have to be added for each new hypothesis, and old facts must be deleted when a clause is deleted.

4.2   Relational Pathfinding

Richards and Mooney (1992) introduced this notion when adding the available literals to a clause does not change its entropy gain. This problem is particularly significant when several predicates are synthesized in parallel. The solution of these authors can be seen as searching, within the graph of the relationships carried by the examples, the sub-graph of the relation carried by the predicate to synthesize. They give a heuristic making the best use of the existing knowledge without attempting to solve the problem of finding all sub-graphs of a given graph. Suppose that, during a proof, it is needed to find a binary predicate P(a, b) where P is to be synthesized and 'a' and 'b' are two instances of the ground facts. Suppose 'a' and 'b' are not linked in the relational graph of the ground facts. They have nevertheless a set {R} of relationships with other instances. As a practical simplification suppose all these relations are binary. Remember that the ground facts are instantiated examples and we call here 'instances' the values of these instantiations, such as names of persons in a family base. The heuristic developed by Richards and Mooney (1992) considers all the paths defined by {R}. If these partial paths create one path linking 'a' and 'b', then this link is made of a sequence of binary predicates in {R}. It has the form $R_i(a, a_i), \ldots, R_j(b_j, b)$. The heuristic supposes that this path is an example of P, that is:

$P(a, b) :- R_i(a, a_i), \ldots, R_j(b_j, b)$. By replacing the instances by variables preserving the links amongst the instances, it easily obtains a predicate

$P(x, y) :- R_i(x, x_i), \ldots, R_j(y_j, y)$ that, once it is added in the knowledge base, enables to prove $P(a, b)$, unless it requests a unary predicate (that cannot be taken into account by this heuristic). This last case is dealt with by yet another heuristic explained in the next section. Anyhow, we have thus obtained a first sequence of the predicates that must be joined in order to increase the entropy gain. If the first set of partial paths do not create a link between 'a' and 'b', this process is repeated on the free nodes which still show a set of relations {R'} with the instances not yet used. If several paths are possible, entropy gain can again be used to choose a best candidate among them. When this heuristic succeeds, it puts in evidence a structure existing in the data, which is important for the completion of the proof. This is what we call an asset. A more general definition will be given in section 6.

4.3   Failure Analysis in ILP

An example of successful failure analysis in the problem of multiple predicates synthesis is provided by Kijsirikul and al. (1992) and Zelle and al. (1994). Suppose that we meet a failure: too many negative examples are covered by a clause C because it misses a predicate P which is available among the ground facts. The problem is choosing which of the available predicates must be added to clause C in order to prevent it to cover so many negative examples. C (as it is) is used in order to prove positive and negative examples. During this proof, we obtain a list of instances, L, belonging to the proof that C covers positive examples, and a list of instances L' belonging to the proof that C covers negative examples. By analyzing the differences between the instances of L and of L' , it is most often possible to single out the predicate that makes these differences, namely P.

4.4   Failure Analysis in PSFS

4.4.1   Some Basic Principles of Recursion Proofs

Let us first recall a few basic principles of the application of recursion proofs to program synthesis. The simplest formula that may be needed to prove has the form $\forall x \exists z, Q(x, z)$. This formula, F1, determines the Skolem function, SF, associated to z, which the value computed by this Skolem function when applied to variable x, that is to say $\forall x, z = SF(x)$. In other words, due to the quantifiers of x and z, the semantics of an input variable are applied to x, and the semantics of an output variable are applied to z. The problem of PSFS is to find 'the' SF which checks F1, which amounts to find a constructive proof of $\forall x \exists z, Q(x, z)$.

Performing a proof recursion consists in the analysis of what are called "base case" and "general case" and to apply to the general case what is called the "induction hypothesis." Since we restrict ourselves here to the natural numbers, the base case is $x = 0$ since the simplest formula we start from includes no conditions on x. The general case is $x = s(a)$ where 's' is the successor function. The induction hypothesis ("if the formula is assumed to be true for n, then we can prove that it is true for n+1") writes $\exists e, Q(a, e) \Rightarrow \exists z, Q(s(a), z)$ or else, using a Skolem function, $Q(a, SF(a)) \Rightarrow Q(s(a), SF(s(a)))$.

We will suppose in the following examples that we are able to prove the base case and that we find, for $x = 0$, that the formula F1 is made true by $z = a_0$, i.e., $SF(0) = a_0$.

4.4.2 The 'Hypothetical' Program

As a first illustration of failure recovery in PSFS, suppose that, in the general case, we can find a function G such that $z = G(e) = G(SF(a))$ at the condition that $P(z)$ is true. This is a condition on the outputs which is a failure of the recursion proof since it is not allowed to impose conditions on an output variable we are not able to compute (otherwise, these conditions should have been included in F1). We propose, in this case, to build a program, F, we qualify of being 'hypothetical', which is inspired by what we already know about G, without including the condition $P(z)$. We hope then that $P(F(x))$ is true for all x. Obviously, if this hope proves false in reality, we still have ways to react. We will now be content with an illustration of the simple case where $\forall x \, P(F(x))$.

This illustration comes from proofs relative to the Ackermann function published in Franova (2008). The reader will have thus to accept that some computation steps are given elsewhere. These computations lead us to try to prove that $\forall x \, \forall y \, \exists z$, Ack $(x, y) = y + z$. In the base case, we have to solve the equation Ack $(0, y) = y + z$. The definition of Ack immediately yields the solution $z = 1$, i.e., $SF(0, y) = 1$. In the general case, and after a rather long computation, we observe that we have to prove the lemma $\exists z$ Ack $(a+1, y) = y + z$. We are thus in position to try an inductive proof on the second input variable, y. In the base case, $y = 0$ and we obtain $SF(a + 1, 0) = 1 + SF(a, 1)$. In the general case, we set $y = b + 1$ and we obtain

$SF(a+1, b+1) + 1 = SF(a + 1, b) + SF(a, b + SF(a + 1, b))$. If we could prove that

$SF(a + 1, b) + SF(a, b + SF(a + 1, b)) > 0$, we would have obtain a recursive way for computing SF. We are not able to do so and we are therefore in front of a failure of the recursion proof. As announced, we build a hypothetical function F (of which we do not know if it is senseless or not) without the condition on the output variable $(z > 0)$. We use all the already obtained results about SF in order to build the function F defined by:

$F(0, y) = 1$
$F(a + 1, 0) = 1 + F(a, 1)$
$F(a+1, b+1) = F(a + 1, b) + F(a, b + F(a + 1, b)) - 1$

We are able to prove that $\exists z, F(x, y) = 1 + z$, that is to say $\forall x \, \forall y, F(x, y) > 0$. It follows that SF which is defined in the same way has the same property.

In this simple case, failure recovery amounts to a proof that this failure does not actually take place. In more complex cases, it often will lead to predicate generation as we shall see.

4.4.3 Dealing with Partially False Formulas

A partially false formula may be used as an indication that a new predicate needs to be built.

Consider the following formula:

(F1) $\forall a\ \forall b,\ b>0,\ \exists z,\ b = a + z$

The analysis of this formula, during an attempt at a recursion proof, shows it is partially false, thus the proof fails. As did Zelle and al. (1994) in ILP, we suppose that there exists a hypothetical predicate P(a, b) with which we will be able to recognize when F1 is true. In order to build P, we use the failure cases of the proof of F1.

In the base case, the condition $b>0$ implies that $b = s(0)$. Considering the formula $\forall a\ \exists z,\ s(0) = a + z$, we observe that it has two partial solutions, namely $z = s(0)$ IF $a = 0$, and $z = 0$ IF $a = s(0)$. These failure conditions provide two features of the hypothetical predicate P:

P(a,s(0)) :- a = 0
P(a,s(0)) :- a = s(0)

In the general case, we have $b=s(c)$, $c>0$, and we have to prove that $\exists e,\ c = a + e \Rightarrow \exists z$ $s(c) = a + z$. Our approach systematically uses a heuristic that urges to study, before using the induction hypothesis, the so-called trivial solutions to the problem $\exists z,\ s(c) = a + z$. We at once obtain two trivial solutions, $z = s(c)$ IF $a = 0$, and $z = 0$ IF $a = s(c)$. This provides us with two new characteristics of P :

P(a,s(c)) :- a = 0
P(a,s(c)) :- a = s(c)

Now, we can use the induction hypothesis and we find $z = s(e)$ without any condition on a. The logics of the recursive features (Peter 1967) tells us to write the recursive feature of P as follows:

P(a,s(c)) :- P(a,c).

We have built or observed five features of P that completely define the predicate P: $\forall a\ \forall b,\ b>0,\ (P(a,b) \Rightarrow \exists z,\ b = a + z$. As you see, we do not obtain a proof of F1 but the conditions at which F1 is true. The proof can then go on by a study of the cases where non P is true. We are able in this way to obtain the largest characterization of the cases where the formal specification we started with can lead to a computable program.

5 Abduction

Kakas and al. (1998) proposed a new method for solving the problem of synthesis of multiple predicates. Their approach modifies the proof technique by using Abductive Logical Programming (ALP) - see Kakas and al. (1998), and Ade and Denecker (1995) instead of ILP. This is not a heuristic but a change of the logical setting. It is thus expected that we have no obvious PSFS equivalent. The definition of ALP given by Kakas and al. (1998) proposes to associate to the usual logical program a set of partially defined predicates, they call "abducible," and a set of integrity constraints insuring that the inferences on the abducibles keep consistency with P. This enables them to replace the classical "negation by failure" used in logical programming, by a "negation by default," defined as follows. To each predicative symbol, *pred*, used in

P, they associate its negation, *non_pred*, and they add to the set of the integrity constraints the supplementary constraint :- *non_pred*, *pred*. The logical derivation is no longer reduced to finding True or False: It becomes, as in classical abduction, a set of predicates that are an "abductive explanation" of the proof. This approach illustrates, at least, the amount of inventiveness the ILP community have shown in order to solve the problem of synthesis from examples of multiple predicates.

When collecting the failures of the proof of the formula

(F1) $\forall a \ \forall b \ b > 0 \ \exists z \ b = a + z,$

we used a similar approach since each of the five features of P 'explains' the failure to prove F1. This heuristics is, however, much less well formalized than the one of Kakas and al. (1998). This formalization work is thus still to be done in the context of PSFS.

6 The Asset Generator

An asset generator defines a human or an automated behavior. It is made of chained actions that are led by strategies, themselves non formally defined. It is thus impossible to give a complete formal definition of an asset generator. We shall give at first an incomplete formal definition we shall then complete in an informal way.

6.1 Definitions

An Incomplete Formal Definition

Given a possibly incomplete undecidable formal theory, and a base theorem (the one to be proven), an **asset** to prove the base theorem is obtained by the analysis of a failure to prove the base theorem. This analysis may detect three different causes of the failure, each leading to a different recovery strategy. Firstly, and quite usually, a 'well-known' and useful part of the domain theory has been forgotten and the asset is a consequence of the theory which has to be retrieved in the existing domain specific literature. Secondly, a very particular non classical lemma is missing and the asset has to be invented as a consequence of the theory. Thirdly, the theory is undecidable, and the asset completes the theory, just enough to be able to prove the base theorem.

Here come now the *informal commentaries* that describe what is an asset generator, given this *definition*.

Let us imagine humans put in the situation of proving a theorem in an undecidable theory. They will at first attempt to directly prove the base theorem from the given theory. Suppose that they are excellent mathematicians and that they fail at finding the proof. They at least know that the proof is going to be either difficult, or impossible. They will thus start a long process by which they will, in principle, combine two strategies. The first strategy looks for the tools missing for completing the proof. They will use their intuition and their knowledge of the constructive proofs, their knowledge of the formal theory, and of the content of the theorem to prove. We call this part of the human behavior the "strategies for choosing the assets" because it enables to find (possibly undecidable) theorems that might possibly be useful in order to prove the base theorem. Our experience, however, is that we are unable to simulate, for the time being, this part of human behavior.

The second of these strategies generates decidable theorems from the theory, and it checks each of these theorems in order to know if they might or not be used some place in a proof of the base theorem. In order to 'give flesh' to this last statement, this paper gives examples of several derivation techniques used in ILP and PSFS.

A Complete Informal Definition

Consider an informally defined proof strategy that does either exist nowhere else than in the mathematician's mind, or it is part of an automatic proving strategy.

Given an undecidable formal theory and a base theorem, we call "asset generator for proving the base theorem," a proof strategy showing the three following features: 1. it generates assets that are defined in the above formal definition. 2. All the generated assets will be easier to prove than the base theorem. 3. If it generates a potentially infinite sequence of assets, then it exists a generalization to this infinite sequence, and this generalization is supposed to be equivalent to the infinite sequence that continues the finite sequence obtained in practice. In the case of an informal specification, condition 3 is formulated rather as excluding infinite sequences: When the asset generator generates an infinite sequence of assets, this amounts to a failure case.

It follows that an asset generator does not insure that we will be able to find a formal proof of the specification. It is only a strategy that gives interesting results in PSFS. It constitutes a strategy which, once it is made fit to the user needs, can help them to better manage the problems of recursion they meet while programming. Given the present state of the art, a real life application of all these techniques seems to us somewhat risky in automatic programming while it should be quite useful as an assistant for dealing with the difficult and counter-intuitive problems.

6.2   Assets and Domain Knowledge

We defined the assets as knowledge invented during a proof. It is however obvious that if some domain knowledge is not given, the problems may become of a very different nature and it may become impossible to discover the necessary assets. A simple ILP example is the one 'favoring' or not recursive calls by including, or not, a process that will systematically attempt to generate a recursive form. In PSFS, section 7.5 will illustrate how, depending of the parameters introduced in the induction hypothesis (knowledge of all the possible forms of the induction hypothesis is part of the domain knowledge), we may fail to find an asset when considering a useless branch of the proof tree.

7   How to Find the Assets in PSFS

The problem dealt with in this last section is to explain in general, and provide two specific examples, of how the assets can be generated in PSFS. From the ILP examples we have seen that humans are in charge to be creative. Creativity is hard to reach, especially in the domain based on a strong theory, such as the one of PSFS. This is why we have been developing a methodology (we called "CM-strategy," see Franova (1985), which can help the humans to know when, where and how their creativity is needed. We do not claim this methodology is able to solve all the problems – we however provide in conclusion a result difficult to reach without using our asset methodology. We do not claim it is easily automated and, as already stated, we see it more as an automated help to prove theorems, though it has been already partially implemented as the CM-strategy, see Franova (1990).

In order to generate assets during our proofs, we need to add two steps to the usual recursion proofs. The one is the management of a kind of stack into which we pile up the conditions needed for solving the problem at hand. We call this process "**introducing abstract arguments**." The second one is a heuristic helping us to generate lemmas such that their proof enables to go on in the proving process. The generation of these special assets is called "**generation of intermediary lemmas**."

7.1 Introducing the 'Abstract Arguments'

We introduce a new type of argument in the predicates a feature of which has to be proven true, we call **abstract arguments**. They are denoted by ξ (or ξ' etc.) in the following. An abstract argument replaces one of the arguments of the base theorem. The first step is choosing which of the arguments of the base theorem will be replaced by an abstract argument, ξ. This argument is known and, in a usual proof, its characteristics are used in order to prove the base theorem. In our approach, we 'forget' for some time these characteristics and we concentrate on studying the features ξ should have so as insuring that the theorem with a substituted argument is true.

In the following, and for the sake of avoiding a too general wording, suppose for example that the formula to prove has two arguments, that is to say that we need to prove that $F(t_1, t_2)$ is true, where F is the base theorem. Suppose further that we have chosen to work with $F(\xi, t_2)$. We shall then look for the features shown by all the ξ such that $F(\xi, t_2)$ is true.

At first, we have to choose which argument will be made abstract. There are two ways to introduce an abstract argument, and we thus start with either $F(t_1, \xi)$ or $F(\xi, t_2)$ since, obviously, $F(\xi, \xi')$, an *a priori* possible choice, would hide all the characteristics of t_1 and t_2.

Supposing we are able to find the features of ξ such that (say) $F(t_1, \xi)$ is true, for all the ξ showing these features, $F(t_1, \xi)$ is true. In other words, calling 'cond' these features and $\{C\}$ the set of the ξ such that $cond(\xi)$ is true, we define $\{C\}$ by $\{C\} = \{\xi \mid cond(\xi)\}$. We can also say that we try to build a 'cond' such that the theorem: $\forall \xi \in \{C\}, F(t_1, \xi)$ is true. It is reasonable to expect that this theorem is much more difficult to prove than $F(t_1, t_2)$. We thus propose a 'detour' that will enable us to prove the theorems that cannot be directly proven, without this 'detour'. Using the characteristics of $\{C\}$ and the definition axioms in order to perform evaluations, and also using the induction hypothesis, we shall build a form of ξ such that $F(t_1, \xi)$. Even though it is still 'ξ' and only for the sake of clarity, let us call ξ_C one of these forms. It is thus such that $F(t_1, \xi_C)$. We are still left with a hard work to perform: Choose the 'good' ξ_C in the set $\{C\}$ and modify it (possibly using hypothetical theorems), in such a way that ξ_C and t_2 will be made identical, which finally completes the proof. Section 7.4 will give a detailed example of this process.

In the following, we will illustrate our methodology by two examples. The simplest of the two is the proof that, for the natural numbers, $2 < 4$. Here F is the predicate '<',

$t_1 = 2$ and $t_2 = 4$. As you will see, we will choose to replace the second argument by ξ and we will study the characteristics of ξ such that $2 < \xi$. As already said, this is a generalization of the base theorem. Note also that we temporarily forget the base theorem and that we focus on the study of the ξ such that $2 < \xi$.

7.1.1 The Sub-problem of the Choice between $F(t_1, \xi)$ and $F(\xi, t_2)$

This choice is done following a heuristic as follows. The predicate F is defined by $F(t_1, t_2)$ and the predicates or functions that define t_1 and t_2. All these definitions are given as a set of axioms and the way these axioms are defined leads to the choice between $F(t_1, \xi)$ and $F(\xi, t_2)$. In the general case, if F defined recursively with respect to one variable, for example the first one, as in

F(s(a), b) IF F(a, b'), then the second argument is replaced by ξ, and we will study $F(t_1, \xi,)$. If the general term of F is defined by a double recursion, then consider the base case, such as for example, $F(0, t_2)$, and the non zero argument is replaced by ξ. If the two terms $F(0, t_2)$ and $F(t_1,0)$ are given in the axioms, then the same kind of considerations are applied to the operators contained in F. If this does not enable to decide, all the possibilities for replacing the arguments by an abstract argument must be studied in turn.

7.1.2 How to Use the Abstract Argument

Suppose we started with $F(t_1, \xi,)$. Since we wish to build solutions enabling us to prove the theorem, the construction process will include checking the likeness of ξ and t_2. The general rule we use during these likeness checks is quite obvious. Suppose that t_2 has the form $f(t'2, t''2)$. Then ξ cannot be matched but with a function as $f(x, y)$ where x and y are variables. Thus, we reach the problem of proving $\exists u \, \exists v, \xi = f(u, v)$. More generally, the failure of a matching between ξ and t_2 leads us to introduce existentially quantified variables that insure the success of the matching of ξ and t_2. The automation of this reasoning step can become very complex and shows many different cases, but each case is quite general and is trivial to solve. For example, suppose that during an evaluation step, we realize that in order to prove the theorem, we have to identify ξ and s(u), then we will make the hypothesis that there indeed exists a 'u' such that $\xi = s(u)$. This relation is looked upon as a condition on ξ: ξ is such that $\exists u, \xi = s(u)$. We shall see more complex examples while going through the examples in 7.3 and 7.4.

7.2 The Generation of Intermediary Lemmas

It is not enough to introduce the existential lemmas we noticed just above. Besides the variables necessary to insure the matching of ξ and t_2, the base theorem included other variables, in particular in t_2. If these variables were universally quantified then what we call an intermediary lemma needs also a universal quantification of these variables. Moreover, the base theorem looks like

[prove that] $\forall x, G(x, \xi)$. By the manipulations ξ undergoes, we discover that the relation G' has to be true, where $\exists u, G'(x, u)$. Since x is universally quantified during all the manipulations we did, the intermediary lemma is thus $\forall x \, \exists u, G'(x, u)$, in which the quantifiers are obviously put in this order. Depending on the conditions on ξ, the intermediary lemmas are not of the same form in the choice of the variables to quantify. We are not yet able to provide a complete list of all the possible forms. The discovery of these forms still relies entirely on human creativity. We shall see, however, that each form can be applied to many particular cases.

7.3 An Illustrative Example Proved without Recursion

In example, and to ease out its reading, we identify each integer with its expression in terms of successor of 0.

Base theorem: Prove that $2 < 4$ by using the following axioms defining the function $<$:

> 1. $0 < s(n)$
> 2. $s(m) < s(n)$ IF $m < n$

In our approach, we shall need to implicitly use the domain knowledge about equality since we start by replacing t_1 or t_2 by ξ. Recall that the axioms defining equality are:

> $x = y \Leftrightarrow y = x$
> $x = x$
> $s(x) = s(y)$ IF $x = y$
> $x = y$ AND $y = z \Rightarrow x = z$
> $s(x) = x$ has the value FALSE (also given as NOT $(s(x) = x)$)
> $s(x) = 0$ has the value FALSE.

This example illustrates the detail of the reasoning steps we perform while applying our methodology. It is however obvious that the simple evaluation classical proof is much faster than ours. We will in reality prove a more general theorem than the one in the base theorem. This example does not illustrate the 'power' of our methodology but the way to use it.

The classical evaluation proof is:

$2 < 4$? By axiom 2. $s(1) < s(3)$ IF $1 < 3$, therefore IF $0 < 2$ and by axiom 1, $0 < s(1)$ is TRUE.

Our proof is a bit more complex. We at first observe that axioms are recursive with respect to the first variable. Thus we replace the second argument, 4, by the abstract argument ξ. We try now to solve the problem to find the conditions on ξ implying $2 < \xi$. The application of axiom 2 demands an auxiliary variable ξ' by which we can suppose that there exists a ξ' such that $\xi = s(\xi')$. Axiom 2 can then be applied to ξ, yielding $[2 < \xi$ IF $[\exists \xi', \xi = s(\xi')]$ AND $[1 < \xi']]$. Axiom 2 can be again applied provided there exists a ξ'' such that $\xi' = s(\xi'')$. Applying it yields: $[1 < s(\xi')$ IF $\exists \xi'']$ AND $[0 < \xi'']$. We can now apply axiom 1 provided we suppose that there exists a ξ''' such that $\xi'' = s(\xi''')$. Replacing ξ'' and ξ' by their values, we thus obtain $[2 < \xi$ IF $\exists \xi''', \xi = s(s(s(\xi''')))]$ which is the condition 'cond' we have spoken of above. This is not yet the answer to the problem at hand. Let $\{C\} = \{\xi \mid \xi = s(s(s(\xi''')))\}$. As announced, we just proved that $[\forall \xi \in \{C\} \ F(t_1, \xi)]$, that is, all numbers equal to or greater than 3 are greater than 2.

We still need to prove that $4 \in \{C\}$, that is $\exists \xi'$, $s(s(s(s(0)))) = s(s(s(\xi')))$, which is trivial in view of the axioms for equality given above, and we can exhibit $\xi' = 1$ that fulfills the condition we looked for.

Looking for a more than necessary complex solution shows a mathematical lack of taste we are quite aware of in the present case. It is however quite common that theorem proving asks for a similar 'detour' as ours when a particular theorem cannot be proven without resorting to a more general one, a quite classical problem.

7.4 An Example Using Recursion

Let us now study yet another simple example needing an inductive proof and one of our 'intermediary lemmas', that is an asset created during the theorem proving process. Besides, in spite of the triviality of the problem, and since we modified one of the axioms defining the predicate $<$, it follows that, to the best of our knowledge, the classical theorem provers fail proving that

$\forall x \, \forall y$, $x - 1 < x + y$ where $x \in Nat - \{0\}$, if they are given the following set of axioms.

Definition of $\neq$: NAT x NAT → BOOL
1. $s(x) \neq s(y)$ IF $x \neq y$
2. $s(x) \neq x$
3. $0 \neq s(x)$

Definition of $+$: NAT → NAT
4. $0 + u = u$
5. $s(u) + v = s(u + v)$

(Note that we do not provide the commutativity of $+$)

Definition of -1 : NAT → NAT
6. $(x - 1)$ is possible IFF $x \neq 0$
7. $s(0) - 1 = 0$
8. $s(y) - 1 = s(y - 1)$ IF $0 \neq y$

Definition of $<$: NAT x NAT → BOOL
9. $0 < u$ IF $0 \neq u$
10. $s(u) < v$ IF $u < v$ AND $s(u) \neq v$

In this case, the 'lack of taste' of our methodology is hidden behind the 'lack of taste' of the clumsy form of axiom 10 with respect to evaluation procedures.

Let us use recursion to prove this theorem. Recall at first that it is part of the 'classical' domain knowledge to know that the proofs can be done after some variables become parameters. A parameter represents any value of the variable it replaces, but this value is fixed while the proof goes on (see, for example, Kleene, 1980). Let us call $E(x, y)$ the formula

$(x - 1 < x + y)$.

The recursive proofs can be done in various ways depending on the choice of the induction variable. This is part of recursive theorem proving domain knowledge, it is not a part of our methodology. We shall choose to perform the recursion proof on the first variable. We thus will have to prove $E(s(0), y)$ in the base case and that $E(a, y) \Rightarrow E(s(a), y)$ in the general case, where y is a parameter.

The induction hypothesis shows, in fact, various possible forms depending on the choice done for replacing the variables by parameters. In the present case, we can use three possible induction hypotheses (Yashuhara, 1971).

The one where 'a' and 'y' are made parameters, it writes: $a - 1 < a + y$. In order to avoid digressing, we shall see how to use this hypothesis in section 7.5, since it leads to a failure.

The one where 'a' only is made parameter, it writes: $\forall u, a - 1 < a + u$. We shall use this one.

For the sake of completeness since we shall not need to use it: the one where 'x' is the only variable that is made parameter, it writes: $\forall t \, \forall q, t < x, [t - 1 < t + q]$.

We shall thus use the induction hypothesis where we hypothesize that

$\forall u, [a - 1 < a + u]$ where, under this hypothesis, we have to prove that $[s(a) - 1 < s(a) + y]$.

We observe that the definition of $<$, which leads us to replace the second argument by ξ in our methodology, that is: $\xi = s(a) + y$. We must study the conditions for $s(a) - 1 < \xi$, that is $\{C\} = \{\xi \mid s(a) - 1 < \xi\}$. Notice at first that by axiom 8, $s(a) - 1 = s(a - 1)$ and that by axiom 10, $s(a - 1) < \xi$ IF $([a - 1 < \xi] \text{ AND } [s(a - 1) \neq \xi])$.

Thus, $\{C\} = \{\xi \mid [a - 1 < \xi] \text{ AND } [s(a - 1) \neq \xi]\}$.

Now, we will try to introduce the induction hypothesis within the definition of $\{C\}$. The matching of $[a - 1 < a + u]$ and $[s(a - 1) \neq \xi]$ is not possible, while we observe it is possible with $[a - 1 < \xi]$ provided $\xi \leftarrow a + u$. In the case where no immediate matching is possible, we generate an intermediary lemma by which this matching might become possible, as we shall now do by building such an 'intermediary lemma'.

We already chose to replace $t_2 = s(a) + y$ by ξ As explained before in section 7.1, we have to find one element in $\{C\}$, we called ξ_C, that can match t_2. Since we do not have the axiom telling us that the matching succeeds if $u \leftarrow s(y)$, we are facing a failure case demanding the generation of an intermediary lemma so as 'inventing' by ourselves that 'u' has to take the value $s(y)$. To that purpose, we choose 'the' ξ_C which is suggested by the induction hypothesis, $\xi_C = a + u$. We thus have to find a 'u' that makes possible the substitution $\xi_C \leftarrow s(a) + y$. We thus face the problem of exhibiting a 'u', such that $[\exists u, s(a) + y = a + u]$. As announced, the lemma to prove has to regain the quantifiers of the variables in the base theorem. We thus have to prove that:

$$\forall a \, \forall y \, \exists u, s(a) + y = a + u$$

The discovery of this lemma is an essential step of creativity and you can check that we discovered it through an analysis of the possible matching of t_2 and ξ_C. Depending on the problem at hand, these matching can generate diverse lemmas and, as already said, we do not have at our disposal (and that will probably never happen) a full battery of all the possible forms. The form we just used, however, is quite standard and can be applied in many cases: We choose to replace t_2 by ξ, then we observe that the induction hypothesis leads us to find a ξ_C which we fail to match to t_2, we then generate the existential feature that, if proven, would enable the matching.

In principle, this lemma is hard to prove because it contains a universal quantifier followed by an existential one. We shall nevertheless prove it by a simple recursion proof, as follows.

In the base case, we have prove that, when $a = 0$, $\exists u$, $s(a) + y = a + u$. We know that $s(a) + y = s(a + y)$, thus that $s(0) + y = s(y)$. Since $0 + u = u$, we obtain, in the base case, that $u = s(y)$.

In the general case, we have to prove that if $[\exists p, s(n + y) = n + p]$ then $[\exists q, s(s(n + y)) = s(n) + q]$. Since $s(n) + y = s(n + y)$ this is equivalent to proving that $[\exists q, s(n + y) = n + q]$, which precisely is the induction hypothesis. From the domain knowledge relative to recursion proofs, we know that the Skolem function of 'u' has the characteristic that $SF(s(b), y) = SF(b, y)$, thus that $u = SF(x, y) = s(y)$.

This completes the proof of the intermediary lemma. We come back to the base theorem and prove that $s(a) - 1 \neq \xi$, where ξ has the value "$a + u$." Using the value of 'u' we just found, that is

$$\forall a \; \forall y, s(a) - 1 \neq a + s(y)$$

This is done easily and completes the proof of the base theorem.

7.5 'Appendix' to 7.4: Failure of the Proof Using [a −1 < a + y]

We attempt to prove that $[a - 1 < a + y] \Rightarrow [s(a) - 1 < s(a) + y]$ where a and y are parameters. As before, we replace $s(a) + y$ by the abstract argument ξ. We thus must study the conditions for $s(a) - 1 < \xi$, that is $\{C\} = \{\xi \mid s(a) - 1 < \xi\}$. As above, we note that by axiom 8, $s(a) - 1 = s(a - 1)$ and by axiom 10, $s(a - 1) < \xi$ IF $[[a - 1 < \xi]$ AND $[s(a - 1) \neq \xi]]$.

Thus, $\{C\} = \{\xi \mid [a - 1 < \xi]$ AND $[s(a - 1) \neq \xi]\}$. We now try to find the induction hypothesis within $\{C\}$. The matching of $[a - 1 < a + y]$ and $[a - 1 < \xi]$ is possible provided $\xi_C \leftarrow a + y$ and we have to solve the equation $s(a) + y = a + y$, that is $s(a + y) = a + y$. This is FALSE by axiom 2.

Note that when we meet an equation we can prove to be false, we do not generate an intermediary lemma since it will be obviously always impossible to prove it. When recovery is possible, such as when finding conditions for a matching to succeed, we generate an 'intermediary lemma' which is a potential asset and this launches the creativity process. A bland failure such as trying to prove FALSE is TRUE should lead to trying something else or giving up.

8   Conclusion

In this paper we provided, at first, some examples of the way human researchers in ILP had to be creative when they wanted to solve the problem of synthesizing from examples multiple and mutually dependent predicates, which are naturally built by the means of mutually recursive definitions. This is an example of human 'computational creativity' in a domain akin to the one in which we want to build a tool for helping creativity, namely theorem proving by recursion. Most of the techniques illustrated here: the 'abstract argument' technique, the 'hypothetical program', the 'intermediary lemma' and even our 'asset generator' may seem a result of simple logical good sense, more than a real forward step in theorem proving. We partly agree since we are convinced that, as a matter of fact, most humans who prove theorems go through one or the other of these techniques. The new result we claim is to have put them in a coherent methodology based on the 'abstract argument' technique, from which all other techniques are called when they are needed. In order to illustrate that we can solve more difficult problems than the simple illustration we gave here, we will give the solution of a problem yet unsolved, as far as we know. The Ackerman function is usually defined by recursion with respect to its first argument. Let us call 'ACK' this definition, it is:

$$ACK(0,n) = n+1$$
$$ACK(m+1,0) = ACK(m,1)$$
$$ACK(m+1,n+1) = ACK(m,ACK(m+1,n))$$

Since this definition is so much accepted in the textbooks, we were curious to know if it is possible find a definition of the Ackerman function where recursion takes place with respect to the second argument. To find such a definition, and following the general idea underlying the generation of the intermediate lemmas, we tried to see if there was a 'z' such that, when exhibited, it could define a second variable recursive Ackerman function, AK. We thus tried to prove the two following theorems, similar to our 'intermediary lemmas':

$$\forall x \, \exists y, \, ACK(x, 0) = z$$
$$\forall x \, \forall y \, \exists z, \, ACK(x, y+1) = ACK(x, y) + z$$

As you see, we did not try to prove directly $\forall x \, \forall y \, [ACK(x,y) = AK(x,y)]$ holds since, anyhow, we ignored the definition of 'AK' which will be invented during the proofs of the above lemmas. The recursive proof of the first lemma asks for an auxiliary function $AUX(x)$ and the proof of the second one asks for another auxiliary function $AUX3(x)$.

The function AK is defined by:

$$AK(x,0) = AUX(x)$$
$$AK(x,y+1) = AK(x,y) + AUX3(x,y)$$

This solution needs two more auxiliary functions as follows:

$$AUX(0) = 1$$
$$AUX(a+1) = AUX(a) + AUX1(a)$$

AUX1 is defined by

$$AUX1(0) = 1$$
$$AUX1(b+1) = AUX2(b,AUX(b+1))$$

AUX2 is defined by

$$AUX2(0,y) = 1$$
$$AUX2(a+1,0) = 1+ AUX2(a,1)$$
$$AUX2(a+1,b+1) = AUX2(a+1,b) + AUX2(a,b+AUX2(a+1,b)) - 1$$

AUX3 is defined by

$$AUX3(0,y) = 1$$
$$AUX3(a+1,y) = AUX2(a,AK(a+1,y))$$

Now that the definition of AK is found, theorem provers can attempt proving $\forall x\ \forall y\ [ACK(x,y) = AK(x,y)]$. Our methodology proved it also, by 'inventing' AK. You notice that the definition of AUX2 is quite similar to the definition of ACK which means that our solution is still strongly biased towards the classical definition. More work would be needed to find other, less obvious, 'intermediate lemmas' leading to other definitions. In spite of this restriction, you see that a really new definition of the Ackerman function has been found. We did not give the complete proof because it would need too much space to be presented. More generally, as we underlined when presenting it, it is clear that our methodology asks for a fair amount of work since many 'intermediary' lemmas can be generated depending on the matching we want to prove to be possible, and each of them has to be disproved before attempting to prove another one. Creativity is expensive, we have to acknowledge this fact. One of us (YK) remembers Ryszard Michalski, in the beginning of the 80's, being very annoyed by the people who were complaining about the computation time needed by his rule generation technique. He would say something like: "Even if it takes months of computation time to find a new rule, unknown to a human specialist, this is a big success!" We cannot more agree with him than by presenting our complicated, cumbersome, however creative methodology.

References

Ade, H., Denecker, M.: AILP: Abductive inductive logic programming. In: Proceedings of the fourteenth International Joint Conference on Artificial Intelligence, pp. 1201–1207 (1995)

Baroglio, C., Botta, M.: Multiple Predicate Learning with RTL. In: Gori, M., Soda, G. (eds.) AI*IA 1995. LNCS, vol. 992, pp. 44–55. Springer, Heidelberg (1995)

Beth, E.: The Foundations of Mathematics, Amsterdam (1959)

Boden, M.: Computational models of creativity. In: Sternberg, R.J. (ed.) Handbook of Creativity, pp. 351–373. Cambridge University Press, Cambridge (1999)

Brazdil, P., Jorge, A.: Learning by Refining Algorithm Sketches. In: ECAI 1994, 11th European Conference of Artificial Intelligence, pp. 427–433 (1994)

De Raedt, L., Lavrac, N., Dzeroski, S.: Multiple Predicate Learning. In: Proceedings of the thirteenth International Joint Conference on Artificial Intelligence, pp. 1037–1042 (1993a)

De Raedt, L., Lavrac, N., Dzeroski, S.: Multiple Predicate Learning. In: Proceedings of the third International Workshop on Inductive Logic Programming, pp. 221–240 (1993b)

De Raedt, L., Lavrac, N.: Multiple Predicate Learning in Two Inductive Logic Programming Settings. J. of the IGPL 4(2), 227–254 (1996)

Esposito, F., Malerba, D., Lisi, F.A.: Multiple Predicate Learning for Document Image Understanding. In: Proceedings of the Fourteenth American Association for Artificial Intelligence Conference, pp. 372–376 (2000)

Fogel, L., Zaverucha, G.: Normal Programs and Multiple Predicate Learning. In: Page, D.L. (ed.) ILP 1998. LNCS, vol. 1446, pp. 175–184. Springer, Heidelberg (1998)

Franova, M.: CM-strategy: A Methodology for Inductive Theorem Proving or Constructive Well-Generalized Proofs. In: Joshi, A.K. (ed.) Proceedings of the Ninth International Joint Conference on Artificial Intelligence, Los Angeles, August 1985, pp. 1214–1220 (1985)

Franova, M.: PRECOMAS - An Implementation of Constructive Matching Methodology. In: Proceedings of ISSAC 1990, Tokyo, Japan, August 20-24, pp. 16–23. ACM, New York (1990)

Franova, M., Kodratoff, Y., Gross, M.: Constructive Matching Methodology: Formally Creative or Intelligent Inductive Theorem Proving? In: Komorowski, J., Raś, Z.W. (eds.) ISMIS 1993. LNCS (LNAI), vol. 689, pp. 476–485. Springer, Heidelberg (1993)

Franova, M.: A Construction of a Definition Recursive with respect to the Second Variable for the Ackermann function. To appear as LRI internal research report, Orsay, France (2008)

Giordana, A., Saitta, L., Baroglio, C.: Learning simple recursive theories. In: Komorowski, J., Raś, Z.W. (eds.) ISMIS 1993. LNCS (LNAI), vol. 689, pp. 425–444. Springer, Heidelberg (1993)

Heitz, T.: Une méthode récursive pour le prétraitement des textes. Thèse, Universite Paris-Sud (2008)

Jorge, A., Brazdil, P.: Architecture for iterative learning of recursive definitions. In: Advances in Inductive Logic Programming, pp. 206–218 (1996)

Kakas, A., Lamma, E., Riguzzi, F.: Learning multiple predicates. In: Giunchiglia, F. (ed.) AIMSA 1998. LNCS (LNAI), vol. 1480, pp. 303–316. Springer, Heidelberg (1998)

Kijsirikul, B., Numao, M., Shimura, M.: Efficient learning of logic programs with non-determinate non-discrimonating literals. In: Kaufmann (ed.) ML 1991 – Machine Learning Conference, pp. 417–421 (1991)

Kijsirikul, B., Numao, M., Shimura, M.: Discrimination-based constructive induction of logic programs. In: Proceedings of the tenth National Joint Conference on Machine Learning, San Jose, CA, pp. 44–49 (1992)

Kleene, S.C.: Introduction to Meta-Mathematics. North-Holland, Amsterdam (1980)

Malerba, D., Esposito, F., Lisi, F.A.: Learning Recursive Theories with ATRE. In: Proceedings of the Thirteenth European Conference on Artificial Intelligence, pp. 435–439 (1998)

Martin, L., Vrain, C.: MULT_ICN: An empirical multiple predicate learner. In: Proc. 5th International Workshop on ILP, pp. 129–144 (1995)

Newell, A., Simon, H.A.: Human Problem Solving. Prentice-Hall, Englewood Cliffs (1972)

Pazzani, M., Kibler, D.: The Utility of Knowledge in Inductive Learning. Machine Learning 9, 57–94 (1992)

Peter, R.: Recursive Functions. Academic Press, New York (1967)

Richards, B.L., Mooney, R.J.: Learning Relations by Pathfinding. In: Proceedings of the Tenth National Conference on Artificial Intelligence. MIT Press, Cambridge (1992)

Yashuhara, A.: Recursive Function Theory and Logic. Academic Press, London (1971)

Zelle, J.M., Mooney, R.J., Konvisser, J.B.: Combining Top-down and Bottom-up Techniques in Inductive Logic Programming. In: Machine Learning: Proceedings of the Eleventh International Conference, pp. 343–351 (1994)

Zhang, X., Numao, M.: MPL-Core: An Efficient Multiple Predicate Learner Based on Fast Failure Mechanism. Journal of Japanese Society for Artificial Intelligence 12(4), 582–590 (1997)

Logical Aspects of the Measures
of Interestingness of Association Rules

Jan Rauch

Faculty of Informatics and Statistics of the University of Economics, Prague*
nám. W. Churchilla 4, 130 67 Prague 3, Czech Republic
`rauch@vse.cz`

Abstract. The relations of the logical calculi of association rules and
of the measures of the interestingness of association rules are studied.
The logical calculi of association rules, 4ft-quantifiers, and known classes
of association rules are briefly introduced. New 4ft-quantifiers and association rules are defined by the application of suitable thresholds to
known measures of interestingness. It is proved that some of the new
4ft-quantifiers are related to known classes of association rules with important properties. It is shown that new interesting classes of association
rules can be defined on the basis of other new 4ft-quantifiers and several
results concerning new classes are proved. Open problems are introduced.

1 Introduction

Association rules are a known form of knowledge studied in the field of KDD
(Knowledge Discovery in Databases). There have been many findings pertaining
to the various properties of association rules. This chapter deals with the logical aspects of association rules, namely with deduction rules concerning pairs
of association rules. We argue for the importance of the study of such deduction rules. An overview of the known findings on such deduction rules is given,
new findings are presented, and unsolved problems are listed. The new findings
concern known measures of interestingness of association rules.

An association rule is understood here not as an implication $X \rightarrow Y$ where
X and Y are conjunctions of simple Boolean attributes, as defined in [8]. An
example of such an association rule is the expression $r \wedge s \rightarrow u \wedge v$ which is
usually understood as an assertion saying that if a market basket contains items
r and s, then it usually also contains items u and v. Two interestingness measures
of the rule $X \rightarrow Y$ are widely used – *confidence* and *support*.

We deal with association rules of the form $\varphi \approx \psi$ where φ and ψ are general
Boolean attributes derived using the propositional connectives $\wedge$, $\vee$, and $\neg$. The

* The work described here has been supported by the grant 201/08/0802 of the Czech
Science Foundation.

rule $\varphi \approx \psi$ means that φ and ψ are associated in the manner given by the symbol $\approx$ that is called a *4ft-quantifier*. Each 4ft-quantifier is a condition derived from a measure of the interestingness of an association rule by the application of a suitable threshold to a value of the measure of interestingness. Important measures of the interestingness of an association rule and their properties are widely studied; see e.g. [1,2,3,4,5]. We can define new interesting 4ft-quantifiers by the application of suitable thresholds to additional known measures of interestingness of an association rule and in this way we get new association rules of the form $\varphi \approx \psi$. The logical properties of such association rules could be useful from various points of view.

We are interested in correct deduction rules of the form $\frac{\varphi \approx \psi}{\varphi' \approx \psi'}$ where φ, ψ, φ', and ψ' are general Boolean attributes. If such a deduction rule is correct, and if the rule $\varphi \approx \psi$ is true in a given data structure, then we can conclude that the rule $\varphi' \approx \psi'$ is also true in this data structure. To study such deduction rules, we need to define logical calculus formulas that correspond to association rules of the form $\varphi \approx \psi$. The logical calculi of association rules are defined and studied in [14]; they are a special case of observational calculi introduced in [9]. There are theoretically interesting and practically important findings concerning deduction rules of the form $\frac{\varphi \approx \psi}{\varphi' \approx \psi'}$. The results are closely related to classes of association rules.

The chapter is organized as follows. The logical calculi of association rules are introduced in Section 2. The reasons for the study of deduction rules of the form $\frac{\varphi \approx \psi}{\varphi' \approx \psi'}$ are discussed in Section 3. An overview of important and long-studied classes of association rules, and of the related results, is in Section 4. The way in which new 4ft-quantifiers are derived from known measures of interestingness of association rules is described in Section 5. The relations of new 4ft-quantifiers to important classes of association rules are presented in Section 6. These relations determine the logical properties of 4ft-quantifiers and corresponding association rules, and thus the related deduction rules $\frac{\varphi \approx \psi}{\varphi' \approx \psi'}$ are also introduced in Section 6. An overview of additional classes defined on the basis of important measures of interestingness of association rules, is in Section 7. Conclusions and open problems for further work are presented in Section 8.

2   Logical Calculi of Association Rules

The logical calculi of association rules studied here belong to the observational calculi defined and studied in [9]. Observational calculi have been defined to be a language in which statements concerning observational data are formulated. Monadic observational predicate calculus is a special case of observational calculus. It is defined by modifications of classical monadic predicate calculi [25]. Modifications consist of adding generalized quantifiers (e.g. 4ft-quantifiers, see below) and in allowing only finite structures as the models in which the formulas are interpreted. The monadic observational predicate calculi were enhanced to calculi with qualitative values in [9]. The calculi with qualitative values were simplified in [20] such that only formulas of the form $\varphi \approx \psi$ corresponding to the association rules mentioned in Sect. 1 are allowed. These calculi can be understood as calculi of association rules. The correctness of important deduction rules of the form $\frac{\varphi \approx \psi}{\varphi' \approx \psi'}$ is proved in these calculi. These calculi are called here the *qualitative calculi of association rules*, and they are informally defined in Section 2.1.

However the theorems on the correctness of the deduction rules of the form $\frac{\varphi \approx \psi}{\varphi' \approx \psi'}$ in qualitative calculi of association rules, are in some respects a bit complex due to dealing with qualitative values. The principles of these theorems can be fully demonstrated in the more simple *predicate calculi of association rules*. The predicate calculi of association rules are (again informally) defined in Section 2.2. We will use them in this chapter to demonstrate the properties of deduction rules of the form $\frac{\varphi \approx \psi}{\varphi' \approx \psi'}$.

2.1 Qualitative Calculi of Association Rules

The Boolean attributes φ and ψ are derived from the columns of the analyzed data matrix $\mathcal{M}$ that contains observation results. The rows of $\mathcal{M}$ correspond to observed objects, and the columns of $\mathcal{M}$ correspond to *attributes* describing observed objects. There is a finite number of possible values for each column of $\mathcal{M}$. Possible values of the attributes (i.e. values of attributes) are called *categories*.

First, the *basic Boolean attributes* are created. The basic Boolean attribute is an expression of the form $A(\alpha)$ where $\alpha \subset \{a_1, \ldots a_k\}$ and $\{a_1, \ldots a_k\}$ is the set of all possible categories of A. The basic Boolean attribute $A(\alpha)$ is true in the row o of $\mathcal{M}$ if it is $A(o) \in \alpha$ where $A(o)$ is the value of the attribute A in the row o. Boolean attributes φ and ψ are derived from basic Boolean attributes using the propositional connectives $\vee$, $\wedge$ and $\neg$ in the usual way. An example of the data matrix with n rows corresponding to the observed objects $o_1, \ldots, o_n$ and with the columns - attributes A, B, C, $\ldots$, Z is the data matrix $\mathcal{M}$ in Fig. 1. There are also examples of the basic Boolean attributes $A(3)$ and $Z(6,8)$ in Fig. 1 (*true* is denoted by 1 and *false* is denoted by 0).

object	A	B	C	D	...	Z	A(3)	Z(6,8)
o_1	3	5	4	9	...	8	1	1
o_2	5	2	7	8	...	1	0	0
$\vdots$	$\vdots$	$\vdots$	$\vdots$	$\vdots$	$\ddots$	$\vdots$	$\vdots$	$\vdots$
o_n	2	1	6	7	...	6	0	1

Fig. 1. Data Matrix $\mathcal{M}$ and the Basic Boolean Attributes $A(3)$, $Z(6,8)$

The rule $\varphi \approx \psi$ is *true in* $\mathcal{M}$, if the condition given by the 4ft-quantifier $\approx$ is satisfied in the four-fold contingency table of φ and ψ in $\mathcal{M}$, otherwise $\varphi \approx \psi$ is *false in* $\mathcal{M}$. The four-fold contingency table $4ft(\varphi, \psi, \mathcal{M})$ (the *4ft table* for short) of φ and ψ in the data matrix $\mathcal{M}$ is the quadruple $\langle a, b, c, d \rangle$ of natural numbers, such that a is the number of rows of $\mathcal{M}$ satisfying both φ and ψ, and b is the number of rows of $\mathcal{M}$ satisfying φ and not satisfying ψ etc.; see Tab. 1.

Table 1. 4ft Table $4ft(\varphi, \psi, \mathcal{M})$ of φ and ψ in $\mathcal{M}$

$\mathcal{M}$	ψ	$\neg\psi$
φ	a	b
$\neg\varphi$	c	d

An example of an association rule is the rule

$$D(7) \wedge A(2,3) \equiv_{p,B} B(6) \vee C(1,4)$$

in which the 4ft-quantifier $\equiv_{p,Base}$ of *founded equivalence* [14] is used. This quantifier is associated to the condition $\frac{a+d}{a+b+c+d} \geq p \wedge a \geq Base$. Thus if the rule $\varphi \equiv_{p,B} \psi$ is true in the data matrix $\mathcal{M}$, then it is true that both, in at least 100 percent of rows of $\mathcal{M}$, have φ and ψ the same value (either both are true or both are false) and moreover there are at least $Base$ rows satisfying both φ and ψ.

Another example of the 4ft-quantifier is the quantifier $\rightarrow_{Cf,Sp}$ that corresponds to the condition $\frac{a}{a+b} \geq Cf \wedge \frac{a}{a+b+c+d} \geq Sp$. The rule $\varphi \rightarrow_{Cf,Sp} \psi$ is the "classical" association rule with confidence Cf and support Sp as defined in [8]. A variant of the 4ft-quantifier $\rightarrow_{Cf,Sp}$ is the quantifier $\Rightarrow_{p,Base}$ of founded implication [9] defined by the condition $\frac{a}{a+b} \geq p \wedge a \geq Base$ for $0 < p \leq 1$ and $Base \geq 0$. Additional examples of 4ft-quantifiers are in Section 6.

2.2 Predicate Calculi of Association Rules

We use informally defined simple predicate calculi of association rules (PCAR for short) to demonstrate the deduction rules concerning association rules. Symbols of the language of PCAR $\mathcal{P}$ are: (unary) predicates $P_1, \ldots, P_K$, 4ft-quantifiers $\approx_1, \ldots, \approx_Q$, logical connectives $\wedge, \vee, \neg$ and additional usual symbols (e.g. brackets). Each predicate corresponds to a *basic Boolean attribute*, see Sect. 2, and *derived Boolean attributes* are created from basic Boolean attributes. A *Boolean attribute* is either a basic Boolean attribute or derived Boolean attribute. An association rule is each expression of the form $\varphi \approx \psi$ where φ and ψ are Boolean attributes and $\approx$ is the 4ft-quantifier. There are no other formulas in $\mathcal{P}$ than association rules. If PCAR $\mathcal{P}$ has K predicates then we say that it is of the type $\langle K \rangle$. An example of association rule in predicate calculus is the rule

$$P_1 \wedge P_3 \equiv_{p,B} P_2 \wedge P_4 .$$

Association rules (i.e., formulas of PCAR $\mathcal{P}$) are interpreted in $\{0,1\}$ matrices that we call *models of PCAR*. If $\mathcal{P}$ is PCAR of the type $\langle K \rangle$ then the model of $\mathcal{P}$ is each $\{0,1\}$ data matrix $\mathcal{M}$ with the columns $f_1,\ldots,f_K$ interpreting $P_1,\ldots,P_K$ respectively. The rows of a data matrix correspond to observed objects, and we denote them $o_1,\ldots,o_n$. An example of such a data matrix is in Fig. 2, where are also examples of derived Boolean attributes. The set $M=\{o_1,\ldots,o_n\}$ is called the *domain of* $\mathcal{M}$, and we write $\mathcal{M}=\langle M;f_1,\ldots,f_K\rangle$. We denote the value of the Boolean attribute φ for the object $o\in M$ as $||\varphi(o)||_{\mathcal{M}}$.

Bool. attributes	P_1	P_2	...	P_K	$\neg P_1$	$P_2\wedge P_n$
object	f_1	f_2	...	f_K		
o_1	1	1	...	1	0	1
$\vdots$	$\vdots$	$\vdots$	$\ddots$	$\vdots$	$\vdots$	$\vdots$
o_n	0	0	...	1	1	0

Fig. 2. An Example of $\{0,1\}$-Data Matrix $\mathcal{M}=\langle M;f_1,\ldots,f_K\rangle$

The value of $\varphi\approx\psi$ in the data matrix $\mathcal{M}$ is denoted as $||\varphi\approx\psi||_{\mathcal{M}}$, it is $||\varphi\approx\psi||_{\mathcal{M}}=\approx\langle a,b,c,d\rangle$ where $\langle a,b,c,d\rangle=4ft(\varphi,\psi,\mathcal{M})$ is the 4ft-table of φ and ψ in $\mathcal{M}$, see Section 2. The association rule $\varphi\approx\psi$ is true in $\mathcal{M}$ if it is $||\varphi\approx\psi||_{\mathcal{M}}=1$, otherwise $\varphi\approx\psi$ is false in $\mathcal{M}$.

3   Why Deduction Rules of Association Rules

We are interested in the deduction rules of the form $\frac{\varphi\approx\psi}{\varphi'\approx\psi'}$ where φ, ψ, φ', and ψ' are Boolean attributes. If such a deduction rule is correct, and if the rule $\varphi\approx\psi$ is true in the data matrix $\mathcal{M}$, then the association rule $\varphi'\approx\psi'$ is also true in the data matrix $\mathcal{M}$. There are at least the following reasons for studying such deduction rules:

- the possibility of reducing the output of a data mining procedure
- the possibility of decreasing the number of actually tested rules
- working with analytical reports.

Reducing of the output of a data mining procedure: If the association rule $\varphi\approx\psi$ is a part of the data mining procedure output (thus it is true in the analyzed data matrix $\mathcal{M}$), and if $\frac{\varphi\approx\psi}{\varphi'\approx\psi'}$ is the correct deduction rule, then it is not necessary to put the association rule $\varphi'\approx\psi'$ into the output; however the deduction rule used must be sufficiently transparent, from the point of view of the user of the data mining procedure. The results of research on transparent deduction rules of this form are in [22].

Decreasing the number of actually tested rules: If the association rule $\varphi\approx\psi$ is true in the analyzed data matrix $\mathcal{M}$, and if $\frac{\varphi\approx\psi}{\varphi'\approx\psi'}$ is the correct deduction rule, then it is not necessary to test $\varphi'\approx\psi'$. This approach is hard to apply in the

Apriori algorithm, but it is applied in the algorithm based on the representation of analyzed data by strings of bits [6].

Working with analytical reports: One of the ways to present the results of data mining to the user is to arrange them into a well-structured analytical report. The core of such a report is a sequence of patterns resulting from the applications of data mining procedures. It could be useful to find a "logical skeleton" of such an analytical report and to deal only with the "logical skeletons", instead of whole reports. By the "logical skeleton" of an analytical report we mean a subset of all the output patterns from which all of the patterns presented in the report can be derived. In this way we can try to communicate the analytical reports even on Semantic Web. Some preliminary considerations on these possibilities are in [26]. Deduction rules (not only of the form $\frac{\varphi \approx \psi}{\varphi' \approx \psi'}$) play an important role in identifying the "logical skeletons" of the analytical reports.

4   Classes of Association Rules

Classes of association rules are defined by classes of 4ft quantifiers. The association rule $\varphi \approx \psi$ belongs to the *class of equivalence rules*, if the 4ft quantifier $\approx$ belongs to the *class of equivalence quantifiers*. We also say that the association rule $\varphi \approx \psi$ is an *equivalence rule* and that the 4ft quantifier $\approx$ is an *equivalence quantifier*. This is the same for additional classes of association rules.

The classes of 4ft-quantifiers are defined using important truth preservation conditions, see section 4.1. There are several important truth preservation conditions that are related to important results both theoretically interesting and practically important, concerning related classes of association rules. These truth preservation conditions are presented in Section 4.2, and related results are listed in Section 4.3.

4.1   Classes of 4ft-Quantifiers and the Truth Preservation Condition

Remember that a condition concerning four-fold tables $\langle a, b, c, d \rangle$ is assigned to each 4ft-quantifier. The rule $\varphi \approx \psi$ is true in the data matrix $\mathcal{M}$ if the condition assigned to the 4ft-quantifier $\approx$ is satisfied in the four-fold contingency $4ft(\varphi, \psi, \mathcal{M}) = \langle a, b, c, d \rangle$; see Section 2.1. We write $\approx (a, b, c, d) = 1$ if the condition assigned to $\approx$ is satisfied for $\langle a, b, c, d \rangle$, otherwise we write $\approx (a, b, c, d) = 0$.

The classes of 4ft-quantifiers (i.e. classes of association rules) are defined using *truth preservation conditions*. Each class $\mathcal{C}$ is defined by the condition $TPC_{\mathcal{C}}(a, b, c, d, a', b', c', d')$ called the *truth preservation condition for $\mathcal{C}$*. This condition concerns two contingency tables, $\langle a, b, c, d \rangle$ and $\langle a', b', c', d' \rangle$. The class $\mathcal{C}$ is defined according to this scheme:

The 4ft-quantifier $\approx$ belongs to the class $\mathcal{C}$ if and only if it satisfies:

$$\approx (a, b, c, d) = 1 \ \wedge \ TPC_{\mathcal{C}}(a, b, c, d, a', b', c', d') \text{ implies } \approx (a', b', c', d') = 1 \ .$$

4.2 Important Classes of Association Rules

Important truth preservation conditions are defined and studied in [9, 14, 17]. They are listed in Tab. 2 together with the classes of association rules they define.

Table 2. Classes of Association Rules Defined by Truth Preservation Conditions

Class		Truth preservation condition
Implicational	$TPC_\Rightarrow$	$a' \geq a \land b' \leq b$
Double implicational	$TPC_\Leftrightarrow$	$a' \geq a \land b' \leq b \land c' \leq c$
Σ-double implicational	$TPC_{\Sigma,\Leftrightarrow}$	$a' \geq a \land b' + c' \leq b + c$
Equivalence	$TPC_\equiv$	$a' \geq a \land b' \leq b \land c' \leq c \land d' \geq d$
Σ-equivalence	$TPC_{\Sigma,\equiv}$	$a' + d' \geq a + d \land b' + c' \leq b + c$

The truth preservation conditions $TPC_\Rightarrow$ for implicational quantifiers and truth preservation conditions $TPC_\equiv$ for equivalence quantifiers are defined in [9]; for the remaining classes see e.g. [14, 17]. Note that the class of equivalence quantifiers is in [9] defined under the name *associational quantifiers*.

An additional important class of association rules is the class of *rules with the F-property* that is defined such that the 4ft quantifier $\approx$ has the *F-property* if it satisfies the conditions F_{bc} and F_{cb}:

F_{bc}: If $\approx (a, b, c, d) = 1$ and $b \geq c - 1 \geq 0$ then $\approx (a, b + 1, c - 1, d) = 1$.
F_{cb}: If $\approx (a, b, c, d) = 1$ and $c \geq b - 1 \geq 0$ then $\approx (a, b - 1, c + 1, d) = 1$,

Note that this definition can be also written in the form

$$\approx (a, b, c, d) = 1 \land TPC_\mathcal{F}(a, b, c, d, a', b', c', d') \text{ implies } \approx (a', b', c', d') = 1$$

where it is $TPC_\mathcal{F} = TPC_{\mathcal{F}_{bc}} \lor TPC_{\mathcal{F}_{cb}}$, (we write only $TPC_\mathcal{F}$ instead of $TPC_\mathcal{F}(a, b, c, d, a', b', c', d')$ etc.) and moreover $TPC_{\mathcal{F}_{bc}}$ is defined as

$$\langle a', b', c', d' \rangle = \langle a, b + 1, c - 1, d \rangle \land b \geq c - 1 \geq 0$$

and $TPC_{\mathcal{F}_{cb}}$ is defined as

$$\langle a', b', c', d' \rangle = \langle a, b - 1, c + 1, d \rangle \land c \geq b - 1 \geq 0 .$$

An overview of the results related to the classes of association rules defined by these truth preservation conditions is in Section 4.3. Namely, of importance are the results concerning the deduction rules for classes of implicational, Σ-double implicational, of Σ-equivalence rules, and for rules with the F-property. These results are provided in more details in Section 6. There are also examples of particular 4ft-quantifiers of these classes in Section 6.

4.3   Results on Classes of Association Rules

There are various results concerning the properties of the logical calculi of association rules published in [9, 12, 13, 14, 16, 17, 18, 18, 19, 24]. They concern the following topics.

- Deduction rules of the form $\frac{\varphi \approx \psi}{\varphi' \approx \psi'}$ where both $\varphi \approx \psi$ and $\varphi' \approx \psi'$ are association rules. The results on such deduction rules are summarized in Section 6.
- The definability of association rules in classical predicate calculi, with an overview of results given in [18].
- The application of the principle of secured extension to the evaluation of association rules in calculi with missing information. This approach was developed in [9], and an overview of related results is in [17].
- Tables of critical frequencies that are tools for avoiding complex computation related to 4ft-quantifiers corresponding to some of statistical hypothesis tests, and an overview of related results is in [17]. Tables of critical frequencies also have a relation to the definability of association rules in classical predicate calculi [18].

5   Measures of Interestingness and New 4ft-Quantifiers

We deal with 4ft-quantifiers defined by the application of suitable thresholds to the measures of interestingness studied in [1, 2, 3, 4, 5]. We use the notation for association rules used in Section 2.1 in [3]. It means that we deal with the association rule $A \approx B$ instead of $\varphi \approx \psi$. Instead of the 4ft table of φ and ψ in the data matrix $\mathcal{M}$, we use the enhanced 4ft table $4ft(A, B, \mathcal{M})$ of the Boolean attributes A and B in the data matrix $\mathcal{M}$, see Tab. 3. Here $r = a + b$ is the number of rows satisfying A, $s = c + d$ is the number of rows not satisfying A, similarly for k, l and B. Moreover $n = a + b + c + d$ is the number of rows of $\mathcal{M}$.

Suppose that each row of data matrix $\mathcal{M}$ corresponds to one market basket (i.e. tuple) and that columns of $\mathcal{M}$ correspond to particular items in the basket. Each column is a Boolean attribute that is true in a row of the data matrix $\mathcal{M}$, if and only if the item corresponding to the column is present in the basket corresponding to the row in question. In Tab. 4 there is another notation [3] used for the frequencies from the enhanced 4ft table $4ft(A, B, \mathcal{M})$ of A and B in $\mathcal{M}$. We see that $a = n(AB)$, $b = n(A\overline{B})$, $r = n(A)$, $k = n(B)$ etc.

In the definitions of interestingness measures, usually used are frequencies (e.g. $n(AB)$, $n(A)$) or probabilities and conditional probabilities [3]. An example of probability is $P(A) = \frac{n(A)}{N}$, which denotes the probability of A. An example of conditional probability is $P(B|A) = \frac{n(AB)}{n(A)}$, which denotes the conditional probability of B, given A. Each interestingness measure $\mathcal{I}$ of the rule $A \approx B$ can also be expressed as the function $\mathcal{I}(a, b, c, d)$ of frequencies a, b, c, d from the enhanced 4ft-table $4ft(A, B, \mathcal{M})$ of A and B in $\mathcal{M}$ see Tab. 3.

Each 4ft-quantifier is defined by the condition concerning the 4ft-table. The 4ft-table is the quadruple $\langle a, b, c, d \rangle$ of natural numbers; see Tab. 1 and 3. The

Table 3. Enhanced 4ft Table $4ft(A, B, \mathcal{M})$ of A and B in $\mathcal{M}$

$\mathcal{M}$	B	$\neg B$	
A	a	b	r
$\neg A$	c	d	s
	k	l	n

Table 4. Basket Rule Frequencies for $A \rightarrow B$

$\mathcal{M}$	B	$\overline{B}$	
A	$n(AB)$	$n(A\overline{B})$	$n(A)$
$\overline{A}$	$n(\overline{A}B)$	$n(\overline{AB})$	$n(\overline{A})$
	$n(B)$	$n(\overline{B})$	N

conditions used in the definitions of 4ft-quantifiers studied in $[9, 12, 13, 14, 16, 17, 18]$ can be composed from simple conditions of the form like

$$\mathcal{I}(a, b, c, d) \geq T$$

concerning functions $\mathcal{I}(a, b, c, d)$ of four natural numbers a, b, c, d where T is a suitable threshold. The used functions $\mathcal{I}(a, b, c, d)$ can be of course understood as additional interestingness measures. Note that some 4ft-quantifiers are defined by conditions like $\mathcal{I}(a, b, c, d) \leq T$.

The 4ft-quantifier $\equiv_{p,B}$ of founded equivalence (see Sect. 2) can be defined using the interestingness measures $\mathcal{I}_{\equiv}$ and $\mathcal{I}_{\odot}$ such that

$$\mathcal{I}_{\equiv}(a, b, c, d) = \frac{a + d}{a + b + c + d} \qquad \text{and} \qquad \mathcal{I}_{\odot}(a, b, c, d) = a .$$

The condition $\frac{a+d}{a+b+c+d} \geq p \wedge a \geq Base$ associated to the 4ft-quantifier $\equiv_{p,B}$ of founded equivalence can be written as $\mathcal{I}_{\equiv} \geq q \wedge \mathcal{I}_{\odot} \geq Base$. The functions $\mathcal{I}_{\equiv}(a, b, c, d)$ and $\mathcal{I}_{\odot}(a, b, c, d)$ are examples of additional interestingness measures derived from the 4ft-quantifiers studied in $[9, 12, 13, 14, 16, 17, 18]$.

We are interested in 4ft-quantifiers derived from the measures of the interestingness of association rules studied in $[1, 2, 3, 4, 5]$ by suitable thresholds. All these 4ft-quantifiers are of the form

$$\mathcal{I}(a, b, c, d) \geq T$$

where $\mathcal{I}(a, b, c, d)$ is the measure of interestingness in question. We are interested in deduction rules of the form $\frac{\varphi \approx \psi}{\varphi' \approx \psi'}$ (i.e., $\frac{A \approx B}{A' \approx B'}$) where $\approx$ is the 4ft-quantifier derived from a measure of interestingness, in the way noted.

The results concerning deduction rules are closely related to classes of association rules. Thus we have to study the relations of defined 4ft-quantifiers to known classes of association rules, or to define and to study new classes containing 4ft-quantifiers not belonging to known classes. There are 20 particular measures of interestingness and corresponding 4ft-quantifiers described in Section 6 together with their relation to known classes of association rules.

6   New 4ft-Quantifiers and Deduction Rules

We deal with 20 measures of interestingness and corresponding 4ft-quantifiers. Most of the measures of interestingness are taken from [3]. Some of them are also defined or studied in additional papers or books, e.g., in [1, 2, 4, 5, 6, 9, 23]. The additional sources are cited only if they use different names of interestingness measures than used in [3]. The citations are used only in the columns *Name* of Tab. 6, Tab. 8, Tab. 10, Tab. 12, and Tab. 13. There are also several interestingness measures defined only in [9], [23] or [6]. They are used to define important or interesting examples of 4ft-quantifiers belonging to various classes of 4ft quantifiers (i.e., classes of association rules). Note that the book [9] is from 1978 and the book [23] is from 1983.

There are two definitions of each measure of interestingness:

- The definition by probabilities as given in [3]; this definition is in the column *Probabilities [3]* of corresponding tables.
- The definition as the function $\mathcal{I}(a, b, c, d)$; it is in the column $\mathcal{I}_{\approx}(a, b, c, d)$ of the corresponding tables. The function corresponding to the quantifier $\approx$ is denoted as $\mathcal{I}_{\approx}(a, b, c, d)$. If the measure of interestingness is defined in [6, 9, 23] then the function $\mathcal{I}_{\approx}(a, b, c, d)$ is taken from these sources.

The 4ft-quantifiers defined by the particular measures of interestingness are described in the columns *Condition* and *Symbol* of corresponding tables. In the column *Condition* there is the condition concerning the function $\mathcal{I}_{\approx}(a, b, c, d)$, this condition defines the 4ft-quantifier. If the 4ft-quantifier is defined in [6,9,23], then the condition is taken from these sources; otherwise, the condition is defined as $\geq Threshold$. All conditions are written in the form $\geq Threshold$, $\leq Threshold$ or $< Threshold$. It means that the whole condition is $\mathcal{I}_{\approx}(a, b, c, d) \geq Threshold$ etc.

The following thresholds are used: $S \in (0; 1\rangle$, integer $B > 0$, $p \in (0; 1\rangle$, real $q > 0$, $\alpha \in (0; 0.5\rangle$, real $\delta > 0$, real r, and χ_{α}^2 is the $(1 - \alpha)$ quantile of the χ^2 distribution function.

In the column *Symbol* there are symbols denoting the 4ft-quantifiers. Each 4ft-quantifier has one or two parameters that are in the lower index of the corresponding symbol. E.g., the quantifier $\Rightarrow_p$ of founded implication has the parameter p; see Tab. 6. Newly defined 4ft-quantifiers are marked by *) in the column *Symbol*.

There are two simple 4ft-quantifiers $\oplus_{Sp}$ of support and $\odot_B$ – base, i.e., absolute support; they are described in Tab. 5, which also has the above-described structure.

The study of classes of association rules was initiated in [9]. An overview of results is in [17]. There are important results concerning deduction rules of the form $\frac{\varphi \approx \psi}{\varphi' \approx \psi'}$ where $\varphi \approx \psi$ and $\varphi' \approx \psi'$ are association rules. The results concern classes of implicational, Σ-double implicational and Σ-equivalence association rules. These results are applicable to all 4ft-quantifiers derived from the known measures of interestingness that belong to some of these classes. Overview of these results and of related 4ft-quantifiers follows.

Table 5. 4ft-quantifiers of Support and Absolute Support

#	Name	Measure of interestingness		4ft-quantifier $\approx$	
		Definition			
		$\mathcal{I}_{\approx}(a,b,c,d)$	Probabilities [3]	Condition	Symbol
1	Support [6]	$\frac{a}{a+b+c+d}$	$P(AB)$	$\geq Sp$	$\oplus_{Sp}$
2	Base [9]	a	–	$\geq Base$	$\odot_{Base}$

6.1 Implicational Rules

There is a theorem proved in [14] that gives a relatively simple criterion of the correctness of the deduction rule $\frac{\varphi \Rightarrow^* \psi}{\varphi' \Rightarrow^* \psi'}$ where $\Rightarrow^*$ is an *interesting implicational quantifier*. Remember that the 4ft-quantifier $\approx^*$ is implicational if

$$\approx (a,b,c,d) = 1 \ \wedge \ a' \geq a \wedge b' \leq b \ \text{ implies } \ \approx (a',b',c,d) = 1 \ ;$$

see Section 4.2. It however means that each implicational quantifier is *c-independent* and *d-independent* (i.e., its value does not depend on either c or d). Thus we write only $\Rightarrow^* (a,b)$ instead of $\Rightarrow^* (a,b,c,d)$ for the implicational quantifier $\Rightarrow^*$.

Interesting implicational quantifiers are defined in [14]. We say that the *implicational quantifier* $\Rightarrow^*$ *is interesting* if both (i) and (ii) are satisfied:

(i) $\Rightarrow^*$ is both *a-dependent* and *b-dependent*
(ii) $\Rightarrow^* (0,0) = 0$

The 4ft-quantifier $\approx$ is *a-dependent* if there are a, a', b, c, d such that $\approx (a,b,c,d) \neq \approx (a',b,c,d)$, and similarly for the *b-dependent* 4ft-quantifier.

We present the correctness criterion of the deduction rule $\frac{\varphi \Rightarrow \psi}{\varphi' \Rightarrow^* \psi'}$ for the *interesting implicational quantifiers* in the *predicate calculus of association rules* defined in Section 2.2. Note that this theorem is a simplified version of an analogous theorem for the *Qualitative Calculi of Association Rules* defined in section 2.1, and proved in [14]. The simplified version is also presented in [12] and its proof can be based on the same principles as the proof of the full version presented in [14].

The presented theorem deals with the notion of *associated propositional formula* to a given Boolean attribute: If ϕ is an attribute, then the *associated propositional formula* $\pi(\phi)$ is the same string of symbols, but the particular basic Boolean attributes are understood as propositional variables. For example: $P_2 \wedge P_n$ is the derived Boolean attribute (see Fig. 2) and $\pi(P_2 \wedge P_n)$ is the propositional formula $P_2 \wedge P_n$ with the propositional variables P_2 and P_n.

The correctness criterion for implicational quantifiers is [14]: Let PCAR $\mathcal{P}$ be the predicate calculus of association rules. In addition let φ, ψ, φ', ψ' be the Boolean attributes of $\mathcal{P}$. If $\Rightarrow^*$ is the interesting implicational quantifier then the deduction rule

$$\frac{\varphi \Rightarrow^* \psi}{\varphi' \Rightarrow^* \psi'}$$

is correct if and only if at least one of the conditions 1) or 2) are satisfied (we use the symbol $\longrightarrow$ for the propositional connective of implication):

1. Both 1.a and 1.b are tautologies of propositional calculus
   1.a $\pi(\varphi \wedge \psi) \longrightarrow \pi(\varphi' \wedge \psi')$
   1.b $\pi(\varphi' \wedge \neg\psi') \longrightarrow \pi(\varphi \wedge \neg\psi)$
2. $\pi(\varphi) \longrightarrow \pi(\neg\psi)$ is a tautology.

There are six important implicational quantifiers defined by the application of suitable thresholds to various measures of interestingness in Tab. 6. Note that the notation used in the column $\mathcal{I}_{\approx}(a, b, c, d)$ is done in Tab. 3 and the notation in the column *Probabilities [3]* is based on Tab. 4; see Section 5.

Table 6. Implicational 4ft-quantifiers

		Measure of interestingness		4ft-quantifier $\approx$	
#	Name	Definition			
		$\mathcal{I}_{\approx}(a, b, c, d)$	Probabilities [3]	Condition	Symbol
1	- Founded implication [9] - Confidence	$\frac{a}{a+b} = \frac{a}{r}$	$P(B\|A)$	$\geq p$	$\Rightarrow_p$
2	Laplace correction	$\frac{a+1}{a+b+2} = \frac{a+1}{r+2}$	$\frac{n(AB)+1}{n(A)+2}$	$\geq p$	$\Rightarrow_p^L$ *)
3	Sebag & Schoenauer	$\frac{a}{b}$	$\frac{P(AB)}{P(A\neg B)}$	$\geq q$	$\Rightarrow_q^S$ *)
4	Example and Counterexample Rate	$1 - \frac{b}{a} = \frac{a-b}{a}$	$1 - \frac{P(A\neg B)}{P(AB)}$	$\geq p$	$\Rightarrow_p^E$ *)
5	Lower critical implication [9]	$\sum_{i=a}^{r} \binom{r}{i}^i (1-p)^{r-i}$	–	$\leq \alpha$	$\Rightarrow_{p,\alpha}^!$
6	Upper critical implication [9]	$\sum_{i=0}^{a} \binom{r}{i}^i (1-p)^{r-i}$	–	$> \alpha$	$\Rightarrow_{p,\alpha}^?$

The quantifiers $\Rightarrow_p$ of founded implication, $\Rightarrow_{p,\alpha}^!$ of lower critical implication, and $\Rightarrow_{p,\alpha}^?$ of upper critical implication (see rows 1, 5 and 6 in Tab. 6) are defined and studied in [9]. It is proved [9] that they are implicational and in [11] that they are interesting implicational quantifiers for $0 < p < 1$ and $0 < \alpha < 0.5$. Definitions of their values are in Tab. 7.

Table 7. Definitions of Values of $\Rightarrow_p$, $\Rightarrow_{p,\alpha}^!$, and $\Rightarrow_{p,\alpha}^?$

#	4ft quantifier $\Rightarrow^*$		$\Rightarrow^* (a,b) = 1$ iff
	Name	$\Rightarrow^*$	
1	Founded implication	$\Rightarrow_p$	$\frac{a}{a+b} \geq p$
5	Lower critical implication	$\Rightarrow_{p,\alpha}^!$	$\sum_{i=a}^{r} \binom{r}{i}^i (1-p)^{r-i} \leq \alpha$
6	Upper critical implication	$\Rightarrow_{p,\alpha}^?$	$\sum_{i=0}^{a} \binom{r}{i}^i (1-p)^{r-i} > \alpha$

There are new quantifiers defined in rows 2–4 of Tab. 6, which are marked by *) in the column *Symbol*. Their definitions and properties are sketched in [19], we describe them here in more details.

The quantifier of *Laplace correction* is described in row 2 of Tab. 6. It is denoted by $\Rightarrow_p^L$ and its value is defined such that

$$\Rightarrow_p^L (a,b) = \begin{cases} 1 \text{ if } \frac{a+1}{a+b+2} \geq p \\ 0 \text{ otherwise.} \end{cases}$$

It is easy to prove that the $\Rightarrow_p^L$ quantifier is an interesting implicational quantifier for $p \in (0.5; 1)$. We have to prove:

(i) $\Rightarrow_p^L$ is implicational i.e. that

$$\Rightarrow_p^L (a,b) = 1 \ \wedge \ a' \geq a \wedge b' \leq b \ \text{ implies } \ \Rightarrow_p^L (a',b') = 1$$

(ii) $\Rightarrow_p^L$ is both *a-dependent* and *b-dependent*
(iii) $\Rightarrow_p^L (0,0) = 0$.

To prove (i) it is enough to show that

$$a' \geq a \ \wedge b' \leq b \ \text{ implies } \ \frac{a'+1}{a'+b'+2} \geq \frac{a+1}{a+b+2} \ .$$

It however follows from several simple inequalities (remember that it is also $a' \geq a \geq 0$ and $b \geq b' \geq 0$):

$$\frac{a'+1}{a'+b'+2} \geq \frac{a+1}{a+b'+2} \geq \frac{a+1}{a+b+2} \ .$$

To prove (ii) we have to find for each $p \in (0.5; 1)$:

(ii.a) a, a', b such that $\Rightarrow_p^L (a',b) \neq \Rightarrow_p^L (a,b)$. We can set $a' = b = 0$ and $a \geq \frac{2p-1}{1-p}$. Then it is $\Rightarrow_p^L (a',b) = 0$ because of $\frac{a'+1}{a'+b'+2} = 0.5 < p$ and $\Rightarrow_p^L (a,b) = 1$ due to $\frac{a+1}{a+b+2} = \frac{a+1}{a+2} \geq \frac{\frac{2p-1}{1-p}+1}{\frac{2p-1}{1-p}+2} = \frac{2p-1+1-p}{1-p} \cdot \frac{1-p}{2p-1+2-2p} = p$.

(ii.b) a, b', b such that $\Rightarrow_p^L (a,b') \neq \Rightarrow_p^L (a,b)$. We can set $a \geq \frac{2p-1}{1-p}$, $b = 0$ and $b' > \frac{a(1-p)+1-2p}{p}$. Then it is $\Rightarrow_p^L (a,b) = 1$, see the previous point (ii.a) and $\Rightarrow_p^L (a,b') = 0$ due to $\frac{a+1}{a+b'+2} < \frac{a+1}{a+\frac{a(1-p)+1-2p}{p}+2} = (a+1)\frac{p}{ap+a(1-p)+1-2p+2p}p = p$.

Proof of (iii) is simple, as it is $\Rightarrow_p^L (0,0) = 0$ due to $\frac{a'+1}{a'+b'+2} = 0.5 < p$; see also point (ii.a). Thus we have shown that the $\Rightarrow_p^L$ quantifier is an interesting implicational quantifier for $p \in (0.5; 1)$.

The quantifier of *Sebag & Schoenauer* is described in row 3 of Tab. 6. It is denoted by $\Rightarrow_q^S$ and its value for $q > 0$ is defined as

$$\Rightarrow_q^S (a,b) = \begin{cases} 1 \text{ if } b > 0 \ \wedge \ \frac{a}{b} \geq q \\ 1 \text{ if } b = 0 \ \wedge \ a > 0 \\ 0 \text{ if } b = 0 \ \wedge \ a = 0 \ . \end{cases}$$

We prove that the $\Rightarrow^S_q$ quantifier is an interesting implicational quantifier for $q > 0$. It is implicational, because of $\frac{a'}{b'} \geq \frac{a}{b'} \geq \frac{a}{b}$ for $a' \geq a$ and $0 < b \leq b'$ and because of $\Rightarrow^S_q (a, 0) = 1$ for $a > 0$.

We choose $a < q$, $a' \geq q$ and $b = 1$, it is $\Rightarrow^S_q (a, 1) = 0$ and $\Rightarrow^S_q (a', 1) = 1$. Note that it also works if $q < 1$, then we have, e.g., $a = 0$, $a' = 1$ and $b = 1$, it is $\Rightarrow^S_q (0, 1) = 0$ and $\Rightarrow^S_q (1, 1) = 1$. It means that $\Rightarrow^S_q$ is a-dependent.

If $q > 1$ then we choose $a > q$, $b = a$ and $b' = 1$, then it is $\Rightarrow^S_q (a, b) = 0$ because of $\frac{a}{b} = 1 < q$ and $\Rightarrow^S_q (a, 1) = 1$ because of $\frac{a}{1} = a > q$. If $q \leq 1$ then we choose $a = 1$, $b > \frac{1}{q}$ and $b' = 1$, then it is $\Rightarrow^S_q (a, b) = 0$ because of $\frac{a}{b} < \frac{a}{\frac{1}{q}} = \frac{1}{\frac{1}{q}} = q$ and $\Rightarrow^S_q (a, b') = 1$ because of $\frac{a}{b} = \frac{1}{1} = 1 \geq q$. It means that $\Rightarrow^S_q$ is b-dependent.

It is also $\Rightarrow^S_q (0, 0) = 0$; see the definition above. It means that the $\Rightarrow^S_q$ quantifier is an interesting implicational quantifier for $q > 0$.

The quantifier of *Example and counterexample rate* is described in row 4 of Tab. 6. It is denoted by $\Rightarrow^E_p$ and its value for $0 < p < 1$ is defined as

$$\Rightarrow^S_q (a, b) = \begin{cases} 1 \text{ if } a > 0 \ \wedge \ 1 - \frac{b}{a} \geq p \\ 0 \text{ otherwise.} \end{cases}$$

We prove that the $\Rightarrow^E_p$ quantifier is an interesting implicational quantifier for $0 < p < 1$. It is implicational, because of $1 - \frac{b'}{a'} \geq 1 - \frac{b'}{a} \geq \frac{b}{a}$ for $a' \geq a > 0$ and $b \leq b'$ and because of $\Rightarrow^E_p (0, b) = 0$.

We choose $a > \frac{1}{1-p}$, $a' = 1$ and $b = 1$, then it is $\Rightarrow^E_q (a', b) = 0$ and $\Rightarrow^S_q (a, b) = 1$ because of $1 - \frac{b}{a} > 1 - \frac{1}{\frac{1}{1-p}} = 1 - (1 - p) = p$. It means that $\Rightarrow^E_p$ is a-dependent. If we set $a = 1$, $b = 1$ and $b' = 0$, then it is $\Rightarrow^E_q (a, b) = 0$ and $\Rightarrow^S_q (a, b') = 1$, thus $\Rightarrow^E_p$ is b-dependent.

We can conclude that the quantifiers $\Rightarrow^L_p$, $\Rightarrow^S_q$, and $\Rightarrow^E_p$ defined in rows 2–4 in Tab. 6, on the basis of known measures of interestingness of association rules, are interesting implicational quantifiers as defined in [14]. It means that the correctness criterion of the deduction rule $\frac{\varphi \Rightarrow^* \psi}{\varphi' \Rightarrow^* \psi'}$ presented at the beginning of this section is valid for these quantifiers.

6.2 Σ-Double Implicational Association Rules

There is an additional theorem presented in [14] that gives a relatively simple correctness criterion of the deduction rule $\frac{\varphi \Leftrightarrow^* \psi}{\varphi' \Leftrightarrow^* \psi'}$ where $\Leftrightarrow^*$ is an *interesting Σ-double implicational quantifier*. Remember that the 4ft-quantifier $\approx^*$ is Σ-double implicational if

$$\approx (a, b, c, d) = 1 \ a' \geq a \ \wedge \ b' + c' \leq b + c \text{ implies } \approx (a', b', c', d') = 1 \ ;$$

see Section 4.2. It however means that each Σ-double implicational quantifier is *d-independent* (i.e., its value does not depend on d). Thus we write only $\Leftrightarrow^* (a, b, c)$ instead of $\Leftrightarrow^* (a, b, c, d)$ for the Σ-double implicational quantifier $\Rightarrow^*$.

Interesting Σ-double implicational quantifiers are defined in [14]. We say that the Σ-*double implicational quantifier* $\Leftrightarrow^*$ *is interesting* if both (i) and (ii) are satisfied:

(i) $\Rightarrow^*$ is both *a-dependent* and *(b+c)-dependent*
(ii) $\Leftrightarrow^* (0,0,0) = 0$

The 4ft-quantifier $\approx$ is *(b+c)-dependent* if there are a, b, c, d, b', c' such that $b + c \neq b' + c'$ and $\approx (a,b,c,d) \neq \approx (a,b',c',d)$.

Again, we present the correctness criterion of the deduction rule $\frac{\varphi \Leftrightarrow^* \psi}{\varphi' \Leftrightarrow^* \psi'}$ for the *interesting Σ-double implicational quantifiers* only in the *predicate calculus of association rules* defined in Section 2.2. This theorem is a simplified version of an analogous theorem for the *Qualitative Calculi of Association Rules* defined in section 2.1 that is presented in [14] and proved in [20].

The correctness criterion for Σ-double implicational quantifiers is [14]: Let PCAR $\mathcal{P}$ be the predicate calculus of association rules. In addition let φ, ψ, φ', ψ' be the Boolean attributes of $\mathcal{P}$. If $\Leftrightarrow^*$ is the interesting Σ-double implicational quantifier, then the deduction rule

$$\frac{\varphi \Leftrightarrow^* \psi}{\varphi' \Leftrightarrow^* \psi'}$$

is correct if and only if at least one of the conditions 1) or 2) are satisfied (we use the symbol $\longrightarrow$ for the propositional connective of implication):

1. Both $\pi(\varphi \wedge \psi) \longrightarrow \pi(\varphi' \wedge \psi')$
   and
   $\pi((\varphi' \wedge \neg\psi') \vee (\neg\varphi' \wedge \neg\psi')) \longrightarrow \pi((\varphi \wedge \neg\psi) \vee (\neg\varphi \wedge \neg\psi))$
   are tautologies
2. $\pi(\varphi) \longrightarrow \pi(\neg\psi)$ or $\pi(\psi) \longrightarrow \pi(\neg\varphi)$ are tautologies.

There are three important Σ-double implicational quantifiers defined by the application of suitable thresholds to the measures of interestingness in Tab. 8.

Table 8. Σ-double Implicational 4ft-quantifiers

#	Name	Measure of interestingness		4ft-quantifier $\approx$	
		Definition			
		$\mathcal{I}_{\approx}(a,b,c,d)$	Probabilities [3]	Condition	Symbol
1	- Jaccard - Founded double implication [23]	$\frac{a}{a+b+c} = \frac{a}{n-d}$	$\frac{P(A\mid B)}{P(A)+P(B)-P(AB)}$	$\geq p$	$\Leftrightarrow_p$
2	Lower critical double implication [23]	$\sum_{i=a}^{n-d} \binom{n-d}{i} p^i (1-p)^{n-d-i}$	$-$	$\leq \alpha$	$\Leftrightarrow^!_{p,\alpha}$
3	Upper critical double implication [23]	$\sum_{i=0}^{a} \binom{n-d}{i} p^i (1-p)^{n-d-i}$	$-$	$> \alpha$	$\Leftrightarrow^?_{p,\alpha}$

These quantifiers are defined in [23] and studied in detail in [20]. It is proved in [20] that the quantifiers $\Rightarrow_p$ of founded double implication $\Rightarrow^!_{p,\alpha}$ of lower critical double implication and $\Rightarrow^?_{p,\alpha}$ of upper critical double implication are interesting Σ-double implicational quantifiers for $0 < p < 1$ and $0 < \alpha < 0.5$. Definitions of their values are in Tab. 9.

Table 9. Definitions of Values of $\Leftrightarrow_p$, $\Leftrightarrow^!_{p,\alpha}$, and $\Leftrightarrow^?_{p,\alpha}$

#	4ft quantifier $\Leftrightarrow^*$		$\Leftrightarrow^* (a,b) = 1$ iff
	Name	$\Leftrightarrow^*$	
1	Founded double implication	$\Leftrightarrow_p$	$\frac{a}{a+b} \geq p$
2	Lower critical double implication	$\Leftrightarrow^!_{p,\alpha}$	$\sum_{i=a}^{n-d} \binom{n-d}{i} p^i (1-p)^{n-d-i} \leq \alpha$
3	Upper critical double implication	$\Leftrightarrow^?_{p,\alpha}$	$\sum_{i=0}^{a} \binom{n-d}{i} p^i (1-p)^{n-d-i} > \alpha$

Note that the quantifier of founded double implication is derived from the measure of the interestingness of association rules that is known as *Jaccard measure* [3]; see row 1 in Tab. 8.

6.3 Σ-Equivalence Association Rules

There is also a theorem presented in [14] that provides a relatively simple correctness criterion of the deduction rule $\frac{\varphi \equiv^* \psi}{\varphi' \equiv^* \psi'}$ where $\equiv^*$ is an *interesting Σ-equivalence quantifier*. Remember that the 4ft-quantifier $\approx^*$ is Σ-equivalence if

$$\approx (a,b,c,d) = 1 \; a' + d' \geq a + d \; \wedge \; b' + c' \leq b + c \text{ implies } \approx (a',b',c',d) = 1 \; ;$$

see Section 4.2.

Interesting Σ-equivalence quantifiers are defined in [14]. We say that the Σ-equivalence quantifier $\equiv^*$ *is interesting* if both (i) and (ii) are satisfied:

(i) $\Rightarrow^*$ is *(a+d)-dependent*
(ii) $\Leftrightarrow^* (0,b,c,0) = 0$ for $b + c > 0$.

The 4ft-quantifier $\approx$ is *(a+d)-dependent* if there are a, b, c, d, a, d such that $a + d \neq a' + d'$ and $\approx (a,b,c,d) \neq \approx (a',b,c,d')$.

Again, we present the correctness criterion of the deduction rule $\frac{\varphi \equiv^* \psi}{\varphi' \equiv^* \psi'}$ for the *interesting Σ-equivalence quantifiers* only in the *predicate calculus of association rules* defined in section 2.2. This theorem is a simplified version of an analogous theorem for the *Qualitative Calculi of Association Rules* defined in section 2.1 that is presented in [14] and proved in [20].

The correctness criterion for Σ-equivalence quantifiers is [14]: Let PCAR $\mathcal{P}$ be the predicate calculus of association rules. In addition let φ, ψ, φ', ψ' be the Boolean attributes of $\mathcal{P}$. If $\equiv^*$ is the interesting Σ-equivalence quantifier then the deduction rule

$$\frac{\varphi \equiv^* \psi}{\varphi' \equiv^* \psi'}$$

is correct if and only if the formula

$$\pi((\varphi \wedge \psi) \vee (\neg\varphi \wedge \neg\psi)) \longrightarrow \pi((\varphi' \wedge \psi') \vee (\neg\varphi' \wedge \neg\psi'))$$

is a tautology (we use the symbol $\longrightarrow$ for the propositional connective of implication).

There are three important Σ-equivalence quantifiers defined by the application of suitable thresholds to measures of interestingness in Tab. 10. Again, the notation used in the column $\mathcal{I}_{\approx}(a, b, c, d)$ is in accordance with Tab. 3 and the notation in the column *Probabilities [3]* is based on Tab. 4; see Section 5.

Table 10. Σ-equivalence 4ft-quantifiers

#	Name	Measure of interestingness		4ft-quantifier $\approx$	
		Definition			
		$\mathcal{I}_{\approx}(a, b, c, d)$	Probabilities [3]	Condition	Symbol
1	- Founded equivalence [23] - Accuracy - Success rate [1]	$\frac{a+d}{a+b+c+d} = \frac{a+d}{n}$	$P(AB) + P(\neg A \neg B)$	$\geq p$	$\equiv_p$
2	Lower critical equivalence [23]	$\sum_{i=a+d}^{n} \binom{n}{i} p^i (1-p)^{n-i}$	$-$	$\leq \alpha$	$\equiv_{p,\alpha}^{!}$
3	Upper critical Equivalence [23]	$\sum_{i=0}^{a+d} \binom{n}{i} p^i (1-p)^{n-i}$	$-$	$> \alpha$	$\equiv_{p,\alpha}^{?}$

These quantifiers are defined in [23] and studied in detail in [20]. It is proved in [20] that the quantifiers $\equiv_p$ of founded equivalence, $\equiv_{p,\alpha}^{!}$ of lower critical equivalence, and $\equiv_{p,\alpha}^{?}$ of upper critical equivalence are interesting Σ-equivalence quantifiers for $0 < p < 1$ and $0 < \alpha < 0.5$. Definitions of their values are in Tab. 11.

Table 11. Definitions of the Values of $\equiv_p$, $\equiv_{p,\alpha}^{!}$, and $\equiv_{p,\alpha}^{?}$

#	4ft quantifier $\equiv^*$		$\equiv^* (a, b) = 1$ iff
	Name	$\equiv^*$	
1	Founded equivalence	$\equiv_p$	$\frac{a+d}{a+b+c+d} \geq p$
2	Lower critical equivalence	$\equiv_{p,\alpha}^{!}$	$\sum_{i=a+d}^{n} \binom{n}{i} p^i (1-p)^{n-i} \leq \alpha$
3	Upper critical equivalence	$\equiv_{p,\alpha}^{?}$	$\sum_{i=0}^{a+d} \binom{n}{i} p^i (1-p)^{n-i} > \alpha$

Note that the quantifier of founded double implication is derived from the measure of the interestingness of association rules that is known as *Accuracy* [3] or *Success rate* [1]; see row 1 in Tab. 10.

6.4   Association Rules with the F-Property

The results on association rules with the F-property are presented in detail in [16]. There is also a theorem presented in [14] that gives a (not excessively simple) correctness criterion of the deduction rule $\frac{\varphi \approx^* \psi}{\varphi' \approx^* \psi'}$ where $\approx^*$ is a 4ft-quantifier belonging to an important subclass of the class of 4ft-quantifiers with the F-property. Remember that the 4ft-quantifier $\approx^*$ has the F-property if it satisfies the conditions F_{bc} and F_{cb}; see Section 4.2:

F_{bc}: If $\approx (a, b, c, d) = 1$ and $b \geq c - 1 \geq 0$ then $\approx (a, b+1, c-1, d) = 1$.
F_{cb}: If $\approx (a, b, c, d) = 1$ and $c \geq b - 1 \geq 0$ then $\approx (a, b-1, c+1, d) = 1$;

see Section 4.2. The presented correctness criterion of deduction rules concerns strong symmetrical quantifiers with the F^+ property. We say that the 4ft-quantifier $\approx$ is *strong symmetrical* if it is satisfied:

$$\approx (a, b, c, d) \;\; = \;\; \approx (a, c, b, d) = \;\; \approx (d, b, c, a) \;.$$

The 4ft-quantifier $\approx$ *has the F^+ property*, if it has the F property and if it satisfies the following conditions F_1^+ and F_2^+:

F_1^+: It is $\approx (0, b, c, d) = 0$ for each 4ft-table $\langle 0, b, c, d \rangle$ and it is $\approx (a, b, c, 0) = 0$ for each 4ft-table $\langle a, b, c, 0 \rangle$.
F_2^+: There are 4ft-tables $\langle a, b, c, d \rangle$ and $\langle a, b', c', d \rangle$ such that $a + b + c + d = a + b' + c' + d$, $\approx (a, b, c, d) = 1$ and $\approx (a, b', c', d) = 0$.

The following correctness criterion is proved in [11] and it concerns the predicate calculus $\mathcal{P}$ of association rules, as defined in section 2.2:

Let $\approx_F$ be a strong symmetrical equivalence quantifier with the F^+ property. Then the deduction rule $\frac{\varphi \approx_F \psi}{\varphi' \approx_F \psi'}$ where φ, ψ, φ' ψ' are the Boolean attributes of $\mathcal{P}$ is correct, if and only if at least one of the conditions a) – g) is satisfied. (The symbol $\longrightarrow$ denotes the propositional connective of implication.)

a) The following formulas are tautologies:
   $\pi(\varphi \wedge \psi) \longrightarrow \pi(\varphi' \wedge \psi')$, $\pi(\varphi' \wedge \neg\psi') \longrightarrow \pi(\varphi \wedge \neg\psi)$,
   $\pi(\neg\varphi' \wedge \psi') \longrightarrow \pi(\neg\varphi \wedge \psi)$, $\pi(\neg\varphi \wedge \neg\psi) \longrightarrow \pi(\neg\varphi' \wedge \neg\psi')$.
b) At least one of the formulas $\pi(\varphi \longrightarrow \neg\psi)$ and $\pi(\neg\varphi \longrightarrow \psi)$ are tautologies.
c) The following formulas are tautologies
   $\pi(\varphi \wedge \psi) \longrightarrow \pi(\varphi' \wedge \psi')$, $\pi(\varphi' \wedge \neg\psi') \longrightarrow \pi(\neg\varphi \wedge \psi)$,
   $\pi(\neg\varphi' \wedge \psi') \longrightarrow \pi(\varphi \wedge \neg\psi)$, $\pi(\neg\varphi \wedge \neg\psi) \longrightarrow \pi(\neg\varphi' \wedge \neg\psi')$.
d) The following formulas are tautologies
   $\pi(\varphi \wedge \psi) \longrightarrow \pi(\neg\varphi' \wedge \neg\psi')$, $\pi(\varphi' \wedge \neg\psi') \longrightarrow \pi(\varphi \wedge \neg\psi)$,
   $\pi(\neg\varphi' \wedge \psi') \longrightarrow \pi(\neg\varphi \wedge \psi)$, $\pi(\neg\varphi \wedge \neg\psi) \longrightarrow \pi(\varphi' \wedge \psi')$.
e) The following formulas are tautologies
   $\pi(\varphi \wedge \psi) \longrightarrow \pi(\neg\varphi' \wedge \neg\psi')$,
   $\pi(\varphi' \wedge \neg\psi') \longrightarrow \pi(\neg\varphi \wedge \psi)$,
   $\pi(\neg\varphi' \wedge \psi') \longrightarrow \pi(\varphi \wedge \neg\psi)$, $\pi(\neg\varphi \wedge \neg\psi) \longrightarrow \pi(\varphi' \wedge \psi')$.
f) Both of the following two formulas are tautologies
   $\pi(\varphi \wedge \psi) \longrightarrow \pi(\varphi' \wedge \psi')$, $\pi(\neg\varphi \wedge \neg\psi) \longrightarrow \pi(\neg\varphi' \wedge \neg\psi')$,
   and at least one of the formulas
   $\pi(\varphi') \longrightarrow \pi(\psi')$, $\pi(\neg\varphi') \longrightarrow \pi(\neg\psi')$ are tautologies.

g) Both of the following two formulas are tautologies

$$\pi(\varphi \wedge \psi) \longrightarrow \pi(\neg\varphi' \wedge \neg\psi'), \ \pi(\neg\varphi \wedge \neg\psi) \longrightarrow \pi(\varphi' \wedge \psi'),$$

and at least one of the formulas

$$\pi(\varphi') \longrightarrow \pi(\psi'), \ \pi(\neg\varphi') \longrightarrow \pi(\neg\psi') \text{ are tautologies.}$$

There are eight important quantifiers with the F-property defined by the application of suitable thresholds to measures of interestingness in Tab. 12 and Tab. 13. Again, the notation used in the column $\mathcal{I}_\approx(a,b,c,d)$ is in accordance with Tab. 3 and the notation in the column *Probabilities [3]* is based on Tab. 4; see Section 5.

Table 12. 4ft-quantifiers with F-property – Definitions by $\mathcal{I}_\approx(a,b,c,d)$

		Measure of interestingness	4ft-quantifier $\approx$	
#	Name	$\mathcal{I}_\approx(a,b,c,d)$	Condition	Symbol
1	- Simple deviation [9] - Odds ratio	$\frac{ad}{bc}$	$> e^\delta$	$\sim_\delta$
2	Fisher [9]	$\sum_{i=a}^{\min(r,k)} \frac{\binom{k}{i}\binom{n-k}{r-i}}{\binom{r}{n}}$	$\leq \alpha$	$\approx_\alpha$
3	χ^2 [9]	$\frac{(ad-bc)^2}{rkls}n$	$\geq \chi_\alpha^2$	$\sim_\alpha^2$
4	AA [6]	$\frac{ad-bc}{(a+b)(a+c)}$	$\geq q$	$\Rightarrow_q^+$
5	Yule's Q	$\frac{ad-bc}{ad+bc}$	$\geq p$	$\approx_p^Q$ *)
6	Yule's Y	$\frac{\sqrt{ad}-\sqrt{bc}}{\sqrt{ad}+\sqrt{bc}}$	$\geq p$	$\approx_p^Y$ *)
7	Lift / interest	$\frac{an}{(a+b)(a+c)} = \frac{a(a+b+c+d)}{(a+b)(a+c)}$	$\geq q$	$\Rightarrow_q^{L+}$ *)
8	Information gain	$\log(\frac{an}{(a+b)(a+c)})$	$\geq q$	$\Rightarrow_q^{I+}$ *)

Table 13. 4ft-quantifiers with the F-property – Definitions by Probabilities [3]

		Measure of interestingness	4ft-quantifier $\approx$	
#	Name	Probabilities [3]	Condition	Symbol
1	- Simple deviation [9] - Odds ratio	$\frac{P(AB)P(\neg A\neg B)}{P(\neg AB)P(A\neg B)}$	$> e^\delta$	$\sim_\delta$
2	Fisher [9]	–	$\leq \alpha$	$\approx_\alpha$
3	χ^2 [9]	–	$\geq \chi_\alpha^2$	$\sim_\alpha^2$
4	AA [6]	–	$\geq q$	$\Rightarrow_q^+$
5	Yule's Q	$\frac{P(AB)P(\neg A\neg B)-P(A\neg B)P(\neg AB)}{P(AB)P(\neg A\neg B)+P(A\neg B)P(\neg AB)}$	$\geq p$	$\approx_p^Q$ *)
6	Yule's Y	$\frac{\sqrt{P(AB)P(\neg A\neg B)}-\sqrt{P(A\neg B)P(\neg AB)}}{\sqrt{P(AB)P(\neg A\neg B)}+\sqrt{P(A\neg B)P(\neg AB)}}$	$\geq p$	$\approx_p^Y$ *)
7	Lift / interest	$\frac{P(B\vert A)}{P(B)}$ or $\frac{P(AB)}{P(A)P(B)}$	$\geq q$	$\Rightarrow_q^{L+}$ *)
8	Information gain	$\log(\frac{P(AB)}{P(A)P(B)})$	$\geq q$	$\Rightarrow_q^{I+}$ *)

The quantifiers $\sim_\delta$ of simple deviation, $\approx_\alpha$, i.e., Fisher's quantifier, and χ_α^2 i.e. χ^2 quantifier (see rows $1-3$ in Tab. 12 and Tab. 13) are defined and studied in [9]. It is proved in [11] that they are strong symmetrical equivalence quantifiers with the F^+ property for $\delta \geq 0$ and $0 < \alpha < 0.5$. The only problem is to prove that that the Fisher's quantifier has the F-property; the core of the proof is published in [10]. Note that the the Fisher's quantifier $\approx_\alpha$ is defined in the book [9] by the condition $\sum_{i=a}^{\min(r,k)} \frac{\binom{k}{i}\binom{n-k}{r-i}}{\binom{r}{n}} \leq \alpha \wedge ad > bc$. Here we omit the partial condition $ad > bc$ because we are interested in the measure of interestingness. The definitions of the values of these quantifiers are in Tab. 14. Note that χ_α^2 is the $(1 - \alpha)$ quantile of the χ^2 distribution function.

Table 14. Definitions of the Values of $\sim_\delta$, $\approx_\alpha$, and $\sim_\alpha^2$

#	4ft quantifier $\approx_F$		$\approx_F (a,b) = 1$ iff
	Name	$\approx_F$	
1	Simple deviation	$\sim_\delta$	$\frac{ad}{bc} > e^\delta$
2	Fisher	$\approx_\alpha$	$\sum_{i=a}^{\min(r,k)} \frac{\binom{k}{i}\binom{n-k}{r-i}}{\binom{r}{n}} \leq \alpha$
3	χ^2	$\sim_\alpha^2$	$\frac{(ad-bc)^2}{rkls} n \geq \chi_\alpha^2$

AA quantifier

The *AA quantifier* is described in row 4 of Tab. 12. It is denoted by $\Rightarrow_q^+$ and its value for $q > 0$ is defined as

$$\Rightarrow_q^+ (a,b,c,d) = \begin{cases} 1 \text{ if } \frac{ad-bc}{(a+b)(a+c)} \geq q \\ 0 \text{ otherwise.} \end{cases}$$

The condition $\frac{ad-bc}{(a+b)(a+c)} \geq q$ is equivalent to the condition $\frac{a}{r} \geq (1+q)\frac{s}{n}$; see also Tab. 3. The condition $\frac{a}{r} \geq (1+q)\frac{s}{n}$ means $\frac{a}{r} * \frac{n}{s} - 1 \geq q$ and it is $\frac{a}{r} * \frac{n}{s} - 1 = \frac{a}{a+b} * \frac{a+b+c+d}{a+c} - 1 = \frac{ad-bc}{(a+b)(a+c)}$. The condition $\frac{a}{r} \geq (1+q)\frac{s}{n}$ means that if the rule $\varphi \Rightarrow_q^+ \psi$ is true in the data matrix $\mathcal{M}$, then the relative frequency of rows satisfying ψ among rows satisfying φ (i.e., $\frac{a}{r}$) is at least $100q$ per cent higher than the relative frequency of rows satisfying ψ among all the rows of the analyzed data matrix (i.e., $\frac{s}{n}$).

It is easy to prove that the AA quantifier $\Rightarrow_q^+$ has the property F. We prove only the condition F_{bc} from the two conditions F_{bc} and F_{cb} defining the F property above. The proof of the condition F_{cb} is analogous. We have to prove:

If $\Rightarrow_q^+ (a,b,c,d) = 1 \wedge b \geq c - 1 \geq 0$ then $\Rightarrow_q^+ (a,b+1,c-1,d) = 1$.

We suppose $\frac{ad-bc}{(a+b)(a+c)} \geq q$ and we also have to prove that $\frac{ad-(b+1)(c-1)}{(a+b+1)(a+c-1)} \geq q$. Thus it is enough to prove $\frac{ad-(b+1)(c-1)}{(a+b+1)(a+c-1)} \geq \frac{ad-bc}{(a+b)(a+c)}$. To finish the proof we show that

$$[ad - (b+1)(c-1)](a+b)(a+c) \geq (ad - bc)(a+b+1)(a+c-1).$$

We denote $X = ad - bc$, $Y = (a+b)(a+c)$ and $Z = b - (c-1)$. Simple algebraic operations lead to the equivalent condition

$$(X + Z)Y \geq X(Y - Z) \text{ and thus to } ZY \geq -XZ \ .$$

We suppose $b \geq c - 1$, thus $Z \geq 0$ and it remains to prove $Y \geq -X$ that is equivalent to $(a + b)(a + c) \geq bc - ad$.

We also suppose $\Rightarrow^+_q (a, b, c, d) = 1$ and $q \geq 0$ that means $\frac{ad-bc}{(a+b)(a+c)} \geq q$ and thus $ad - bc \geq 0$ that implies $bc - ad \leq 0$. Thus $(a + b)(a + c) \geq 0 \geq bc - ad$ and this finishes the proof of the condition F_{bc}. We can conclude that the AA quantifier $\Rightarrow^+_q$ has the F-property.

The AA quantifier $\Rightarrow^+_q$ however is not strong symmetrical. We show it for $\Rightarrow^+_1$, and for additional parameters the proof is similar. Let be $\langle a, b, c, d \rangle = \langle 1, 1, 1, 10 \rangle$, then the condition $\approx (a, b, c, d) = \approx (d, b, c, a)$ is not satisfied: We have $\Rightarrow^+_1 (1, 1, 1, 10) = 1$ because of $\frac{1*10-1*1}{(1+1)(1+1)} = \frac{9}{4} > 1$ while we have $\Rightarrow^+_1 (10, 1, 1, 1) = 0$ because of $\frac{10*1-1*1}{(10+1)(10+1)} = \frac{9}{121} < 1$. It means that the above-presented correctness theorem of the deduction rules $\frac{\varphi \approx_F \psi}{\varphi' \approx_F \psi'}$ is not applicable for the AA quantifier $\Rightarrow^+_q$.

It is however clear that the AA quantifier $\Rightarrow^+_q$ is symmetrical [9]. The 4ft-quantifier $\approx$ is *symmetrical* if it is satisfied:

$$\approx (a, b, c, d) \ = \ \approx (a, c, b, d) \ .$$

Note that if the 4ft-quantifier $\approx$ is symmetrical, then the rule $\varphi \approx \psi$ is true, if and only if the rule $\psi \approx \varphi$ is also true.

Yule's Q Quantifier

The *Yule's Q quantifier* is described in row 5 of Tab. 12 and of Tab. 13. It is denoted by $\approx^Q_p$ and its value for $0 < p < 1$ is defined as

$$\approx^Q_p (a, b, c, d) = \begin{cases} 1 \text{ if } \frac{ad-bc}{ad+bc} \geq p \\ 0 \text{ otherwise.} \end{cases}$$

It is easy to prove that the Yule's Q quantifier $\approx^Q_p$ has the property F. Again, we prove only the condition F_{bc} from the two conditions F_{bc} and F_{cb} defining the F property above. We have to prove:

$$\text{If } \approx^Q_p (a, b, c, d) = 1 \ \wedge \ b \geq c - 1 \geq 0 \ \text{ then } \ \approx^Q_p (a, b+1, c-1, d) = 1 \ .$$

We assume that $\frac{ad-bc}{ad+bc} \geq q$ and we have to prove that also $\frac{ad-(b+1)(c-1)}{ad+(b+1)(c-1)} \geq p$. Thus it is enough to prove $\frac{ad-(b+1)(c-1)}{ad+(b+1)(c-1)} \geq \frac{ad-bc}{ad+bc}$. To finish the proof we show that

$$[ad - (b + 1)(c - 1)](ad + bc) \geq (ad - bc)[ad + (b + 1)(c - 1)] \ .$$

We denote $X = ad - bc$, $Y = ad + bc$ and $Z = b - (c - 1)$. Simple algebraic operations lead to the equivalent condition

$$(X + Z)Y \geq X(Y - Z) \text{ and thus to } ZY \geq -XZ.$$

We assume that $b \geq c - 1$, thus $Z \geq 0$ and it remains to prove $Y \geq -X$ that is equivalent to $ad + bc) \geq bc - ad$ and also to $2ad \geq 0$.

We also assume that $\approx_p^Q (a, b, c, d) = 1$ and $p > 0$ that means $\frac{ad - bc}{ad - bc} \geq p > 0$ and thus it must be $ad \geq 0$ that finishes the proof. We can conclude that the Yule's Q quantifier $\approx_p^Q$ has the F-property.

It is evident that

$$\approx_p^Q (a, b, c, d) = \approx_p^Q (a, c, b, d) = \approx_p^Q (d, b, c, a)$$

and thus the Yule's Q quantifier $\approx_p^Q$ is strong symmetrical. We prove that the Yule's Q quantifier $\approx_p^Q$ has the F^+ property for $0 < p < 1$. It remains to prove the conditions F_1^+ and F_2^+ defined above:

F_1^+: It is $\approx_p^Q (0, b, c, d) = 0$ for each 4ft-table $\langle 0, b, c, d \rangle$ because of $\frac{0*d - bc}{ad + bc} \leq 0 < p$, similarly it is $\approx_p^Q (a, b, c, 0) = 0$ for each 4ft-table $\langle a, b, c, 0 \rangle$.

F_2^+: There are 4ft-tables $\langle a, b, c, d \rangle = \langle 1, 1, 1, 1 \rangle$ and $\langle a, b', c', d \rangle = \langle 1, 2, 0, 1 \rangle$ such that $a+b+c+d = a+b'+c'+d$, $\approx_p^Q (1, 2, 0, 1) = 1$ because of $\frac{1*1 - 2*0}{1*1 + 2*0} = 1 > p$ and $\approx_p^Q (1, 1, 1, 1) = 0$ because of $\frac{1*1 - 1*1}{1*1 + 1*1} = 0 < p$.

We can conclude that Yule's Q quantifier $\approx_p^Q$ is strong symmetrical and it has the F^+ property for $0 < p < 1$.

Yule's Y quantifier

The *Yule's Y quantifier* is described in row 6 of Tab. 12 and of Tab. 13. It is denoted by $\approx_p^Y$ and its value for $0 < p < 1$ is defined as

$$\approx_p^Y (a, b, c, d) = \begin{cases} 1 \text{ if } \frac{\sqrt{ad} - \sqrt{bc}}{\sqrt{ad} + \sqrt{bc}} \geq p \\ 0 \text{ otherwise.} \end{cases}$$

The Yule's Y quantifier $\approx_p^Y$ is strong symmetrical and it has the F^+ property for $0 < p < 1$. The proof is similar to the proof of the same assertion for the Yule's Q quantifier $\approx_p^Q$ that is given above.

Lift/Interest Quantifier

The *lift / interest quantifier* is described in row 7 of Tab. 12 and of Tab. 13. It is denoted by $\Rightarrow_q^{L+}$ and its value for $q > 0$ is defined as

$$\Rightarrow_q^{L+} (a, b, c, d) = \begin{cases} 1 \text{ if } \frac{a(a+b+c+d)}{(a+b)(a+c)} \geq q \\ 0 \text{ otherwise.} \end{cases}$$

The lift / interest quantifier has similar properties to those of the AA quantifier. It means that it has the F-property but it is not strong symmetrical and thus the above-given correctness criterion of the deduction rules $\frac{\varphi \approx_F \psi}{\varphi' \approx_F \psi'}$ is not applicable for the lift / interest quantifier $\Rightarrow_q^{L+}$.

We show that $\Rightarrow_{10}^{L+}$ is not strong symmetrical, and for additional parameters, the proof is similar. Let $\langle a, b, c, d \rangle = \langle 1, 1, 1, 10 \rangle$; then the condition $\approx$ $(a, b, c, d) = \approx (d, b, c, a)$ is not satisfied: It is $\Rightarrow_1^{L+} (1, 1, 1, 10) = 0$ because of $\frac{1*(1+1+1+10)}{(1+1)(1+1)} = \frac{13}{4} < 10$ while we have $\Rightarrow_1^{L+} (10, 1, 1, 1) = 1$ because of $\frac{10*(1+1+1+10)}{(1+1)(1+1)} = \frac{10*13}{4} > 10$.

We also prove that the lift / interest quantifier $\Rightarrow_q^{L+}$ satisfies the condition F_{bc} from the two conditions F_{bc} and F_{cb} defining the F property. The proof for the condition F_{cb} is analogous. We have to prove:

If $\quad \Rightarrow_q^{L+} (a, b, c, d) = 1 \;\wedge\; b \geq c - 1 \geq 0 \quad$ then $\quad \Rightarrow_q^{L+} (a, b + 1, c - 1, d) = 1$.

We suppose $\frac{a(a+b+c+d)}{(a+b)(a+c)} \geq q$ and we have to prove $\frac{a(a+(b+1)+(c-1)+d)}{(a+b+1)(a+c-1)} \geq q$. Thus it is enough to prove $\frac{a(a+(b+1)+(c-1)+d)}{(a+b+1)(a+c-1)} \geq \frac{a(a+b+c+d)}{(a+b)(a+c)}$. To finish the proof we show that

$$(a + b)(a + c) \geq (a + b + 1)(a + c - 1) .$$

This is equivalent to $0 \geq -b + c - 1$ and it follows from the assumption $b \geq c - 1$. It finishes the proof.

Information Gain Quantifier

The *information gain quantifier* is described in row 8 of Tab. 12 and of Tab. 13. It is denoted by $\Rightarrow_q^{I+}$ and its value for $q > 0$ is defined as

$$\Rightarrow_q^{I+} (a, b, c, d) = \begin{cases} 1 \text{ if } \log(\frac{an}{(a+b)(a+c)}) \geq q \\ 0 \text{ otherwise.} \end{cases}$$

The information gain quantifier $\Rightarrow_q^{I+}$ is similar to the lift / interest quantifier $\Rightarrow_q^{L+}$. It means that it has the F-property but it is not strong symmetrical, and thus the above-given correctness criterion of the deduction rules $\frac{\varphi \approx_F \psi}{\varphi' \approx_F \psi'}$ is not applicable for the information gain quantifier $\Rightarrow_q^{L+}$. The proof is analogous to the proof for the lift / interest quantifier $\Rightarrow_q^{L+}$.

7 Additional Classes of Association Rules

We can distinguish two types of classes of association rules - $\mathcal{M}-$ *independent classes* and $\mathcal{M}-$ *dependent classes*. These notions were introduced in [19] in relation to the effort to solve some problems related to deduction rules, concerning the "classical association rule" $X \rightarrow Y$ defined, e.g., in [8]. Such rules are verified using the interestingness measures confidence and support. There are both $\mathcal{M}$–independent and $\mathcal{M}$– dependent classes among the classes of association rules mentioned in Section 4.

Both $\mathcal{M}$–independent and $\mathcal{M}$– dependent classes of association rules are introduced in Section 7.1. Reasons leading to the definition of new $\mathcal{M}$– dependent classes are explained in Section 7.2, and examples of new $\mathcal{M}$– dependent classes and of their properties are in Section 7.3. There are also new $\mathcal{M}$–independent classes of association rules inspired by several interestingness measures; see Section 7.4.

7.1 $\mathcal{M}$ – Independent and $\mathcal{M}$ – Dependent Classes of Rules

Remember that the classes of 4ft-quantifiers (i.e., classes of association rules) are defined using truth preservation conditions TPC; see Section 4.1. Each class $\mathcal{C}$ is defined by the truth preservation condition $TPC_{\mathcal{C}}(a, b, c, d, a', b', c', d')$ that concerns two contingency tables $\langle a, b, c, d\rangle$ and $\langle a', b', c', d'\rangle$. The class $\mathcal{C}$ is defined according to this scheme:

The 4ft-quantifier $\approx$ belongs to the class $\mathcal{C}$ if and only if it satisfies:

$$\approx (a, b, c, d) = 1 \ \wedge \ TPC_{\mathcal{C}}(a, b, c, d, a', b', c', d') \text{ implies } \approx (a', b', c', d') = 1 \ .$$

The condition $TPC_{\mathcal{C}}(a, b, c, d, a', b', c', d')$ is usually simple, typically it is based on two or three inequalities. Examples of TPC's are in Tab. 2 in Section 4.2. Note that neither of these TPC's include the condition $a + b + c + d = a' + b' + c' + d'$ or allow $a + b + c + d = a' + b' + c' + d'$ to be derived from them.

This is, however, not true for the $TPC_{\mathcal{F}} = TPC_{\mathcal{F}_{bc}} \vee TPC_{\mathcal{F}_{cb}}$ (see Section 4.2) where $TPC_{\mathcal{F}_{bc}}$ is defined as

$$\langle a', b', c', d'\rangle = \langle a, b+1, c-1, d\rangle \ \wedge \ b \geq c - 1 \geq 0$$

and $TPC_{\mathcal{F}_{cb}}$ is defined as

$$\langle a', b', c', d'\rangle = \langle a, b-1, c+1, d\rangle \ \wedge \ c \geq b - 1 \geq 0 \ .$$

Both $TPC_{\mathcal{F}_{bc}}$ and $TPC_{\mathcal{F}_{cb}}$ imply $a + b + c + d = a' + b' + c' + d'$.

The condition $a + b + c + d = a' + b' + c' + d'$ could be interpreted such that the 4ft-tables $\langle a, b, c, d\rangle$ and $\langle a', b', c', d'\rangle$ concern the one data matrix $\mathcal{M}$ (or at least that $\langle a, b, c, d\rangle$ and $\langle a', b', c', d'\rangle$ concern the two data matrices $\mathcal{M}$ and $\mathcal{M}'$ with the same number of rows). Thus if the condition $TPC_{\mathcal{C}}(a, b, c, d, a', b', c', d')$ includes the condition $a + b + c + d = a' + b' + c' + d'$ or if it is possible to deduce that, then we say that the class $\mathcal{C}$ is $\mathcal{M}$–dependent. Otherwise we say that the class $\mathcal{C}$ is $\mathcal{M}$–independent.

7.2   Why New $\mathcal{M}$ – Dependent Classes

The "classical" association rule is the implication $X \rightarrow Y$, where X and Y are conjunctions of simple Boolean attributes. As mentioned in Section 2.1, the association rule $X \to Y$ with confidence Cf and support Sp can be understood as a formula of $X \to_{Cf,Sp} Y$ of suitable calculus of association rules, where the 4ft-quantifier $\to_{Cf,Sp}$ corresponds to the condition $\frac{a}{a+b} \geq Cf \wedge \frac{a}{a+b+c+d} \geq Sp$ concerning the 4ft-table $4ft(X, Y, \mathcal{M})$ of X and Y in the data matrix $\mathcal{M}$ in question.

The 4ft-quantifier $\to_{p,Sp}$ with confidence p and support Sp is a composition of two 4ft-quantifiers: the quantifier $\Rightarrow_p$ of founded implication (i.e., of confidence; see row 1 in Tab. 6) and the 4ft-quantifier $\oplus_{Sp}$ of support (see row 1 in Tab. 5). The problem is that the 4ft-quantifier $\oplus_{Sp}$ is not equivalence (i.e., not associational in the sense of [9]). To prove it, we have to find for each $Sp \in (0; 1)$ two quadruples $\langle a, b, c, d \rangle$ and $\langle a', b', c', d' \rangle$ of natural numbers such that $a' \geq a \wedge b' \leq b \wedge c' \leq c \wedge d' \geq d$, $\frac{a}{a+b+c+d} \geq Sp$, and $\frac{a'}{a'+b'+c'+d'} < Sp$. It is easy to verify that this is satisfied for $\langle a, b, c, d \rangle = \langle 1, 0, 0, 0 \rangle$, and $\langle a', b', c', d' \rangle = \langle 1, 0, 0, d' \rangle$ where $d' > \frac{1-Sp}{Sp}$.

Each 4ft-quantifier belonging to one of the classes of implicational quantifiers, double implicational quantifiers, Σ-double implicational quantifiers, and Σ-equivalence quantifiers is also equivalence (i.e., associational) quantifier [9,16]. The 4ft-quantifier $\oplus_{Sp}$ is not equivalence, and thus it is not implicational. It means that the correctness criterion of deduction rules $\frac{\varphi \Rightarrow^* \psi}{\varphi' \Rightarrow^* \psi'}$ for the interesting implicational quantifiers presented in section 6.1 is not applicable to "classical" association rules $\varphi \to_{Cf,Sp} \psi$.

It is, however, evident that the quantifier $\oplus_{Sp}$ *is* $\mathcal{M}$*-dependent equivalence* where the class of $\mathcal{M}$-dependent equivalence quantifiers is defined by the truth preservation condition $TPC_{\equiv}^{\mathcal{M}}$:

$$a + b + c + d = a' + b' + c' + d' \ \wedge \ a' \geq a \wedge b' \leq b \wedge c' \leq c \wedge d' \geq d \ .$$

Note that we can write $TPC_{\equiv}^{\mathcal{M}} = TPC^{\mathcal{M}} \wedge TPC_{\equiv}$ where the truth preservation condition $TPC^{\mathcal{M}}$ is defined as $TPC^{\mathcal{M}} = a + b + c + d = a' + b' + c' + d'$. Thus it is natural to ask what would happen if we derive new $\mathcal{M}$-dependent classes by adding the truth preservation condition $TPC^{\mathcal{M}}$ to already known $\mathcal{M}$-independent classes.

7.3 New $\mathcal{M}$ – Dependent Classes

Three $\mathcal{M}$-dependent classes of association rules (i.e., classes of 4ft-quantifiers) derived by adding the truth preservation condition $TPC^{\mathcal{M}}$ to known $\mathcal{M}$-independent classes are briefly introduced in this section. Their overview is in Tab. 15, and their detailed definitions are in Tab. 16.

There are six quantifiers belonging to the class of implicational quantifiers listed in Tab. 7. The 4ft-quantifier $\to_{p,Sp}$ with confidence p and support Sp corresponding to the "classical" associational rule mentioned in Section 7.2 is a composition of the quantifiers $\Rightarrow_p$ of founded implication and the 4ft-quantifier $\oplus_{Sp}$ of support, and it is thus defined by the condition $\frac{a}{a+b} \geq p \wedge \frac{a}{a+b+c+d} \geq Sp$. It is easy to prove that the 4ft-quantifier $\to_{p,Sp}$ is $\mathcal{M}$-dependent implicational. We can derive thereby a new $\mathcal{M}$-dependent implicational quantifier from each

Table 15. Overview of New $\mathcal{M}$ – dependent Classes of 4ft-quantifiers

Class	Truth preservation condition	
	Symbol	Definition
$\mathcal{M}$-dependent implicational	$TPC^{\mathcal{M}}_{\Rightarrow}$	$TPC^{\mathcal{M}} \wedge TPC_{\Rightarrow}$
$\mathcal{M}$-dependent Σ-double implicational	$TPC^{\mathcal{M}}_{\Sigma,\Leftrightarrow}$	$TPC^{\mathcal{M}} \wedge TPC_{\Sigma,\Leftrightarrow}$
$\mathcal{M}$-dependent Σ-equivalence	$TPC^{\mathcal{M}}_{\Sigma,\equiv}$	$TPC^{\mathcal{M}} \wedge TPC_{\Sigma,\equiv}$

Table 16. Definitions of New $\mathcal{M}$ – dependent Classes of 4ft-quantifiers

Symbol	Truth preservation condition
	Detail of definition
$TPC^{\mathcal{M}}_{\Rightarrow}$	$a + b + c + d = a' + b' + c' + d' \ \wedge \ a' \geq a \wedge b' \leq b$
$TPC^{\mathcal{M}}_{\Sigma,\Leftrightarrow}$	$a + b + c + d = a' + b' + c' + d' \ \wedge \ a' + c' \geq a + c \wedge b' \leq b$
$TPC^{\mathcal{M}}_{\Sigma,\equiv}$	$a + b + c + d = a' + b' + c' + d' \ \wedge \ a' + d' \geq a + d \wedge b' + c' \leq b + c$

of the implicational quantifiers listed in Tab. 7. It means, e.g., that the $\mathcal{M}$-dependent implicational quantifier $\to^{L}_{p,Sp}$ of $\mathcal{M}$-*dependent Laplace correction* is defined by the condition $\frac{a+1}{a+b+2} \geq p \wedge \frac{a}{a+b+c+d} \geq Sp$ etc.

We can define, in an analogous way, additional $\mathcal{M}$-dependent Σ-double implicational quantifiers from the Σ-double implicational quantifiers listed in Tab. 9, and additional $\mathcal{M}$-dependent Σ-equivalence quantifiers from the Σ-equivalence quantifiers listed in Tab. 11. Note that we can define in the same way additional $\mathcal{M}$-dependent classes for the $\mathcal{M}$-independent classes introduced in Sect. 7.4.

However there are no corresponding results concerning the correctness of deduction rules of the form $\frac{\varphi \approx \psi}{\varphi' \approx \psi'}$ where φ, ψ for the $\mathcal{M}$ – dependent classes of association rules introduced in Tab. 15. The only result is the following theorem concerning $\mathcal{M}$-dependent implicational quantifiers.

The theorem concerns interesting $\mathcal{M}$-dependent implicational quantifiers. The $\mathcal{M}$-*dependent implicational quantifier* $\to^{*}$ *is interesting* if it is both *a-dependent* and *b-dependent* and if $\to^{*} (0,0,c,d) = 0$ for each couple $\langle c, d \rangle$ of natural numbers.

Let PCAR $\mathcal{P}$ be the calculus of association rules and let φ, ψ, φ', ψ' be the Boolean attributes of $\mathcal{P}$. If $\to^{*}$ is the interesting $\mathcal{M}$-dependent implicational quantifier then the deduction rule $\frac{\varphi \to^{*} \psi}{\varphi' \to^{*} \psi'}$ is correct, if at least one of the conditions 1) or 2) are satisfied (again, $\longrightarrow$ is the propositional connective of implication):

1. Both 1.a and 1.b are tautologies of propositional calculus
   1.a: $\pi(\varphi \wedge \psi) \longrightarrow \pi(\varphi' \wedge \psi')$
   1.b: $\pi(\varphi' \wedge \neg\psi') \longrightarrow \pi(\varphi \wedge \neg\psi)$
2. $\pi(\varphi) \longrightarrow \pi(\neg\psi)$ is a tautology.

The proof of this theorem is outlined in [19]. Note that this theorem does not says "*is correct if and only if*" but only "*is correct if*", however, it seems that the "*is correct if and only if*" variant of this theorem could be formulated and proved [19] as well as the corresponding theorems for the additional $\mathcal{M}$ – dependent classes of association rules introduced in Tab. 15.

7.4 New $\mathcal{M}$ – Independent Classes

There are numerous measures of interestingness defined and studied in [1, 2, 3, 4, 5], and we did not mention all of them. Some of such measures of interestingness can be used to define additional $\mathcal{M}$ – independent classes of association rules. We introduce two such measures of interestingness; see, e.g., [3]. Both of them are somehow analogous to a founded implication (i.e. confidence), see row 1 in Tab. 6. We use the notation in accordance with Tab. 3.

The first one is the measure of *recall* defined as $\frac{a}{a+c}$ that leads to the 4ft-quantifier $\Rightarrow_p^R$ of recall defined for $0 < p \leq 1$ by the condition $\frac{a}{a+c} \geq p$. We can define the class of *recall-like quantifiers* by the truth preservation condition TPC_{recc} defined as $a' \geq a \land c' \leq c$.

The second one is the measure of *specificity* defined as $\frac{d}{b+d}$ that leads to the 4ft-quantifier $\Rightarrow_p^S$ of specificity defined for $0 < p \leq 1$ by the condition $\frac{d}{b+d} \geq p$. We can define the class of *specificity-like quantifiers* by the truth preservation condition TPC_{spec} defined as $d' \geq d \land b' \leq b$.

It is relatively easy to formulate and proof the corresponding correctness theorems of deduction rules of the form $\frac{\varphi \approx \psi}{\varphi' \approx \psi'}$ for the *recall - like quantifiers* and *specificity - like quantifiers*. Of course, the corresponding $\mathcal{M}$ – dependent classes of rules can be defined on the basis of the truth preservation conditions TPC_{recc} and TPC_{spec}.

8 Conclusions and Further Work

We introduced the logical calculi of association rules together with 4ft-quantifiers and known classes of 4ft-quantifiers and of association rules. We argued for their usefulness, and we emphasized deduction rules between association rules; see Sect. 4.3. We defined new 4ft-quantifiers by the application of suitable thresholds to several known measures of interest. We proved that some of new 4ft-quantifiers belong to known classes of rules with important properties. We have also shown that new interesting classes of rules can be defined on the basis of the other new 4ft-quantifiers, and we mentioned some interesting results concerning the new classes. Due to the limited space of the chapter, some of the results are presented only very briefly.

Moreover, there are various open questions concerning the relation of the measures of interestingness of association rules and logical calculi of association rules. Some of them are listed below, and they are subject of current research.

- What are the relations of the additional tens of interestingness measures defined and studied, e.g., in [1,2,3,4,5] to known classes of association rules?
- Are there some additional new classes of association rules with interesting properties?

– Are there some unknown useful deduction rules concerning known or new classes of rules?
– What are the relations of the properties of the interestingness measures (see, e.g., Table V in [3]) to classes of association rules?
– What about definability of association rules corresponding to not yet studied interestingness measures in classical predicate calculus with equality; see [18]?

References

1. Hébert, C., Crémilleux, B.: A Unified View of Objective Interestingness Measures. In: Perner, P. (ed.) MLDM 2007. LNCS (LNAI), vol. 4571, pp. 533–547. Springer, Heidelberg (2007)
2. Hilderman, R., Hamilton, H.: Knowledge Discovery and Measures of Interest, p. 162. Kluwer Academic Publishers, Dordrecht (2001)
3. Geng, L., Hamilton, H.J.: Interestingness Measures for Data Mining: A survey. ACM Computing Surveys 38(33) (2006)
4. Pang-Ning, T., Kumar, V., Srivastava, J.: Selecting the Right Objective Measure for Association Analysis. Information Systems 29(4), 293–313 (2004)
5. Piatetski-Shapiro, G.: Discovery, Analysis, and Presentation of Strong Rules. In: Knowledge Discovery in Databases, pp. 229–248. AAI/MIT Press (1991)
6. Rauch, J., M. Šimunek, M.: An Alternative Approach to Mining Association Rules. In: Lin, T., et al. (eds.) Data Mining: Foundations, Methods, and Applications, pp. 219–238. Springer, Heidelberg (2005)
7. Ralbovský, M., Kuchař, T.: Using Disjunctions in Association Mining. In: Perner, P. (ed.) ICDM 2007. LNCS (LNAI), vol. 4597, pp. 339–351. Springer, Heidelberg (2007)
8. Agraval, R., Imielinski, T., Swami, A.N.: Mining Association Rules between Sets of Items in Large Databases. In: Proc. of the ACM-SIGMOD 1993 Int. Conference on Management of Data, Washington, D.C., pp. 207–216 (1993)
9. Hájek, P., Havránek, T.: Mechanising Hypothesis Formation - Mathematical Foundations for a General Theory. Springer, Heidelberg (1978)
10. Rauch, J.: Ein Beitrag zu der GUHA Methode in der dreiwertigen Logik. Kybernetika 11, 101–113 (1975)
11. Rauch, J.: Logical Foundations of Hypothesis Formation from Databases. Mathematical Institute of the Czechoslovak Academy of Sciences, Prague, Czech Republic, Dissertation (1986) (in Czech)
12. Rauch, J.: Logical Calculi for Knowledge Discovery in Databases. In: Komorowski, J., Żytkow, J.M. (eds.) PKDD 1997. LNCS, vol. 1263, pp. 47–57. Springer, Heidelberg (1997)
13. Rauch, J.: Classes of Four-Fold Table Quantifiers. In: Zytkow, J., Quafafou, M. (eds.) PKDD 1998. LNCS, vol. 1510, pp. 203–211. Springer, Heidelberg (1998)
14. Rauch, J.: Logic of Association Rules. Applied Intelligence 22, 9–28 (2005)
15. Rauch, J.: Definability of Association Rules in Predicate Calculus. In: Lin, T.Y., Ohsuga, S., Liau, C.J., Hu, X. (eds.) Foundations and Novel Approaches in Data Mining, pp. 23–40. Springer, Heidelberg (2005)
16. Rauch, J.: Observational Calculi, Classes of Association Rules and F-property. In: Granular Computing 2007, pp. 287–293. IEEE Computer Society Press, Los Alamitos (2007)

17. Rauch, J.: Classes of Association Rules - an Overview. In: Lin, T., et al. (eds.) Datamining: Foundations and Practice. Studies in Computational Intelligence, vol. 118, pp. 283–297. Springer, Heidelberg (2008)

18. Rauch, J.: Definability of Association Rules and Tables of Critical Frequencies. In: Lin, T., et al. (eds.) Datamining: Foundations and Practice. Studies in Computational Intelligence, vol. 118, pp. 299–321. Springer, Heidelberg (2008)

19. Rauch, J.: Measures of Interestingness and Classes of Association Rules. Accepted for Foundations of Data Mining, ICDM 2008 Workshop (2008)

20. Rauch, J.: Contribution to Logical Foundations of KDD. Faculty of Informatics and Statistics, University of Economics Prague, Czech Republic, Assoc. Prof. Thesis (1998) (in Czech)

21. Burian, J.: Data Mining and AA (Above Average) Quantifier. In: Svátek, V. (ed.) Proceedings of Znalosti 2003. TU Ostrava, Ostrava (2003) (in Czech)

22. Chrz, M.: Transparent Deduction Rules for the GUHA Procedures Diploma thesis. Faculty of Mathematics and Physics. Charles University in Prague, p. 63 (2007)

23. Hájek, P., Havránek, T., Chytil, M.: GUHA Method. Academia, Prague (1983) (in Czech)

24. Hájek, P., Holeňa, M., Rauch, J.: The GUHA Method and Foundations of (Relational) Data Mining. In: de Swart, H., Orłowska, E., Schmidt, G., Roubens, M., et al. (eds.) Theory and Applications of Relational Structures as Knowledge Instruments. LNCS, vol. 2929, pp. 17–37. Springer, Heidelberg (2003)

25. Mendelson, E.: Introduction to Mathematical Logic. Princeton, D. Van Nostrand Company, Inc. (1964)

26. Rauch, J., Šimunek, M.: Semantic Web Presentation of Analytical Reports from Data Mining – Preliminary Considerations. In: Web Intelligence, pp. 3–7. IEEE Computer Society, Los Alamitos (2007)

Part III
Text and Web Mining

Clustering the Web 2.0

Katharina Morik and Michael Wurst

Artificial Intelligence Unit, Technische Universität Dortmund, Germany

Abstract. Ryszard Michalski has been the pioneer of Machine Learning. His conceptual clustering focused on the understandability of clustering results. It is a key requirement if Machine Learning is to serve users successfully. In this chapter, we present two approaches to clustering in the scenario of Web 2.0 with a special concern of understandability in this new context. In contrast to semantic web approaches which advocate ontologies as a common semantics for homogeneous user groups, Web 2.0 aims at supporting heterogeneous user groups where users annotate and organize their content without a reference to a common schema. Hence, the semantics is not made explicit. It can be extracted by Machine Learning, though, hence providing users with new services.

1 Introduction

From its very beginning, Machine Learning aimed at helping people in doing their job more effectively. Ryszard Michalski has always stressed the *service to people* as a motivation for Machine Learning. Early on, classifier learning received a lot of attention [18], and subsequently eased the development of knowledge-based systems that support experts [15], [17]. Since the need of classified training examples could become a bottleneck for applications, clustering approaches became attractive because, there, no expert is necessary, who classifies the observations. However, statistical clustering approaches often lack the understandability of the clusters. The use of logic expressions for clustering turned precise cluster descriptions into easily understandable conditions for an observation to belong to the cluster [14]. The *understandability* of learning results is a strong factor of its success. It is a necessity for many applications. There is a large variety of understandable representations: depending on a user's education and school of thinking, logic, visual, or numerical representations ease understanding. Since interpretations depend on the user's background, it is quite difficult to supply heterogeneous user groups with one adequate representations. Instead, the representation has to cope with the heterogeneity of users, presenting information to a user in the way he or she prefers.

In contrast to the systems supporting a user in professional performance, the World Wide Web (WWW) offers documents, music, videos to a large variety of users, no longer restricted to one profession but offered to and by most different cultures and communities. The approach to organize the collections according to ontologies (the semantic web approach) has failed possibly just because of this heterogeneity of users. The view of the WWW as a space for collaboration

J. Koronacki et al. (Eds.): Advances in Machine Learning II, SCI 263, pp. 207–223.

among diverse users who not only seek but also deliver information (the *Web 2.0* approach) takes into account that users do not want to obey a given semantic but use their own one without explicit declaration.

Integrating Machine Learning capabilities into the WWW hence demands to cope with diverse representations. No longer, the same annotation of an object can be considered to have the same meaning: for different users the same label may have completely different meanings. Turning it the other way around, different annotations may mean the same. Machine Learning approaches normally use a fixed feature set and – in supervised learning – a fixed set of class labels. In the Web 2.0 context, this is no longer appropriate. Instead, the intensional meaning of a feature or label can only be determined on the basis of its extension regarding one particular user. We call this the *locality* requirement. The similarity of extensions can then be used to determine the similarity of features or labels given by different users, and, hence, the similarity of users. We shall investigate this issue in Section 2[1].

Moreover, the heterogeneity of users also challenges the design of user interfaces. Users differ in the way they like to look at a collection of items (i.e., pictures, music,...). Some prefer to navigate in a step-wise deepening manner with a small number of selectable nodes at each level. Others prefer to have many choices and few levels. For the developer of an interface it is not at all clear how to best organize a collection. Here, Machine Learning can be of help, if it constructs all possible and optimal structures and offers these to the user who can then choose the one he or she likes. The construction of optimal structures in one learning run technically means to learn the Pareto-optimal structures. Methodologically, our learning algorithm can be classified as an instance of *multi-strategy learning*. It exploits frequent set mining and evolutionary optimization in order to cluster a collection on the basis of annotations (tag sets) [13]. We shall describe our approach to this problem in Section 3[2].

2   Collaboratively Structuring Collections

Media collections in the internet have become a commercial success. Companies provide users with texts, music, and videos for downloading and even offer to store personal collections at the company's site. The structuring of large media collections has thus become an issue. Where general ontologies are used to structure the central archive, personal media collections are locally structured in very different ways by different users. The level of detail, the chosen categories, and the extensions even of categories with the same name can differ completely from user to user. In a one year project, a collection of music was structured by our students [10]. We found categories of mood, time of day, instruments, occasions (e.g., "when working" or "when chatting"), and memories (e.g., "holiday songs from 2005" or "songs heard with Sabine"). The *level of detail*

---

[1] The work on collaborative clustering is based in the Ph D thesis of Michael Wurst [22].
[2] The work on tagset clustering is based on the diploma thesis of Andreas Kaspari [12].

depends on the interests of the collector. Where some students structure instruments into electronic and unplugged, others carefully distinguish between string quartet, chamber orchestra, symphony orchestra, requiem, opera. A specialist of jazz designs a hierarchical clustering of several levels, each with several nodes, where a lay person considers jazz just one category. Where the most detailed structure could become a general taxonomy, from which less finely grained, local structures can easily be computed, categories under headings like "occasions" and "memories" cannot become a general structure for all users. Such categories depend on the personal attitude and life of a user. They are truly *local* to the user's collection. Moreover, the classification into a structure is far from being standardized. This is easily seen when thinking of a node "favorite songs". Several users' structures show completely different songs under this label, because different users have different favorites. The same is true for the other categories. We found that even the general genre categories can *extensionally vary* among users, the same song being classified, e.g., as "rock'n roll" in one collection and "blues" in another one. Hence, even if (part of) the collections' structure looks the same, their extensions can differ considerably [6]. In summary, structures for personal collections differ in the level of detail, the chosen categories, and the extensions for even the same labels.

Can Machine Learning be of help also for structuring personal collections? Since users do not want to have their hand-made structures overwritten, one could deny the benefit of automatic structuring. While users like to classify some songs into their own structure, they would appreciate it, if a learning system would clean-up their collection "accordingly". Moreover, users like to exchange songs (or pictures, or videos) with others. The success of Amazon's collaborative recommendations shows that users appreciate to share preferences. A structure of another user might serve as a blueprint for refining or enhancing the own structure. The main objective seems to be that users are given a choice among alternatives instead of providing them with just one result. To summarize, the requirements of a learning approach are, that it should

- not overwrite hand-made structures,
- not aim at a global model, but enhance a local one,
- add structure where a category's extension has become too large,
- take structures of others into account in a collaborative manner,
- recommend objects which fit nicely into the local structure, and
- should deliver several alternatives among which the user can choose.

2.1   The Learning Task of Localized Alternative Cluster Ensembles

The requirements listed above actually pose a new learning task. We characterize learning tasks by their inputs and outputs. Let X denote the set of all possible objects. A function $\varphi : S \rightarrow G$ is a function that maps objects $S \subseteq X$ to a (finite) set G of groups. We denote the domain of a function φ with D_φ. In cases where we have to deal with overlapping and hierarchical groups, we denote the set of groups as 2^G. The input for a learning task is a finite set of functions

$I \subseteq \{\varphi | \varphi : S \rightarrow G\}$. The same holds for the output $O \subseteq \{\varphi | \varphi : S \rightarrow G\}$. We consider the structuring of several users u_i, each described by $\varphi_i : S_i \rightarrow G_i$. A user with the problem of structuring her left-over objects S might now exploit the cluster models of other users in order to enhance the own structure. Cluster ensembles are almost what we need [3], [19], [20]. However, there are three major drawbacks: first, for cluster ensembles, all input clusterings must be defined at least on S. Second, the consensus model of cluster ensembles does not take the locality of S into account. Finally, merging several heterogenous user clusterings by a global consensus does not preserve the user's hand-made structuring. Hence, we have defined a new learning task [23].

Definition 1 (Localized Alternative Cluster Ensembles). *Given a set* $S \subseteq X$, *a set of input functions* $I \subseteq \{\varphi_i : S_i \rightarrow G_i\}$, *and a quality function*

$$q : 2^{\Phi} \times 2^{\Phi} \times 2^{S} \rightarrow R \tag{1}$$

with R *being partially ordered,* LOCALIZED ALTERNATIVE CLUSTERING ENSEMBLES *delivers the output functions* $O \subseteq \{\varphi_i | \varphi_i : S_i \rightarrow G_i\}$ *so that* $q(I, O, S)$ *is maximized and for each* $\varphi_i \in O$ *it holds that* $S \subseteq D_{\varphi_i}$.

Note that in contrast to cluster ensembles, the input clusterings can be defined on any subset S_i of X. Since for all $\varphi_i \in O$ it must hold that $S \subseteq D_{\varphi_i}$, all output clusterings must at least cover the objects in S.

2.2   The LACE Algorithm

The algorithm LACE derives a new clustering from existing ones by extending and combining them such, that each covers a subset of objects in S. We need two more definitions in order to describe the algorithm.

Definition 2 (Extended function). *Given a function* $\varphi_i : S_i \rightarrow G_i$, *the function* $\varphi_i' : S_i' \rightarrow G_i$ *is the* EXTENDED FUNCTION *for* φ_i, *if* $S_i \subset S_i'$ *and* $\forall x \in S_i : \varphi_i(x) = \varphi_i'(x)$.

Definition 3 (Bag of clusterings). *Given a set* I *of functions. A* BAG OF CLUSTERINGS *is a function*

$$\varphi_i(x) = \begin{cases} \varphi_{i1}'(x), & \text{if } x \in S_{i1}' \\ \vdots & \vdots \\ \varphi_{ij}'(x), & \text{if } x \in S_{ij}' \\ \vdots & \vdots \\ \varphi_{im}'(x), & \text{if } x \in S_{im}' \end{cases} \tag{2}$$

where each φ_{ij}' *is an extension of a* $\varphi_{ij} \in I$ *and* $\{S_{i1}', \ldots, S_{im}'\}$ *is a partitioning of* S.

Now, we can define the quality for the output, i.e. the objective function for our bag of clusterings approach to local alternative clustering ensembles.

Definition 4 (Quality of an output function). *The* QUALITY OF AN INDI-VIDUAL OUTPUT FUNCTION *is measured as*

$$q^*(I, \varphi_i, S) = \sum_{x \in S} \max_{x' \in S_{ij}} sim(x, x') \text{ with } j = h_i(x) \tag{3}$$

where sim *is a similarity function* sim $: X \times X \to [0, 1]$ *and* h_i *assigns each example to the corresponding function in the bag of clusters* $h_i : S \to \{1, \ldots, m\}$ *with*

$$h_i(x) = j \Leftrightarrow x \in S'_{ij}. \tag{4}$$

The QUALITY OF A SET OF OUTPUT FUNCTIONS *now becomes*

$$q(I, O, S) = \sum_{\varphi_i \in O} q^*(I, \varphi_i, S). \tag{5}$$

Besides this quality function, we want to cover the set S with a bag of clusterings that contains as few clusterings as possible.

The main task is to cover S by a bag of clusterings φ. The basic idea of this approach is to employ a sequential covering strategy. In a first step, we search for a function φ_i in I that best fits the set of query objects S. For all objects not sufficiently covered by φ_i, we search for another function in I that fits the remaining points. This process continues until either all objects are sufficiently covered, a maximal number of steps is reached, or there are no input functions left that could cover the remaining objects. All objects that could not be covered are assigned to the input function φ_j containing the object which is closest to the one to be covered. Alternative clusterings are produced by performing this procedure several times, such that each input function is used at most once.

When is an object sufficiently covered by an input function so that it can be removed from the query set S? We define a threshold based criterion for this purpose. Let Z_{φ_i} be the set of objects delivered by φ.

Definition 5. *A function* φ SUFFICIENTLY COVERS *an object* $x \in S$ *(written as* $x \sqsubseteq_\alpha \varphi$ *), iff* $x \sqsubseteq_\alpha \varphi :\Leftrightarrow \max_{x' \in Z_\varphi} sim(x, x') > \alpha$.

This threshold allows to balance the quality of the resulting clustering and the number of input clusters. A small value of α allows a single input function to cover many objects in S. This, on average, reduces the number of input functions needed to cover the whole query set.

Turning it the other way around: when do we consider an input function to fit the objects in S well? First, it must contain at least one similar object for each object in S. This is essentially what is stated in the quality function q^*. Second, it should cover as few additional objects as possible. This condition follows from the locality demand. Using only the first condition, the algorithm would not distinguish between input functions which span a large part of the data space and those which only span a small local part. This distinction, however, is essential

$$O = \emptyset$$
$$I' = I$$
while $(|O| < max_{alt})$ **do**
    $S' = S$
    $B = \emptyset$
    $step = 0$
    while $((S' \neq \emptyset) \wedge (I' \neq \emptyset) \wedge (step < max_{steps}))$ **do**
        $\varphi_i = \arg\max_{\varphi \in J} q_f^*(Z_\varphi, S')$
        $I' = I' \setminus \{\varphi_i\}$
        $B = B \cup \{\varphi_i\}$
        $S' = S' \setminus \{x \in S' | x \sqsubset_\alpha \varphi_i\}$
        $step = step + 1$
    end while
    $O = O \cup \{bag(B, S)\}$
end while

Fig. 1. The sequential covering algorithm finds bags of clusterings in a greedy manner. max_{alt} denotes the maximum number of alternatives in the output, max_{steps} denotes the maximum number of steps that are performed during sequential covering. The function bag constructs a bag of clusterings by assigning each object $x \in S$ to the function $\varphi_i \in B$ that contains the object most similar to x. Z_{φ_i} is the set of objects delivered by φ.

for treating the locality appropriately. The situation we are facing is similar to that in information retrieval. The target concept S – the ideal response – is approximated by φ delivering a set of objects – the retrieval result. If all members of the target concept are covered, the retrieval result has the highest recall. If no objects in the retrieval result are not members of S, it has the highest precision. We want to apply precision and recall to characterize how well φ covers S. We can define

$$precision(Z_{\varphi_i}, S) = \frac{1}{|Z_{\varphi_i}|} \sum_{z \in Z_{\varphi_i}} max\{sim(x, z) | x \in S\} \qquad (6)$$

and

$$recall(Z_{\varphi_i}, S) = \frac{1}{|S|} \sum_{x \in S} max\{sim(x, z) | z \in Z_{\varphi_i}\}. \qquad (7)$$

Please note that using a similarity function which maps identical objects to 1 (and 0 otherwise) leads to the usual definition of precision and recall. The fit between an input function and a set of objects now becomes

$$q_f^*(Z_{\varphi_i}, S) = \frac{(\beta^2 + 1)recall(Z_{\varphi_i}, S)precision(Z_{\varphi_i}, S)}{\beta^2 recall(Z_{\varphi_i}, S) + precision(Z_{\varphi_i}, S)}. \qquad (8)$$

Recall directly optimizes the quality function q^*, precision ensures that the result captures local structures adequately. The fitness $q_f^*(Z_{\varphi_i}, S)$ balances the two criteria.

Deciding whether φ_i fits S or whether an object $x \in S$ is sufficiently covered requires to compute the similarity between an object and a cluster. Remember that Z_{φ_i} is the set of objects delivered by φ. If the cluster is represented by all of its objects ($Z_{\varphi_i} = S_i$, as usual in single-link agglomerative clustering), this central step becomes inefficient. If the cluster is represented by exactly one point ($|Z_{\varphi_i}| = 1$, a centroid in k-means clustering), the similarity calculation is very efficient, but sets of objects with irregular shape, for instance, cannot be captured adequately. Hence, we adopt the representation by "well scattered points" Z_{φ_i} as representation of φ_i [8], where $1 < |Z_{\varphi_i}| < |S_i|$. These points are selected by stratified sampling according to G.

We can now compute the fitness q_f^* of all $Z_{\varphi_i} \in I$ with respect to a query set S in order to select the best φ_i for our bag of clusterings.

The whole algorithm works as depicted in Figure 1. We start with the initial set of input functions I and the set S of objects to be clustered. In a first step, we select an input function that maximizes $q_f^*(Z_{\varphi_i}, S)$. φ_i is removed from the set of input functions leading to a set I'. For all objects S' that are not sufficiently covered by φ_i, we select a function from I' with maximal fit to S'. This process is iterated until either all objects are sufficiently covered, a maximal number of steps is reached, or there are no input functions left that could cover the remaining objects. All input functions selected in this process are combined to a bag of clusters, as described above. Each object $x \in S$ is assigned to the input function containing the object being most similar to x. Then, all input functions are extended accordingly (cf. definition 2). We start this process anew with the complete set S and the reduced set I' of input functions until the maximal number of alternatives is reached.

2.3 Results of the LACE Algorithm

The LACE algorithm has been successfully applied to the collection of student structures for the music collection. Leaving out one clustering, the learning task was to again structure the now unstructured music. We compared our approach with single-link agglomerative clustering using cosine measure, top down divisive clustering based on recursively applying kernel k-means [5] (K k-means), and with random clustering. Localized Alternative Cluster Ensembles were applied using cosine similarity as inner similarity measure. The parameters for all algorithms were chosen globally optimal. In our experiments we used $\alpha = 0.1$. For β, the optimal value was 1. Kernel k-means and random clustering were started five times with different random initializations.

Table 1 shows the results. As can be seen, the local alternative cluster ensembles approach LACE performs best (see [23] for more details on the evaluation).

The application opportunities of LACE are, on the one hand, to structure collections automatically but personalized and, on the other hand, to recommend items mass-tailored to a user's needs. If all not yet classified instances fit into a hand-made structure, the structure is not changed. If, however, some instances do not fit or a cluster has become too large, a set S is formed and other structures

Table 1. The results for different evaluation measures

Method	Correlation	Absolute distance	FScore
LACE	0.44	0.68	0.63
K k-means audio	0.19	2.2	0.51
K k-means ensemble	0.23	2.5	0.55
single-link audio	0.11	9.7	0.52
single-link ensemble	0.17	9.9	0.60
random	0.09	1.8	0.5

φ_i are exploited. This leads to some new structures and is accompanied by recommendations.

Let us look at some examples for illustration.

- In the cluster "pop" of user A there might be too many instances. Now, a structure φ_B dividing similar songs into "rock", "metal", and "disco" might be found in the collection of user B under a different title, say "dance-floor". Integrating these clusters is accompanied by new instances from user B. These recommendations are specific to user A's "pop" cluster (i.e., local).
- Another example starts with several music plays S, which user A did not find the time to structure. He might receive a structure φ_1 according to instruments and another one, φ_2, with funny titles like "songs for Sabine", "poker nights", and "lonely sundays". When listening to music in the latter clusters, user A might like them and change the tags to "romance", "dinner background", and "blues".
- Of course, memory categories (e.g., "holidays 2005") will almost never become tasks for automatic structuring, because their extension is precisely determined and cannot be changed. However, for somebody else, also these categories can be useful and become (under a different name) a cluster for a larger set of instances.

The localized alternative cluster ensembles offer new services to heterogeneous user groups. The approach opens up applications in the Web 2.0 context ranging from content sharing to recommendations and community building, where the locality requirement is essential. Machine Learning, here, does not support professional performance but the computer use for leisure. It supports the many small groups of users, which together are more than the main stream group. Both from a commercial and ethical point of view, this is challenging service.

3   Structuring Tagged Collections

Collaborative tagging of resources has become popular. Systems like DEL.ICIO.US[3], LAST.FM[4], or FLICKR[5] allow users to annotate all resources by

---

[3] http://del.icio.us/
[4] http://last.fm/
[5] http://flickr.com

freely chosen tags, without a given obligatory taxonomy of tags (ontology). Since the users all together develop the "taxonomy" of the Web 2.0 by their tags, the term "folksonomy" came up.

Definition 6 (Folksonomy). *A* FOLKSONOMY *is a tuple* $\mathcal{F} = (U, T, R, Y)$, *where U is the set of users, T the set of tags, and R the set of resources. $Y \subseteq U \times T \times R$ is a relation of taggings. A tuple $(u, t, r) \in Y$ means, that user u has annotated resource r with tag t.*

The popular view of folksonomies is currently the tag cloud, like the one shown in Figure 2. Tags are understandable although not precisely defined. While tag clouds support users to hop from tag to tag, inspecting the resources is cumbersome. When selecting a tag, the user finds all the resources annotated with this tag. There is no extensionally based hierarchy guiding the exploration of resources. For instance, a user cannot move from all photos to those being tagged as "art", as well. Hence, navigation in folksonomies is particularly restricted. However, the data for a more appropriate structure are already given – they just need to be used. A resource, which has been tagged as, e.g., $\{art, photography\}$ is linked with the termsets $\{art\}, \{photography\}, \{art, photography\}$ by a function $g : T \times R \to \mathbb{N}$. The user may now refine the selection of resources, e.g., from all photos to those being annotated as "art", as well. As is easily seen, the power set of tags, together with union and intersection forms a lattice (see Figure 3). If we choose cluster sets within this lattice, these no longer need to form a lattice. Only quite weak assumptions about valid cluster sets are necessary:

Definition 7 (Frequent Termset Clustering Conditions). *A cluster set* $\mathcal{C} \subseteq \mathcal{P}(T)$ *is valid, if it fulfills the following constraints:*

$$\emptyset \in \mathcal{C} \tag{9a}$$

$$\forall D \in \mathcal{C} \ \text{with} \ D \neq \emptyset : \exists C : C \prec D \tag{9b}$$

$$\forall C \in \mathcal{C} : \exists r \in R : r \nabla C. \tag{9c}$$

Condition (9a) states that the empty set must be contained in each cluster set. Condition (9b) ensures that there is a path from each cluster to the empty set

Fig. 2. Tag cloud view of an excerpt of the DEL.ICIO.US tags. More frequently used tags are printed in a bolder style.

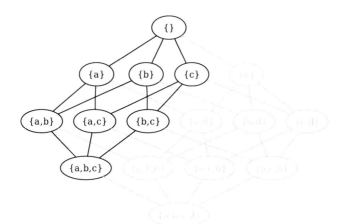

Fig. 3. The frequent termsets with the lattice of possibly frequent termsets grayed

(thus the cluster set is a connected graph). Condition (9c) ensures that each cluster contains at least one resource.

The link to the resources is given by the cover relation based on the function $g:\nabla \subseteq R \times \mathcal{P}(T)$ for which the following holds:

$$r\nabla C \equiv \forall t \in C : g(t,r) > 0 \tag{10}$$

A resource is covered by a termset, if all terms in the termset are assigned to the resource. The *support* of a termset is defined as the fraction of resources it covers. The frequency of tag sets can either be defined as the number of resources it covers, or as the number of users annotating resources with it, or as the number of tuples $U \times R$. Our method works with any of these frequencies. Using FPgrowth of [9], we find the frequent sets which only need to obey the clustering conditions 9a, 9b, and 9c.

3.1   Learning Pareto-optimal Clusterings from Frequent Sets

Having found all frequent tag sets, the task of structuring them according to the preference of a user is to select a cluster set $\mathcal{C}$ among all possible valid clusterings Γ. Since we do not know the preference of the user, we do not tweak heuristics, such that these select a preferred clustering, as done in [21], [1], [7]. In contrast, we decompose the selection criteria into two orthogonal criteria and apply multi-objective optimization. Given orthogonal criteria, multi-objective optimization [2] finds all trade-offs between the criteria such that the solution cannot be enhanced further with respect to one criterion without destroying the achievements with respect to the other criteria – it is *Pareto-optimal*. The user may then explore the Pareto-optimal solutions or select just one to guide her through a media collection.

Several algorithms were proposed for multi-objective optimization, almost all of them based on evolutionary computation [2, 24]. In this work, we use the genetic algorithm NSGA-2 [4] to approximate the set of Pareto-optimal solutions. We used the operators implemented within the RapidMiner system, formerly known as YALE [16]. The population consists of cluster sets. These individuals are represented by binary vectors. A mapping from cluster sets to binary vectors is defined such that each element of the set of frequent termsets corresponds to one position in the vector. The result of the initial frequent termset clustering corresponds to a vector where each component has the value 1. This hierarchy of frequent sets is traversed in breadth-first manner when mapping to the vector components. We are looking for optimal solutions which are part of the frequent sets result, i.e., vectors where some of the components have the value 0. As a illustration, Figure 4 shows the binary encoding of a cluster set, where the frequent sets $\{a\}, \{b\}, \{c\}, \{a, b\}, \{b, c\}$ had been found. In the course of mutation and cross-over, the NSGA-2 algorithm may create vectors that correspond to invalid cluster sets. These are repaired by deleting those clusters that are not linked to a path which leads to the root cluster. Hence, our cluster conditions are enforced by post-processing each individual.

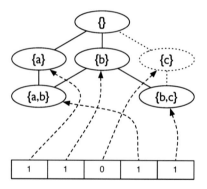

Fig. 4. Each cluster corresponds to a component of the vector. The ordering corresponds to breadth-first search in the result of the frequent termset clustering, i.e. the complete search space.

The algorithm approximates the set $\Gamma^* \subseteq \Gamma$ of Pareto-optimal cluster sets.

$$\Gamma^* = \{\mathcal{C} \in \Gamma \mid \not\exists \mathcal{D} \in \Gamma : f(\mathcal{D}) \succeq f(\mathcal{C})\} \tag{11}$$

where $f(\mathcal{D}) \succeq f(\mathcal{C})$ states that there is no cluster set $\mathcal{D}$ that Pareto-dominates the cluster set $\mathcal{C}$ with respect to the fitness functions f. Thus Γ^* contains all non-dominated cluster sets.

3.2   Resulting Navigation Structures

We have applied our multistrategy approach which combines frequent sets with multi-objective optimization to data from the Bibsonomy system ([11]), which is a collaborative tagging system allowing users to annotate web resources and publications. Our data set contains the tag assignments of about 780 users. The number of resources tagged by at least one user is about 59.000. The number of tags used is 25.000 and the total number of tag assignments is 330.000. Since all Pareto-optimal clusterings are found in one learning run, it does not make sense to compare the found clusterings to a one delivered by an alternative approach. Either the one clustering is part of the Pareto front, or it is not Pareto-optimal.

The clusterings in Figure 5 show some navigation structures from a Pareto front minimizing overlap and maximizing coverage. In the picture, only the depth of a node is indicated. Each node is a set of resources labeled by (a set of) tags. Since users regularly navigate by tags, these labels are easily understood. Also the structure is easily understood so that users can, indeed, select the structure they like. Other approaches combine the two criteria into one heuristic and deliver just one of the shown clusterings. A more detailed analysis of the individual results shows the following:

- Cluster sets that fulfill the overlap criterion well are quite narrow and do not cover many resources.

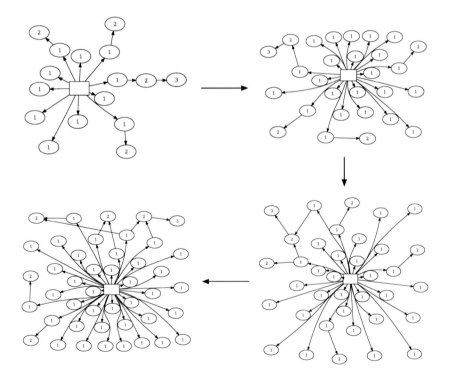

Fig. 5. Some cluster sets from the Pareto front optimizing overlap vs. coverage: starting with a small overlap (upper left) moving to a high coverage (lower left)

- Cluster sets that fulfill the coverage criterion well are quite broad and contain a lot of overlap. Note, however, that overlap might be desired for navigation: the same resource can then be retrieved using different tags or tag sets.
- All resulting cluster structures are very shallow, as neither of the criteria forces the selection of deep clusters. Both, high coverage and low overlap can be achieved with clusters of level one.

Multi-objective optimization requires the criteria to be orthogonal. Hence, it has to be verified for each pair of criteria that they are negatively correlated. Actually, on artificially generated data, the correlations of criteria has been investigated [12]. In addition to the usual clustering criteria of overlap and coverage, we have defined criteria which take into account the hierarchy of clusters and the given frequent tag sets, namely childcount and completeness. Since navigation is often performed top-down, we use the number of child nodes at the root and each inner node as indicator of the complexity of a cluster set.

Definition 8 (Child count). *Given a cluster set* $\mathcal{C}$, *we define* $\mathrm{succ} \colon \mathcal{C} \to \mathcal{P}(\mathcal{C})$ *as* $\mathrm{succ}\,(C) = \{D \in \mathcal{C} \mid C \prec D\}$ *and thus the set* $\mathcal{C}' \subseteq \mathcal{C}$ *as* $\mathcal{C}' = \{C \in \mathcal{C} \mid\mid \mathrm{succ}(C) \mid > 0\}$. *Based on this, we can define*

$$childcount_{max}(\mathcal{C}) = max_{C \in \mathcal{C}'} \mid succ(C) \mid \tag{12}$$

Thus the complexity of a cluster structure is given as the most complex node. The maximal child count will usually increase with increasing coverage, since to cover more resources often means to add clusters.

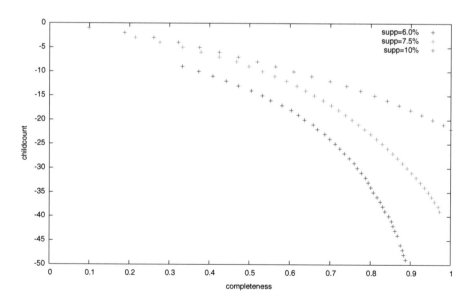

Fig. 6. Pareto front for child count vs. completeness for different minimal supports

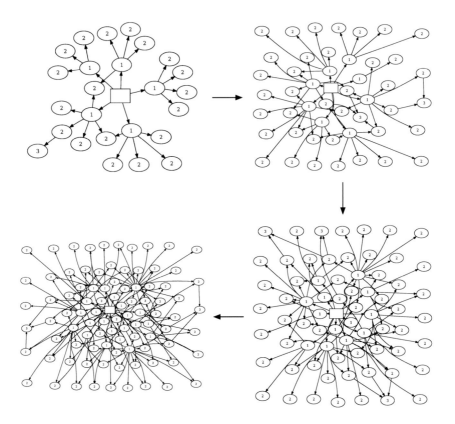

Fig. 7. Some cluster sets from the Pareto front optimizing child count vs. completeness: starting with a small child count (upper left) moving to a high completeness (lower left)

The criterion of completeness is similar to coverage, but tailored to the frequent termsets that are the starting point of our clustering. The idea of completeness is, that the selected clusters should represent the given frequent termsets as completely as possible.

Definition 9 (Completeness). *Given two cluster sets C and C_{ref}. We assume $C \subset C_{ref}$. Then the function complete : $\Gamma \times \Gamma \to$ is defined as:*

$$complete(C, C_{ref}) = \frac{|C|}{|C_{ref}|} \qquad (13)$$

Thus, the more of the original frequent termsets are contained in the final clustering, the higher the completeness. This combines coverage and cluster depth in one straightforward criterion. Figure 6 shows the Pareto front when minimizing childcount and maximizing completeness. Figure 7 shows some clusterings along the Pareto front when optimizing child count vs. completeness.

Inspecting the Pareto optimal results more closely yields the following:

- Clusterings with a small maximum child count are narrow, but deep. This effect can be explained, as deep clustering yield on average a higher completeness.
- Clusterings with high completeness are broader, but still deep and contain much of the heterogeneity contained in the original data. They also show a very high coverage.

In this way, we can actually optimize three criteria at once: the simplicity of the cluster structure, its detailedness in terms of cluster depth, and the coverage of resources. These criteria are furthermore not biased to remove heterogeneity from the data, which is essential in many explorative applications.

Tag set clustering by multi-objective optimization on the basis of frequent set mining is a multistrategy learning approach which supports the personalized access to large collections in the Web 2.0. The decomposition of clustering criteria into orthogonal ones allows to compute all Pareto-optimal solutions in one learning run. Again, the service is to heterogeneous user groups. Machine Learning helps to automatically build Human Computer Interfaces that correspond to a user's preferences and give him or her a choice.

4   Conclusion

In this chapter, two approaches of Machine Learning serving the Web 2.0 have been shown. Both deliver sets of clusterings in order to allow users to choose. Both are labeled by user given tag (sets) without a formally defined semantics but understood in the way natural language is understood. In order to serve heterogeneous user groups, we have focused on the locality in the LACE algorithm and on delivering all Pareto-optimal solutions in the tagset clustering. Since Ryszard Michalski has always been open to new approaches and, at his last visit at Dortmund, was curious about our then emerging attempts to apply Machine Learning to the Web 2.0, we are sorry that we cannot discuss matters with him, anymore. Even though our methods are different from his, the impact of his ideas about understandability, service of Machine Learning, and multistrategy learning is illustrated also by this moderate work.

References

[1] Beil, F., Ester, M., Xu, X.: Frequent term-based text clustering. In: Proceedings of the International Conference on Knowledge Discovery and Data Mining, KDD (2002)
[2] Coello Coello, C.A.: A comprehensive survey of evolutionary-based multiobjective optimization techniques. Knowledge and Information Systems 1(3), 129–156 (1999)
[3] Datta, S., Bhaduri, K., Giannella, C., Wolff, R., Kargupta, H.: Distributed data mining in peer-to-peer networks. IEEE Internet Computing, special issue on distributed data mining (2005)

[4] Deb, K., Agrawal, S., Pratab, A., Meyarivan, T.: A fast elitist non-dominated sorting genetic algorithm for multi-objective optimization: NSGA-II. In: Deb, K., Rudolph, G., Lutton, E., Merelo, J.J., Schoenauer, M., Schwefel, H.-P., Yao, X. (eds.) PPSN 2000. LNCS, vol. 1917, Springer, Heidelberg (2000)

[5] Dhillon, I.S., Guan, Y., Kulis, B.: Kernel k-means: spectral clustering and normalized cuts. In: Proc. of the conference on Knowledge Discovery and Data Mining (2004)

[6] Flasch, O., Kaspari, A., Morik, K., Wurst, M.: Aspect-based tagging for collaborative media organization. In: Berendt, B., Hotho, A., Mladenic, D., Semeraro, G. (eds.) WebMine 2007. LNCS (LNAI), vol. 4737, pp. 122–141. Springer, Heidelberg (2007)

[7] Fung, B.C.M., Wang, K., Ester, M.: Hierarchical document clustering using frequent itemsets. In: Proceedings of the SIAM International Conference on Data Mining (2003)

[8] Guha, S., Rastogi, R., Shim, K.: CURE: an efficient clustering algorithm for large databases. In: Proc. of ACM SIGMOD International Conference on Management of Data, pp. 73–84 (1998)

[9] Han, J., Pei, J., Yin, Y.: Mining frequent patterns without candidate generation. In: Proceedings of the ACM SIGMOD International Conference on Management of Data (2000)

[10] Homburg, H., Mierswa, I., Möller, B., Morik, K., Wurst, M.: A benchmark dataset for audio classification and clustering. In: Proceedings of the International Conference on Music Information Retrieval (2005)

[11] Hotho, A., Jäschke, R., Schmitz, C., Stumme, G.: BibSonomy: A social bookmark and publication sharing system. In: Proceedings of the Conceptual Structures Tool Interoperability Workshop at the International Conference on Conceptual Structures (2006)

[12] Kaspari, A.: Maschinelle Lernverfahren für kollaboratives tagging. Master's thesis, Technische Univ. Dortmund, Computer Science, LS8 (2007)

[13] Michalski, R.S., Kaufman, K.: Intelligent evolutionary design: A new approach to optimizing complex engineering systems and its application to designing heat exchangers. International Journal of Intelligent Systems 21(12) (2006)

[14] Michalski, R.S., Stepp, R., Diday, E.: A recent advance in data analysis: Clustering objects into classes characterized by conjunctive concepts. In: Kanal, L., Rosenfeld, A. (eds.) Progress in Pattern Recognition, pp. 33–55. North-Holland, Amsterdam (1981)

[15] Michalski, R.S., Chilausky, R.: Knowledge acquisition by encoding expert rules versus computer induction from examples: A case study involving soybean pathology. International Journal for Man-Machine Studies (12), 63–87 (1980)

[16] Mierswa, I., Wurst, M., Klinkenberg, R., Scholz, M., Euler, T.: Yale: Rapid prototyping for complex data mining tasks. In: Ungar, L., Craven, M., Gunopulos, D., Eliassi-Rad, T. (eds.) KDD 2006: Proceedings of the 12th ACM SIGKDD international conference on Knowledge discovery and data mining, pp. 935–940. ACM, New York (2006)

[17] Morik, K.: Balanced cooperative modeling. In: Michalski, R., Tecuci, G. (eds.) Machine Learning - A Multistrategy Approach, pp. 295–318. Morgan Kaufmann, San Francisco (1994)

[18] Ryszard, R.S., Michalski, S.: A variable decision space approach for implementing a classification system. In: Proceedings of the Second International Joint Conference on Pattern Recognition, Copenhagen, Denmark, pp. 71–75 (1974)

[19] Strehl, A., Ghosh, J.: Cluster ensembles – a knowledge reuse framework for combining partitionings. In: Proceedings of the AAAI (2002)

[20] Topchy, A.P., Jain, A.K., Punch, W.F.: Combining multiple weak clusterings. In: Proceedings of the International Conference on Data Mining, pp. 331–338 (2003)

[21] Wang, K., Xu, C., Liu, B.: Clustering transactions using large items. In: Proceedings of the International Conference on Information and Knowledge Management (1998)

[22] Wurst, M.: Distributed Collaborative Structuring – A Data Mining Approach to Information Management in Loosely-Coupled Domains. PhD thesis, Technische Univ. Dortmund, Computer Science, LS8 (2008)

[23] Wurst, M., Morik, K., Mierswa, I.: Localized alternative cluster ensembles for collaborative structuring. In: Fürnkranz, J., Scheffer, T., Spiliopoulou, M. (eds.) ECML 2006. LNCS (LNAI), vol. 4212, pp. 485–496. Springer, Heidelberg (2006)

[24] Zitzler, E., Thiele, L.: Multiobjective evolutionary algorithms: a comparative case study and the strength Pareto approach. IEEE Transactions on Evolutionary Computation 3(4), 257–271 (1999)

Induction in Multi-Label Text Classification Domains

Miroslav Kubat, Kanoksri Sarinnapakorn, and Sareewan Dendamrongvit

Department of Electrical & Computer Engineering University of Miami Coral Gables,
FL 33146, U.S.A.
mkubat@miami.edu, ksarin@miami.edu, s.dendamrongvit@umiami.edu

Abstract. Automated classification of text documents has two distinctive aspects. First, each training or testing example can be labeled with more than two classes at the same time—this has serious consequences not only for the induction algorithms, but also for how we evaluate the performance of the induced classifier. Second, the examples are usually described by great many attributes, which makes induction from hundreds of thousands of training examples prohibitively expensive. Both issues have been addressed by recent machine-learning literature, but the behaviors of existing solutions in real-world domains are still far from satisfactory. Here, we describe our own technique and report experiments with a concrete text database.

Keywords: Classifier induction, text classification, multi-label examples, information fusion, Dempster-Shafer theory.

1 Introduction

The last two decades were marked by significant progress in the studies of digital libraries, some of them containing millions of documents. The fast development of corpuses of this size places a heavy burden on the indexing mechanisms that provide access to the stored information. Ideally, we would like the user to be able to identify the documents of interest either by selecting from hierarchically organized (pre-defined) categories or by entering well-chosen keywords. In response, the system would return those files that are relevant to the user's request.

The indexing scheme is easy to create as long as the number of documents is manageable: a human operator simply labels every instance in the collection with the category it belongs to, and—if needed—furnishes additional information such as the language, source, or intended audience (e.g., a scientific versus a popular paper). But manual annotation is impractical and prohibitively expensive in the case of collections that contain millions of documents, especially if thousands of new instances are added on a daily basis. In such domains, we would appreciate a computer program capable of automating the process. In the scenario considered in this paper, the human expert labels a small fraction of the documents, thus creating a training set from which the computer induces a classifier to be used to label the rest.

This looks like a standard job for machine learning, but we must also be aware of some traits that set this task apart. First, the number of classes can be so high as to render the use of certain types of classifiers problematic; second, the tens of thousands

J. Koronacki et al. (Eds.): Advances in Machine Learning II, SCI 263, pp. 225–244.
springerlink.com   © Springer-Verlag Berlin Heidelberg 2010

of attributes that typically describe a document impose on induction a heavy computational burden; and third, while classical machine learning assumes one class for each example, a text document often belongs to two or more (even many) classes at the same time. Therefore, we need new induction algorithms, and we also have to re-think the ways to evaluate the performance of the resulting classifiers. As indicated by the literature survey in the next section, this specific strand of research has been pursued by several research groups, but the results are still far from satisfactory. The difficulty of the involved problems—and their economic and practical importance—is bound to keep the research alive for years to come.

The work reported in this paper has been inspired by the challenges posed by EU-ROVOC, a particularly large multilingual collection, created by virtue of close collaboration of the European Parliament, the European Commission's Publications Office, and the national organizations of the EU member states. The documents in this collection come from such diverse fields as law, economics, trade, education, communications, medicine, agriculture, and many others. The corpus is expected to grow to millions of documents, each described by about 100,000 numeric features (each specifying the relative frequency of a different word) and known to belong into a subset of more than 5,000 different classes. The classes are hierarchically organized, with 30 different labels at the top level [9, 17].

Very soon, we learned that existing techniques would be impractical. For illustration: having experimented with one of the best-known methods—*ADTree* [10]—we were astonished to observe that several hours were needed to process even a highly simplified version of the data: 20,000 documents described by 100 features. Worse still, when we increased the number of features from the clearly inadequate 100, we realized that the computation time grew supralinearly—induction from examples described by 1,000 features incurred much more than ten times the CPU time needed when only 100 features were involved. Later, we learned that another technique, *AdaBoost.MH*, was faster than *ADTree*, but it still suffered from the superlinear nature of the computational costs; these disqualified its straightforward application.

This last observation suggested where to look for an improvement. Suppose we divide the feature set into N subsets. If T is the time needed to induce a classifier from *all* features, and T_i the time needed to induce a classifier from the i-th feature subset, we have $\Sigma T_i \ll T$. Figure 1 illustrates the point: running the same classifier on data described by 50% of the features may take about 10% of the time consumed when all 100% are used; induction of two classifiers, each from a different set of 50% features will then take about 20% time. We thus surmised that a lot of computation might be saved by running a *baseline induction algorithm* (BIA) repeatedly on the same training examples, each time described by a different feature subset, inducing N subclassifiers. To classify a document, we submit the feature vector describing it to all subclassifiers in parallel, and then merge ("fuse") their outputs.

We put this idea to test in the course of our work on the EUROVOC problem—we experimented with a few state-of-the-art fusion techniques, and then developed a new one, built around certain ideas borrowed from the Dempster-Shafer theory of evidence combination. This paper summarizes the results of experiments designed to investigate our system's behavior under different conditions.

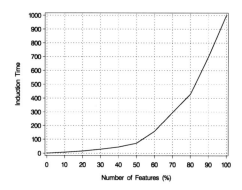

Fig. 1. Computational costs of induction often grow supralinearly, even exponentially, in the number of features describing the training examples

After a brief survey of related work (in the next section) we formally state our goal and the requisite performance criteria in Section 3. Then, in Section 4, we outline *AdaBoost.MH*, the BIA of our choice. Section 5 describes the fusion mechanisms, including the relevant background from the Dempster-Shafer theory. Section 6 reports experiments that indicate that the approach indeed achieves computational savings that are not outbalanced by impaired classification performance, and that the approach compares favorably with another recently proposed technique, induction of multi-label decision trees.

2   Previous Work

In the field of automated document classification, it is quite common that a document simultaneously belongs to two or more categories—this is certainly the case of a multidisciplinary scientific paper, a patent, or a commentary on a controversial article in an international-law magazine. Among the text-categorization domains with multi-label examples, the best known is perhaps the Reuters-21578 dataset, consisting of documents from the 1987 Reuters newswire [30]; but the same is true for the EUROVOC domain [24].

The fact that traditional machine learning expects each example to have one and only one class label motivates the most straightforward solution: for each class, induce a binary classifier that for each example returns "yes" or "no," depending on whether the example is or is not regarded as a representative of the class; a testing example is then submitted to all these binary classifiers in parallel. Literature has reported several successful attempts to employ this approach. For instance, mechanisms based on the Bayesian decision theory were studied by [20], [13], and [22], the behavior of the instance-based rule (nearest-neighbor classifiers) was explored by [21], and the currently popular support vector machines were addressed by [18] and [19]. Somewhat expectedly, experiments reported in these papers indicate that the support vector machines tend to outperform more conventional approaches, but usually at the price of very high computational costs.

The main defect of the idea to induce a separate classifier for each class is that the mutual relations between classes are thus neglected. By way of a rectification, [15] developed a method that trains all classifiers simultaneously (although the discriminant function they worked with was still based on individual classes), while [16] resorted to "stacking": at the lower-level, they used one SVM for each class, and then fed the outputs of these lower-level SVMs into a higher-level SVM that decided about the final class labels. The entropy-maximizing technique proposed by [34] explicitly considered inter-class correlations, and the authors were able to show that their method, while expensive, outperformed earlier attempts in certain domains; however, their assumption of the mutual independence of the estimated errors in real-world domains is rarely satisfied. Finally, trying to avoid the need to induce a separate classifier for each label, [7] extended the methodology of decision tree induction to make induction from multi-label examples more natural.

The third strand of research focusing on induction from multi-label examples relies on the "boosting" method that was developed by [26]. The idea is to combine classifiers that have each been induced from somewhat different training data. Perhaps the best known from the many versions of this approach is *AdaBoost* [11, 12]. For domains with multi-label examples, [27] developed two extensions, *AdaBoost.MH* and *AdaBoost.MR*. Both of them return not only class labels, but also the classifiers' levels of confidence in each label. Four extended versions of these algorithms have been included in the system *BoosTexter* [27].

In summary, the literature we have reviewed has approached the problem of multi-label examples from two directions: one induces a binary classifier for each class, the other induces a general classifier that handles all combinations of classes. Either way, what all these approaches share is the extreme computational intensity resulting from the circumstance that thousands of features are needed to describe each document. It is thus only natural that computational costs constitute a critical criterion to be considered during the selection of the most appropriate technique. Without significant reduction of these costs, full-scale applications in large-scale real-world settings are all but impossible.

3   Problem Statement and Performance Criteria

Let $\mathscr{R}^p$ be an instance space, let $\mathscr{X} \subset \mathscr{R}^p$ be a finite set of documents, and let $\mathscr{Y}$ be a finite set of classes such that each $x_i \in \mathscr{X}$ belongs to its subset, $Y_i \subseteq \mathscr{Y}$. The features describing the documents have been obtained from the relative frequencies of words or terms. Given a set of training examples, $S = \{(x_1, Y_1), \ldots, (x_n, Y_n)\}$, the goal is to find a classifier to carry out the mapping $g : \mathscr{X} \rightarrow 2^{\mathscr{Y}}$ in a way that optimizes classification performance. Moreover, the induction of the classifier has to be accomplished in a realistic time.

To obtain reasonable criteria to measure classification performance, let us start with those employed by the field of *information retrieval* for domains where only two class labels are permitted: positive examples and negative examples. Let us denote by TP (true positives) the number of correctly classified positive examples; by FN (false negatives), the number of positive examples misclassified as negative; by FP (false positives), the number of negative examples misclassified as positive ones; and by TN (true negatives), the number of correctly classified negative examples. Let us now use these four quantities to define *precision, Pr,* and *recall, Re,* by the following simple formulas:

$$Pr = \frac{TP}{TP + FP} \qquad\qquad Re = \frac{TP}{TP + FN} \qquad (1)$$

Observing that the user often wants to maximize both criteria, while balancing their values, [23] proposed to combine *precision* and *recall* in a single formula, F_β, parameterized by the user-specified $\beta \in [0, \infty)$ that quantifies the relative importance ascribed to either criterion:

$$F_\beta = \frac{(\beta^2 + 1) \times Pr \times Re}{\beta^2 \times Pr + Re} \qquad (2)$$

It is easy to see that $\beta > 1$ gives more weight to *recall* and $\beta < 1$ gives more weight to *precision*; that F_β converges to *recall* if $\beta \to \infty$, and to *precision* if $\beta = 0$. The situation where *precision* and *recall* are deemed equally relevant is reflected by the value $\beta = 1$, in which case F_1 degenerates to the following formula:

$$F_1 = \frac{2 \times Pr \times Re}{Pr + Re} \qquad (3)$$

Based on these preliminaries, [31] proposed two alternative ways how to generalize these criteria for domains with multi-label examples: (1) *macro-averaging*, where *precision* and *recall* are first computed for each category and then averaged; and (2) *micro-averaging*, where *precision* and *recall* are obtained by summing over all individual decisions. The formulas are summarized in Table 1 where $Pr_i, Re_i, TP_i, FN_i, FP_i$, and TN_i stand for the *precision, recall,* and the four above-mentioned variables for the i-th class.

Table 1. The macro-averaging and micro-averaging versions of the *precision* and *recall* performance criteria for domains with multi-label examples

	Precision	Recall	F_1
Macro	$Pr^M = \frac{\sum_i^k Pr_i}{k}$	$Re^M = \frac{\sum_i^k Re_i}{k}$	$F_1^M = \frac{\sum_i^k F_{1,i}}{k}$
Micro	$Pr^\mu = \frac{\sum_{i=1}^k TP_i}{\sum_{i=1}^k (TP_i + FP_i)}$	$Re^\mu = \frac{\sum_{i=1}^k TP_i}{\sum_{i=1}^k (TP_i + FN_i)}$	$F_1^\mu = \frac{2 \times Pr^\mu \times Re^\mu}{Pr^\mu + Re^\mu}$

For the sake of completeness, we have to mention that other performance metrics have been recommended (and used) for performance evaluation in multi-label domains, such as *One_error, Coverage, Average Precision, Hamming loss,* and *Ranking loss* [28, 33]. To keep things simple, we will not employ them here, although we did report their use in [24].

4   Baseline Induction Algorithm

The first step in our research was to choose an appropriate Baseline Induction Algorithm (BIA). We were interested in techniques whose induction time grows supralinearly in the number of attributes (because this is what motivated our use of information fusion in the first place), but also technique also had to induce classifiers with high classification performance. Experience of several authors was suggesting the use of the boosting algorithm [6, 8, 28] that had been used in text categorization tasks before. Our own early experiments [25] indicated that, in the EUROVOC domain, good results might be achieved by the use of *AdaBoost.MH* [27], an extension of the classical boosting approach to multi-label domains. An important aspect of this program is that it returns not only class labels, but also confidence in each label. Later, the reader will realize how we exploit this circumstance by our fusion technique.

For the sake of completeness, Figure 2 summarizes the principle of this technique. Let $\mathcal{X}$ be a set of documents, and let $\mathcal{Y}$ be a set of class labels such that each $x \in \mathcal{X}$ is assigned some $Y \subseteq \mathcal{Y}$. The algorithm maintains a distribution, D_t, over the training examples and labels. The distribution, uniform at the beginning, is updated after each boosting round (indexed by t). The learner selects the next training subset randomly according to D_t, and induces from this training set the next "weak hypothesis," h_t, that in our experiments has the form of a single-attribute test. After the training session, *AdaBoost.MH* outputs a ranking function, $f : \mathcal{X} \times \mathcal{Y} \to \mathcal{R}$ (where $\mathcal{R}$ is the set of real values). Label l is assigned to x only if $f(x,l) > 0$.

This algorithm (together with some others) is included in the software package *Boos-Texter* implemented by [27]. Being primarily interested in the fusion mechanisms, we will regard *AdaBoost.MH* as a black box invoked by a master algorithm that will run this BIA on specially designed versions of the training data. The reader will kindly

Given: $(x_1, Y_1), \ldots, (x_n, Y_n)$ where $x_i \in \mathcal{X}$, $Y_i \subseteq \mathcal{Y}$
Initialize: $D_1(i,l) = 1/(nk)$, where k is the size of $\mathcal{Y}$

For $t = 1, \ldots, T$:
1. Pass distribution D_t to a "weak learner"
2. Induce a "weak hypothesis," $h_t : \mathcal{X} \times \mathcal{Y} \to R$, in the form of a single-attribute test
3. Choose $\alpha_t \in R$ and update the distribution:

$$D_{t+1}(i,l) = \frac{D_t(i,l) \exp(-\alpha_t Y_i\{l\} h_t(x_i,l))}{Z_t}$$

where Z_t is a normalization factor and α_t is a parameter
Obtain the final hypothesis by a weighted voting over the weak hypotheses:

$$f(x,l) = \sum_{t=1}^{T} \alpha_t h_t(x,l).$$

Fig. 2. The essence of the *AdaBoost.MH* algorithm

remember that each subclassifier induced by this BIA returns for each document x, and for each class label l, a number $f(x,l) \in (-\infty, \infty)$ that quantifies the subclassifier's confidence that x should be labeled with l.

5   Information Fusion

A subclassifier induced by our BIA returns for each document x, and for each class label l, a number $f(x,l) \in (-\infty, \infty)$ that quantifies the subclassifier's confidence that x should be labeled with l. Receiving these values for each document-label pair from each subclassifier, we need a mechanism to fuse all these "opinions" into one final decision about the classes with which x is to be labeled.

Figure 3 depicts the overall schema of our system from the classification point of view. The classification process is triggered by the submission of a feature vector that describes a document whose class labels are to be establish. Each the subclassifiers induced by the BIA works with a different subset of these features, and returns a different set of (ranking) functions, $f(x,l)$. From these values, the system computes so-called basic-belief assignments (BBA) the nature of which we will explain shortly. Finally, the BBAs are fused by a formula derived from the Dempster-Shafer theory of evidence combination.

To make the paper self-contained, we first briefly summarize the relevant notion from the Dempster-Shafer theory, and only then proceed to the description of our own solution.

5.1   Elements of the Dempster-Shafer Theory

Consider a set of mutually exclusive and exhaustive propositions, $\Theta = \{\theta_1, \ldots, \theta_k\}$, referred to as the *frame of discernment* (FoD). In our context, θ_i states that "document x belongs to class θ_i." Dempster-Shafer Theory (DST) operates with the *Basic Belief Assignment* (BBA) that assigns to any set, $A \subseteq \Theta$, a numeric value $m(A) \in [0,1]$, called a *mass function*, that quantifies the evidence that supports the proposition that the given document belongs to A and only A. The mass function has to satisfy the following conditions [29]:

$$m(\emptyset) \quad = \quad 0; \qquad\qquad \sum_{A \subseteq \Theta} m(A) = 1 \qquad\qquad (4)$$

Any A such that $m(A) > 0$ is called a *focal element*. If $\bar{A}$ is the complement of A, then $m(A) + m(\bar{A}) \leq 1$. A *belief function*, $Bel(A) \in [0,1]$, assigns to every nonempty subset $A \subseteq \Theta$ the degree of support for the claim that the document's classes are all contained in A. This is why the system's belief in A is calculated as the sum of the masses of all A's subsets:

$$Bel(A) \quad = \quad \sum_{B \subseteq A} m(B) \qquad\qquad (5)$$

Note that $Bel(A) = m(A)$ if A is a singleton. The DST rule of combination makes it possible to arrive at a new BBA by fusing the information from several BBAs that

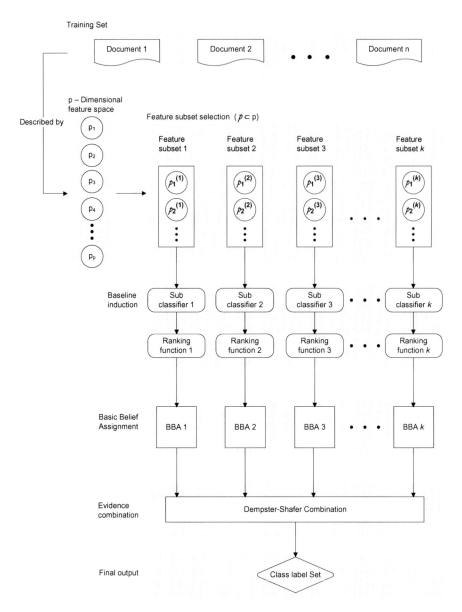

Fig. 3. Classifier induction system where different feature subsets are used in generating subclassifiers induced by a baseline induction algorithm and Dempster-Shafer theory is used as a master algorithm to combine subclassifiers' evidence

span the same FoD. Let $\Im(\Theta) = \{A \subseteq \Theta : m(A) > 0\}$ be the set of focal elements within a given body of evidence, BoE. Consider two bodies of evidence, $\{\Theta, \Im_1, m_1\}$ and $\{\Theta, \Im_2, m_2\}$. That is, m_1 and m_2 are BBAs for the same Θ with focal elements $\Im F_1$ and $\Im F_2$, respectively. The normalization constant,

$$K_{12} = 1 - \sum_{\substack{B_i \in \mathfrak{I}_1; C_j \in \mathfrak{I}_2; \\ B_i \cap C_j = \emptyset}} m_1(B_i) m_2(C_j),$$

tells us how much m_1 and m_2 are conflicting. If $K_{12} > 0$, then the two BoEs are said to be compatible, and the two masses, m_1 and m_2, can be combined to obtain the overall m that for any $A \neq \emptyset$ is calculated as follows:

$$m(A) \equiv (m_1 \oplus m_2)(A)$$

$$= \sum_{\substack{B_i \in \mathfrak{I}_1; C_j \in \mathfrak{I}_2; \\ B_i \cap C_j = A}} m_1(B_i) m_2(C_j) \div K_{12}.$$

The idea of applying DST to classification problems is not new. [2] reported their experimental evaluation of five ensemble methods from which the one based on DST gave encouraging results under realistic circumstances. [1] then described their own DST-based classifier combination technique that outperformed other classifier combination methods on three different domains. In the experiments of [3], the performance of the best combination of different classifiers on ten benchmark domains slightly outperformed the best individual method. A successful attempt to combine—with the help of DST—sets of rules used in text categorization was also reported by [4]. However, all these approaches focus on domains with single-label examples. The contribution of our paper is the use of DST-fusion in domains with multi-label examples.

5.2   Fusion Mechanisms

Each run of the BIA induces a subclassifier that for document x and class label l returns the value of a function, $f(x,l) \in (-\infty, \infty)$, that quantifies the subclassifier's confidence in l—higher $f(x,l)$ indicates higher confidence. To combine the returned labels and confidence values of multiple subclassifiers, we need an appropriate fusion mechanism.

To be able to use DST, we first have to convert $f(x,l)$ into what can be treated as a mass function that (the reader will recall) must have values from the interval $[0,1]$. This conversion can only be accomplished heuristically because literature on uncertainty processing does not offer any straightforward analytical technique.[1]

Our proposed solution is summarized in Figure 4: for a given l, it calculates the difference between $f(x,l)$ and the minimum confidence of the subclassifier in any label, and then divides the result by the maximum difference observed in the outputs of this subclassifier. The idea is to make sure that the converted values can be treated as degrees of belief where values close to 1 indicate strong belief in l, and values close to 0 indicate strong disbelief in l. In the same spirit, if a subclassifier returns $f(x,l) > 0$ for all class labels, then all converted values should fall into the interval $[0.5,1]$ (case "a"), whereas if the subclassifier returns $f(x,l) < 0$ for all class labels, then all converted values should fall into the interval $[0,0.5]$ (case "b"). But of course, most common is the case where the subclassifier returns $f(x,l) > 0$ for some labels and $f(x,l) < 0$ for others (case "c").

---

[1] See the extensive discussion in the chapter "Attribute Data Fusion" of the monograph [5].

1. When $f(x,l) \geq 0$ for all labels,

$$f^*(x,l) = 0.5 + \frac{f(x,l) - \min_{l'} f(x,l')}{2 \times (\max_{l'} f(x,l') - \min_{l'} f(x,l'))}$$

2. When $f(x,l) < 0$ for all labels,

$$f^*(x,l) = \frac{f(x,l) - \min_{l'} f(x,l')}{2 \times (\max_{l'} f(x,l') - \min_{l'} f(x,l'))}$$

3. When some $f(x,l) \geq 0$ and some $f(x,l) < 0$,

$$f^*(x,l) = \begin{cases} -\dfrac{f(x,l) - \min\limits_{l'} f(x,l')}{2 \times \min\limits_{l'} f(x,l')} & \text{,if } f(x,l) < 0 \\[3mm] 0.5 + \dfrac{f(x,l)}{2 \times \max\limits_{l'} f(x,l')} & \text{,if } f(x,l) \geq 0 \end{cases}$$

Fig. 4. Converting the ranking function of each label to a degree of belief

The next step is summarized in Figure 5: the converted confidence values are used in the calculations of basic belief assignments (BBAs) for different class labels. Let us denote by $\bar{l}$ the negation of l (meaning that the document is *not* to be labeled with l). From the DST perspective, a classifier can for each class and each document return one out of four different labellings: $\Theta = \{l, \bar{l}, [l, \bar{l}], \emptyset\}$ (recall that $[l, \bar{l}]$ is in the DST interpreted as meaning that both l and $\bar{l}$ are possible). The mass has to be calculated for each of these outcomes separately. The formal proof that the masses thus calculated satisfy the condition from Equation 4 was presented in [24].

Finally, the formulas obtained by the procedure from Figure 6 employ the Dempster-Shafer rule of combination to fuse the mass values returned by the individual subclassifiers for each of the four possibilities of each class label. Once this step has been accomplished, the beliefs in the individual classes are calculated as $Bel(l) = m(l)$ and $Bel(\bar{l}) = m(\bar{l})$. Note that l and $\bar{l}$ are singletons. The following classification rule is then used:

Assign label l to the document if $Bel(l) > Bel(\bar{l})$.

Put another way, the system assigns label l to the document if its belief in l is higher than its belief in $\bar{l}$. In the rest of the paper, the approach just described will be referred to as *DST-Fusion*.

Weighted sum. Let us now briefly mention that another recently published fusion method relies on a so-called "weighted sum" approach [14], where the confidence scores are weighed by the accuracies of the individual subclassifiers. To be more concrete, let $f_i^*(x,l)$ denote the normalized confidence score (see above) that subclassifier i assigns

1. Compute classification error probabilities for each label.

$$P(\bar{l}|l) = \frac{\text{\#of class } l \text{ documents that are not classified as class } l}{\text{\#of class } l \text{ documents}}$$

$$P(l|\bar{l}) = \frac{\text{\#of documents not in class } l \text{ that are classified as class } l}{\text{\#of documents that are not class } l}$$

2. Estimate prior probability of each label.

$$P(l) = \frac{\text{\#of class } l \text{ documents}}{\text{Total\#of documents}}$$

$$P(\bar{l}) = 1 - P(l)$$

3. Compute a set of BBAs associated with each label.

$$m(\Theta) = P(l|\bar{l})P(\bar{l}) + P(\bar{l}|l)P(l) \tag{6}$$

$$m(l) = f^*(x,l) * \{1 - P(l|\bar{l})P(\bar{l}) - P(\bar{l}|l)P(l)\} \tag{7}$$

$$m(\bar{l}) = f^*(x,\bar{l}) * \{1 - P(l|\bar{l})P(\bar{l}) - P(\bar{l}|l)P(l)\} \tag{8}$$

$$= (1 - f^*(x,l)) * \{1 - P(l|\bar{l})P(\bar{l}) - P(\bar{l}|l)P(l)\} \tag{9}$$

Fig. 5. Computing the Dempster-Shafer masses for class labels

For a given l, combine two sets of BBAs, m_1 and m_2, associated with l:

$$m(A) = \begin{cases} 0, & \text{if } A = \emptyset \\[2mm] (m_1(l)m_2(l) + m_1(l)m_2(\Theta) \\ +m_1(\Theta)m_2(l)) \div K_{12}, & \text{if } A = l \\[2mm] (m_1(\bar{l})m_2(\bar{l}) + m_1(\bar{l})m_2(\Theta) \\ +m_1(\Theta)m_2(\bar{l})) \div K_{12}, & \text{if } A = \bar{l} \\[2mm] (m_1(\Theta)m_2(\Theta)) \div K_{12}, & \text{if } A = \Theta \end{cases}$$

where

$$K_{12} = 1 - \{m_1(l)m_2(\bar{l}) + m_1(\bar{l})m_2(l)\}.$$

For more classifiers, continue combining the resulting BBA, m, from previous combination with BBA from the next using the same formula.

Fig. 6. Evidence combination

to class label l for example x. The accuracy of subclassifier i in predicting class l is obtained by the following formula:

$$Acc_i(l) = P(l|l)P(l) + P(\bar{l}|\bar{l})P(\bar{l}).$$

For the case of N subclassifiers, the weighted-sum approach assigns to document x label l if the following condition is satisfied:

$$w(l) = \frac{\sum\limits_{i=1}^{N} Acc_i(l) * f_i^*(x,l)}{\sum\limits_{i=1}^{N} Acc_i(l)} > 0.5.$$

6   Experiments

The size of the original database makes systematic performance evaluation rather impractical. Given that each single experimental run on the complete data takes many days, it is impossible to go through the hundreds of experiments needed for statistically justified conclusions. As the next-best solution, we decided to work with a simplified database: 10,000 documents described by 4,000 features, and labeled with only the 30 classes. We used all class labels from the top-most level of the class hierarchy, and we selected the features by the *Document Frequency* criterion, an unsupervised feature selection method recommended for text categorization by [32]—in principle, we picked randomly 4,000 features from those that appeared in more than 50 documents.

We will report two sets of experiments. The first was designed to tell us more about the behavior of DST-fusion under diverse circumstances: different numbers of features used by the individual subclassifiers, and different numbers of subclassifiers. In particular, we wanted to know how the classification performance, as well as computational costs, vary with the changing values of these parameters. In the second set of experiments, meant to place DST-fusion in the context of related research, we compared our technique with a completely different approach known from recent literature: the *Multi-Label C4.5* [7]. We suspected that the superiority of the one or the other may depend on the number of features used to describe the document to be classified.

6.1   Fusion's Performance

The basic motivation for the subclassifier combination was to reduce computational costs. At the same time, we needed to make sure that the speed-up has not been achieved at the cost of seriously curtailed classification performance. Finally, we wanted to learn how DST-fusion's behavior was affected by different sizes of the feature subsets, and we needed to ascertain whether some general guideline could be devised for the choice of an optimum number of subclassifiers.

In the first experiment, we randomly selected from the (reduced) feature set 500, 1,000, 2,000, and 4,000 features, respectively. For each of these selections, we created five equally-sized overlapping feature subsets, each with 25% randomly selected features, thus obtaining sets of 125, 250, 500, and 1,000 features, respectively. Using the training examples described by these feature vectors, we compared the performance of *AdaBoost.MH* when run on all features (in our graphs, this case is labeled with "NoFusion") with that of *AdaBoost.MH* when run on feature subsets with subsequent fusion. The number of boosting rounds used by *AdaBoost.MH* was set at 10% of the number of

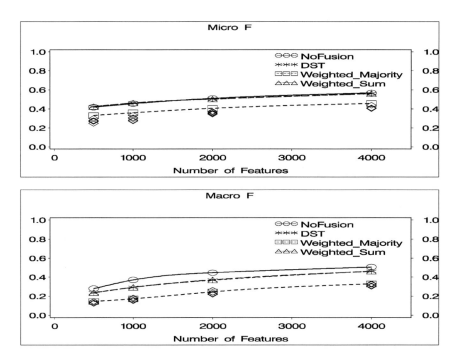

Fig. 7. Average classification performance of fused sublcassifiers as compared to the original "NoFusion" method: varying the number of features

features (e.g., 50 rounds for the experiment involving 500 features). Apart from the pure DST-Fusion, we also worked with the "weighted sum" mechanism (both approaches were described in Section 5.2). For the sake of comparison, we experimented also with two other classifier combination methods: plain majority voting, and weighted majority voting. In the latter, the vote of each classifier supporting class l was weighted by the accuracy, $P(l|l)P(l) + P(\bar{l}|\bar{l})P(\bar{l})$. All graphs in this section were obtained as averages from 5-fold crossvalidation.

The graphs in Figure 7 compare the performance of the "NoFusion" case with that of DST-Fusion, weighted majority voting, and "weighted sum." We omitted plain majority voting because it almost always underperformed the other approaches; therefore, the inclusion of its chart would only reduce clarity. As for performance criteria, we used the micro- and macro-averaging of *precision*, *recall*, and F_1. The graphs show that the performances of DST-Fusion and "weighted sum" are comparable to "NoFusion" in terms of micro-averaging F_1.[2] A closer look reveals that DST-Fusion and "weighted sum" have the lowest micro-averaging *precision* and the highest micro-averaging *recall* among the methods, while the two majority-voting schemes have poor micro-averaging

---

[2] When subjected to the t-test, the differences in the performance of DST-Fusion and "weighted sum" turn out to be statistically insignificant. However, this does not mean that they both tended to label the documents with the same classes. We observed that each of them misclassified different documents, but they both committed about the same number of errors.

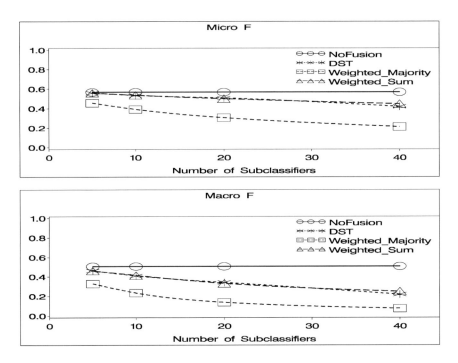

Fig. 8. Average classification performance of fusion methods as compared to the original "NoFusion" method: varying the number of subclassifiers

recall and F_1. In terms of macro-averaging, "NoFusion" outperforms the other methods. In a separate set of experiments (not detailed here) we ascertained that that DST-Fusion, "weighted sum," and "NoFusion" outperformed each single subclassifier, but the weighted majority voting occasionally failed to outperform the best subclassifier.

The relatively small reduction of DST-fusion's classification performance as compared to the "NoFusion" case is particularly encouraging in view of the convincing savings in computation time. The latter are plotted in the left part of Figure 9 that shows the exponential growth in the time needed to induce the "NoFusion" classifier as compared to the (more or less) linear growth in the time needed to induce the DST-fusion system. For the case of 4,000 features, the "NoFusion" approach needs six times more time than DST-fusion.

In the second round of experiments, we wanted to know how the classification performance varied with the number of subclassifiers. To this end, we divided the 4,000 features in four different ways into overlapping subsets: 5 sets of 1,000 features each, 10 sets of 500 features each, 20 sets of 250 features each, and 40 sets of 125 features each. For each of these divisions, we compared the performance of DST-Fusion to that of a single classifier induced from all 4,000 features. As before, the number of boosting rounds was set to 10% of the number of features. The results are shown in the graphs in Figure 8. The reader can see that the performance of DST-Fusion and "weighted sum" along these criteria deteriorated with the growing number of subclassifiers,

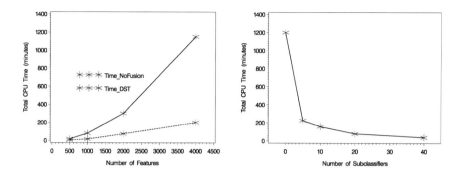

Fig. 9. CPU time as plotted against the growing number of features (left) and growing number of subclassifiers (right)

although they still clearly outperformed the other voting methods. We did not observe any discernible difference between DST-Fusion and "weighted sum." Moving from the "NoFusion" case to the extreme case of the fusion of 40 subclassifiers incurred performance loss of 24%, clearly indicating that the use of 125 features was insufficient for the induction of useful subclassifiers.

The right part of Figure 9 shows that the induction time of DST-fusion steeply drops with the growing number of subclassifiers. Evidently, the trade-off between classification performance and computation costs deserves closer attention in our future research. the concrete choice of how many subclassifiers are to be induced (and combined) will probably depend on the specific needs of a concrete application.

6.2   Comparing BoosTexter and Multi-Label C4.5

Our final task was to compare the behavior of the "fused" *AdaBoost.MH* with that of the perhaps more traditional *Multi-Label C4.5*, a program that generalizes classical induction of decision trees into the multi-label case. We will be interested in two performance criteria: classification accuracy, and induction time; the former was evaluated in terms of precision and recall; the latter, in minutes of CPU time. In our experiments, we wanted to know how the classification performance and computational costs change as we vary the number of features used to describe the individual documents.

Again, we ran both programs on a simplified version of the database, consisting of 10,000 documents described by 4,000 features. Only the 30 top-level class labels we used. To achieve acceptable statistical reliability of the results, we followed the methodology of 5-fold crossvalidation. This means that in each run a classifier was induced from 8,000 documents and then tested on the remaining 2,000 documents. We repeated this experiment for different numbers of features, running from 500 to all 4,000 features used in the "reduced" database.

Figure 10 summarizes the results of these experiments. The first thing to observe is that *AdaBoost.MH* systematically outperformed *Multi-Label C4.5* in terms of both Macro-F_1 and Micro-F_1, especially when a larger number of features was employed.

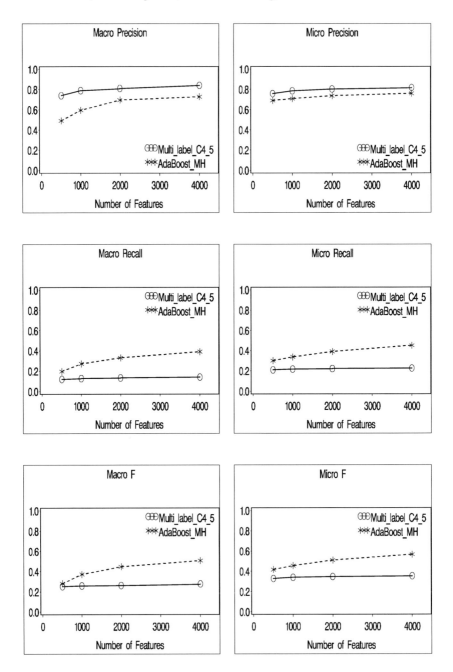

Fig. 10. Classification performance of the two approaches along different criteria, as measured on independent testing data. The *AdaBoost.MH* is labeled by the name of the entire package of which it is a part, *BoosTexter*. The reader can see that *Multi-Label C4.5* is better along *precision*, whereas *AdaBoost.MH* is better along *recall* and F_1. *AdaBoost.MH* seems to gain an edge with the growing number of features employed.

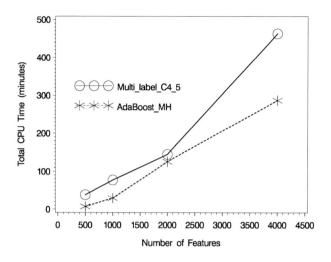

Fig. 11. Induction time measured in minutes. The time indicated in the graph is always the sum total of all five runs of the 5-fold crossvalidation procedure.

Perhaps more interestingly, a closer look reveals that each method displayed a somewhat different behavior along the component criteria of F_1: *AdaBoost.MH* turned out to be better in terms of *recall* (both micro and macro), whereas *Multi-Label C4.5* turned out to be better in terms of *precision*, especially in situations where only a relatively small subset of features was used. The experiments seem to indicate that the decision-tree based *Multi-Label C4.5* is able to get the most from even a very small feature set. This may be due to the fact that *AdaBoost.MH* only isolated features or linear combinations of features, whereas decision trees allow for more flexible representation.

The disparate behavior of the two techniques along *precision* and *recall* needs to be properly understood before choosing the induction method. For instance, users of automated recommender systems are discouraged when offered a wrong document, even if this happens very rarely. Ability to minimize such cases is measured by *precision*, and this is why the decision-tree based system will in domains of this kind be preferred. On the other hand, *recall* is important when we want to make sure that all (or almost all) documents of the requested class have been returned. Then, the *recall* criterion will be critical, which means that *AdaBoost* will be given preference. At the same time, we have to be aware of the circumstance that our experiments indicate that a growing number of features seems to mitigate the difference between the two systems' behavior.

The next round experiments was designed to illustrate the computational complexity of the algorithms. The results are summarized by Figure 11 that plots the CPU times consumed by the two programs for growing numbers of features. The reader can see that *Multi-Label C4.5* is clearly more expensive than the competing program: note how fast the costs grow with the increasing number of features. We conclude that the practical utility of multi-label decision trees in the complete EUROVOC domain (with hundreds of thousand of documents and tens of thousands of features) is limited.

7   Conclusion

A popular research strand in machine learning focuses on how to induce a set of subclassifiers—each for a somewhat different version of the training set—and then how to combine ("fuse") their outputs. In this paper, we concentrated on the problems posed by the classification of text documents. Two critical circumstances deserve our attention, in this context: first, text documents are typically described by very long feature vectors; second, each document can belong to two or more classes at the same time. The former aspect motivates our intention to work with several features subsets, as it is common in certain version of the boosting algorithm. The second aspect necessitates a somewhat more sophisticated fusion mechanism than those that have been common in the machine learning literature.

In our proposed technique, we used a well-known boosting algorithm, *AdaBoost.MH*, as a "baseline induction algorithm" to be used for the induction of a set of subclassifiers, each from the same training set that, however, uses a different feature subset. As for the fusion, we did consider the use of classical voting schemes, but—feeling that these schemes might not give full justice to all the information (such as confidence values) returned by the subclassifiers—we developed our own fusion method around the principles of the Dempster-Shafer Theory. We call this technique *DST-fusion*.

Experiments with a simplified version of our application domain indicate that (1) DST-fusion can lead to impressive savings in the computational time without seriously impairing the classification performance, and (2) that DST-Fusion and "weighted sum" systematically outperformed the more traditional methods of plain voting and weighted majority voting. We did not observe statistically significant difference between the performance of *DST-Fusion* and "weighted sum," but since *DST-Fusion* is easier to implement, we may be inclined to favor the "weighted sum" approach. Yet, closer inspection (not detailed in this paper) revealed that, although each of the two winning techniques committed about the same number of errors, each erred on different documents. In this sense, we think it premature to discard one in favor of the other. More systematic analysis is needed.

When comparing *DST-Fusion* with a more traditional approach, we observed that, most of the time, our approach compares favorably with one based on induction of decision trees, namely, the program known as *Multi-Label C4.5*. The exception is the case when the user wants to make sure that the vast majority of the returned documents are relevant to the query, even if many other relevant documents have been overlooked. Then, *Multi-Label C4.5* might be a better choice. Even so, the high computational costs incurred in similar domains by decision-tree induction represent a major detriment.

Acknowledgements. The research was partly supported by the NSF grant IIS-0513702.

References

[1] Al-Ani, A., Deriche, M.: A new technique for combining multiple classifiers using the Dempster-Shafer theory of evidence. Journal of Artificial Intelligence Research 17, 333–361 (2002)

[2] Bahler, D., Navarro, L.: Methods for combining heterogeneous sets of classifiers. In: Proc. Natl. Conf. on Artificial Intelligence (AAAI) Workshop on New Research Problems for Machine Learning (2000), citeseer.ist.psu.edu/470241.html

[3] Bi, Y., Bell, D., Wang, H., Guo, G., Greer, K.: Combining multiple classifiers using Dempster's rule of combination for text categorization. In: Torra, V., Narukawa, Y. (eds.) MDAI 2004. LNCS (LNAI), vol. 3131, pp. 127–138. Springer, Heidelberg (2004)

[4] Bi, Y., McClean, S., Anderson, T.: Improving classification decisions by multiple knowledge. In: Proc. IEEE Int'l Conf. on Tools with Artificial Intelligence (ICTAI 2005), pp. 340–347 (2005)

[5] Blackman, S., Popoli, R.: Design and Analysis of Modern Tracking Systems. Artech House, Norwood (1999)

[6] Chen, J., Zhou, X., Wu, Z.: A multi-label Chinese text categorization system based on boosting algorithm. In: Proc. IEEE Int'l Conf. on Computer and Information Technology (CIT 2004), pp. 1153–1158 (2004)

[7] Clare, A., King, R.D.: Knowledge discovery in multi-label phenotype data. In: Siebes, A., De Raedt, L. (eds.) PKDD 2001. LNCS (LNAI), vol. 2168, p. 42. Springer, Heidelberg (2001)

[8] Diao, L., Hu, K., Lu, Y., Shi, C.: Boosting simple decision trees with Bayesian learning for text categorization. In: Proc. World Congress on Intelligent Control and Automation, Shanghai, P.R.China, pp. 321–325 (2002)

[9] European Communities (2005), http://europa.eu.int/celex/eurovoc

[10] Freund, Y., Mason, L.: The alternating decision tree learning algorithm. In: Proc. Int'l Conf. on Machine Learning (ICML 1999), pp. 124–133. Morgan Kaufmann, San Francisco (1999), citeseer.ist.psu.edu/freund99alternating.html

[11] Freund, Y., Schapire, R.E.: Experiments with a new boosting algorithm. In: Proc. Int'l Conf. on Machine Learning (ICML 1996), pp. 148–156 (1996), citeseer.ist.psu.edu/freund96experiments.html

[12] Freund, Y., Schapire, R.E.: A decision-theoretic generalization of on-line learning and an application to boosting. Journal of Computer and System Sciences 55(1), 119–139 (1997)

[13] Friedman, N., Geiger, D., Goldszmidt, M.: Bayesian network classifiers. Machine Learning 29(2-3), 131–163 (1997), citeseer.ist.psu.edu/friedman97bayesian.html

[14] Fürnkranz, J.: Hyperlink ensembles: A case study in hypertext classification. Information Fusion 3(4), 299–312 (2002), citeseer.ist.psu.edu/578531.html

[15] Gao, S., Wu, W., Lee, C.H., Chua, T.S.: A MFoM learning approach to robust multiclass multi-label text categorization. In: Proc. Int'l Conf. on Machine Learning (ICML 2004), pp. 329–336 (2004)

[16] Godbole, S., Sarawagi, S.: Discriminative methods for multi-labeled classification. In: Dai, H., Srikant, R., Zhang, C. (eds.) PAKDD 2004. LNCS (LNAI), vol. 3056, pp. 22–30. Springer, Heidelberg (2004)

[17] Institutional Exploratory Research Project JRC – IPSC (2005), http://www.jrc.cec.eu.int/langtech/eurovoc.html

[18] Joachims, T.: Text categorization with support vector machines: learning with many relevant features. In: Nédellec, C., Rouveirol, C. (eds.) ECML 1998. LNCS, vol. 1398, pp. 137–142. Springer, Heidelberg (1998), citeseer.ist.psu.edu/article/joachims98text.html

[19] Kwok, J.T.: Automated text categorization using support vector machine. In: Proc. Int'l Conf. on Neural Information Processing (ICONIP 1998), Kitakyushu, JP, pp. 347–351 (1998), citeseer.ist.psu.edu/kwok98automated.html

[20] Langley, P., Iba, W., Thompson, K.: An analysis of Bayesian classifiers. In: Natl. Conf. on Artificial Intelligence, pp. 223–228 (1992),
`citeseer.ist.psu.edu/article/langley92analysis.html`

[21] Li, B., Lu, Q., Yu, S.: An adaptive k-nearest neighbor text categorization strategy. ACM Trans. on Asian Language Information Processing (TALIP) 3, 215–226 (2004)

[22] McCallum, A., Nigam, K.: A comparison of event models for naive Bayes text classification. In: Proc. Workshop on Learning for Text Categorization, AAAI 1998 (1998),
`citeseer.ist.psu.edu/mccallum98comparison.html`

[23] van Rijsbergen, C.J.: Information Retrieval, 2nd edn. Butterworths, London (1979),
`http://www.dcs.gla.ac.uk/~iain/keith/index.htm`

[24] Sarinnapakorn, K., Kubat, M.: Combining subclassifiers in text categorization: A dst-based solution and a case study. IEEE Transactions on Knowledge and Data Engineering 19(12), 1638–1651 (2007)

[25] Sarinnapakorn, K., Kubat, M.: Induction from multilabel examples in information retrieval systems. Applied Artificial Intelligence 22(5), 407–432 (2008)

[26] Schapire, R.E.: The strength of weak learnability. Machine Learning 5(2), 197–227 (1990)

[27] Schapire, R.E., Singer, Y.: Improved boosting using confidence-rated predictions. Machine Learning 37(3), 297–336 (1999),
`citeseer.ist.psu.edu/schapire99improved.html`

[28] Schapire, R.E., Singer, Y.: BoosTexter: A boosting-based system for text categorization. Machine Learning 39(2/3), 135–168 (2000),
`citeseer.ist.psu.edu/schapire00boostexter.html`

[29] Shafer, G.: A mathematical theory of evidence. Princeton University Press, Princeton (1976)

[30] Vilar, D., Castro, M.J., Sanchis, E.: Multi-label text classification using multinomial models. In: Vicedo, J.L., Martínez-Barco, P., Muñoz, R., Saiz Noeda, M. (eds.) EsTAL 2004. LNCS (LNAI), vol. 3230, pp. 220–230. Springer, Heidelberg (2004)

[31] Yang, Y.: An evaluation of statistical approaches to text categorization. Information Retrieval 1(1/2), 69–90 (1999),
`citeseer.ist.psu.edu/article/yang98evaluation.html`

[32] Yang, Y., Pedersen, J.O.: A comparative study on feature selection in text categorization. In: Fisher, D.H. (ed.) Proceedings of ICML1997, 14th International Conference on Machine Learning, pp. 412–420. Morgan Kaufmann Publishers, San Francisco (1997), `citeseer.ist.psu.edu/yang97comparative.html`

[33] Zhang, M.L., Zhou, Z.H.: A k-nearest neighbor based algorithm for multi-label classification. In: The 1st IEEE Int'l Conf. on Granular Computing (GrC 2005), Beijing, China, vol. 2, pp. 718–721 (2005),
`cs.nju.edu.cn/people/zhouzh/zhouzh.files/publication/grc05.pdf`

[34] Zhu, S., Ji, X., Xu, W., Gong, Y.: Multi-labelled classification using maximum entropy method. In: Proc. ACM SIGIR Conf. Research and Development in Information Retrieval, pp. 274–281 (2005)

Cluster-Lift Method for Mapping Research Activities over a Concept Tree

Boris Mirkin[1], Susana Nascimento[2], and Luís Moniz Pereira[2]

[1] School of Computer Science, Birkbeck University of London,
London, UK WC1E 7HX
and
Department of Data Analysis and Artificial Intelligence,
State University–School of Higher Economics, Moscow, Russia
mirkin@dcs.bbk.ac.uk

[2] Computer Science Department and Centre for Artificial Intelligence (CENTRIA)
Faculdade de Ciências e Tecnologia, Universidade Nova de Lisboa,
2829-516 Caparica, Portugal
snt@di.fct.unl.pt, lmp@di.fct.unl.pt

Abstract. The paper builds on the idea by R. Michalski of inferential concept interpretation for knowledge transmutation within a knowledge structure taken here to be a concept tree. We present a method for representing research activities within a research organization by doubly generalizing them. To be specific, we concentrate on the Computer Sciences area represented by the ACM Computing Classification System (ACM-CCS). Our cluster-lift method involves two generalization steps: one on the level of individual activities (clustering) and the other on the concept structure level (lifting). Clusters are extracted from the data on similarity between ACM-CCS topics according to the working in the organization. Lifting leads to conceptual generalization of the clusters in terms of "head subjects" on the upper levels of ACM-CCS accompanied by their gaps and offshoots. A real-world example of the representation is provided.

Keywords: Cluster-lift method, additive clustering, concept generalization, concept tree, knowledge transmutation.

1 Introduction: Inductive Generalization for Concept Interpretation

In his work on inferential learning theory [5,6], R. Michalski pointed out the importance of knowledge transmutation defined as the process of deriving desirable knowledge from a given input and background knowledge. He envisioned that knowledge transmutation can be performed in terms of pairs of operations such as

> selection vs. generation, replication vs. removal, reformulation vs. randomization, abstraction vs. concretion, similization vs. dissimilization, and generalization vs. specialization.

[6], p. 3. In this paper, we would like to draw attention to the possibility of formalizing the generalization step within the framework of knowledge represented by a concept

J. Koronacki et al. (Eds.): Advances in Machine Learning II, SCI 263, pp. 245–257.
springerlink.com   © Springer-Verlag Berlin Heidelberg 2010

tree, such as decision trees advocated by R. Michalski in the framework of conceptual clustering [7,17]. Concept trees currently are well recognized knowledge structures being important part of ontologies, taxonomies and other forms of knowledge representation.

Consider, for example, ACM Computing Classification System (ACM-CCS), a conceptual four-level classification of the Computer Science subject area, built to reflect the vast and changing world of computer oriented writing. This classification was first published in 1982 and then thoroughly revised in 1998, and it is being updated since [1]. ACM-CCS comprises eleven major partitions (first-level subjects):

A. *General Literature*
B. *Hardware*
C. *Computer Systems Organization*
D. *Software*
E. *Data*
F. *Theory of Computation*
G. *Mathematics of Computing*
H. *Information Systems*
I. *Computing Methodologies*
J. *Computer Applications*
K. *Computing Milieux*

These are subdivided into 81 second-level subjects. For example, item *I. Computing Methodologies* consists of eight subjects:

I.0 GENERAL
I.1 SYMBOLIC AND ALGEBRAIC MANIPULATION
I.2 ARTIFICIAL INTELLIGENCE
I.3 COMPUTER GRAPHICS
I.4 IMAGE PROCESSING AND COMPUTER VISION
I.5 PATTERN RECOGNITION
I.6 SIMULATION AND MODELING (G.3)
I.7 DOCUMENT AND TEXT PROCESSING (H.4, H.5)

which are further subdivided into third-layer topics as, for instance, *I.5 PATTERN RECOGNITION* which consists of seven topics:

I.5.0 General
I.5.1 Models
I.5.2 Design Methodology
I.5.3 Clustering
    Algorithms
    Similarity measures
I.5.4 Applications
I.5.5 Implementation (C.3)
I.5.m Miscellaneous

These are further subdivided in unlabeled subtopics such as those two shown for topic *I.5.3 Clustering*.

As can be seen from the examples above, there are a number of collateral links between topics both on the second and the third layers - they are in the parentheses in the ends of some topics such as *I.6*, *I.7*, and *I.5.5* above.

Concept tree structures such as the ACM-CCS are used, mainly, as devices for annotation and search for documents or publications in collections such as that on the ACM portal [1]. However, being adequate domain ontologies, concept trees can and should be used for other tasks as well. For example, the ACM-CCS tree has been applied as:

- A gold standard for ontologies derived by web mining systems such as the CORDER engine [18];
- A device for determining the semantic similarity in information retrieval [9] and e-learning applications [21];
- A device for matching software practitioners' needs and software researchers' activities [2].

Here we concentrate on yet another application of ACM-CCS, mapping research activities in Computer Sciences. However, our method can be utilized in other knowledge domains as well. The method works for any domain if its structure has been represented with a concept tree. We propose the use of concept tree structures for representing activities of research organizations by using a two-stage generalization of individual research topics over the tree topology.

A concept tree such as the ACM-CCS taxonomy can be seen as a representative generic ontology, with its explicitly expressed hierarchical subsumption relation between subject classes along with the collateral relation of association between different nodes. The art of representation of various items on an ontology is of interest in many areas such as text analysis, web mining, bioinformatics and genomics. In web mining, representations are extracted from domain ontologies: the ontologies are used to automatically characterize usage profiles by describing user's interests and preferences for web personalisation [20]. There are also recommender systems for on-line academic research papers [8], which extract user profiles based on an ontology of research topics. In bioinformatics several clustering techniques have been successfully applied in the analysis of gene expression profiles and gene function prediction by incorporating gene ontology information into clustering algorithms [4].

However, this line of thinking has never been applied for representing the activities of research organizations. The very idea of representing the activities of a research organization as a whole may seem rather odd because conventionally it is only the accumulated body of results that does matter in the sciences, and these always have been and still are provided by individual efforts. The assumption of individual research efforts implicitly underlies systems for reviewing and comparing different research departments in countries such as the United Kingdom in which scientific organizations are subject to regular comprehensive review and evaluation practices. The evaluation is based on the analysis of individual researchers' achievements, leaving the portrayal of a general picture to subjective declarations by the departments [14]. Such an evaluation provides for the assessment of relative strengths among different departments, which is good for addressing funding issues. Yet there exists another aspect, that of the integral portrayal rather than comparative analysis of the developments. This aspect is important for decisions regarding long-term or wide-range issues of scientific development such as national planning or addressing the so-called 'South–North divide' between developed and underdeveloped countries. The latter would require comparing between integral systems of scope and capabilities of scientific organizations and university departments in both the South and North (see, for instance, The United Nations Millennium Project task force web-site [19]).

Representation of activities over the ACM-CCS concept tree can be used for:

1. Overviewing scientific subjects that are being developed in the organization.
2. Positioning the organization over ACM-CCS.
3. Overviewing scientific disciplines being developed in organizations over a country or other territorial unit, with a quantitative assessment of controversial subjects, for example, those in which the level of activity is not sufficient or the level of activities by far exceeds the level of results.
4. Assessing the scientific issues in which the character of activities in organizations does not fit well onto the classification; these can be potentially the growth points or other breakthrough developments.
5. Planning research restructuring and investment.

Similar lists of objectives can be drawn for the analysis of other activities.

2   Cluster – Lift Method

We represent a research organization by clusters of ACM-CCS topics to reflect communalities between activities of members or teams working on these topics. Each of the clusters is mapped to the ACM-CCS tree and then lifted in the tree to express its general tendencies. The clusters are found by analyzing similarities between topics which are derived from either automatic analysis of documents posted on web by the teams or by explicitly surveying the members of the department. The latter option is especially convenient in situations in which the web contents do not properly reflect the developments. If such is the case, a tool for surveying research activities of the members and teams is needed.

Accordingly, this work involves developing:

1. e-screen based ACM-CCS topic surveying device,
2. method for deriving similarity between ACM-CCS topics,
3. method for finding possibly overlapping topic clusters from similarity data, and
4. method for parsimoniously lifting topic clusters on ACM-CCS.

In the following subsections, we describe these four.

2.1   E-Screen Survey Tool

An interactive survey tool has been developed to provide two types of functionality: i) data collection about the research results of individual members, described in terms of the ACM-CCS topics; ii) statistical analysis and visualization of the data and results of the survey. The period of research activities comprises the survey year and the previous four years. This is supplied with interactive "focus + context" navigation functionalities [16]. The respondent is asked to select up to six topics among the leaf nodes of the ACM-CCS tree and assign each with a percentage expressing the proportion of the topic in the total of the respondent's research activity. Figure 1 shows a screenshot of the interface for a respondent who has chosen six ACM-CCS topics during his/her survey session. Another, "research results" form allows to make a more detailed assessment in terms of individual research results of the respondent in categories such as refereed publications, funded projects, and theses supervised.

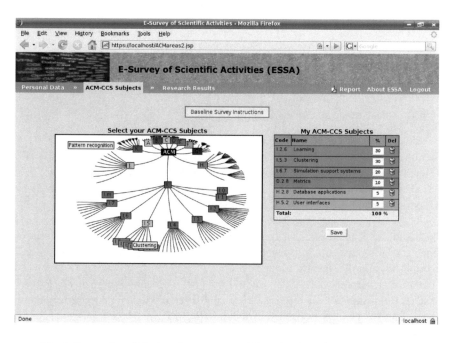

Fig. 1. Screenshot of the interface survey tool for selection of ACM-CCS topics

The leaf nodes of the ACM-CCS tree are populated thus by the respondent supplied weights, which can be interpreted as fuzzy membership degrees of the respondent's activity with respect to ACM-CCS topics.

2.2   Deriving Similarity between ACM-CCS Topics

We define similarity between ACM-CCS topics i and j as the weighted sum of individual similarities. The individual similarity is just the product of weights f_i and f_j assigned by the respondent to the topics. Clearly, topics that are left outside of the individual's list, have zero similarities with other topics.

We assign weights to the surveyed individuals too. An individual's weight is inversely proportional to the number of subjects they selected in the survey. This smoothes out the differences between topic weights imposed by the selection sizes.

It is not difficult to see that the resulting topic-to-topic similarity matrix $A = (a_{ij})$ is positive semidefinite.

2.3   Finding Overlapping Clusters

The topic clusters are to be found over similarity matrix $A = (a_{ij})$ with no conventional mandatory nonoverlapping condition imposed.

We employ the data recovery approach described in [10,11] for the case of crisp clustering and in [13] for the case of fuzzy clustering. We consider only the crisp clustering case in this paper. We find clusters one by one as subsets of ACM-CCS leaf topics S maximizing criterion

$$g(S) = s^T A s / s^T s = a(S)|S|. \tag{1}$$

where

1. $s = (s_i)$ denotes a binary membership vector corresponding to subset S so that $s_i = 1$ if $i \in S$ and $s_i = 0$, otherwise;
2. $a(S)$ is the average similarity a_{ij} within S and
3. $|S|$ is the number of topics in S.

Criterion (1) can be considered as a compromise between two contradicting criteria: (a) maximizing the within-cluster similarity and (b) maximizing the cluster size. When squared, the criterion expresses the proportion of the similarity data scatter, which is taken into account by cluster S according to the data recovery model described in [10,11].

It should be pointed out that this criterion emerges not only in the data recovery framework but it also fits into some other frameworks such as (i) maximum density subgraphs [3] and (ii) spectral clustering [15].

We use a version of ADDI-S algorithm from [10] for locally optimizing criterion (1) that starts from singleton $S = \{i\}$ for a topic $i \in I$. Then the algorithm iteratively finds an entity j to move in or remove from S by maximizing $g(S \pm j)$ where $S \pm j$ stands for $S + j$ if $j \notin S$ or $S - j$ if $j \in S$. It appears that this can be done easily - just by comparing the average similarity between j and S, $a(j, S)$, with the threshold $\pi = a(S)/2$; the greater the difference, the better the j. The process stops when the change of the state of j with respect to S is not beneficial anymore, that is, if π is greater than $a(j, S)$ if $j \notin S$, or smaller than $a(j, S)$ if $j \in S$. In this way, by starting from each $i \in I$, ADDI-S produces a number of potentially overlapping or even coinciding locally optimal clusters S_i – of which that with the highest contribution is taken as the algorithm's output S.

Thus produced S is rather tight because each $j \in S$ has a high degree of similarity with S, greater than half the average similarity within S, and it is also well separated from the rest, because for each entity $j \notin S$, its average similarity with S is less than that.

Next cluster can be found with the same procedure applied to residual similarity matrix $A' = A - a(S)ss^T$. Its contribution to the initial data scatter is computed as g^2 where g is defined in (1) by using the residual matrix A' rather than A. More clusters can be extracted in a similar manner by using residual matrices obtained by subtraction of all the previously found clusters [10].

2.4   Parsimonious Lifting Method

To generalise the main contents of a cluster of topics, we translate it to higher layers of the taxonomy by lifting it according to the principle: if all or almost all children of a node in an upper layer belong to the cluster, then the node itself is taken to represent the cluster on a higher level of ACM-CCS taxonomy. Such a lift can be done differently leading to different portrayals of the cluster on ACM-CCS tree depending on the relative weights of accompanying events, "gaps" and "offshoots", as described below.

A cluster can fit quite well into the classification or not (see Fig. 2), depending on how much its topics are dispersed among the tree nodes.

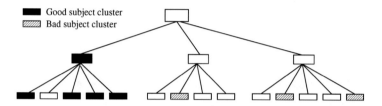

Fig. 2. Two clusters of second-layer topics, presented with checked and diagonal-lined boxes, respectively. The checked box cluster fits within one first-level category (with one gap only), whereas the diagonal line box cluster is dispersed among two categories on the right. The former fits the classification well; the latter does not fit at all.

The best possible fit would be when all topics in the subject cluster fall within a parental node in such a way that all the siblings are covered and no gap occurs. The parental tree node, in this case, can be considered as the head subject of the cluster. A second best case is when one of the children does not belong to the cluster (a gap) or when one of the children is covered by a different parent (an offshoot). A few gaps, that is, head subject's children topics that are not included in the cluster, although diminish the fit, still leave the head subject unchanged. A larger misfit occurs when a cluster is dispersed among two or more head subjects (see Fig. 3). It is not difficult to see that the gaps and offshoots are determined by the head subjects specified in a lift.

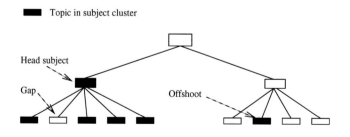
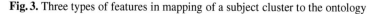

Fig. 3. Three types of features in mapping of a subject cluster to the ontology

The total count of head subjects, gaps and offshoots, each type weighted accordingly, can be used for scoring the extent of the cluster misfit needed for lifting a grouping of research topics over the classification tree as illustrated on Fig. 4. The smaller the score, the more parsimonious the lift and the better the fit. When the topics under consideration relate to deeper levels of classification, such as the third layer of ACM-CCS, the scoring may allow some tradeoffs between different gap-offshoot configurations at different head subject structures. In the case illustrated on Fig. 4, the subject cluster of third-layer topics presented by checked boxes, can be lifted to two head subjects as in (A) or, just one, the upper, category in (B), with the "cost" of three more gap nodes and one offshoot less. Depending on the relative weighting of gaps, offshoots and multiple head subjects, either lifting can minimize the total misfit.

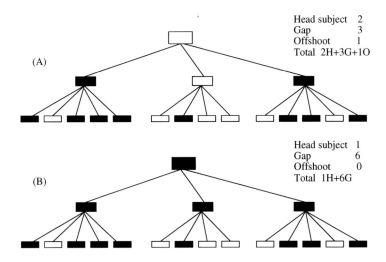

Fig. 4. Tradeoff between different liftings of the same subject cluster: mapping (B) is more parsimonious than (A) if gaps are much cheaper than additional head subjects

Altogether, the set of topic clusters, their head subjects, offshoots and gaps constitutes what can be referred to as a profile of the organization under consideration. Such a representation can be easily accessed and expressed as an aggregate. It can be further elaborated by highlighting those subjects in which members of the organization have been especially successful (i.e., publication in best journals or award) or distinguished by a special feature (i.e., industrial product or inclusion to a teaching program). Multiple head subjects and offshoots, when persist at subject clusters in different organizations, may show some tendencies in the development of the science, that the classification has not taken into account yet.

A parsimonious lifting of a subject cluster can be achieved by recursively building a parsimonious scenario for each node of the ACM-CCS tree based on parsimonious scenarios for its children. In this, we assume that any head subject is automatically present at each of the nodes it covers, unless they are gaps (as presented on Fig. 4 (B)). This assumption allows us to set the algorithm as a recursive procedure.

The procedure determines, at each node of the tree, sets of head gain, gap and off-shoot events to iteratively raise them to those of the parents, under each of two different assumptions that specify the situation at the parental node. One assumption is that the head subject has been inherited at the parental node from its own parent, and the second assumption is that it has not been inherited but gained in the node only. In the latter case the parental node is labeled as a head subject. Consider the parent-children system as shown in Fig. 5, with each node assigned with sets of offshoot, gap and head gain events under the above two inheritance of head subject assumptions.

Let us denote the total number of events, to be minimized, under the inheritance and non-inheritance assumptions by e_i and e_n, respectively. A lifting result at a given node is defined by a triplet of sets (H, G, O), representing the tree nodes at which events of head gains, gaps and offshoots, respectively, have occurred in the subtree rooted at the node. We use (Hi, Gi, Oi) and (Hn, Gn, On) to denote lifting results under the inheritance and non-inheritance assumptions, respectively. The algorithm computes parsimonious scenarios for parental nodes according to the topology of the tree, proceeding from the leaves to the root in the manner which is similar to that described in [12].

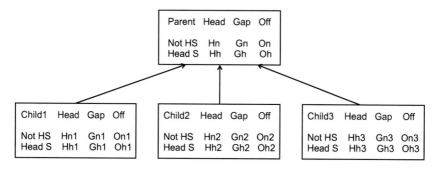

Fig. 5. Events in a parent-children system according to a parsimonious lifting scenario; HS and Head S stand for Head subject

At a leaf node the six sets Hi, Gi, Oi, Hn, Gn and On are empty, except that Hn $=\{S\}$ if the given leaf belongs to topic cluster S or Gi $=\{S\}$ if not. The algorithm then will compute parsimonious scenarios for parental nodes according to the topology of the tree, proceeding from the leaves to the root. Let us, for the sake of simplicity, consider the case when the penalty for an offshoot is taken to be zero while penalties for the head gain and gap events are specified by arbitrary positive h and g, respectively. Then, in a parsimonious scenario, the total score of events, weighted by h and g, can be derived from those of its children (indicated by subscripts 1, 2 and 3 for the case of three children on Fig. 5) as $e_i = \min(e_{n1} + e_{n2} + e_{n3} + g, \ e_{i1} + e_{i2} + e_{i3})$ or $e_n = \min(e_{i1} + e_{i2} + e_{i3} + h, \ e_{n1} + e_{n2} + e_{n3})$, under the inheritance or non-inheritance assumption, respectively; the proof given in [12] for the binary tree case can be easily extended to an arbitrary rooted tree.

3   An Example of Implementation

Let us illustrate the approach by using the data from a survey conducted at the Department of Computer Science, Faculty of Science & Technology, New University of Lisboa (DI-FCT-UNL). The survey involved 49 members of the academic staff of the department.

For simplicity, we use only data of the second level of ACM-CCS, each coded in the format V.v where V=A,B,...,K, and v =1,..,mK, with mK being the number of second level topics. Each member of the department supplied three ACM subjects most relevant to their current research. Altogether, these comprise 26 of the 59 topics at the second level in ACM-CCS. (Two subjects of the second level, General and Miscellaneous, occurred in every first-level division, are omitted because they do not contribute to the representation.)

With the algorithm ADDI-S sequentially applied to the 26×26 similarity matrix, the following six sequentially extracted clusters have been obtained:

1. Cl1 (contribution 27.08%, intensity 2.17), 4 items: *D.3, F.1, F.3, F.4*;
2. Cl2 (contribution 17.34%, intensity 0.52), 12 items: *C.2, D.1, D.2, D.3, D.4, F.3, F.4, H.2, H.3, H.5, I.2, I.6*;
3. Cl3 (contribution 5.13%, intensity 1.33), 3 items: *C.1, C.2, C.3*;
4. Cl4 (contribution 4.42%, intensity 0.36), 9 items: *F.4, G.1, H.2, I.2, I.3, I.4, I.5, I.6, I.7*;
5. Cl5 (contribution 4.03%, intensity 0.65), 5 items: *E.1, F.2, H.2, H.3, H.4*;
6. Cl6 (contribution 4.00%, intensity 0.64), 5 items: *C.4, D.1, D.2, D.4, K.6*.

The next 7th cluster's contribution is just 2.5%, on par with the contributions of each of the 26 individual topics, which justifies halting the process at this point.

The six found clusters lifted in the ACM-CCS are presented on Fig. 6 along with the relevant first-level categories.

The lifting results show the following:

- The department covers, with a few gaps and offshoots, six head subjects shown with pentagons filled in by different patterns;
- The most contributing cluster, with the head subject *F. Theory of Computation*, comprises a very tight group of a few second level topics;
- The next contributing cluster has two, not one, head subjects, *D* and *H*, and offshoots to every other head subject in the consideration, which shows that this cluster currently is the structure underlying the unity of the department;
- Moreover, the two head subjects of this cluster come on top of two other clusters, each pertaining to just one of the head subjects, *D. Software* or *H. Information Systems*. This means that the two-headed cluster signifies a new direction in Computer Sciences, that combines *D* and *H* into a single direction, which seems a feature of the current developments in Computer Sciences indeed; this should eventually get reflected in an update of the ACM classification (by raising *D.2 Software Engineering* to the level 1?);
- There are only three offshoots outside the department's head subjects: *E.1 Data Structures* — from *H. Information Systems*, *G.1 Numerical Analysis* — from *I.*

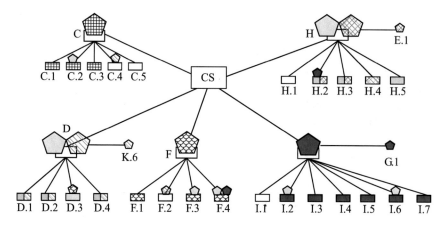

Fig. 6. Six subject clusters in the DI-FCT-UNL represented over the ACM-CCS ontology. Head subjects are shown with differently patterned pentagons. Topic boxes shared by different clusters are split-patterned.

Computing Methodologies, and *K.6 Management of Computing and Information Systems* — from *D. Software*. All three seem natural and should be reflected in the list of collateral links between different parts of the classification tree if supported by similar findings at other departments.

4   Concluding Remarks

We have described a method in the area of knowledge transmutation – for representing aggregated research activities over a concept tree. The method involves two generalization steps: (a) clustering research topics according to their similarities in terms of the efforts by individuals involved, with no relation to the concept tree in question, and (b) generalization of clusters mapped to a concept tree by lifting them to more general categories - this is done over the tree only. Therefore, the generalization steps cover both sides of the representing process.

This work is part of the research project *Computational Ontology Profiling of Scientific Research Organization* (COPSRO), whose main goal is to develop a methodology to represent a Computer Science organization such as a University department over the ACM-CCS classification tree. Such an approach involves the following steps:

1. surveying the members of ACM-CCS topics they are working on; this can be supplemented with indication of the degree of success achieved (publication in a good journal, award, etc.);
2. deriving similarity between ACM-CCS topics resulting from the survey and clustering them;
3. mapping clusters to the ACM-CCS taxonomy and lifting them in a parsimonious way by minimizing the weighted sum of counts of head subjects, gaps and offshoots;

4. aggregating results from different clusters and, potentially, different organizations by means of the taxonomy;
5. interpretation of the results and drawing conclusions.

Current research work includes a survey that is being conducted over several C.S. departments in Universities in Portugal and the U.K., the exploration of fuzzy similarity measures between research topics of the ACM-CCS tree according to the weighted choices of the respondents, and the extension of the additive clustering model to a fuzzy additive version in the framework of the data recovery approach.

In principle, the approach can be extended to other areas of science or engineering, provided that such an area has been systematized into a comprehensive concept tree representation. Potentially, this approach could lead to a useful instrument for visually feasible comprehensive representation of developments in any field of human activities.

Acknowledgements. The authors are grateful to DI-FCT-UNL members that participated in the survey. Igor Guerreiro is acknowledged for developing the interface shown in Fig. 1. This work has been supported by the grant PTDC/EIA/69988/2006 from the Portuguese Foundation for Science & Technology.

References

1. ACM Computing Classification System (1998), http://www.acm.org/about/class/1998 (Cited 9 September 2008)
2. Feather, M., Menzies, T., Connelly, J.: Matching software practitioner needs to researcher activities. In: Proc. of the 10th Asia-Pacific Software Engineering Conference (APSEC 2003), p. 6. IEEE, Los Alamitos (2003)
3. Gallo, G., Grigoriadis, M.D., Tarjan, R.E.: A fast parametric maximum flow algorithm and applications. SIAM Journal on Computing 18(1), 30–55 (1989)
4. Liu, J., Wang, W., Yang, J.: Gene ontology friendly biclustering of expression profiles. In: Proc. of the IEEE Computational Systems Bioinformatics Conference, pp. 436–447. IEEE, Los Alamitos (2004)
5. Michalski, R.S.: Two-tiered concept meaning, inferential matching and conceptual cohesiveness. In: Vosniadou, S., Ortony, A. (eds.) Similarity and Analogical Reasoning. Cambridge University Press, N.Y (1989)
6. Michalski, R.S.: Inferential learning theory: A conceptual framework for characterizing learning processes. Reports on the Machine Learning and Inference Laboratory, MLI 91–9. George Mason University (1991)
7. Michalski, R.S., Stepp, R.E.: Learning from observation: Conceptual clustering. In: Michalski, R.S., Carbonell, J.G., Mitchell, T.M. (eds.) Machine Learning: An Artificial Intelligence Approach, pp. 331–363. Morgan Kauffmann, San Mateo (1983)
8. Middleton, S., Shadbolt, N., Roure, D.: Ontological user representing in recommender systems. ACM Trans. on Inform. Systems 22(1), 54–88 (2004)
9. Miralaei, S., Ghorbani, A.: Category-based similarity algorithm for semantic similarity in multi-agent information sharing systems. In: IEEE/WIC/ACM Int. Conf. on Intelligent Agent Technology, pp. 242–245 (2005), doi:10.1109/IAT.2005.50
10. Mirkin, B.: Additive clustering and qualitative factor analysis methods for similarity matrices. Journal of Classification 4(1), 7–31 (1987)
11. Mirkin, B.: Clustering for Data Mining: A Data Recovery Approach, 276 p. Chapman & Hall /CRC Press, Boca Raton (2005)

12. Mirkin, B., Fenner, T., Galperin, M., Koonin, E.: Algorithms for computing parsimonious evolutionary scenarios for genome evolution, the last universal common ancestor and dominance of horizontal gene transfer in the evolution of prokaryotes. BMC Evolutionary Biology 3(2) (2003), doi:10.1186/1471-2148-3-2

13. Nascimento, S., Mirkin, B., Moura-Pires, F.: Modeling proportional membership in fuzzy clustering. IEEE Transactions on Fuzzy Systems 11(2), 173–186 (2003)

14. RAE2008: Research Assessment Exercise (2008), http://www.rae.ac.uk/ (Cited 9 September 2008)

15. Shi, J., Malik, J.: Normalized cuts and image segmentation. IEEE Transactions on Pattern Analysis and Machine Intelligence 22(8), 888–905 (2000)

16. Spence, R.: Information Visualization, 206 p. Addison-Wesley, ACM Press (2000)

17. Stepp, R., Michalski, R.S.: Conceptual clustering of structured objects: A goal-oriented approach. Artificial Intelligence 28(1), 43–69 (1986)

18. Thorne, C., Zhu, J., Uren, V.: Extracting domain ontologies with CORDER. Tech. Report kmi-05-14. Open University, pp. 1–15 (2005)

19. The United Nations Millennium Project Task Force,
    http://www.cid.harward.edu/cidtech (Cited 1 September 2006)

20. Weiss, S.M., Indurkhya, N., Zhang, T., Damerau, F.J.: Text Mining: Predictive Methods for Analyzing Unstructured Information, 237 p. Springer, Heidelberg (2005)

21. Yang, L., Ball, M., Bhavsar, V., Boley, H.: Weighted partonomy-taxonomy trees with local similarity measures for semantic buyer-seller match-making. Journal of Business and Technology 1(1), 42–52 (2005)

On Concise Representations of Frequent Patterns Admitting Negation

Marzena Kryszkiewicz[1], Henryk Rybiński[1], and Katarzyna Cichoń[2]

[1] Institute of Computer Science, Warsaw University of Technology,
Nowowiejska 15/19, 00-665 Warsaw, Poland
{mkr,hrb}@ii.pw.edu.pl
[2] Institute of Electrical Apparatus, Technical University of Lodz,
Stefanowskiego 18/22, 90-924 Lodz, Poland
cichon@p.lodz.pl

Abstract. The discovery of frequent patterns is one of the most important issues in the data mining area. An extensive research has been carried out for discovering positive patterns, however, very little has been offered for discovering patterns with negation. One of the main difficulties concerning frequent patterns with negation is huge amount of discovered patterns. It exceeds the number of frequent positive patterns by orders of magnitude. The problem can be significantly alleviated by applying concise representations that use generalized disjunctive rules to reason about frequent patterns, both with and without negation. In this paper, we examine three types of generalized disjunction free representations and derive the relationships between them. We also present two variants of algorithms for building such representations. The results obtained on a theoretical basis are verified experimentally.

1 Introduction

The problem of discovering frequent patterns in large databases is nowadays one of the most important issues in the area of data mining. Introduced in [1] for a sales transaction database, it became a standard approach to knowledge discovery in many data discovering tasks, e.g. for telecom providers [31], or text mining tasks such as mining grammatical patterns [11], hierarchical clustering of the sets of documents [32], discovering synonyms or homonyms [34, 39, 40], as well as in medical areas [44]. Frequent patterns, as defined there, are sets of items that co-occur more often than a given threshold. Frequent patterns are commonly used for building association rules. For example, an association rule may state that 30% of sentences with the term "apple" have also a term "iphone". The rule may result from a commercial action of a new Apple product. Patterns and association rules can be generalized by admitting negations. A sample rule with negation could state that 90% of sentences that contain words "apple" and "pie" do not contain the word "motherboard".

Already a number of frequent "positive" patterns is usually huge; admitting negation results in exponential explosion of mined patterns, which makes analysis of the discovered knowledge hardly difficult, if possible at all. It is thus preferable to discover and store a possibly small fraction of patterns, from which one can derive all

J. Koronacki et al. (Eds.): Advances in Machine Learning II, SCI 263, pp. 259–289.
springerlink.com   © Springer-Verlag Berlin Heidelberg 2010

other significant patterns whenever required. The problem of concise representing frequent positive patterns of various types was considered in [5-10, 14-30, 33, 35-38, 41-43, 45]. In particular, the representations have been considered in the context of closed sets [15], generators [15], simple disjunctive rules [5-6, 15], and generalized disjunctive rules [7-9, 17, 19, 21-23, 28]. A disjunction-free set representation, as offered in [6], uses disjunctive rules with one or 2 items in the rule consequent for reasoning about supports of patterns. In [28] and in [19], more concise representations (called GDFGR and GDFSR, respectively) were offered. For reasoning about supports of patterns they use disjunctive rules with an unlimited number of items in the rule consequent. Both representations consist of main components and borders. All elements of GFDGR are generators. It was shown in [19] that the main component and infrequent part of the border of GDFSR are the same as in the case of GDFGR, and the frequent part of the border of GDFSR contains the frequent part of the border of GDFGR and possibly some non-generators. Similar representation, called non-redundant itemsets, was offered independently in [8].

The problem of concise representing of frequent patterns admitting negation was addressed in [24-26]. As in the case of GDFGR and GDFSR, the representations discussed there, were based on generalized disjunctive rules. In [25], a representation (GDFSRN) of all frequent patterns, both with and without negation was presented, as a particular variant of GDFSR. In [24], a refined version of GDFSRN, namely GDFLR, was offered as a lossless representation. Although GDFLR represents both frequent positive patterns and frequent patterns with negation, it consists of only positive patterns. The k-GDFLR representation was offered in [26] as a variant of GDFLR, which represents frequent patterns admitting at most k negated items.

Recently the use of concise representations of the frequent patterns with negation in practical application has started, especially for text mining, and a possibility to dynamically change support in the mining process. For example in [34], it is shown how such representations can be successfully used for discovering dominated meanings from WEB. In this paper, we concentrate on three concise representations, namely GDFSR, GDFSRN and GDFLR, which in the sequel will be called briefly GDF representations. We will analyze relationships between them in detail.

In the literature, the algorithms for computing concise representations for GDFGR (*GDFGR-Apriori*, [17, 28]) and GDFLR (*GDFLR-Apriori* [24], and *GDFLR-SO-Apriori* [27]) have been proposed. As shown in [27], *GDFLR-SO-Apriori* is much faster than *GDFLR-Apriori*, which results from a proposed order of calculating supports of both positive patterns, and the ones containing negations. We provide in the paper *GDFSR-Apriori* as a special case of *GDFLR-Apriori*. In addition, we offer a modification of *GDFSR-Apriori* (called here *GDFSR-SO-Apriori*), and perform a number of experiments to check its efficiency. It turns out to be faster in building the GDFSR representation. In spite of the fact that GDFSR contains only positive patterns, the new algorithm determines supports of not only positive, but also non-positive patterns, which makes possible finding if there are useful generalized disjunctive rules associated with candidate patterns or not. In *GDFSR-SO-Apriori* the same order of calculating supports is applied, as in the case of *GDFLR-SO-Apriori*. A simple way or calculating GDFSRN will be shown.

The layout of the paper is as follows: Subsections 2.1-2.5 recall basic notions of frequent positive patterns and patterns with negation, as well as methods of inferring frequencies (or supports) of patterns from frequencies of other patterns by means of generalized disjunctive rules. In Subsection 2.6, we provide new theorems related to patterns admitting negated items. Section 3 recalls the three concise representations, i.e. GDFSR, GDFSRN, GDFLR, and their properties. In addition, a new theorem related to derivable patterns in the GDFLR representation is provided in Subsection 3.3. In Section 4, we analyze in detail relationships between the representations. All the results presented there are new. In Section 5, we recall the original algorithm for finding the GDFLR representation, and provide algorithms for finding the GDFSR and GDFSRN representations, whereas in Section 6 we present the "support oriented" algorithms: *GDFLR-SO-Apriori*, as well as new algorithms *GDFSR-SO-Apriori* and *GDFSRN-SO-Apriori*. The experimental results are presented in Section 7. The paper ends with the conclusions (Section 8).

2 Basic Notions

2.1 Itemsets, Frequent Itemsets

Let $I = \{i_1, i_2, ..., i_m\}$, $I \neq \emptyset$, be a set of distinct *items*. In the case of a transactional database, a notion of an item corresponds to a sold product, while in the case of a relational database an item will be a pair (*attribute, value*). Any set of items is called an *itemset*. An itemset consisting of k items will be called a *k-itemset*. Let $\mathcal{D}$ be a set of transactions (or tuples, respectively), where each transaction (tuple) T is a subset of I. Without any loss of generality, we will restrict further considerations to transactional databases. *Support* of itemset X is denoted by $sup(X)$ and is defined as the number (or percentage) of transactions in $\mathcal{D}$ that contain X. Itemset X is called *frequent* if its support is greater than some user-defined threshold *minSup*, where $minSup \in [0, |\mathcal{D}|]$. The set of all frequent itemsets will be denoted by $\mathcal{F}$:

$$\mathcal{F} = \{X \subseteq I | \ sup(X) > minSup\}.$$

Property 2.1.1 [2]

a) Let $X, Y \subseteq I$. If $X \subset Y$, then $sup(X) \geq sup(Y)$.
b) If $X \in \mathcal{F}$, then $\forall Y \subset X \ Y \in \mathcal{F}$.
c) If $X \notin \mathcal{F}$, then $\forall Y \supset X \ Y \notin \mathcal{F}$.

2.2 Generalized Disjunctive Sets and Generalized Disjunction-Free Sets

In this section, we recall the notion of generalized disjunctive sets and generalized disjunction-free sets and their properties [17, 28]. Informally speaking, generalized disjunctive sets enable reasoning about supports of their supersets. To the contrary, the supports of generalized disjunction-free sets are not derivable. A key concept in the definitions of both types of sets is the notion of a generalized disjunctive rule.

Let $Z \subseteq I$. $X \rightarrow a_1 \vee ... \vee a_n$ is defined a *generalized disjunctive rule based on* Z (and Z is the *base of* $X \rightarrow a_1 \vee ... \vee a_n$) if $X \subset Z$ and $\{a_1, ..., a_n\} = Z \backslash X$.

In the sequel, $\vee A$, where $A = \{a_1, ..., a_n\}$, will denote $a_1 \vee ... \vee a_n$. One can easily note that $\{Z \backslash A \rightarrow \vee A) | \ \emptyset \neq A \subseteq Z\}$ is the set of all distinct generalized disjunctive rules based on Z. Hence, there are $2^{|Z|-1}$ distinct generalized disjunctive rules based on Z.

Support of $X \rightarrow a_1 \vee \ldots \vee a_n$, is denoted by $sup(X \rightarrow a_1 \vee \ldots \vee a_n)$ and is defined as the number (or percentage) of transactions in $\mathcal{D}$ in which X occurs together with a_1 or a_2, or ... or a_n. Please note that $sup(X \rightarrow a) = sup(X \cup \{a\})$.

Error of $X \rightarrow a_1 \vee \ldots \vee a_n$, is denoted by $err(X \rightarrow a_1 \vee \ldots \vee a_n)$ and is defined as the number (or percentage) of transactions containing X that do not contain any item in $\{a_1, \ldots, a_n\}$; that is,

$$err(X \rightarrow a_1 \vee \ldots \vee a_n) = sup(X) - sup(X \rightarrow a_1 \vee \ldots \vee a_n).$$

$X \rightarrow a_1 \vee \ldots \vee a_n$ is defined a *certain rule* if $err(X \rightarrow a_1 \vee \ldots \vee a_n) = 0$.

Thus, $X \rightarrow a_1 \vee \ldots \vee a_n$ is certain if each transaction containing X contains also a_1 or a_2, or ... or a_n.

An itemset X is defined as a *generalized disjunctive set* if there is a certain generalized disjunctive rule based on X; that is, if $\exists a_1, \ldots, a_n \in X$ such that $X \setminus \{a_1, \ldots, a_n\} \rightarrow a_1 \vee \ldots \vee a_n$ is a certain rule. Otherwise, X is defined a *generalized disjunction-free set*. Let us note that $\varnothing$ is a generalized disjunction-free set.

The set of all generalized disjunction-free sets will be denoted by *GDFree*; that is,

$$GDFree = \{X \subseteq I| \ \forall a_1, \ldots, a_n \in X \ err(X \setminus \{a_1, \ldots, a_n\} \rightarrow a_1 \vee \ldots \vee a_n) > 0, n \geq 1\}.$$

Example 2.2.1. Let us consider database $\mathcal{D}$ from Table 1. Table 2 presents all generalized disjunctive rules based on $\{ab\}$. None of these rules is certain. Hence, $\{ab\}$ is generalized disjunction-free. Table 3 presents all generalized disjunctive rules based on $\{abc\}$, some of which are certain. Thus, $\{abc\}$ is generalized disjunctive. □

Table 1. Sample database $\mathcal{D}$

Id	Transaction
T_1	$\{abce\}$
T_2	$\{abcef\}$
T_3	$\{abceh\}$
T_4	$\{abe\}$
T_5	$\{aceh\}$
T_6	$\{bce\}$
T_7	$\{h\}$

Table 2. Generalized disjunctive rules based on $\{ab\}$

$r: X \rightarrow a_1 \vee \ldots \vee a_n$	$sup(X)$	$sup(r)$	$err(r)$	certain?
$\{a\} \rightarrow b$	5	4	1	no
$\{b\} \rightarrow a$	5	4	1	no
$\varnothing \rightarrow a \vee b$	7	6	1	no

Table 3. Generalized disjunctive rules based on $\{abc\}$

$r: X \rightarrow a_1 \vee \ldots \vee a_n$	$sup(X)$	$sup(r)$	$err(r)$	certain?
$\{ab\} \rightarrow c$	4	3	1	no
$\{ac\} \rightarrow b$	4	3	1	no
$\{bc\} \rightarrow a$	4	3	1	no
$\{a\} \rightarrow b \vee c$	5	5	0	yes
$\{b\} \rightarrow a \vee c$	5	5	0	yes
$\{c\} \rightarrow a \vee b$	5	5	0	yes
$\varnothing \rightarrow a \vee b \vee c$	7	6	1	no

Of course, if each occurrence of X implies the occurrence of item a_1 or ... or a_n, then each occurrence of a superset of X implies the occurrence of a_1 or ... or a_n too.

Property 2.2.1. If $X \rightarrow a_1 \vee ... \vee a_n$ is certain, then $\forall Z \supset X\ Z \rightarrow a_1 \vee ... \vee a_n$ is certain.

The next property is an immediate consequence of Property 2.2.1 and states that supersets of a generalized disjunctive set are generalized disjunctive. In consequence, all subsets of a generalized disjunction-free set are generalized disjunction-free.

Property 2.2.2. Let $X \subseteq I$.

a) If $X \notin GDFree$, then $\forall Y \supset X\ Y \notin GDFree$.
b) If $X \in GDFree$, then $\forall Y \subset X\ Y \in GDFree$.

It was proved in [17, 28] that each certain generalized disjunctive rule determines an errorless method of calculating the support of the base of that rule from the supports of proper subsets of this base. In addition, it was shown there that the information on the support of a generalized disjunctive set X and the supports of all proper subsets of X is sufficient to reconstruct all certain generalized disjunctive rules based on X. As a result, the information on the supports of all generalized disjunction-free sets and the supports of all minimal generalized disjunctive sets is sufficient to derive the supports of all other itemsets in 2^I. Beneath we recall these results related to determining supports of rules and itemsets that will be used in the paper.

Property 2.2.3. Let $X, Y, \{a\} \subset I$, $Y \neq \emptyset$, and $X \rightarrow \vee Y \vee a$ be a generalized disjunctive rule. Then:

$$sup(X \rightarrow \vee Y \vee a) = sup(X \rightarrow \vee Y) + sup(X \rightarrow a) - sup(X \cup \{a\} \rightarrow \vee Y).$$

The main property related to calculating the support of a generalized disjunctive rule, which can be derived inductively from Property 2.2.3, is provided beneath:

Property 2.2.4. Let $X, Y \subset I$ and $X \rightarrow \vee Y$ be a generalized disjunctive rule. Then:

$$sup(X \rightarrow \vee Y) = \Sigma_{\emptyset \neq Z \subseteq Y} (-1)^{|Z|-1} \times sup(X \cup Z).$$

It follows from the presented property that the support of $X \rightarrow \vee Y$ depends on the supports of all itemsets that belong to the left-open interval $(X, X \cup Y]$.

Corollary 2.2.1. Let $X, Y \subset I$ and $X \rightarrow \vee Y$ be a generalized disjunctive rule. The error of $X \rightarrow \vee Y$ is derivable from the supports of itemsets in the interval $[X, X \cup Y]$:

$$err(X \rightarrow \vee Y) = \Sigma_{Z \subseteq Y} (-1)^{|Z|} \times sup(X \cup Z).$$

Therefore, $X \rightarrow \vee Y$ is a certain generalized disjunctive rule iff $\Sigma_{Z \subseteq Y} (-1)^{|Z|} \times sup(X \cup Z)] = 0$. After transforming this equation, one obtains what follows:

Property 2.2.5. Let $X, Y \subset I$ and $X \rightarrow \vee Y$ be a generalized disjunctive rule. Then:

$$err(X \rightarrow \vee Y) = 0 \text{ iff } sup(X \cup Y) = (-1)^{|Y|-1} \times [\Sigma_{Z \subset Y} (-1)^{|Z|} \times sup(X \cup Z)].$$

Thus, if $X \rightarrow \vee Y$ is a certain generalized disjunctive rule, then $sup(X \cup Y)$ is determinable from the supports of itemsets in the right-open interval $[X, X \cup Y)$.

Example 2.2.2. Let the set of all items $I = \{abcefh\}$ and $\mathcal{D}$ be the database from Table 1. Table 4 presents all generalized disjunction-free sets and all minimal generalized disjunctive sets found in this database. Values provided in square brackets in the subscript denote supports of itemsets.

Table 4. *GDFree* and minimal generalized disjunctive sets in $\mathcal{D}$ from Table 1

k	k-itemsets in *GDFree*	minimal generalized disjunctive k-itemsets
3		$\{abc\}_{[3]}$, $\{ach\}_{[2]}$
2	$\{ab\}_{[4]}$, $\{ac\}_{[4]}$, $\{ah\}_{[2]}$, $\{bc\}_{[4]}$, $\{ch\}_{[2]}$, $\{fh\}_{[0]}$	$\{ae\}_{[5]}$, $\{af\}_{[1]}$, $\{be\}_{[5]}$, $\{bf\}_{[1]}$, $\{bh\}_{[1]}$, $\{ce\}_{[5]}$, $\{cf\}_{[1]}$, $\{ef\}_{[1]}$, $\{eh\}_{[2]}$
1	$\{a\}_{[5]}$, $\{b\}_{[5]}$, $\{c\}_{[5]}$, $\{e\}_{[6]}$, $\{f\}_{[1]}$, $\{h\}_{[3]}$	
0	$\varnothing_{[7]}$	

Let us demonstrate how the information on *GDFree* and minimal generalized disjunctive sets can be used to determine supports of remaining itemsets in 2^I. Let $\{ace\}$ be an itemset for which we wish to calculate the support. We note that $\{ace\}$ has subsets $\{ae\}$ and $\{ce\}$ among minimal generalized disjunctive sets (see Table 4). Thus, by Property 2.2.2a, $\{ace\}$ is also generalized disjunctive. By Property 2.2.1, any certain generalized disjunctive rule based on $\{ae\}$ or $\{ce\}$ determines a certain generalized disjunctive rule based on $\{ace\}$. Let us identify an arbitrary certain generalized disjunctive rule based on $\{ae\}$. The only rules based on $\{ae\}$ are as follows: $\{a\}\rightarrow e$, $\{e\}\rightarrow a$, and $\varnothing\rightarrow a \vee e$. Let us start with the evaluation of the first rule: $err(\{a\}\rightarrow e) = $ /* by Corollary 2.2.1 */ $= sup(\{a\}) - sup(\{ae\}) = $ /* see Table 4 */ $= 5 - 5 = 0$. Thus, $\{a\}\rightarrow e$ is certain. Having found a certain rule based on $\{ae\}$, we can use it to determine a certain rule based on $\{ace\}$. Otherwise, we would continue the evaluation of other rules based on $\{ae\}$. Since $\{a\}\rightarrow e$ is certain, then rule $\{ac\}\rightarrow e$, which is based on $\{ace\}$, is also certain; that is, $err(\{ac\}\rightarrow e) = 0$. Hence, by Property 2.2.5, $sup(\{ace\}) = sup(\{ac\}) = $ /* see Table 4 */ $= 4$. Please, see database $\mathcal{D}$ from Table 1 to verify the calculated support value for $\{ace\}$.   □

2.3 Sets Admitting Negated Items

In Subsections 2.3-2.5, we will introduce the notions related to patterns with negated items and their properties based on [24]. Let $L = I \cup \{-a \,|a\in I\}$. Each element in L will be called a *literal*. Elements in $L \setminus I$ will be called *negative literals*. By analogy, items in I will be also called *positive literals*. Each pair of literals a and $-a$ in L is called *contradictory*.

For the sake of convenience, we will apply the following notation: if l stands for a literal, then $-l$ will stand for its contradictory literal.

A *literal set* (or briefly *liset*) is defined as a set consisting of non-contradictory literals in L. A *liset* is called *positive* if all literals contained in it are positive. A *liset* is called *negative* if all literals contained in it are negative.

Lisets X and Y are called *contradictory* if $|X| = |Y|$ and for each literal in X there is a contradictory literal in Y. A liset contradictory to X will be denoted by $-X$.

Support of liset X is denoted by $sup(X)$ and defined as the number (or percentage) of transactions in $\mathcal{D}$ that contain all positive literals in X and do not contain any negative literal from X.

Liset X is called *frequent* if its support is greater than *minSup*. The set of all frequent lisets will be denoted by *FL*.

Instead of an original database $\mathcal{D}$, it is sometimes convenient to consider an extended database $\mathcal{D}$' in which each transaction T in $\mathcal{D}$ is extended with all negative literals contradictory to the items that do not occur in T. Table 5 is such an extended version of the database from Table 1. Clearly, all transactions in the extended database will be of the same size equal to $|I|$. Using the extended database, the support of liset X can be calculated as the number (or percentage) of transactions containing all literals (both positive and negative) in X. Though we do not recommend this method for evaluating lisets, this interpretation allows us to infer Property 2.3.1 related to supports of subsets and supersets of lisets by analogy to Property 2.1.1.

Table 5. Extended version $\mathcal{D}$' of database $\mathcal{D}$ from Table 1

Id	Transaction
T_1	$\{(\ a)(\ b)(\ c)(\ e)(-f)(-h)\}$
T_2	$\{(\ a)(\ b)(\ c)(\ e)(\ f)(-h)\}$
T_3	$\{(\ a)(\ b)(\ c)(\ e)(-f)(\ h)\}$
T_4	$\{(\ a)(\ b)(-c)(\ e)(-f)(-h)\}$
T_5	$\{(\ a)(-b)(\ c)(\ e)(-f)(\ h)\}$
T_6	$\{(-a)(\ b)(\ c)(\ e)(-f)(-h)\}$
T_7	$\{(-a)(-b)(-c)(-e)(-f)(\ h)\}$

Property 2.3.1

a) Let X,Y be lisets. If $X \subset Y$, then $sup(X) \geq sup(Y)$.
b) If $X \in FL$, then $\forall Y \subset X\ Y \in FL$.
c) If $X \notin FL$, then $\forall Y \supset X\ Y \notin FL$.

Let X be a liset. A *canonical variation of X* (denoted by $cv(X)$) is defined as an itemset obtained from X by replacing all negative literals in X by contradictory literals; that is,

$$cv(X) = P \cup (-N),$$

where P is the set of all positive literals in X and N is the set of all negative literals in X. Clearly, for any liset X, $cv(X) = \bigcup_{x \in X} cv(\{x\})$. In addition, if X is a positive liset, then $cv(X) = X$.

All lisets having the same canonical variation as liset X are denoted by $\mathcal{V}(X)$; that is,

$$\mathcal{V}(X) = \{Y \subseteq L | cv(Y) = cv(X)\}.$$

Each liset in $\mathcal{V}(X)$ is called a *variation of X*.

$\mathcal{V}(X)$ contains only one positive liset, which is $cv(X)$, and only one negative liset, namely $-cv(X)$. Clearly, the number of all variations of X equals $2^{|X|}$ and the sum of the supports of all variations of X equals $|\mathcal{D}|$.

Property 2.3.2. Let X be a liset.

a) $\mathcal{V}(Z) = \mathcal{V}(X)$ for any $Z \in \mathcal{V}(X)$.
b) $|\mathcal{V}(X)| = 2^{|X|}$.
c) $\Sigma_{Z \in \mathcal{V}(X)}\, sup(Z) = |\mathcal{D}|$.

2.4 Generalized Disjunctive Lisets and Generalized Disjunction-Free Lisets

Let us start with definitions analogous to those from Section 2.2.

Let Z be a liset. $X \rightarrow a_1 \vee \ldots \vee a_n$ is defined a *generalized disjunctive rule based on liset* Z if $X \subset Z$ and $\{a_1, \ldots, a_n\} = Z \backslash X$.

Support of rule $X \rightarrow a_1 \vee \ldots \vee a_n$ *based on a liset* is denoted by $sup(X \rightarrow a_1 \vee \ldots \vee a_n)$ and defined as the number (or percentage) of transactions in the extended database $\mathcal{D}'$ in which X occurs together with a_1 or a_2, or $\ldots$ or a_n. *Error of rule* $X \rightarrow a_1 \vee \ldots \vee a_n$ *based on a liset* is defined in usual way: $err(X \rightarrow a_1 \vee \ldots \vee a_n) = sup(X) - sup(X \rightarrow a_1 \vee \ldots \vee a_n)$. A rule r based on a liset is defined *certain* if $err(r) = 0$.

Liset X is defined as a *generalized disjunctive liset* if there is a certain generalized disjunctive rule based on X. Otherwise, X is defined a *generalized disjunction-free liset*.

The set of all generalized disjunction-free lisets will be denoted by *GDFreeL*; i.e.,

$$GDFreeL = \{X \subseteq L|\ \forall a_1, \ldots, a_n \in X\ err(X \backslash \{a_1, \ldots, a_n\} \rightarrow a_1 \vee \ldots \vee a_n) > 0, n \geq 1\}.$$

By analogy to Properties 2.2.1-2.2.2, one can easily observe:

Property 2.4.1. Let $X \cup \{a_1, \ldots, a_n\}$ be a liset.

If $X \rightarrow a_1 \vee \ldots \vee a_n$ is certain, then $\forall Z \supset X\ Z \rightarrow a_1 \vee \ldots \vee a_n$ is certain.

Property 2.4.2. Let X be a liset.

a) If $X \notin GDFreeL$, then $\forall Y \supset X\ Y \notin GDFreeL$.
b) If $X \in GDFreeL$, then $\forall Y \subset X\ Y \in GDFreeL$.

Proof: Follows immediately from Property 2.4.1.   □

2.5 Errors of Generalized Disjunctive Rules and Supports of Liset Variations

In Section 2.4, we have provided the definition of an error of a generalized disjunctive rule $X \rightarrow \vee A$ based on a liset. Here we will provide its equivalent interpretation, which will be more suitable for determining the relationship between the errors of rules for a given liset and the supports of its variations.

Property 2.5.1. Let A, X and Z be lisets such that A, $X \subset Z$ and $Z \backslash X = A$ and $A \neq \emptyset$. The error of rule $X \rightarrow \vee A$, which is based on Z, equals the number of transactions in the extended database $\mathcal{D}'$ in which X occurs and no literal from the set A occurs; that is,

$$err(X \rightarrow VA) = sup(X \cup (-A)).$$

Example 2.5.1. By Property 2.5.1,

$err(\{ab\} \rightarrow c \vee d) = sup(\{ab(-c)(-d)\})$, $err(\{ab\} \rightarrow (-c) \vee (-d)) =$
$sup(\{abcd\})$, $err(\{(-a)b\} \rightarrow (-c) \vee d) = sup(\{(-a)bc(-d)\})$. □

By Property 2.5.1, the antecedent (or alternatively, consequent) of a generalized disjunctive rule r based on a liset Z uniquely determines the variation V of Z, $V \neq Z$, the support of which equals the error of r. Hence, we may conclude that the set of the supports of all distinct variations of liset Z that are different from Z equals the set of the errors of all distinct generalized disjunctive rules based on Z (see Theorem 2.5.1).

Theorem 2.5.1. Let Z be a liset.

a) $\{sup(V) | V \in \mathcal{V}(Z) \wedge V \neq Z\} = \{err(X \rightarrow V(Z \backslash X)) | X \subset Z\}$.
b) $\{sup(V) | V \in \mathcal{V}(Z)\} = \{err(X \rightarrow V(Z \backslash X)) | X \subset Z\} \cup \{sup(Z)\}$.

2.6   Generalized Disjunctive Sets and Generalized Disjunction-Free Sets versus Supports of Liset Variations

In this section, we will examine more closely the relationship between supports and types (generalized disjunctive/generalized disjunction-free) of liset variations. We claim that for any liset Z with zero-support all its variations $V \neq Z$ are generalized disjunctive.

Theorem 2.6.1. Let Z be a liset. If $sup(Z) = 0$, then $\forall Y \in \mathcal{V}(Z)$ $(Y \neq Z \Rightarrow Y \notin GDFreeL)$.

Proof: Let $sup(Z) = 0$ and Y be any liset such that $Y \in \mathcal{V}(Z)$ and $Y \neq Z$. Let $X = Z \cap Y$ and $A = Y \backslash Z$ (hence, $Z \backslash Y = -A$). Thus, $Y = X \cup A$ and $Z = X \cup (-A)$. By Property 2.5.1, $err(X \rightarrow VA) = sup(X \cup (-A)) = sup(Z) = 0$. Since, $err(X \rightarrow VA) = 0$, then $Y = X \cup A$ is a generalized disjunctive liset. □

Now, we are able to derive the equivalence between the existence of a generalized disjunctive variation of liset Z and the existence of a zero-support variation of Z.

Theorem 2.6.2. Let Z be a non-empty liset.

a) $(\exists X \in \mathcal{V}(Z) \ X \notin GDFreeL)$ iff $(\exists Y \in \mathcal{V}(Z) \ sup(Y) = 0)$.
b) $(\forall X \in \mathcal{V}(Z) \ X \in GDFreeL)$ iff $(\forall Y \in \mathcal{V}(Z) \ sup(Y) \neq 0)$.

Proof: Ad a) ($\Rightarrow$) Follows immediately from Property 2.5.1 and Property 2.3.2a.

($\Leftarrow$) Follows immediately from Theorem 2.6.1 and Property 2.3.2a.
Ad b) Follows immediately from Theorem 2.6.2a. □

As follows from Theorem 2.6.2b, all variations of liset Z are generalized disjunction-free if and only if all variations of Z have zero supports.

Let Z be generalized disjunction-free. Since all variations of Z are supported by non-empty disjoint subsets of transactions in the database $\mathcal{D}$, the number of the variations, which equals $2^{|Z|}$, cannot exceed the number of transactions in $\mathcal{D}$. Thus, $2^{|Z|} \leq |\mathcal{D}|$.

Corollary 2.6.1. Generalized disjunction-free lisets contain at most $\lfloor \log_2 |\mathcal{D}| \rfloor$ literals. Minimal generalized disjunctive lisets contain at most $\lfloor \log_2 |\mathcal{D}| \rfloor + 1$ literals.

Finally, we claim that all variations of a generalized disjunction-free liset with non-zero support are generalized disjunction-free lisets.

Theorem 2.6.3. Let Z be a liset such that $sup(Z) \neq 0$ and $Z \in GDFreeL$. Then $\forall Y \in \mathcal{V}(Z)$ $Y \in GDFreeL$.

Proof: Follows from Theorem 2.5.1b and Theorem 2.6.2b.   □

3   Generalized Disjunction-Free Set Representations

3.1   Representing Frequent Positive Patterns with Generalized Disjunction-Free Set Representation (GDFSR)

A generalized disjunction-free representation of itemsets (denoted by GDFSR, [19] is defined as consisting of the following components:

- *the main component Main* = $\{X \subseteq I|$ $(X \in \mathcal{F}) \wedge (X \in GDFree)\}$ enriched by the information on the support for each itemset in *Main*;
- *the infrequent border IBd*$^-$ = $\{X \subseteq I|$ $(X \notin \mathcal{F}) \wedge (\forall Z \subset X$ $Z \in Main)\}$;
- *the generalized disjunctive border DBd*$^-$ = $\{X \subseteq I|$ $(X \in \mathcal{F}) \wedge (X \notin GDFree) \wedge (\forall Y \subset X$ $Y \in Main)\}$ enriched by the information on the support (and/or a certain generalized disjunctive rule) for each itemset in *DBd*$^-$.

Thus, the *Main* component contains frequent generalized disjunction-free itemsets. The infrequent border *IBd*$^-$ contains infrequent itemsets all proper subsets of which belong to *Main*. The generalized disjunctive border *DBd*$^-$ consists of frequent generalized disjunctive itemsets all proper subsets of which belong to *Main*.

Theorem 3.1.1

a) All itemsets in *Main* have non-zero support.
b) If $X \in Main$, then $\forall Y \subset X$ $Y \in Main$.
c) Itemsets that do not belong to GDFSR have at least one proper subset in the border $DBd^- \cup IBd^-$.

The GDFSR representation is a lossless representation of all frequent itemsets; namely, it is sufficient to determine for any itemset, whether it is frequent, and if so, enables determining its support.

Example 3.1.1. Let us consider database $\mathcal{D}$ from Table 1, which consists of seven transactions. For $minSup = 3$, the following GDFSR representation will be found:

$Main = \{\varnothing_{[7]}, \{a\}_{[5]}, \{b\}_{[5]}, \{c\}_{[5]}, \{e\}_{[6]}, \{ab\}_{[4]}, \{ac\}_{[4]}, \{bc\}_{[4]}\};$
$IBd^- = \{\{f\}, \{h\}, \{abc\}\};$
$DBd^- = \{\{ae\}_{[5, a \Rightarrow e]}, \{be\}_{[5, a \Rightarrow e]}, \{ce\}_{[5, c \Rightarrow e]}\}.$

Now we will illustrate how to determine whether itemsets are frequent or not, and how to calculate the support of a frequent itemset. Let us consider itemset $\{aef\}$. We note that $\{aef\}$ has a subset (here: $\{f\}$) in the infrequent border IBd^-. This means that $\{aef\}$ is infrequent. Now we will consider itemset $\{abe\}$. It does not have any subset in IBd^-, but has a subset, e.g., $\{ae\}_{[5,\ a\ \Rightarrow\ e]}$, in DBd^-. Since, $\{a\}\rightarrow e$ is certain, then $\{ab\}\rightarrow e$, which is based on $\{abe\}$, is also certain. So, $err(\{ab\}\rightarrow e) = 0$. Hence, by Property 2.2.5, $sup(\{abe\}) = sup(\{ab\})$. Since $\{ab\}_{[4]} \in Main$, its support is already known (here: equal to 4). Finally, $sup(\{abe\}) = sup(\{ab\}) = 4$.   □

3.2 Representing Frequent Positive and Negative Patterns with Generalized Disjunction-Free Set Representation (GDFSRN)

A generalized disjunction-free representation of sets admitting negated items (denoted by GDFSRN, [25]) is defined as consisting of the following components:

- *the main component Main* $= \{X\subseteq L|\ (X\in \mathcal{F}) \wedge (X\in GDFreeL)\}$ enriched by the information on the support for each liset in *Main*;
- *the infrequent border IBd*$^-$ $= \{X\subseteq L|\ (X\notin \mathcal{F}) \wedge (\forall Z\subset X\ Z\in Main)\}$;
- *the generalized disjunctive border DBd*$^-$ $= \{X\subseteq L|\ (X\in \mathcal{F}) \wedge (X\notin GDFreeL) \wedge (\forall Y\subset X$ $Y\in Main)\}$ enriched by the information on the support (and/or a certain generalized disjunctive rule) for each liset in DBd^-.

Thus, GDFSRN is defined in analogical way as GDFSR except that GDFSRN is built from lisets, while GDFSR is built only from itemsets.

Theorem 3.2.1

a) All lisets in *Main* have non-zero support.
b) If $X\in Main$, then $\forall Y\subset X\ Y\in Main$.
c) Lisets that do not belong to GDFSRN have at least one proper subset in the border $DBd^- \cup IBd^-$.

GDFSRN is a lossless representation of all frequent lisets and it enables determining for any liset, whether it is frequent, and if so, enables determining its support.

Example 3.2.1. Let us consider database $\mathcal{D}$ from Table 5, which is an extended version of Table 1, so that the transactions from Table 1 are in Table 5 with added the missing items. We obtain the following

GDFSRN for $minSup = 3$:

$Main = \{\varnothing_{[7]}, \{(-f)\}_{[6]}, \{(-h)\}_{[4]}, \{a\}_{[5]}, \{b\}_{[5]}, \{c\}_{[5]}, \{e\}_{[6]}, \{ab\}_{[4]}, \{ac\}_{[4]}, \{bc\}_{[4]}\};$
$IBd^- = \{\{(-a)\}, \{(-b)\}, \{(-c)\}, \{(-e)\}, \{f\}, \{h\}, \{a(-h)\}, \{c(-h)\}, \{(-f)(-h)\}, \{abc\}\};$
$DBd^- = \{\{ae\}_{[5,\ a\Rightarrow e]}, \{be\}_{[5,\ b\Rightarrow e]}, \{ce\}_{[5,\ c\Rightarrow e]}, \{a(-f)\}_{[4,\ \varnothing\Rightarrow a\vee(-f)]}, \{b(-f)\}_{[4,\ \varnothing\Rightarrow b\vee(-f)]},$
$\{b(-h)\}_{[4,\ (-h)\Rightarrow b]}, \{c(-f)\}_{[4,\ \varnothing\Rightarrow c\vee(-f)]}, \{e(-f)\}_{[5,\ \varnothing\Rightarrow e\vee(-f)]}, \{e(-h)\}_{[4,\ (-h)\Rightarrow e]}\}.$

If we wish to evaluate liset $\{ab(-f)(-h)\}$, we find that it has a subset, e.g. $\{a(-h)\}$, in IBd^-, so it is infrequent. Now, let us consider liset $\{be(-h)\}$. It does not have any subset in IBd^-, but has a subset $\{b(-h)\}_{[4,\ (-h)\ \Rightarrow\ b]}$ in DBd^-. Since, $\{(-h)\}\rightarrow b$ is certain, then $\{e(-h)\}\rightarrow b$, which is based on $\{be(-h)\}$, is also certain. So, $err(\{e(-h)\}\rightarrow b) = 0$. Hence, by Property 2.2.5, $sup(\{be(-h)\}) = sup(\{e(-h)\})$. Since $\{e(-h)\}_{[4,\ (-h)\ \Rightarrow\ e]} \in DBd^-$, its support is already known (here: equal to 4). Finally, $sup(\{be(-h)\}) = sup(\{e(-h)\}) = 4$. $\square$

3.3   Representing Frequent Positive and Negative Patterns with Generalized Disjunction-Free Literal Representation (GDFLR)

A generalized disjunction-free representation of lisets (denoted by GDFLR, [24]) is defined as consisting of the following components:

- the main component *Main* = $\{X\subseteq I|\ (\exists Z\in \mathcal{V}(X)\ Z\in FL) \wedge (\forall V\in \mathcal{V}(X)\ V\in GDFreeL)\}$ enriched by the information on the support for each set in *Main*;
- the infrequent border *IBd⁻* = $\{X\subseteq I|\ (\forall Z\in \mathcal{V}(X)\ Z\notin FL) \wedge (\forall Z\subset X\ Z\in Main)\}$;
- the generalized disjunctive border *DBd⁻* = $\{X\subseteq I|\ (\exists Z\in \mathcal{V}(X)\ Z\in FL)\ (\exists V\in \mathcal{V}(X)\ V\notin GDFreeL) \wedge (\forall Y\subset X\ Y\in Main)\}$ enriched by the information on the support (and/or a certain generalized disjunctive rule) for each itemset in *DBd⁻*.

Thus, the *Main* component contains itemsets for which all variations are generalized disjunction-free (i.e., by Theorem 2.6.2b, have non-zero supports) and at least one is frequent. The infrequent border *IBd⁻* contains itemsets for which all variations are infrequent, and all proper subsets belong to *Main*. The generalized disjunctive border *DBd⁻* consists of itemsets each of which has at least one frequent variation and has at least one generalized disjunctive variation (i.e., by Theorem 2.6.2a, has at least one variation with zero support) and all proper subsets of which belong to *Main*.

Theorem 3.3.1

a) All variations of all itemsets in *Main* have non-zero support.
b) If $X\in Main$, then $\forall Y\subset X\ Y\in Main$.
c) Itemsets that do not belong to GDFLR have at least one proper subset in the border $DBd^- \cup IBd^-$.

GDFLR is a lossless representation of all frequent lisets and enables determining for any liset, whether it is frequent, and if so, enables determining its support.

Example 3.3.1. Let us consider database $\mathcal{D}$ from Table 1. For *minSup* = 3, the following GDFLR representation will be found:

$Main = \{\varnothing_{[7]}, \{a\}_{[5]}, \{b\}_{[5]}, \{c\}_{[5]}, \{e\}_{[6]}, \{f\}_{[1]}, \{h\}_{[3]}, \{ab\}_{[4]}, \{ac\}_{[4]}, \{bc\}_{[4]}\};$
$IBd^- = \{\{ah\}, \{ch\}, \{fh\}, \{abc\}\};$
$DBd^- = \{\{ae\}_{[5,\ a\ \Rightarrow\ e]}, \{af\}_{[1,\ f\Rightarrow a]}, \{be\}_{[5,\ b\ \Rightarrow\ e]}, \{bf\}_{[1,\ f\Rightarrow b]}, \{bh\}_{[1,\ h\ \Rightarrow\ b]},$
$\quad\quad \{ce\}_{[5,\ c\ \Rightarrow\ e]}, \{cf\}_{[1,\ f\Rightarrow c]}, \{ef\}_{[1,\ f\Rightarrow e]}, \{eh\}_{[2,\ \varnothing\ \Rightarrow\ e\vee h]}\}.$

Let us illustrate how to determine if some lisets are frequent or not, and how to calculate the support of a frequent liset. Let us consider liset $\{(-a)e(-h)\}$. We note that its canonical variation $\{aeh\}$ has a subset $\{ah\}$ in the infrequent border *IBd⁻*. Since $\{ah\}$ and all its variations including $\{(-a)(-h)\}$ are infrequent, then $\{(-a)e(-h)\}$ is also infrequent as a superset of infrequent liset $\{(-a)(-h)\}$. Now, let us consider liset $\{e(-f)\}$. Its canonical version $\{ef\}$ does not have any subset in *IBd⁻*, but has a subset (namely, $\{ef\}_{[1,\ f\ \Rightarrow\ e]}$ itself) in *DBd⁻*. Since the supports of $\{ef\}$ and all its subsets are already known, we can determine the support of its variation $\{(e)(-f)\}$: $sup(\{e(-f)\}) = $ /* by Property 2.5.1 */ $= err(\{e\} \rightarrow f) = $ /* by Corollary 2.2.1 */ $= sup(\{e\}) - sup(\{ef\}) = $ /* see *Main* and DBd^- */ $= 6 - 1 = 5$.

Please, see Table 1 or Table 5 to verify these results.   □

Beneath we formulate and prove a new theorem related to DBd^- of GDFLR.

Theorem 3.3.2. Frequent variations of elements of DBd^- are generalized disjunctive.

Proof: Let itemset P belong to DBd^- of GDFLR. Then there is a variation of P which is generalized disjunctive, i.e. which is a base of a generalized disjunctive rule with zero-error. Hence, by Property 2.5.1, there exists a variation of P, say P', with support equal to 0. Clearly, P' is infrequent. In addition, by Theorem 2.6.1, all variations of P that are different from P' are generalized disjunctive. Hence, all frequent variations of P are generalized disjunctive.   □

3.4   Common Properties of GDF Representations

It has been proved in [22, 24] that upper bounds on the length of main and border elements in the GDF representations depend logarithmically on the number of transaction in the database as follows:

Theorem 3.4.1

a) $\forall Z \in Main, |Z| \leq \lfloor \log_2(|\mathcal{D}| - minSup) \rfloor$.
b) $\forall Z \in DBd^- \cup IBd^-, |Z| \leq \lfloor \log_2(|\mathcal{D}| - minSup) \rfloor + 1$.

Another common feature of each GDF representation is that its border elements are all and only sets that do not belong to *Main* component, but have all their proper subsets in *Main*.

Theorem 3.4.2 [19, 24]

a) $DBd^- \cup IBd^- = \{X \mid (X \notin Main) \wedge (\forall Y \subset X\ Y \in Main)\}$.
b) $IBd^- = \{X \mid (X \notin Main) \wedge (\forall Y \subset X\ Y \in Main)\} \setminus DBd^-$.

By Theorem 3.4.2a, the set being the set-theoretical union of DBd^- and IBd^- can be reconstructed based only on the knowledge of the *Main* component. By Theorem 3.4.2b, IBd^- can be determined based only on the knowledge of the *Main* and DBd^- components. As a result, all GDF representations will remain lossless representations of respective frequent patterns, even after discarding IBd^- component [19]. More advanced techniques for reducing borders were offered in [9] and in [21]. As follows from the experiments in [21], the combination of these two techniques reduces the border by up to two orders of magnitude. Finally, a simple technique for lossless reduction of the main component was proposed in [23].

4   Relationships between the GDF Representations

In this section, we will compare all three GDF representations built for a same $minSup$ value. We will use the following notation: $Main_{GDFSR}$, $Main_{GDFSRN}$, $Main_{GDFLR}$,

DBd^-_{GDFSR}, DBd^-_{GDFSRN}, DBd^-_{GDFLR}, IBd^-_{GDFSR}, IBd^-_{GDFSRN}, IBd^-_{GDFLR}, to indicate respective representation to which a considered component belongs to.

Let us start with a trivial relationship between the GDFSR and GDFSRN representations, namely:

- $Main_{GDFSR} \subseteq Main_{GDFSRN}$;
- $DBd^-_{GDFSR} \subseteq DBd^-_{GDFSRN}$;
- $IBd^-_{GDFSR} \subseteq IBd^-_{GDFSRN}$.

In the following subsections, we will investigate the relationships between GDFLR and GDFSRN, as well as between GDFLR and GDFSR. A special attention will be paid to a particular case, when $minSup = 0$. We will show that in this case the elements of GDFLR are identical with the elements of GDFSR, and GDFSRN contains all variations of each element of GDFLR (and by this, also of each element of GDFSR).

4.1 Relationship between GDFLR and GDFSRN

We start our analysis of the relationship between GDFLR and GDFSRN by examining the correspondence between lisets in the $Main$ and DBd^- components of both representations.

Lemma 4.1.1. If itemset P belongs to $Main_{GDFLR}$, then all frequent variations of P belong to $Main_{GDFSRN}$, and there is at least one such a variation of P.

Proof: Let itemset P belong $Main_{GDFLR}$. Then all variations of P are generalized disjunction-free and there is a variation of P, which is frequent. Hence, all frequent variations of P are generalized disjunction-free and thus belongs to $Main_{GDFSRN}$.   □

Lemma 4.1.2. If itemset P belongs to DBd^-_{GDFLR}, then all frequent variations of P belong to DBd^-_{GDFSRN}, and there is at least one such a variation of P.

Proof: Let itemset P belong to DBd^-_{GDFLR}. Then at least one variation of P is frequent. Clearly, all subsets of all frequent variations of P are frequent (*). In addition, by Theorem 3.3.2, all frequent variations of P are generalized disjunctive (**). Since P belongs to DBd^-_{GDFLR}, all variations of all proper subsets of P are generalized disjunction-free as belonging to $Main_{GDFLR}$. Hence, all proper subsets of all frequent variations of P are generalized disjunction-free (***). By (*), (**) and (***), all frequent variations of P belong to DBd^-_{GDFSRN}.   □

Now, based on the fact that the sum of the supports of all variations of any liset equals the number of all transactions, we will propose a simple estimation of the maximal number of frequent variations of a liset. Clearly, for $minSup \geq 50\%$, it can be only one frequent variation of a liset, as the supports of all other variations of the liset do not exceed $minSup$. Now, let us consider $minSup$ equal to, say, 34%. Then the liset may have at most two frequent variations. Eventually, we generalize these observations as follows:

Proposition 4.1.1. A liset may have at most $\lceil 100\% \, / \, minSup \rceil - 1$ frequent variations.

The corollary beneath follows from Lemmas 4.1.1-2 and Proposition 4.1.1:

Corollary 4.1.1

a) Each itemset in $Main_{GDFLR}$ represents at most $\lceil 100\% \,/\, minSup\rceil - 1$ of its variations in $Main_{GDFSRN}$.

b) Each itemset in DBd^-_{GDFLR} represents at most $\lceil 100\% \,/\, minSup\rceil - 1$ of its variations in DBd^-_{GDFSRN}.

c) If $minSup \geq 50\%$, then $|Main_{GDFLR}| = |Main_{GDFSRN}|$ and $|DBd^-_{GDFLR}| = |DBd^-_{GDFSRN}|$.

Table 6. Another database

Id	Transaction
T_1	{abce}
T_2	{abcef}
T_3	{abch}
T_4	{abe}
T_5	{acfh}
T_6	{bef}
T_7	{h}
T_8	{af}

Table 7. GDFLR and GDFSRN for database from Table 6 and $minSup = 2$

k	GDFLR			GDFSRN		
	$|Main_k|$	$|DBd^-_k|$	$|IBd^-_k|$	$|Main_k|$	$|DBd^-_k|$	$|IBd^-_k|$
3	0	2	3	0	2	1
2	9	3	3	11	5	39
1	6	0	0	11	0	1
0	1	0	0	1	0	0
0-3	16	5	6	23	7	40

Now, we will investigate, if an element from IBd^-_{GDFLR} has its variation in GDFSRN. Let us consider the database from Table 6. Table 7 shows the cardinalities of the representations GDFLR and GDFSRN found from Table 6 for $minSup = 2$. As follows from Table 7, each component of GDFSRN is more numerous than the respective component of GDFLR. Nevertheless, GDFLR contains five 3-itemsets (three out of which belong to IBd^-_{GDFLR}), while GDFSRN contains three 3-lisets (one out of which belongs to IBd^-_{GDFSRN}). This means that at least two 3-itemsets in IBd^-_{GDFLR} do not have any of their variations in GDFSRN. Lemma 4.1.2 generalizes this observation.

Lemma 4.1.3. An itemset that belongs to IBd^-_{GDFLR} is not guaranteed to have any of its variations in GDFSRN.

Let us now examine the relationship between GDFLR and GDFSRN for $minSup=0$.

Lemma 4.1.4. Let $minSup = 0$. Itemset P belongs to $Main_{GDFLR}$ iff all variations of P are frequent generalized disjunction-free sets.

Proof: P belongs to $Main_{GDFLR}$ iff all variations of P are generalized disjunction-free and at least one variation of P has support greater than 0 iff /* by Theorem 2.6.2b */ all variations of P are frequent generalized disjunction-free sets. □

Lemma 4.1.5 is an immediate consequence of Lemma 4.1.4.

Lemma 4.1.5. Let $minSup = 0$. Itemset P belongs to $Main_{GDFLR}$ iff all variations of P belong to $Main_{GDFSRN}$.

Lemma 4.1.6. Let $minSup = 0$ and $|\mathcal{D}| > 0$. Then $IBd^-_{GDFLR} = \{\,\}$.

Proof: Since $|\mathcal{D}| > 0$, then $sup(\varnothing) > 0$ and $\varnothing \notin IBd^-_{\text{GDFLR}}$. Let P be any non-empty liset. Then P contains more than one variation and the sum of the supports of all variations of P equals $|\mathcal{D}|$, which is greater than 0. So, at least one variation of P has support different from 0. Hence, $P \notin IBd^-_{\text{GDFLR}}$. We have thus shown that neither $\varnothing$ nor any other liset belongs to IBd^-_{GDFLR}. So, $IBd^-_{\text{GDFLR}} = \{\}$.   □

Lemma 4.1.7. Let $minSup = 0$ and $|\mathcal{D}| > 0$. If $P \in DBd^-_{\text{GDFLR}}$, then $IBd^-_{\text{GDFSRN}} \neq \{\}$, $DBd^-_{\text{GDFSRN}} \neq \{\}$, and all variations of P belong to the border of GDFSRN.

Proof: Let P be an element DBd^-_{GDFLR}. Thus, all proper subsets of P belong to $Main_{\text{GDFLR}}$. Hence, by Lemma 4.1.5, all variations of all proper subsets of P belong to $Main_{\text{GDFSRN}}$ (*). Since P belongs to DBd^-_{GDFLR}, then at least one variation of P is generalized disjunctive and at least one variation of P is frequent. So, by Theorem 2.6.2a, at least one variation of P, say V, has support equal to 0 and at least one variation of P has support greater than 0. Thus, by Theorem 2.6.1, all variations of P that are different from V, are generalized disjunctive. V and all variations of P that have support equal to 0 are infrequent and by (*) all their proper subsets belong to $Main_{\text{GDFSRN}}$. Hence, these variations of P belong to IBd^-_{GDFSRN}. The remaining variations of P are frequent generalized disjunctive and by (*) all their proper subsets belong to $Main_{\text{GDFSRN}}$. Thus, these variations of P belong to DBd^-_{GDFSRN}. Hence, all variations of P belong to the border of GDFSRN, which consists of two non-empty parts.   □

Lemma 4.1.8. Let $minSup = 0$ and $|\mathcal{D}| > 0$. If pattern P belongs to the border of GDFSRN, then the canonical variation of P belongs to the border of GDFLR.

Proof (by contradiction): Let us assume that P belongs to the border of GDFSRN and $cv(P)$ does not belong to the border of GDFLR. Then $cv(P)$ either 1) belongs to $Main_{\text{GDFLR}}$, or 2) does not belong to GDFLR.

Case 1: Since $cv(P) \in Main_{\text{GDFLR}}$, then by Lemma 4.1.1, $P \in Main_{\text{GDFSRN}}$, which contradicts the assumption.

Case 2: By Theorem 3.3.1c, there is a proper subset, say X, of $cv(P)$ in the border of GDFLR. By Lemma 4.1.7, all variations of X belong to the border of GDFSRN. Clearly, P has a proper subset among all variations of X. Thus, there is a proper subset of P that does not belong to $Main_{\text{GDFSRN}}$. Hence P does not belong to the border of GDFSRN, which contradicts the assumption.   □

Now, we will consider the case, when $\mathcal{D}$ contains no transaction.

Lemma 4.1.9. Let $minSup = 0$ and $\mathcal{D}$ contains no transaction. Then:

a) $Main_{\text{GDFLR}} = Main_{\text{GDFSRN}} = \{\}$;
b) $DBd^-_{\text{GDFLR}} = DBd^-_{\text{GDFSRN}} = \{\}$;
c) $IBd^-_{\text{GDFLR}} = IBd^-_{\text{GDFSRN}} = \{\varnothing\}$.

Proof: Since $\mathcal{D}$ contains no transaction, then $sup(\varnothing) = 0$. Hence, $\varnothing$, which is its own and only variation, is a minimal infrequent set. Thus, $\varnothing \in IBd^-_{\text{GDFLR}}$ and $\varnothing \in$

IBd^-_{GDFSRN}. Therefore, $IBd^-_{GDFLR} = IBd^-_{GDFSRN} = \{\varnothing\}$, $Main_{GDFLR} = Main_{GDFSRN} = \{\ \}$, and $DBd^-_{GDFLR} = DBd^-_{GDFSRN} = \{\ \}$. $\qquad\qquad\square$

Lemmas 4.1.5-9 allow us to conclude:

Theorem 4.1.1. Let $minSup = 0$. Pattern P belongs to GDFLR iff all (i.e. $2^{|P|}$) variations of P belong to GDFSRN.

4.2   Relationship between GDFLR and GDFSR

Now we will examine the relationship between the GDFLR and GDFSR representations.

Lemma 4.2.1. If itemset P belongs to $Main_{GDFSR}$, then P belongs to $Main_{GDFLR}$.

Proof: Let $P \in Main_{GDFSR}$. Then P is a frequent generalized-disjunction free itemset having support greater than 0. Thus, by Theorem 2.6.3, all variations of P are generalized-disjunction free sets. Hence, $P \in Main_{GDFLR}$. $\qquad\qquad\square$

Lemma 4.2.2. If pattern P belongs to DBd^-_{GDFSR}, then P belongs to DBd^-_{GDFLR}.

Proof: Let P belong to DBd^-_{GDFSR}. Then P is a minimal generalized disjunctive itemset, which has a frequent variation (namely, itself) (*) and all its proper subsets in $Main_{GDFSR}$. Thus, by Lemma 4.2.1, all proper subsets of P belong to $Main_{GDFLR}$ (**). By (*) and (**), P belongs to DBd^-_{GDFLR}. $\qquad\qquad\square$

Lemma 4.2.3. If pattern P belongs to IBd^-_{GDFSR}, then P belongs to GDFLR.

Proof: Let $P \in IBd^-_{GDFSR}$. Then P is a minimal infrequent itemset, such that its all proper subsets belong to $Main_{GDFSR}$. Thus, by Lemma 4.2.1, all proper subsets of P belong to $Main_{GDFLR}$. Having this in mind, we will consider three mutually exclusive cases: 1) all variations of P are infrequent, 2) at least one variation of P is frequent and at least one variation of P is generalized disjunctive, 3) at least one variation of P is frequent and no variation of P is generalized disjunctive.

Case 1: In this case, $P \in IBd^-_{GDFLR}$.

Case 2: In this case, $P \in DBd^-_{GDFLR}$.

Case 3: In this case, $P \in Main_{GDFLR}$. $\qquad\qquad\square$

Theorem 4.2.1. If pattern P belongs to GDFSR, then P belongs to GDFLR.

Now, we will examine the relationship between GDFLR and GDFSR in the case when $minSup = 0$.

Lemma 4.2.4. Let $minSup = 0$. If itemset $P \in Main_{GDFLR}$, then $P \in Main_{GDFSR}$.

Proof: Let $P \in Main_{GDFLR}$. Then all variations of P are generalized-disjunction free sets, so by Theorem 2.6.2b, the supports of all variations of P are greater than 0. Thus, P is a frequent generalized-disjunction free itemset. Hence, $P \in Main_{GDFSR}$. $\qquad\square$

Lemma 4.2.5. Let $minSup = 0$. If $P \in DBd^-_{GDFLR}$, then $P \in DBd^-_{GDFSR} \cup IBd^-_{GDFSR}$.

Proof: Let $P \in DBd^-_{\text{GDFLR}}$. Then P is a minimal itemset having all proper subsets in $Main_{\text{GDFLR}}$ and either 1) $sup(P) = 0$ or 2) $sup(P) > 0$ and P is a generalized disjunctive set. Since all proper subsets of P belong to $Main_{\text{GDFLR}}$, then by Lemma 4.2.4, all proper subsets of P belong to $Main_{\text{GDFSR}}$. Hence, if $sup(P) = 0$, then $P \in IBd^-_{\text{GDFSR}}$. Otherwise, $P \in DBd^-_{\text{GDFSR}}$.   □

Lemma 4.2.6. Let $minSup = 0$. If itemset $P \in IBd^-_{\text{GDFLR}}$, then $P \in IBd^-_{\text{GDFSR}}$.

Proof: Let $P \in IBd^-_{\text{GDFLR}}$. Hence, all variations of P are infrequent, so the supports of all variations of P are equal to 0. Thus, $|\mathcal{D}| = 0$. By Lemma 4.1.9, $IBd^-_{\text{GDFLR}} = \{\varnothing\}$. Hence, $P = \varnothing$ is the only minimal infrequent itemset. Thus, $IBd^-_{\text{GDFSR}} = \{\varnothing\} = \{P\}$.   □

The theorem beneath follows from Theorem 4.2.1 and Lemmas 4.2.4-6:

Theorem 4.2.2. Let $minSup = 0$. Itemset P belongs to GDFLR iff P belongs to GDFSR.

5   Rule Error Oriented Algorithms

5.1   Building GDFLR with *GDFLR-Apriori*

In this section, we recall the *GDFLR-Apriori* algorithm [24], which builds GDFLR. The process of creating candidate elements and the calculation of their supports in *GDFLR-Apriori* are assumed to be carried out as in the *Apriori*-like algorithms [2] that discover all frequent positive patterns. However, *GDFLR-Apriori* differs from them by introducing additional tests classifying candidates to one of the three components *Main*, *IBd⁻* or *DBd⁻* respectively. In the algorithm, we apply the following notation:

Notation for *GDFLR-Apriori*
• X_k – candidate k-itemsets;
• $X.sup$ – the support field of itemset X;

First the *GDFLR-Apriori* algorithm initializes the three GDFLR components. Then it checks if the number of transactions in the database $\mathcal{D}$ is greater than $minSup$. If so, then $\varnothing$, being a generalized disjunction-free set, is frequent, so it is inserted into $Main_0$. Next, the set X_1 is assigned all items that occur in the database $\mathcal{D}$. Now, the following steps are performed level-wise for all k item candidates, for $k \geq 1$:

• Supports of all candidates in X_k are determined during a pass over the database.
• For each candidate X in X_k, the errors *Errs* of all generalized disjunctive rules based on X are calculated from the supports of respective subsets of X, according to Corollary 2.2.1 Since $\{X.sup\} \cup Errs$ equals the set of the supports of all variations of X (by Theorem 2.5.1b), with the condition $max(\{X.sup\} \cup Errs) \leq minSup$ it is checked if all variations of X are infrequent. If so, X is classified as an element of the infrequent border *IBd⁻*. Otherwise, at least one variation of X is frequent. Next, the condition $min(\{X.sup\} \cup Errs) = 0$ is checked if X is generalized disjunctive or its support equals 0. If so, X is assigned to the border *DBd⁻*. Otherwise, it is assigned to the *Main* component.

- After assigning all the candidates from X_k to the respective components of GDFLR, candidates for frequent patterns, longer by one item, are created as X_{k+1}. As all proper subsets of each element in GDFLR must belong to *Main*, the creation of $k+1$ item candidates is limited to merging[1] appropriate pairs of k item patterns from *Main*. In addition, the newly created candidates that have missing k item subsets in *Main* are not valid GDFLR elements, and thus are removed from X_{k+1}.

Algorithm. *GDFLR-Apriori*(database $\mathcal{D}$, support threshold *minSup*);

$Main = \{\}$; $DBd^- = \{\}$; $IBd^- = \{\emptyset\}$;   // initialize GDFLR
if $|\mathcal{D}| > minSup$ **then begin**
  $\emptyset.sup = |\mathcal{D}|$;
  move $\emptyset$ from IBd^- to $Main_0$;
  $X_1 = \{1\text{-itemsets}\}$;
  for $(k = 1; X_k \neq \emptyset; k++)$ **do begin**
    calculate the supports of itemsets in X_k within one scan of database $\mathcal{D}$;
    forall candidates $X \in X_k$ **do begin**
      $Errs = Errors\text{-}of\text{-}rules(X, Main)$;
      if $max(\{X.sup\} \cup Errs) \leq minSup$ **then** // by Th. 2.5.1b, all variations of X are infrequent
        add X to IBd^-_k
      elseif $min(\{X.sup\} \cup Errs) = 0$ **then** //by Th. 2.6.2a, there is a gen. dis. variation of X
        add X to DBd^-_k
      else
        add X to $Main_k$
      endif
    endfor;
    /* create new $(k+1)$-candidates by merging k-itemsets in *Main* (by Ths. 3.3.1b & 3.4.2a) */
    $X_{k+1} = \{X \subseteq I|\ \exists Y,Z \in Main_k\ (|Y \cap Z| = k-1 \wedge X = Y \cup Z)\}$;
    /* remain only those candidates that have all k-subsets in *Main* (by Ths. 3.3.1b & 3.4.2a) */
    $X_{k+1} = X_{k+1} \setminus \{X \in X_{k+1}|\ \exists Y \subseteq X\ (|Y| = k \wedge Y \notin Main_k\}$
  endfor;
endif;
return $<\cup_k Main_k, \cup_k DBd^-_k, \cup_k IBd^-_k>$;

The algorithm ends when there are no more candidates to evaluate, i.e. $X_{k+1} = \emptyset$. Please note that the algorithm will iterate no more than $\lfloor \log_2(|\mathcal{D}| - minSup) \rfloor + 1$ times (by Theorem 3.4.1). Hence, database $\mathcal{D}$ will be read no more than $\lfloor \log_2(|\mathcal{D}| - minSup) \rfloor + 1$ times.

Function. *Errors-of-rules*(itemset X, component *Main*);

$Errs = \{\}$;
  forall non-empty subsets $Y \subseteq X$ **do begin**
    calculate $err(X \setminus Y \rightarrow \vee Y)$ from the supports of subsets of X;   // by Corollary 2.2.1
    insert $err(X \setminus Y \rightarrow \vee Y)$ in $Errs$;
    /* optionally break if $max(\{X.sup\} \cup Errs) > minSup$ and $min(\{X.sup\} \cup Errs) = 0$ */
  endfor;
return $Errs$;

---

[1] As in the *Apriori* algorithm [2].

Let us note that the most critical operation in the *GDFLR-Apriori* algorithm is the calculation of errors of a given candidate pattern X. As follows from Corollary 2.2.1, the error of rule $x_1 \dots x_m \to x_{m+1} \lor \dots \lor x_n$ built from X, which has n items in consequent, requires the knowledge of the supports of X and its $2^n - 1$ proper subsets. Taking into account that one can built $\binom{|X|}{n}$ distinct rules from X that have n items in their consequents and a rule consequent may have from 1 to $|X|$ items, the calculation of the errors of all rules based on X requires $\Sigma_{n=1..|X|} \binom{|X|}{n} (2^n - 1) = 3^{|X|} - 2^{|X|}$ accesses to proper subsets of X. In fact, the calculation of the errors of generalized disjunctive rules based on a candidate X can be broken as soon as we identify a frequent variation of X (i.e. when $max(\{X.sup\} \cup Errs) > minSup$) and a generalized disjunctive variation of X (i.e. when $min(\{X.sup\} \cup Errs) = 0$). Satisfaction of these two conditions guarantees that X is an element of the generalized disjunctive border.

5.2 Building GDFSR and GDFSRN with *GDFSR-Apriori* and *GDFSRN-Apriori*

For building the GDFSR representation, we provide here the algorithm *GDFSR-Apriori*, as a simpler version of *GDFLR-Apriori*. In particular, *GDFSR-Apriori* does not calculate errors of rules for the infrequent candidates. Additionally, in the *GDFSR-Apriori* algorithm, the errors of rules built from a given frequent candidate are calculated until a first rule with a zero error is identified instead of calculating the errors of all rules based on the candidate. The main differences between the two algorithms are shadowed in the code below (see the Algorithm *GDFSR-Apriori*).

Given the *GDFSR-Apriori*, we can easily modify it to the form (we call it *GDFSRN-Apriori*) which builds the GDFSRN representation. Namely, in the prologue we extend each transaction from the database with negations of the items missing in the transaction, and then run *GDFSR-Apriori* on the extended database. Schematically it is shown in the code below. Let us note that this simple mode of building GDFSRN, does not guarantee high efficiency, as it works on much larger transactions (built on much larger set of items).

Algorithm. *GDFSR-Apriori*(database $\mathcal{D}$, support threshold *minSup*);

```
Main = { }; DBd⁻ = { }; IBd⁻ = {∅};                    // initialize GDFSR
if |D| > minSup then begin
    ∅.sup = |D|;
    move ∅ from IBd⁻ to Main₀;
    X₁ = {1-itemsets};
    for (k = 1; Xₖ ≠ ∅; k++) do begin
        calculate the supports of itemsets in Xₖ within one scan of database D;
        forall candidates X∈Xₖ do begin
            if X.sup ≤ minSup then              // X is infrequent
                add X to IBd⁻ₖ
            else                                // X is frequent
                /* search a generalized disjunctive rules based on X with the error = 0*/
```

    forall non-empty subsets $Y \subseteq X$ **do**
      /* calculate the errors *Errs* of generalized disjunctive rules based on X */
      /* from the supports of subsets of X (according to Corollary 2.2.1)   */
      /* until $min(\{X.sup\} \cup Errs) = 0$ */
      endfor;
      if $min(\{X.sup\} \cup Errs) = 0$ **then**   // Is X a base of a certain gen. disjunctive rule?
        add X to DBd^-_k
      else add X to $Main_k$ **endif**
    endif
    endfor;
    /* create new $(k+1)$-candidates by merging k item patterns in *Main* */
    $\mathcal{X}_{k+1} = \{X \subseteq I | \exists Y,Z \in Main_k (|Y \cap Z| = k-1 \wedge X = Y \cup Z)\}$;
    /* remain only those candidates that have all k item subsets in *Main* */
    $\mathcal{X}_{k+1} = \mathcal{X}_{k+1} \setminus \{X \in \mathcal{X}_{k+1} | \exists Y \subseteq X (|Y| = k \wedge Y \notin Main_k\}$
  endfor
endif;
return $<\cup_k Main_k, \cup_k DBd^-_k, \cup_k IBd^-_k>$;

Algorithm. *GDFSRN-Apriori*(database $\mathcal{D}$, support threshold *minSup*);

$\mathcal{D}' = \mathcal{D}$;

extend all transactions in $\mathcal{D}'$ with negations of missing items;
return *GDFSR-Apriori*($\mathcal{D}'$, *minSup*);

---

6 Support Oriented Algorithms

In this section we will present possibilities of speeding up the process of building the concise representations GDFLR, GDFSR and GDFSRN. Our goal is to re-use efficiently the information of the supports of subsets (and their variations) when calculating the errors of rules built from a candidate pattern. Since the calculation of the errors of rules built from the pattern X is equivalent to the determination of the supports of X's variations with negation, we will focus only on the latter task. First, we will recall an ordering of X's variations, as proposed in our earlier work [27]. Based on this ordering, we will present a method of calculating the support of each variation from the supports of two patterns. We will recall here the "support oriented" algorithm *GDFLR-SO-Apriori* [27], and propose two new algorithms *GDFSR-SO-Apriori*, and *GDFSRN-SO-Apriori*, which are based on the ideas from *GDFLR-SO-Apriori*.

6.1 Efficient Calculation of Supports of Pattern Variations

We start with defining an ordering of the variations of the pattern X:

Let $0 \le n < 2^{|X|}$. The nth *variation of pattern* X ($\mathcal{V}_n(X)$) is defined as this variation of X that differs from X on all and only the bit positions having value 1 in the binary representation of n. For the variation $\mathcal{V}_n(X)$, n is called its *(absolute) ordering number*.

Let $0 \le i < |X|$. The ith *cluster* ($C_i(X)$) for the pattern X is defined as the set of all variations of X, such that i is the leftmost bit position with value 1 in the binary representation of their ordering numbers. Please note that X, which is 0th variation of X, does not belong to any cluster $C_i(X)$, $0 \le i < |X|$, since the binary representation of its ordering number does not contain any bit position with value 1.

Below we illustrate these concepts.

Example 6.1.1. Let $X = \{abc\}$. $\mathcal{V}_5(X) = \mathcal{V}_{2^2+2^0}(X) = \mathcal{V}_{(101)_2}(X) = \{(-a)b(-c)\}$; that is, 5th variation of X differs from X on the positions 2 and 0. Table 8 enumerates all the variations of X. The variations of X that are different from X can be split to $|X| = 3$ clusters: $C_0(X) = \{\mathcal{V}_{(001)_2}(X)\}$; $C_1(X) = \{\mathcal{V}_{(010)_2}(X), \mathcal{V}_{(011)_2}(X)\}$; $C_2(X) = \{\mathcal{V}_{(100)_2}(X), \mathcal{V}_{(101)_2}(X), \mathcal{V}_{(110)_2}(X), \mathcal{V}_{(111)_2}(X)\}$. Note that the ordering numbers of variations in cluster $C_i(X)$, $i \in \{0, ..., |X|\text{-}1\}$, can be expressed as $2^i + j$, where $j \in \{0, ..., 2^i - 1\}$ (see Table 8). □

Let $0 \le i < |X|$ and $j \in \{0, ..., 2^i - 1\}$. A jth *variation of pattern X in cluster* $C_i(X)$ is defined as $\mathcal{V}_{2^i + j}(X)$. For variation $\mathcal{V}_{2^i + j}(X)$, j is called its *ordering number in cluster* $C_i(X)$ (or *relative ordering number*).

Table 8. Absolute and relative ordering of variations of pattern $X = \{abc\}$

variation V of pattern X	ordering number n of variation V	$\|X\|$ bit binary representation of n	cluster $C_i(X)$ including variation V	j - ordering number of variation V in $C_i(X)$	binary representation of j	absolute versus rel. ordering of variation V
$\{(a)(b)(c)\}$	0	$(000)_2$	–	–	$(000)_2$	–
$\{(a)(b)(-c)\}$	1	$(001)_2$	$C_0(X)$	0	$(000)_2$	$1 = 2^0 + 0$
$\{(a)(-b)(c)\}$	2	$(010)_2$	$C_1(X)$	0	$(000)_2$	$2 = 2^1 + 0$
$\{(a)(-b)(-c)\}$	3	$(011)_2$	$C_1(X)$	1	$(001)_2$	$3 = 2^1 + 1$
$\{(-a)(b)(c)\}$	4	$(100)_2$	$C_2(X)$	0	$(000)_2$	$4 = 2^2 + 0$
$\{(-a)(b)(-c)\}$	5	$(101)_2$	$C_2(X)$	1	$(001)_2$	$5 = 2^2 + 1$
$\{(-a)(-b)(c)\}$	6	$(110)_2$	$C_2(X)$	2	$(010)_2$	$6 = 2^2 + 2$
$\{(-a)(-b)(-c)\}$	7	$(111)_2$	$C_2(X)$	3	$(011)_2$	$7 = 2^2 + 3$

Corollary 6.1.1. Let X be a pattern. The set of all variations of X consists of X and all variations in the clusters $C_i(X)$, where $i \in \{0, ..., |X| - 1\}$:

$$\mathcal{V}(X) = \{X\} \cup \bigcup_{i = 0..|X|-1} C_i(X) = \{X\} \cup \bigcup_{i = 0..|X|-1, j = 0..2^i-1} \{\mathcal{V}_{2^i + j}(X)\}.$$

Note that two variations $\mathcal{V}_j(X)$ and $\mathcal{V}_{2^i + j}(X)$, $j \in \{0, ..., 2^i-1\}$, of a non-empty pattern X differ only on the position i; namely, the item on i^{th} position in $\mathcal{V}_{2^i + j}(X)$ is negation of the item on i^{th} position in $\mathcal{V}_j(X)$. In addition, $\mathcal{V}_j(X)$ and $\mathcal{V}_{2^i + j}(X)$ do not differ from X on positions greater than i. Thus, we can formulate Theorem 6.1.1.

Theorem 6.1.1. Let X be a non-empty pattern, $i \in \{0, ..., |X| - 1\}$ and $j \in \{0, ..., 2^i-1\}$. Then the following holds:

$$sup(\mathcal{V}_{2^i + j}(X)) = sup(\mathcal{V}_j(X \setminus \{X[i]\})) - sup(\mathcal{V}_j(X)).$$

Table 9. Calculation of supports of consecutive variations of $X = \{abc\}$

i – X's cluster no.	$X \setminus \{X[i]\}$	j – rel. ordering number of X's variation in $C_i(X)$	support calculation for j^{th} variation of X in cluster $C_i(X)$ (that is, for variation $\mathcal{V}_{2^i+j}(X)$)
0	$\{ab\}$	0	$sup(\mathcal{V}_{2^0+0}(X)) = sup(\mathcal{V}_0(X \setminus X[0])) - sup(\mathcal{V}_0(X))$ /* $sup(\{ab(-c)\}) = sup(\{ab\}) - sup(\{abc\})$ */
1	$\{ac\}$	0	$sup(\mathcal{V}_{2^1+0}(X)) = sup(\mathcal{V}_0(X \setminus X[1])) - sup(\mathcal{V}_0(X))$ /* $sup(\{a(-b)c\}) = sup(\{ac\}) - sup(\{abc\})$ */
		1	$sup(\mathcal{V}_{2^1+1}(X)) = sup(\mathcal{V}_1(X \setminus X[1])) - sup(\mathcal{V}_1(X))$ /* $sup(\{a(-b)(-c)\}) = sup(\{a(-c)\}) - sup(\{ab(-c)\})$ */
2	$\{bc\}$	0	$sup(\mathcal{V}_{2^2+0}(X)) = sup(\mathcal{V}_0(X \setminus X[2])) - sup(\mathcal{V}_0(X))$ /* $sup(\{(-a)bc\}) = sup(\{bc\}) - sup(\{abc\})$ */
		1	$sup(\mathcal{V}_{2^2+1}(X)) = sup(\mathcal{V}_1(X \setminus X[2])) - sup(\mathcal{V}_1(X))$ /* $sup(\{(-a)b(-c)\}) = sup(\{b(-c)\}) - sup(\{ab(-c)\})$ */
		2	$sup(\mathcal{V}_{2^2+2}(X)) = sup(\mathcal{V}_2(X \setminus X[2])) - sup(\mathcal{V}_2(X))$ /* $sup(\{(-a)(-b)c\}) = sup(\{(-b)c\}) - sup(\{a(-b)c\})$ */
		3	$sup(\mathcal{V}_{2^2+3}(X)) = sup(\mathcal{V}_3(X \setminus X[2])) - sup(\mathcal{V}_3(X))$ /* $sup(\{(-a)(-b)(-c)\}) = sup(\{(-b)(-c)\}) - sup(\{a(-b)(-c)\})$ */

Corollary 6.1.2. Let X be a non-empty pattern, and $i \in \{0, ..., |X| - 1\}$. The support of each variation in $C_i(X)$ can be calculated from the support of a variation of $X \setminus \{X[i]\}$, and the support of either X or its variation belonging to a cluster $C_l(X)$, where $l < i$.

Table 9 illustrates how the supports of consecutive variations of a pattern X can be calculated based on Theorem 6.1.1 and Corollary 6.1.2. Please note that given the support of X, and the supports of all variations of all proper $|X|-1$ item subsets of X one can calculate the supports of all the variations of X.

6.2 Building GDFLR with the Algorithm *GDFLR-SO-Apriori*

In this section, we recall the algorithm *GDFLR-SO-Apriori*. The main (and only) difference between *GDFLR-SO-Apriori* and the original *GDFLR-Apriori* is that the first one determines and uses the supports of variations instead of the errors of rules built from candidate patterns. The difference is highlighted in the code below.

Additional notation for *GDFLR-SO-Apriori*

- $X.Sup$ – table storing supports of all variations of pattern X; note: $|X.Sup| = 2^{|X|}$.
  Example: Let $X = \{ab\}$, then:
  $X.Sup[0] = X.Sup[(00)_2] = sup(\{(\ a)(\ b)\})$; $X.Sup[1] = X.Sup[(01)_2] = sup(\{(\ a)(-b)\})$;
  $X.Sup[2] = X.Sup[(10)_2] = sup(\{(-a)(\ b)\})$; $X.Sup[3] = X.Sup[(11)_2] = sup(\{(-a)(-b)\})$.

Algorithm. *GDFLR-SO-Apriori*(support threshold *minSup*);

$Main = \{\}$; $DBd^- = \{\}$; $IBd^- = \{\varnothing\}$; // initialize GDFLR
if $|\mathcal{D}| > minSup$ **then begin**

  $\varnothing.Sup[0] = |\mathcal{D}|$; move $\varnothing$ from IBd^- to $Main_0$; $X_1 = \{1$ item patterns$\}$;
  for $(k = 1; X_k \neq \varnothing; k$++$)$ **do begin**
    calculate supports of patterns in X_k within one scan of $\mathcal{D}$

```
  forall candidates X∈ X_k do begin
    Calculate-supports-of-variations(X, Main_{k-1});
    if max({X.Sup[l]| l = 0..2^k−1}) ≤ minSup then add X to IBd⁻_k
    elseif min({X.Sup[l]| l = 0..2^k−1}) = 0 then add X to DBd⁻_k
    else add X to Main_k endif
  endfor;
  /* create new (k+1)-candidates by merging k item patterns in Main */
  X_{k+1} = {X⊆I| ∃Y,Z ∈ Main_k (|Y∩Z| = k−1 ∧ X = Y∪Z)};
  /* remain only those candidates that have all k item subsets in Main */
  X_{k+1} = X_{k+1} \ {X∈ X_{k+1}| ∃Y⊂X (|Y| = k ∧ Y∉Main_k}
  endfor
endif;
return <∪_k Main_k, ∪_k DBd⁻_k, ∪_k IBd⁻_k>;
```

As one can observe, the function *Errors-of-rules* is replaced here by the procedure *Calculate-supports-of-variations*, which determines the supports of variations of candidate pattern X in two loops: the external one iterates over clusters of variations of X, whereas the internal one iterates over variations within a current cluster. Below a code of the procedure *Calculate-supports-of-variations* is presented. The supports of variations are calculated according to Theorem 6.1.1. Let us note that the procedure requires only $|X|$ accesses to the proper subsets of a given candidate pattern X, instead of $3^{|X|} - 2^{|X|}$ accesses, which would be performed by the equivalent function *Errors-of-rules* in the original algorithm *GDFLR-Apriori*.

```
procedure. Calculate-supports-of-variations(k-pattern X, Main_{k-1});
/* assert 1: X.Sup[0] stores the support of pattern X       */
/* assert 2: all k−1 item subsets of X are in Main_{k-1} and   */
/*           the supports of all their variations are known */
for (i = 0; i < k; i++) do begin          // focus on cluster C_i(X)
  Y = X \ {X[i]};  find Y in Main_{k-1};    // Y⊂X is accessed once per cluster
  for (j = 0; j < 2^i; j++) do begin
    X.Sup[2^i + j] = Y.Sup[j] − X.Sup[j];   // calculate support of j^{th} variation in cluster C_i(X)
    /* optionally break if max{X.Sup[l]| l=0..2^i + j}>minSup and min{X.Sup[l]| l=0..2^i + j}=0 */
  endfor;
  /* optionally break if max{X.Sup[l]| l=0..2^i + j}>minSup and min{X.Sup[l]| l=0..2^i + j}=0 */
endfor;
return;
```

Let us note, that in order to create and evaluate the $(k+1)$-item candidates *GDFLR-SO-Apriori* requires storing the k-itemsets of the *Main* component in GDFLR, along with the supports of all their variations, whereas the *GDFLR-Apriori* requires storing all i-itemsets from *Main*, $i=0...k$, along with their supports, and does not require storing the supports of the variations of the *Main* elements.

The *Apriori-GDFLR* algorithm can be further optimized by breaking the calculation of the supports of the variations of a candidate X as soon as the

Calculate-supports-of-variations procedure identifies a frequent variation of X (i.e. when $max(\{X.Sup[l]|\ l = 0..\ 2^i + j\}) > minSup$) and a generalized disjunctive variation of X (i.e. when $min(\{X.Sup[l]|\ l = 0..\ 2^i + j\}) = 0$). Satisfaction of these two conditions guarantees that X is an element of the generalized disjunctive border.

6.3 Building GDFSR and GDFSRN with *GDFSR-SO-Apriori* and *GDFSRN-SO-Apriori*

Now we introduce the *GDFSR-SO-Apriori* algorithm, which calculates the GDFSR representations based on the supports of the variations of currently maximal elements from *Main*, rather than on errors of rules. *GDFSR-SO-Apriori* can be obtained from *GDFSR-Apriori* in an analogical way, as *GDFLR-SO-Apriori* was obtained from *GDFLR-Apriori*. The introduced changes are shadowed in the code below.

Algorithm. *GDFSR-SO-Apriori*(database $\mathcal{D}$, support threshold *minSup*);

```
Main = {}; DBd⁻ = {}; IBd⁻ = {∅};                          // initialize GDFSR
if |𝒟| > minSup then begin
  ∅.sup = |𝒟|;
  move ∅ from IBd⁻ to Main₀;
  𝒳₁ = {1-itemsets};
  for (k = 1; 𝒳ₖ ≠ ∅; k++) do begin
    calculate the supports of itemsets in 𝒳ₖ within one scan of database 𝒟;
    forall candidates X∈𝒳ₖ do begin
      if X.sup ≤ minSup then add X to IBd⁻ₖ     // X is infrequent
      else begin                                // X is frequent
        /* calculate the supports of variants of X (according to Theorem 6.1.1) */
        /* until X's variation with zero support is identified or              */
        /* the supports of all variations of X are calculated;                */
        if there is variation of X with zero support then // X is a base of a certain gen. dis. rule
          add X to DBd⁻ₖ
        else add X to Mainₖ  endif
      endif
    endfor;
    /* create new (k+1)-candidates by merging k item patterns in Main */
    𝒳ₖ₊₁ = {X⊆I| ∃Y,Z ∈ Mainₖ (|Y∩Z| = k–1 ∧ X = Y∪Z)};
    /* remain only those candidates that have all k item subsets in Main */
    𝒳ₖ₊₁ = 𝒳ₖ₊₁ \ {X∈𝒳ₖ₊₁| ∃Y⊆X (|Y| = k ∧ Y∉Mainₖ}
  endfor
endif;
return <∪ₖ Mainₖ, ∪ₖ DBd⁻ₖ, ∪ₖ IBd⁻ₖ>;
```

Analogically to the modification we did in Subsection 5.2, in *GDFSR-Apriori* in order to obtain *GDFSRN-Apriori*, now we transform *GDFSR-SO-Apriori* to *GDFSRN-SO-Apriori*, which builds the GDFSRN representation with the use of an extended database.

Algorithm. *GDFSRN-SO-Apriori*(database $\mathcal{D}$, support threshold *minSup*);

$\mathcal{D}' = \mathcal{D}$;

extend all transactions in $\mathcal{D}'$ with negations of missing items;

return *GDFSR-SO-Apriori*($\mathcal{D}'$, *minSup*);

---

7 Experimental Results

In order to carry out experiments, we have expanded our tool [10], so that it contains the implementations of all the algorithms presented in this paper. A specific feature of our implementation is an application of hashing maps as proposed in [10], which guarantees a very efficient usage of main memory when searching patterns.

The experiments producing the GDFLR and GDFSR representations were carried out on the benchmark *mushroom* and *connect-4* data sets (accessible from http://www.ics.uci.edu/~mlearn/MLRepository.html). The *mushroom* dataset contains 8124 transactions; each of which consists of 23 items; the total number of distinct items in this data set is 119. The *connect-4* dataset contains 67557 transactions of length 43 items; the total number of distinct items is 129. The GDFSRN representation was built on extended with negations of missing items copies of the *mushroom* and *connect-4* datasets.

The experimental results are presented in Tables 10-13. The notation "N/A" used in the tables means that a respective representation was not built due to insufficient capacity of main memory or unacceptable long duration time of the calculations. Tables 10-11 provide the cardinalities of all three GDF representations w.r.t. *minSup* threshold value, which confirm the results obtained on a theoretical basis in Section 4. In particular, we observe that $|Main_{GDFLR}| = |Main_{GDFSRN}|$ and $|DBd^-_{GDFLR}| = |DBd^-_{GDFSRN}|$ for *minSup* $\geq 50\%$. On the other hand, as follows from Table 10, for very low *minSup* values (below 1%) the cardinality of GDFSR, which represents only frequent positive patterns in the *mushroom* dataset, becomes similar to the cardinality of GDFLR, which represents both positive patterns and patterns with negated items. In addition, the cardinality of GDFSRN is by an order of magnitude greater than the cardinality of GDFLR for *minSup* = 2% and this gap increases when lowering the value of the *minSup* threshold.

Table 10. Cardinalities of GDF representations for the *mushroom* dataset

Representation	GDFSR				GDFSRN				GDFLR			
minSup	|Main|	|DBd⁻|	|IBd⁻|	Total	|Main|	|DBd⁻|	|IBd⁻|	Total	|Main|	|DBd⁻|	|IBd⁻|	Total
0.012%	220 696	30 533	22 341	273 570	N/A	N/A	N/A	N/A	229 382	44 253	0	273 635
1%	33 822	15 430	22 092	71 344	3 429 417	411 625	450 517	4 291 559	229 382	44 253	0	273 635
2%	18 336	10 128	14 042	42 506	2 355 207	337 582	289 817	2 982 606	229 382	44 253	0	273 635
5%	7 040	4 786	6 208	18 034	1 178 303	223 955	153 541	1 555 799	229 246	44 253	90	273 589
10%	2 542	1 923	2 410	6 875	559 469	135 521	82 876	777 866	216 881	43 945	3 799	264 625
20%	645	528	905	2 078	187 647	61 534	36 655	285 836	134 916	37 708	14 974	187 598
30%	207	124	369	700	74 724	30 280	17 563	122 567	66 795	24 767	13 097	104 659
40%	89	57	233	379	31 366	16 531	9 059	56 956	30 417	15 337	8 137	53 891
50%	30	20	149	199	13 755	9 163	4 291	27 209	13 755	9 163	4 173	27 091
60%	15	6	121	142	6 195	5 864	2 156	14 215	6 195	5 864	2 038	14 097
70%	9	3	114	126	2 715	4 090	1 173	7 978	2 715	4 090	1 055	7 860
80%	8	3	115	126	1 262	2 782	610	4 654	1 262	2 782	492	4 536
90%	5	1	117	123	417	1 739	527	2 683	417	1 739	409	2 565
99%	1	1	118	120	35	172	285	492	35	172	167	374

Table 11. Cardinalities of GDF representations for the *connect-4* dataset

Representation	GDFSR				GDFSRN				GDFLR			
minSup	|Main|	|DBd⁻|	|IBd⁻|	Total	|Main|	|DBd⁻|	|IBd⁻|	Total	|Main|	|DBd⁻|	|IBd⁻|	Total
10%	19 471	9 696	8 388	37 555	N/A	N/A	N/A	N/A	N/A	N/A	N/A	N/A
20%	4 728	2 845	2 304	9 877	N/A	N/A	N/A	N/A	3 595 988	1 964 380	692 205	6 252 573
30%	1 625	1 595	890	4 110	1 609 513	845 398	461 542	2 916 453	1 541 747	819 120	427 568	2 788 435
40%	819	1 247	493	2 559	590 095	328 087	201 338	1 119 520	586 324	326 662	198 503	1 111 489
50%	462	935	382	1 779	208 774	122 113	83 740	414 627	208 774	122 113	83 611	414 498
60%	265	629	355	1 249	68 991	49 439	30 127	148 557	68 991	49 439	29 998	148 428
70%	161	384	245	790	21 968	21 815	10 569	54 352	21 968	21 815	10 440	54 223
80%	83	265	199	547	5 939	6 996	5 401	18 336	5 939	6 996	5 272	18 207
90%	22	177	141	340	646	2 170	975	3 791	646	2 170	846	3 662
99%	9	16	133	158	41	283	313	637	41	283	184	508

Table 12. Runtime of calculating 1) supports of variations by *GDFxx-SO-Apriori* algorithms and 2) errors of rules by *GDFxx-Apriori* algorithms for *mushroom*

	runtime [s]						runtime ratio		
minSup	GDFSR-SO-Apriori	GDFSR-Apriori	GDFSRN-SO-Apriori	GDFSRN-Apriori	GDFLR-SO-Apriori	GDFLR-Apriori	GDFSR-Apriori / GDFSR-SO-Apriori	GDFSRN-Apriori / GDFSRN-SO-Apriori	GDFLR-Apriori / GDFLR-SO-Apriori
0.012%	4,68	N/A	N/A	N/A	5,32	N/A	N/A	N/A	N/A
1%	0,59	13,46	113,65	N/A	5,26	1 099,57	22,7	N/A	209,1
2%	0,31	5,73	73,07	N/A	5,27	1 099,05	18,4	N/A	208,5
5%	0,13	1,62	34,46	5 699,26	5,24	1 097,43	12,9	165,4	209,3
10%	0,03	0,50	15,15	1 938,22	4,93	881,03	15,6	128,0	178,7
20%	0,02	0,06	4,43	356,27	3,06	291,30	3,9	80,4	95,3
30%	0,00	0,02	1,56	82,01	1,44	80,98		52,5	56,4
40%	0,00	0,00	0,58	21,55	0,59	22,84		37,3	38,6
50%	0,00	0,00	0,22	6,46	0,27	7,05		29,6	26,6
60%	0,00	0,00	0,09	1,70	0,09	2,04		18,3	22,0
70%	0,00	0,00	0,03	0,50	0,05	0,62		15,6	13,3
80%	0,00	0,00	0,03	0,17	0,02	0,19		5,7	12,5
90%	0,00	0,00	0,00	0,03	0,00	0,08			

Table 13. Runtime of calculating 1) supports of variations by *GDFxx-SO-Apriori* algorithms and 2) errors of rules by *GDFxx-Apriori* algorithms for *connect-4*

	runtime [s]						runtime ratio		
minSup	GDFSR-SO-Apriori	GDFSR-Apriori	GDFSRN-SO-Apriori	GDFSRN-Apriori	GDFLR-SO-Apriori	GDFLR-Apriori	GDFSR-Apriori / GDFSR-SO-Apriori	GDFSRN-Apriori / GDFSRN-SO-Apriori	GDFLR-Apriori / GDFLR-SO-Apriori
10%	0,31	4,24	N/A	N/A	N/A	N/A	13,6	N/A	N/A
20%	0,06	0,51	N/A	N/A	N/A	11 354,27	8,2	N/A	N/A
30%	0,05	0,14	57,14	2 875,78	55,85	1 146,58	3,1	50,33	20,5
40%	0,02	0,06	18,86	438,64	19,72	271,81	4,1	23,3	13,8
50%	0,00	0,05	5,90	96,32	6,60	63,61		16,3	9,6
60%	0,00	0,03	1,78	20,16	1,94	13,44		11,3	6,9
70%	0,00	0,02	0,56	4,63	0,58	3,05		8,3	5,3
80%	0,00	0,00	0,11	0,72	0,14	0,56		6,6	4,0
90%	0,00	0,00	0,02	0,05	0,03	0,03		2,9	1,0

The contents of Tables 12-13 allow us to compare the impact of applying support oriented- versus error oriented algorithms building the GDF representations. We have compared only the duration time of the operations that are different and crucial for the performance of the algorithms; namely, in the case of *GDFSR-SO-Apriori*, *GDFSRN-SO-Apriori*, and *GDFLR-SO-Apriori* algorithms, we report the runtime of calculating supports of variations of candidate sets, while in the case of *GDFSR-Apriori*, *GDFSRN-Apriori*, and *GDFLR-Apriori* algorithms, we report the runtime of calculating errors of rules based on candidate sets. The speed-up obtained by applying support oriented versions is particularly significant in the case of the algorithms building GDFSRN and GDFLR, and increases when lowering *minSup* value. As follows from Table 12, *GDFLR-SO-Apriori* is at least 200 times faster than *GDFLR-Apriori* for *minSup* ≤ 5%. Nevertheless, insufficient capacity of the main memory may disable the usage of this method. In the case of the *connect-4* dataset (see Table 13), we were not able to build the GDFSRN and GDFLR representations with support oriented algorithms for *minSup* ≤ 10%.

8 Summary and Conclusions

We have reviewed three GDF representations of frequent patterns, namely: GDFSR that represents all frequent positive patterns, and GDFSRN, as well as, GDFLR that represent all frequent patterns including those with negated items. GDFSR and GDFLR consist of only some positive patterns (itemsets), while GDFSRN consists of some patterns including those with negated items (literal sets). The GDF representations use generalized disjunctive rules to derive the supports of patterns. The GDF representations store some generalized disjunction-free patterns (i.e. *Main* component), some minimal generalized disjunctive patterns (i.e. *DBd⁻* component), and optionally some minimal infrequent patterns (i.e. *IBd⁻* component). In the paper, we have formulated and proved new theorems related to generalized disjunctive sets and generalized disjunction-free sets versus supports of patterns variations. We have found that all frequent variations of elements of the *DBd⁻* component of the GDFLR representation are generalized disjunctive. Eventually, we have found and proved a number of relationships between all three GDF representations. In particular, we have proved that for any given *minSup* value:

• $Main_{GDFSR} \cup DBd^-_{GDFSR} \cup IBd^-_{GDFSR} \subseteq Main_{GDFSRN} \cup DBd^-_{GDFSRN} \cup IBd^-_{GDFSRN}$;

• $Main_{GDFSR} \cup DBd^-_{GDFSR} \cup IBd^-_{GDFSR} \subseteq Main_{GDFLR} \cup DBd^-_{GDFLR} \cup IBd^-_{GDFLR}$;

• $|Main_{GDFSR}| \leq |Main_{GDFLR}| \leq |Main_{GDFSRN}|$ and $|DBd^-_{GDFSR}| \leq |DBd^-_{GDFLR}| \leq |DBd^-_{GDFSRN}|$.

It follows from the last dependency that the cardinality of the lossless reduced GDFLR representation obtained by discarding redundant IBd^-_{GDFLR} component does not exceed the cardinality of the reduced GDFSRN representation obtained by discarding its IBd^-_{GDFSRN} component. In addition, we have proved that for *minSup* = 0, the elements of GDFLR are identical with the elements of GDFSR, and GDFSRN contains all variations of each element of GDFLR and GDFSR. This implies that for *minSup* = 0, the cardinality of GDFSRN is in practice by orders of magnitude greater than the cardinalities of GDFLR and GDFSR, since any element *X* of GDFLR and GDFSR has $2^{|X|}$ variations in GDFSRN.

In the paper, we provided a complete set of error oriented and support oriented versions of algorithms for calculating each GDF representation. *GDFSR-Apriori* and *GDFSRN-Apriori* are adapted versions of the *GDFLR-SO-Apriori* algorithm, while *GDFSR-SO-Apriori* and *GDFSRN-SO-Apriori* are adapted versions of the *GDFLR-SO-Apriori* algorithm. The experiments prove that the support oriented versions of the algorithms, which apply a particular order of calculating the supports of variations of candidates, are faster than respective error oriented versions by up to two orders of magnitude for low support threshold values. The experimental results confirm also the properties related to cardinalities of the GDF representations that we found on a theoretical basis.

References

1. Agrawal, R., Imielinski, T., Swami, A.: Mining Associations Rules between Sets of Items in Large Databases. In: Proceedings of the ACM SIGMOD, Washington, USA, pp. 207–216 (1993)
2. Agrawal, R., Mannila, H., Srikant, R., Toivonen, H., Verkamo, A.I.: Fast Discovery of Association Rules. In: Advances in KDD, pp. 307–328. AAAI, Menlo Park (1996)
3. Baptiste, J., Boulicaut, J.-F.: Constraint-Based Discovery and Inductive Queries: Application to Association Rule Mining. In: Proceedings of Pattern Detection and Discovery, ESF Exploratory Workshop, London, UK, pp. 110–124 (2002)
4. Baralis, E., Chiusano, S., Garza, P.: On Support Thresholds in Associative Classification. In: Proceedings of SAC 2004 ACM Symposium on Applied Computing, Nikosia, Cyprus, pp. 553–558 (2004)
5. Boulicaut, J.-F., Bykowski, A., Rigotti, C.: Approximation of Frequency Queries by Means of Free-Sets. In: Zighed, D.A., Komorowski, J., Żytkow, J.M. (eds.) PKDD 2000. LNCS (LNAI), vol. 1910, pp. 75–85. Springer, Heidelberg (2000)
6. Bykowski, A., Rigotti, C.: A Condensed Representation to Find Frequent Patterns. In: Proceedings of PODS 2001 ACM SIGACT-SIGMOD-SIGART Symposium on Principles of Database Systems, Santa Barbara, USA, pp. 267–273 (2001)
7. Calders, T.: Axiomatization and Deduction Rules for the Frequency of Itemsets, Ph.D. Thesis, University of Antwerp (2003)
8. Calders, T., Goethals, B.: Mining All Non-Derivable Frequent Itemsets. In: Elomaa, T., Mannila, H., Toivonen, H. (eds.) PKDD 2002. LNCS (LNAI), vol. 2431, pp. 74–85. Springer, Heidelberg (2002)
9. Calders, T., Goethals, B.: Minimal k-free representations of frequent sets. In: Lavrač, N., Gamberger, D., Todorovski, L., Blockeel, H. (eds.) PKDD 2003. LNCS (LNAI), vol. 2838, pp. 71–82. Springer, Heidelberg (2003)
10. Cichon K.: Fast Discovering Representation of Frequent Patterns with Negation and Reducts of Decision Tables, Ph.D. Thesis, Warsaw University of Technology (2006) (in Polish)
11. Gołębski M.: Inducing Grammair Rules for Polish, Ph.D. Thesis, Warsaw University of Technology (2007) (in Polish)
12. Han, J., Kamber, M.: Data Mining: Concepts and Techniques. Morgan Kaufmann Publishers, San Francisco (2000)
13. Harms, S.K., Deogun, J., Saquer, J., Tadesse, T.: Discovering Representative Episodal Association Rules from Event Sequences using Frequent Closed Episode Sets and Event Constraints. In: Proceedings of ICDM 2001 IEEE International Conference on Data Mining, San Jose, California, USA, pp. 603–606 (2001)
14. Kryszkiewicz, M.: Closed Set Based Discovery of Representative Association Rules. In: Hoffmann, F., Adams, N., Fisher, D., Guimarães, G., Hand, D.J. (eds.) IDA 2001. LNCS, vol. 2189, pp. 350–359. Springer, Heidelberg (2001)
15. Kryszkiewicz, M.: Concise Representation of Frequent Patterns Based on Disjunction–Free Generators. In: Proceedings of ICDM 2001 IEEE International Conference on Data Mining, San Jose, California, USA, pp. 305–312 (2001)

16. Kryszkiewicz, M.: Inferring Knowledge from Frequent Patterns. In: Bustard, D.W., Liu, W., Sterritt, R. (eds.) Soft-Ware 2002. LNCS, vol. 2311, pp. 247–262. Springer, Heidelberg (2002)
17. Kryszkiewicz, M.: Concise Representations of Frequent Patterns and Association Rules. Publishing House of Warsaw University of Technology, Warsaw (2002)
18. Kryszkiewicz, M.: Concise Representations of Association Rules. In: Proceedings, Pattern Detection and Discovery, ESF Exploratory Workshop, London, UK, pp. 92–109 (2002)
19. Kryszkiewicz, M.: Reducing Infrequent Borders of Downward Complete Representations of Frequent Patterns. In: Proceedings of The First Symposium on Databases, Data Warehousing and Knowledge Discovery, Baden-Baden, Germany, pp. 29–42 (2003)
20. Kryszkiewicz, M.: Closed set based discovery of maximal covering rules. International Journal of Uncertainty, Fuzziness and Knowledge-Based Systems 11(Supplement-1), 15–29 (2003)
21. Kryszkiewicz, M.: Reducing Borders of k-disjunction Free Representations of Frequent Patterns. In: Proceedings, SAC 2004 ACM Symposium on Applied Computing, Nikosia, Cyprus, pp. 559–563 (2004)
22. Kryszkiewicz, M.: Upper Bound on the Length of Generalized Disjunction Free Patterns. In: Proceedings of SSDBM 2004 International Conference on Scientific and Statistical Database Management, Santorini, Greece, pp. 31–40 (2004)
23. Kryszkiewicz, M.: Reducing Main Components of k-disjunction Free Representations of Frequent Patterns. In: Proceedings of IPMU 2004 International Conference in Information Processing and Management of Uncertainty in Knowledge-Based Systems, Perugia, Italy, pp. 1751–1758 (2004)
24. Kryszkiewicz, M.: Generalized Disjunction-Free Representation of Frequent Patterns with Negation, pp. 63–82. JETAI, Taylor & Francis Group, UK (2005)
25. Kryszkiewicz, M.: Reasoning about Frequent Patterns with Negation. In: Encyclopedia of Data Warehousing and Mining, pp. 941–946. Information Science Publishing, Idea Group (2005)
26. Kryszkiewicz, M.: Generalized Disjunction-Free Representation of Frequents Patterns with at Most k Negations. In: Ng, W.-K., Kitsuregawa, M., Li, J., Chang, K. (eds.) PAKDD 2006. LNCS (LNAI), vol. 3918, pp. 468–472. Springer, Heidelberg (2006)
27. Kryszkiewicz, M., Cichon, K.: Support Oriented Discovery of Generalized Disjunction-Free Representation of Frequent Patterns with Negation. In: Ho, T.-B., Cheung, D., Liu, H. (eds.) PAKDD 2005. LNCS (LNAI), vol. 3518, pp. 672–682. Springer, Heidelberg (2005)
28. Kryszkiewicz, M., Gajek, M.: Concise Representation of Frequent Patterns Based on Generalized Disjunction-Free Generators. In: Chen, M.-S., Yu, P.S., Liu, B. (eds.) PAKDD 2002. LNCS (LNAI), vol. 2336, pp. 159–171. Springer, Heidelberg (2002)
29. Kryszkiewicz, M., Gajek, M.: Why to Apply Generalized Disjunction-Free Generators Representation of Frequent Patterns? In: Hacid, M.-S., Raś, Z.W., Zighed, D.A., Kodratoff, Y. (eds.) ISMIS 2002. LNCS (LNAI), vol. 2366, pp. 383–392. Springer, Heidelberg (2002)
30. Kryszkiewicz, M., Rybiński, H., Gajek, M.: Dataless Transitions between Concise Representations of Frequent Patterns. Journal of Intelligent Information Systems 22(1), 41–70 (2004)
31. Kryszkiewicz, M., Rybinski, H., Muraszkiewicz, M.: Data Mining Methods for Telecom Providers. In: MOST 2002 Conference, Warsaw (2002)
32. Kryszkiewicz, M., Skonieczny, Ł.: Hierarchical Document Clustering Using Frequent Closed Sets. In: Advances in Soft Computing, pp. 489–498. Springer, Heidelberg (2006)

33. Mannila, H., Toivonen, H.: Multiple Uses of Frequent Sets and Condensed Representations. In: Proceedings of KDD 1996, Portland, USA, pp. 189–194 (1996)
34. Nykiel, T., Rybinski, H.: Word Sense Discovery for Web Information Retrieval. In: MCD Workshop 2008, Piza, ICDM (2008)
35. Pasquier, N.: Data mining: Algorithmes d'extraction et de Réduction des Règles d'association dans les Bases de Données, Ph.D. thesis, Université Blaise Pascal – Clermont–Ferrand II (2000)
36. Pasquier, N., Bastide, Y., Taouil, R., Lakhal, L.: Efficient Mining of Association Rules Using Closed Itemset Lattices. Journal of Information Systems 24(1), 25–46 (1999)
37. Pei, J., Dong, G., Zou, W., Han, J.: On Computing Condensed Frequent Pattern Bases. In: Proceedings of ICDM 2002 IEEE International Conference on Data Mining, Maebashi, Japan, pp. 378–385 (2002)
38. Phan-Luong, V.: Representative Bases of Association Rules. In: Proceedings of ICDM 2001 IEEE International Conference on Data Mining, San Jose, California, USA, pp. 639–640 (2001)
39. Rybinski, H., Kryszkiewicz, M., Protaziuk, G., Jakubowski, A., Delteil, A.: Discovering Synonyms based on Frequent Termsets. In: Kryszkiewicz, M., Peters, J.F., Rybiński, H., Skowron, A. (eds.) RSEISP 2007. LNCS (LNAI), vol. 4585, pp. 516–525. Springer, Heidelberg (2007)
40. Rybinski, H., Kryszkiewicz, M., Protaziuk, G., Kontkiewicz, A., Marcinkowska, K., Delteil, A.: Discovering Word Meanings Based on Frequent Termsets. In: Raś, Z.W., Tsumoto, S., Zighed, D.A. (eds.) MCD 2007. LNCS (LNAI), vol. 4944, pp. 82–92. Springer, Heidelberg (2008)
41. Toivonen, H.: Discovery of Frequent Patterns in Large Data Collections. Ph.D. Thesis, Report A-1996-5, University of Helsinki (1996)
42. Saquer, J., Deogun, J.S.: Using Closed Itemsets for Discovering Representative Association Rules. In: Ohsuga, S., Raś, Z.W. (eds.) ISMIS 2000. LNCS (LNAI), vol. 1932, pp. 495–504. Springer, Heidelberg (2000)
43. Savinov, A.: Mining Dependence Rules by Finding Largest Itemset Support Quota. In: Proceedings of SAC 2004 ACM Symposium on Applied Computing, Nikosia, Cyprus, pp. 525–529 (2004)
44. Tsumoto, S.: Discovery of Positive and Negative Knowledge in Medical Databases Using Rough Sets. In: Arikawa, S., Shinohara, A. (eds.) Progress in Discovery Science. LNCS (LNAI), vol. 2281, pp. 543–552. Springer, Heidelberg (2002)
45. Zaki, M.J., Hsiao, C.J.: CHARM: An Efficient Algorithm For Closed Itemset Mining. In: Proceedings of SIAM 2002 International Conference on Data Mining, Arlington, USA (2002)

Part IV
Classification and Beyond

A System to Detect Inconsistencies between a Domain Expert's Different Perspectives on (Classification) Tasks

Derek Sleeman[1], Andy Aiken[1], Laura Moss[1], John Kinsella[2], and Malcolm Sim[2]

[1] Department of Computing Science, University of Aberdeen,
Aberdeen, AB24 3UE
{d.sleeman,a.aiken,lmoss}@abdn.ac.uk
[2] Department of Anaesthesia, Glasgow Royal Infirmary, University of Glasgow,
Glasgow, G31 2ER
j.kinsella@clinmed.gla.ac.uk

Abstract. This paper discusses the range of knowledge acquisition, including machine learning, approaches used to develop knowledge bases for Intelligent Systems. Specifically, this paper focuses on developing techniques which enable an expert to detect inconsistencies in 2 (or more) perspectives that the expert might have on the same (classification) task. Further, the INSIGHT system has been developed to provide a tool which supports domain experts exploring, and removing, the inconsistencies in their conceptualization of a task. We report here a study of Intensive Care physicians reconciling 2 perspectives on their patients. The high level task which the physicians had set themselves was to classify, on a 5 point scale (A-E), the hourly reports produced by the Unit's patient management system. The 2 perspectives provided to INSIGHT were an annotated set of patient records where the expert had selected the appropriate category to describe that snapshot of the patient, and a set of rules which are able to classify the various time points on the same 5-point scale.

Inconsistencies between these 2 perspectives are displayed as a confusion matrix; moreover INSIGHT then allows the expert to revise both the annotated datasets (correcting data errors, and/or changing the assigned categories) and the actual rule-set. Each expert achieved a very high degree of consensus between his refined knowledge sources (i.e., annotated hourly patient records and the rule-set).

Further, the consensus between the 2 experts was ~95%. The paper concludes by outlining some of the follow-up studies planned with both INSIGHT and this general approach.

1 Introduction

Contemporary knowledge-based systems, as their expert systems predecessors (Buchanan & Shortliffe 1984), have 2 principal components, namely, a task-specific inference engine, and the corresponding associated domain-specific knowledge base. If the area of interest is both large and complex then it is likely that knowledge engineers will

J. Koronacki et al. (Eds.): Advances in Machine Learning II, SCI 263, pp. 293–314.
springerlink.com   © Springer-Verlag Berlin Heidelberg 2010

spend a great deal of time and effort producing the appropriate knowledge base (KB), and so various efforts have been made to reuse existing knowledge bases whenever possible, (Corsar & Sleeman 2007). This paper surveys a number of methods by which KBs can be produced from scratch including: traditional interviewing, computer-based tools which have incorporated classical psychological approaches such as card sort, systems to acquire information to support a particular problem solver (PS), the use of machine learning in knowledge acquisition / capture, as well as more recent attempts to infer information from data sets produced by large numbers of users of systems like Open Mind, (Singh et al, 2002).

A central problem, since the inception of Expert Systems, is how to deal with the uncertainty inherent in such knowledge bases (Buchanan & Shortliffe, 1984). EMYCIN (Buchanan & Shortliffe, 1984) associated certainty factors with particular pieces of information (both facts & rules) and evolved a calculus which allows the uncertainty associated with decisions to be calculated, and then reported to the user. Bayesian Networks have developed these ideas further, so that it is possible for decision support systems to identify a range of possible decisions and to associate each with strength of belief, (Pearl, 1988). Both these approaches provide pragmatic approaches to the handling of uncertainty associated with expertise. Clearly, however there are different types of uncertainty associated with pieces of knowledge including the fact that even experts retain incorrect information, and further they can also misapply information. Developing techniques for capturing and refining expertise is an important sub-activity at the intersection of Cognitive Psychology & Artificial Intelligence.

The focus of the work reported here is an attempt to get experts to provide 2 perspectives on a classification task, and then to provide a system / tool which enables the domain expert to appreciate when a particular entity has been classified differently by the 2 perspectives. Further, the tool provides the expert with support in revising one or both of the knowledge sources until a consensus is reached (or the expert abandons that particular task). As usual we believe it is vital that this activity is grounded in a real-world task and we have chosen the classification of hourly Intensive Care Unit (ICU) patient records; specifically the domain expert's task was to classify records (which can contain up to 60 pieces of information) on a 5-point A-E scale where E is severely ill.

The rest of the paper is structured as follows: section 2 gives an overview of ICU patient management systems, and the types of information which they produce; additionally, patient scoring systems are discussed. Section 3 gives an overview of the cognitive science literature on expertise, on knowledge acquisition / capture including the important role which machine learning has played in these activities; thirdly we review cooperative knowledge acquisition and knowledge refinement systems. Section 4 provides a conceptual overview of the INSIGHT system which takes 2 perspectives on an expert's classification knowledge, detects inconsistencies between them, and allows the domain expert to revise both knowledge sources to see if a consensus on the current task can be reached. Section 5 describes the use of INSIGHT by experts to reconcile 2 perspectives of their knowledge about ICU patients; namely a set of annotated patient records and a rule-set which covers each of the 5 categories (A-E). A high level of consensus was achieved by both experts. Section 6 outlines several of the contributions of this work. Section 7 concludes the paper by outlining some planned follow-up studies.

2   Overview of Patient Management Systems Used in Intensive Care Units (ICUs)

This section gives an overview of patient monitoring systems which are used in Intensive Care Units (ICUs), together with examples of parameters collected. We also discuss the need for patient scoring systems, and outline a 5-point qualitative scale which we have developed.

Many ICUs have patient management systems which collect the patients' physiological parameters, records nursing activities, and other interventions (such as the administration of drugs and boluses of fluids). This information is typically collected at specified time periods say every minute or hour, is recorded on a data base associated with the patient monitoring system, and is continuously available on a monitor at the patient's bedside where it is usually displayed as a conventional chart; this is the form of the information which clinicians use when they attend patients. Thus many ICUs are now paperless. Often this information is not systematically analysed subsequently for trends or inconsistencies in the data sets. This is the focus of an aspect of our work which has led us to produce the ACHE (Architecture for Clinical Hypothesis Evaluation) infrastructure, (Moss et al., 2008). That paper also outlines one preliminary study which we have undertaken with ACHE to identify the occurrence of Myocardial Infarctions in this group of ICU patients.

The patient management system used at Glasgow Royal Infirmary (GRI), a Phillips CareVue, records up to 60 parameters. Table 1 lists the principal parameters, and lists the frequency of recording in the current data set. It should be noted that the data sets which we analyse are extracted from the patient database, de-identified, and output as a spreadsheet; the spreadsheet is then input to the ACHE system & different analyses are performed on the data "off-line".

2.1 Patient Scoring System

For a variety of reasons it would be helpful to clinicians if they were able to obtain a regular summary of each patient's overall condition. Such information would be useful to determine whether there has been any appreciable progress / deterioration, would be a useful summary for the next shift of clinical staff, and could be included as a component of a discharge summary. To date the APACHE-2 scale (Knaus *et al.*, 1991) is widely used in ICUs in the western world, but the APACHE score is created only once during a patient's ICU stay, usually 24 hours after admission. Additionally this scoring system does not take into account the effect of interventions on a patient. For example if a patient has a very low blood pressure this is clearly a very serious condition, but it is even more serious if the patient has this blood pressure despite having received a significant dose of a drug like Adrenaline.[1]

---

[1] Adrenaline normally raises a patient's blood pressure through its inotropic effect.

Table 1. Parameters used in the study

Parameter	Recorded Interval
Heart Rate	Hourly
Temperature	Hourly
Mean Arterial Pressure (MAP)	Hourly
Diastolic	Hourly
Systolic	Hourly
FiO_2	Hourly
SpO_2	Hourly
Urine output	Hourly
Central Venous Pressure (CVP)	Hourly
LiDCO (If applicable to patient)	Hourly
Drug Infusions (eg Adrenaline, Noradrenaline)	As applicable
Fluid Infusions	As applicable
Dialysis Sessions	As applicable

The clinical authors of this paper (JK & MS) have been addressing this issue for some while. More recently we have produced a 5-point (high-level) qualitative description of ICU patients, which can be summarized as follows:

E   Patient is highly unstable with say a number of his physiological parameters (e.g., blood pressure, heart rate) having extreme values (either low or high).

D   Patient more stable than patients in category E but is likely to be receiving considerable amounts of support (e.g., fluid boluses, drugs such as Adrenaline, & possible high doses of oxygen)

C   Either more stable than patients in category D or the same level of stability but on lower levels of support (e.g., fluids, drugs & inspired oxygen)

B   Relatively stable (i.e., near normal physiological parameters) with low levels of support

A   Normal physiological parameters *without* use of drugs like Adrenaline, only small amounts of fluids, and low doses of inspired oxygen

For more details on the descriptions, please see Appendix A.

The objective of the study is to derive a series of rules which can be used with a high degree of consistency, to classify the hourly patient reports produced by the patient management system. The top-level outline of the study is:

- The administrator of the patient management system produced listings (in spreadsheet format) for 10 patients' complete stays in the ICU (the number of days varied from 1-23 days)
- One of the clinical investigators (MS) annotated each of the hourly records (nearly 3000 records in all) with his assessment of the patient's status on the 5-point qualitative scale on the basis of the information provided by the Phillips CareVue system i.e., that contained in the spreadsheets
- Further we asked the same clinician to articulate rules to describe each of the 5 categories, (i.e., A-E).

- We used the INSIGHT tool (described below) to help this clinician make this data set & his rule set more consistent by modifying, as he saw fit, either the annotations in the date set, his rule-set, or both.
- The second clinician (JK) annotated 3 of the patients' data sets, again using the same qualitative scale (A-E)
- We used the INSIGHT tool to help the second clinician (JK) make his data set consistent with the rule set produced by the first clinician. This clinician was, of course, allowed to modify both his annotations of the data set & the actual rules).

More details of this study are given in section 5.

3 Literature Overview

This section gives a Cognitive Science perspective on the acquisition of expertise (section 3.1), provides an overview of knowledge acquisition (including machine learning) approaches in section 3.2, and discusses cooperative knowledge acquisition and knowledge refinement systems in section 3.3.

3.1 The Cognitive Science Perspective on the Acquisition of Expertise

The classic book on Protocol Analysis by Ericsson & Simon (1993) argues that to acquire a person's genuine expertise it is essential that one does not get the expert to articulate what they do *in the abstract*, but one should essentially observe what they do when solving an *actual task*. In the case of protocol analysis they further argue that the process of verbalizing the steps of problem solving does not perturb the expert's actual problem solving processes. Effectively, Ericsson & Simon introduced the distinction between "active" knowledge which is used to solve tasks as opposed to "passive" knowledge which is used to discuss tasks / a domain.

This has been a recurrent theme / perspective in much of cognitive science and in the study of expertise since that time, as is illustrated by the very nice study reported by Johnson (1983). This investigator attended a medical professor's lectures on diagnosis where he explained the process. The investigator then accompanies the professor's ward round (with a group of medical students) & noticed a difference in his procedures. When challenged about these differences the medical professor said:

> *"Oh, I know that, but you see I don't know how I do diagnosis, and yet I need to teach things to students. I create what I think of as plausible means for doing tasks and hope students will be able to convert them into effective ones."*

Thus the essential "rule" of expertise / knowledge acquisition (KA) is that one should ask an expert to solve specific task(s), and (preferably) explain what s/he is doing as the task proceeds; one should **not** normally ask a domain expert to discuss their expertise in the abstract (this includes asking an expert to articulate rules and procedures they use to solve tasks).

3.2   Summary of Knowledge Acquisition Including Uses of Machine Learning to Extract Domain Knowledge in a Number of Domains

In a recent overview at the K-CAP 2007 conference, Sleeman (Sleeman et al, in press) argued that Knowledge Acquisition (KA) is "a broad church" and consists of a very wide range of approaches including:

- Interviewing of domain experts by Knowledge Engineers: an approach which was dominant in the early development of Expert Systems (Buchanan & Shortliffe, 1984).
- Techniques, including card sort, repertory grids, laddering, which had originally been developed by psychologists as "manual" techniques which Computer Scientists redeveloped as a series of Computer-based systems, (Diaper, 1989).
- Problem-Solving Method (PSM) driven systems such as MOLE, MORE, SALT acquire more focussed information which is sufficient to satisfy a *particular* type of problem solver / PSM. The use of these systems is less demanding for the domain expert as the information collected is generally less, and the purpose of the information collected is usually more apparent, (Marcus & McDermott, 1989.)
- Machine-learning approaches have played an important role of transforming sets of usually labelled instances into knowledge (usually rule sets). Given the context of this volume I provide some more detail of these approaches below.
- Natural Language techniques (specifically Information Extraction approaches) have now matured to the point where they have been successfully applied to a number of textual sources & have extracted useful information (Etzioni et al, 2005).
- Captalizing on greater connectivity & the willingness of some people to provide samples of texts, and to complete sentences in meaningful ways. Systems like OpenWorld have collected vast corpora which they have then analysed using statistical techniques to extract some very interesting concepts & associations (Singh et al, 2002). Similarly, von Ahn has exploited peoples' enthusiasm for online game playing, von Ahn (2006).

Michalski and Chilausky (1980) had a notable early success using Machine Learning approaches to extracting knowledge / rules from instances, in the domain of crop disease. The soya bean crop is of major importance to the state of Illinois, and so it employed a number of plant pathologists to advise farmers on crop diseases. Michalski and Chilausky studied the standard reference book on the subject, and also spent 40 or so hours interviewing an expert. This allowed them to determine, what they believed were, the appropriate set of descriptors for soya bean diseases. Subsequently, they developed a questionnaire to illicit, from farmers, examples of actual crop diseases which they had experienced; in fact, they obtained nearly 700 such cases. They then trained a version of the ID3 program (Quinlan, 1986) with 307 instances, and used the trained system to classify 376 test cases. The performance of the trained program was impressive; it only misclassified 2 instances whereas humans following the information given in the standard textbook misclassified 17% of the cases. A final step in this project was to extract a series of IF-THEN-ELSE rules from the ID3 tree, for day-to-day use by the plant pathologists and farmers.

For a more recent survey of the application of Machine Learning approaches to real-world tasks, see Langley & Simon (1995).

3.3   Cooperative Knowledge Acquisition and Knowledge Refinement Systems

Building large knowledge bases is a demanding task; and particularly so if one is working in a domain where the knowledge / information is still "fluid". When one attempts to use such knowledge bases in conjunction with an appropriate inference engine to solve real-world tasks, one often finds that information is missing (and hence needs to be acquired), or the system gives answers to tasks which the domain expert says are incorrect (and hence the knowledge base needs to be refined). Again, if the domain is at the cutting edge of human knowledge then it is not possible to draw on an existing source of knowledge to support the processes of acquisition and refinement noted above, and hence one must use a well-chosen domain expert to act as the oracle. For obvious reasons the systems which have been built, by our group and others, to fulfil this role are often referred to as *Cooperative Knowledge Acquisition and Knowledge Refinement systems.* See Sleeman (1994) for a review of such systems. Over the last decade or more we have implemented systems which are able to refine knowledge bases (KBs) in a variety of formalisms including rules, cases, taxonomies, and causal graphs. The family of systems which are most relevant to this discussion are those which are able to refine cases, and they are discussed in the next sub-section.

3.3.1   The REFINER Systems
The REFINER family of programs have been designed to detect inconsistencies in a set of labelled cases. That is, these systems are provided with a set of categories which the domain expert believes are relevant to the domain, a set of descriptors needed to describe the domain, and a set of labelled cases / instances. The descriptors can be of a variety of types including real, integer, string and hierarchical. If the latter, then the system requires some further information about the nature of the taxonomy (for example, Figure 1). Table 2 shows a set of cases including the categories assigned by the domain expert to each case. At the heart of each system is an algorithm which forms a category description from say all the instances of category A, bearing in mind the actual types of the variables. This process is repeated for each of the categories. Table 3 shows the category descriptions which the algorithm infers for this dataset. The systems then check to see whether the set of inferred categories are consistent (i.e., not overlapping with other categories). The set of cases is said to be consistent if each category can be distinguished from the other categories by a particular feature or a particular feature-value pair.

Fig. 1. The hierarchy for the Disease descriptor

Table 2. Sample dataset used to illustrate Refiner++

Case	Heart Rate (HR)	Diastolic Blood Pressure (DBP)	Disease	Category
1	50	90	Disease 1	A
2	56	90	Disease 2	A
3	52	101	Disease 3	A
4	50	95	Disease 3	B
5	56	97	Disease 3	B
6		89	Disease 5	A
7	52	97	Disease 3	B

Table 3. The category descriptions generated by Refiner++

Category	HR	DBP	Disease
A	50 – 56	89 – 101	Any Disease
B	50 – 56	95 – 97	Disease 3

If the set of cases is inconsistent then the algorithm further suggests ways in which the inconsistency(s) might be removed, these include:

- Changing a value of a feature of a case (due perhaps to a typing error)
- Reclassifying a case / instance
- Shelving a case to work on it subsequently
- Adding an additional descriptor to all the cases
- Creating a disjunction by excluding a value or range of values from a category description

Considering the dataset shown in Table 2, the category descriptions are inconsistent (a case with a DBP value in the range 95 – 97 and a Disease value of Disease 3 could not be unambiguously categorised) and so the user would be presented with a set of disambiguation options such as:

- Exclude 95 – 97 from category A's DBP range
- Change the value of DBP in case 4 to 97
- Change the value of Disease in case 3 to Disease 1, Disease 2 or Disease 5
- Add a new descriptor to distinguish between these categories

If, for example, the user opts to create a disjunction, the categories are now distinct. Table 4 shows the updated (non-overlapping) category.

Table 4. Updated category descriptions

Category	HR	DBP	Disease
A	50 – 56	89 – 101, except 95 – 97	Any Disease
B	50 – 56	95 – 97	Disease 3

We have so far effectively implemented 3 systems:

- **REFINER** (Sharma & Sleeman, 1988) was the first system; it was incremental in that it processed a single case / instance and attempted at each stage to remove any inconsistencies detected.
- **REFINER+:** The clear disadvantage of REFINER was that a change made to accommodate an inconsistency associated with case(n) might be reversed when case(n+1) was considered, and so REFINER+ implemented a "batch" algorithm. Namely all the instances were available before any of the category descriptions were created, and hence it was able to avoid much of the unnecessary work done in the initial system.

  When REFINER+ was used with a small number of cases it was quite effective, however the number of inconsistencies noted in a sizable data set could be overwhelming for the expert. To help contain the situation we evolved several heuristics namely:
  - A change which removes a considerable number of inconsistencies is preferred over one which removes a smaller number of inconsistencies;
  - A change which makes a smaller number of changes to the data set is preferred over one which makes more extensive changes
- **REFINER DA:** The essential difference between REFINER DA & its predecessor REFINER+ is that it combined aspects of the two earlier systems. Namely the domain expert is asked to suggest several cases which he/she thought were prototypical of the several categories, from which descriptions of the several categories were inferred as described above. Then this version of REFINER attempted to cover additional cases without causing the set of category descriptors to become inconsistent.

3.3.2 Critique of REFINER DA and Effectively the REFINER Family of Systems

The machine learning algorithm attempts to create, in each version of REFINER, a set of non-overlapping descriptions for the categories; moreover, each of the descriptors is used in each of the categories. Further, the descriptor-value pair which effectively discriminates category A from category B is produced by the machine learning algorithm, and hence is greatly influenced by the set of cases presented to the system. The domain expert's intuitions are not used in guiding this selection of features. So in principle the feature-value pair SpO_2 (96-100) could be used to determine that a patient was in category A (i.e., dischargeable), whereas if that same case had a further feature-value pair of FiO_2: (95-100), this would be clinically described as a very sick patient. So from working with REFINER DA with this data set we made two important observations:

- The feature-value pairs chosen to make a category distinct are often not very intuitive to a domain expert. (The same comment can of course be made of the output from other machine learning algorithms such the decision trees created by C4.5)
- An expert might effectively sub-divide a category like E into a number of subcategories, which he might not initially articulate. (That is a patient can be in

category E for one of several *distinct* reasons: e.g., poor heart rate, or poor oxygen saturation.) If the domain expert does not articulate these sub-categories then category E will be an amorphous category which will influence the descriptions inferred and this in turn will affect the other categories inferred by REFINER. Additionally if the sub-categories are articulated then it is likely that there will only be a small number of examples in each of the sub-categories, which again will mean that the machine learning algorithms will have difficulties in extracting domain-relevant descriptions.

In the next section we outline a further system which we have developed, called INSIGHT, which addresses these issues.

4   Conceptual Design of INSIGHT

Below we give the design criteria for a system, INSIGHT, which we believe addresses (some of) the difficulties noted at the end of the last section.

- Have the experts describe each of the categories & sub-categories in terms of features *which the expert believes are appropriate*. Effectively the expert provides us with a set of classification rules for the domain. (This knowledge source will be considered to be a perspective which the expert holds on this domain.)
- All the REFINER systems require the domain expert to assign a category (a label) to each of the instances. We are continuing with this practise here as it gives us a further perspective on the set of cases / instances.
- Compare the expert's two perspectives on the domain; namely, the rules the expert has articulated for each of the categories versus the annotations he /she has associated with each of the cases.
- We have implemented a system, INSIGHT, which compares these two perspectives. So instead of using a machine learning algorithm as the core of the system we are, in this approach, using a system to check the consistency between the 2 expert-provided perspectives.[2]

As noted above, INSIGHT is a development of the REFINER family of systems, yet incorporates a somewhat different approach. Whereas the REFINER systems are able to infer descriptions of categories from a set of instances and to detect inconsistencies and suggest how they might be resolved, the INSIGHT system highlights discrepancies in two perspectives of an expert on a particular (classification) task, and brings these to the attention of the domain expert. In particular this realization of the checking tool, INSIGHT is able to handle annotated cases where the expert assigns each instance to one of the pre-designated set of categories. The second source of information is a set of rules which are able to classify each of the cases / instances. INSIGHT displays the results of such comparisons as a confusion matrix; an example of a confusion matrix for this domain is shown in Figure 2. The first row of the matrix consists of all the case which have been classified by the domain expert as "A"s whereas the cell (A, B) corresponds to cases which have been annotated by the expert as an

---

[2] We shall see later that one of INSIGHT's modes does use machine learning techniques.

"A" but have been classified by the rule set as a "B". Similarly the cases in the right hand cell in that row, cell (A, E), have been annotated by the expert as "A" but have been classified by the rules set as "E"s. Clearly all the diagonal cells [ie (A, A) (B, B), ...(E, E)] contain instances which have been classified identically by both the expert's annotation & by the rule-set.

	Expected: A	Expected: B	Expected: C	Expected: D	Expected: E	Expected: (none)
Observed: A	90 % 181 of 202	4 % 9 of 202	1 % 3 of 202	(none)	0 % 1 of 202	4 % 8 of 202
Observed: B	0 % 4 of 1053	96 % 1014 of 1053	2 % 22 of 1053	0 % 4 of 1053	0 % 2 of 1053	1 % 7 of 1053
Observed: C	(none)	4 % 22 of 533	87 % 462 of 533	6 % 34 of 533	0 % 1 of 533	3 % 14 of 533
Observed: D	(none)	2 % 6 of 360	8 % 29 of 360	83 % 297 of 360	2 % 6 of 360	6 % 22 of 360
Observed: E	(none)	4 % 20 of 540	2 % 11 of 540	12 % 67 of 540	78 % 423 of 540	4 % 19 of 540
Observed: (none)	(none)	22 % 16 of 73	4 % 3 of 73	8 % 6 of 73	1 % 1 of 73	64 % 47 of 73

Fig. 2. A confusion matrix

INSIGHT provides a range of facilities to enable the expert to view the instances which have been misclassified and to either edit the data set (say to change the annotation of an instance, or correct a clearly incorrect data value) or to revise or enhance the current rule-set.

The Confusion Matrix (CM) seems to be a very intuitive way of presenting the results to experts; so far all the experts who have used it, have had no problem understanding it. Additionally it suggests a procedure for tackling the revision of the discrepancies. Clearly some discrepancies are more surprising than others. For example as all the categories are in a sense ordered, instances in the cell (A, E) can be considered to be more surprising than those only one category away, say those in cell (A, B). Thus this distance measure suggests that the domain expert should be encouraged to consider discrepancies in the following order:

- (A, E) & (E, A); (Distance between categories of 4).
- (A, D), (B, E), (E, B) & (D, A); (Distance between categories of 3)
- (A, C), (B, D), (C, E), (E, C) (D, B) & (C, A); (Distance between categories of 2)
- (A, B) (B, C) (C, D) (D, E) (E, D) (D, C) (C, B) & (B, A); (Distance between categories of 1)

A further strategy which we suggested to the domain experts was for the first period to concentrate on removing the discrepancies from the *data-set* (incorrect annotations & data points) and only at a later stage make changes to the rule-set. This heuristic is based on the perspective that changes to the data set are localized, whereas a change to a rule could, in principle, effect *all* of the instances / cases.

A third strategy suggested was initially to refine each of the patient data sets *individually*, before attempting to refine the complete set of instances.

4.1 The Rule Interpreter

Essentially each rule consists of a set of one or more conjunctive conditions, and a single action which is to assign a particular instance to a category. To date we have implemented only a single conflict resolution strategy, namely the first rule which is satisfied, fires. This means that it is necessary for the domain expert (supported by the analyst) to ensure that the most specific rules are placed at the top of the list, and the more general rules are placed at the bottom of the list. In many situations the rules are mutually exclusive, as they include non-overlapping conditions (or in the extreme case use completely different descriptors) in which case they are order-independent. However if a set of rules has related conditions, then it is important to ensure they are appropriately ordered.

We have kept the format of the rules and the rule interpreter simple for a number of reasons: firstly, this meant the system could be implemented quickly; secondly, the form of the current rules, and the interpreter's decision making appear to be easily understandable by domain experts. (The interpreter and the form of the rules may be enhanced subsequently if there is a clear need.)

4.2 Inferring Rules from Instances

INSIGHT has a mode which infers a rule when it is provided with several instances of a particular category. This mode was added so that an expert would not be forced to specify rules for each of the categories ab initio. However, such rules contain a feature-value pair corresponding to each of the descriptors used to describe instances. Our recent work with INSIGHT has made us aware of the need to select relevant descriptors from the inferred rule, in order to achieve effective distinctions between the categories. So even in this mode, we believe the process will require some involvement by the domain expert who will need to refine each rule by, for example, selecting descriptors from the set inferred by the Machine Learning algorithm.

This mode has still to be used by a domain expert with a demanding application.

5 Use/Evaluation of the INSIGHT System

Section 2 gives an overview of the evaluation to be undertaken; as mentioned in that section, the system's administrators provided us with a spreadsheet which contained the complete ICU stays for 10 patients. Each of these records was de-identified before this information was passed to us. Table 5 gives the code name for each of the patients and the number of recorded time points associated with each patient.

It should be noted that the patients' datasets represent their complete stay in the ICU, and hence it is to be expected that the quality and completeness of the records will not be high at both the beginning and end of the patients' stays. For example, usually when a patient is first admitted to an ICU, they are in need of resuscitation, and as some of this involves manual infusion of drugs, the patient management system does not capture all the actual activities, nor all of the patient's physiological parameters. Thus associated with each patient's stay there may be a number of time points which do not contain all the "core" parameters, and hence, it might be argued, these time points should not be used for this analysis (note that after the first 6 hours in the ICU, a complete set of "core" parameters is normally collected for the patient). It should be noted that some of the descriptors in this dataset (such as urine output and heart rate) were extrapolated to fill in certain missing values; the algorithm used to calculate these missing values was agreed with the clinicians.

Table 5. Patient codes and the number of records provided for each patient; there being in total 2761 patient records

Patient Code	696	705	707	708	720	728	733	738	751	782
Number of time points	129	576	475	40	188	281	396	110	493	73

This section describes a two-stage study conducted with clinician-1 (sections 5.1 & 5.2), and a related 1-stage study undertaken with clinician-2 (section 5.3).

5.1 Review of Study with Clinician-1 (Phase-1)

Clinician-1 (MS) chose initially to concentrate on Patient 705 which has 576 time points (or instances). When he started this session there was a 45.0% (259/576) agreement between his annotations and his initial rule-set, however if the unclassifiable records are ignored that figure becomes 45.7% (258/564).

Further, at the end of the session (with just this one patient) the agreement was 97.0% (559 of 576) or 100% (556 of 556) if we ignore the effects of the (20) unclassifiable records. This session took about 5 hours, and was relatively slow as this was the first time INSIGHT had been used "in anger", and at the beginning of the session it was necessary, for example, to change annotations of instances singly, which was painstaking when the expert wanted to change a group of such values. This and other functionalities have subsequently been added, so the tool is now very much faster to use. One thing which this clinician did at an early stage was to reduce the number of parameters viewed for each instance from the original 41 to just 6; this also speeded up his handling of instances considerably. The parameters which he chose to view were: Adrenaline, FiO_2, HR, Mean Arterial Pressure (MAP), Noradrenaline and SpO_2.

In section 2.1, we outlined the nature of the knowledge available in this domain, and in section 4 we outlined the simple rule interpreter which we have implemented. Here we give some examples of the rules which have been implemented for several categories. For example, the rule associated with category A has the following form:

HR (normal-range) AND BLOOD-PRESSURE (normal-range) AND SPO_2 (normal-range) AND FIO_2 (normal-range) AND ADRENALINE (none) AND NORADRENALINE (none)

Note this is a *conjunctive* rule, and all the conditions have to be satisfied before a time point is classified as an "A".

On the other hand, there are a number of disjunctive rules which represent each of the conditions which correspond to a patient being assigned to category "E", namely:

HR (extremely low) OR HR (extremely high) OR MAP (extremely low) OR MAP (extremely high) OR ADRENALINE (extremely high) OR NORADRENALINE (extremely high)

Table 6. We have used the notation "nn A→B" to indicate that nn items which had been annotated initial by the clinician as an "A", have since been reclassified by the expert as a "B". If the item is followed by a "*" this implies that the changed annotation is now consistent with that predicted by the then-current version of the rule-set. (Remember that when the rule-set changes all the instances are re-evaluated against the revised rule-set.)

Expert	Rule A	Rule B	Rule C	Rule D	Rule E
A		157 A→B* 2 A→C 3 rule edits 2 left as original annotation	11 A→B 7 A→C* 1 A→D	2 A→D* 2 rule edits	3 data edits[3]
B	14 B→A*		1 rule edit 1 B→U 11 B→C*	2 B→C 1 B→D*	1 data edit[2] 1 B→E*
C		12 C→B*		15 C→D* 1 C→E 2 rule edits 4 left as original annotation	1 C→E*
D		13 D→C 4 D→E 3 D→U	11 D→C* 1 D→E 2 D→U 1 rule edit		21 D→E* 1 rule edit 1 data edit[2]
E					

The rules associated with categories B, C & D are also largely disjunctive, and tend to have values on a continuum from those associated with category "A" to those associated to category "E", as section 2.1 suggests.

Clinician-1 (MS) followed roughly the refinement strategy suggested in section 4; note there are no E rows in this CM which means that none of the instances classified by the expert as an "E" was classified as anything else by the rule set. In fact the expert chose to consider cells (A, E), (B, E) (C, E) followed by (B, D), (D, B) & then (A, C) (A, B) (B, A) (D, E) (D, C) (B, C) (D, C) (B, D) (C, B) & (C, D). In the early

---

[3] To remove impermissible values.

stages of the analysis, very obvious inconsistencies were encountered & dealt with, and later it became often an issue of fine-tuning the rule-set and / or the data–set to achieve the classification which the expert wanted between two "adjacent" (1-distance apart) classifications. Table 6 gives a summary of the changes made to the "cells".

Here we provide an overview of the typical decisions made by the domain expert:

- **Inadmissible Readings:** In cell (A, E), the expert considered that 3 of the values given in the data-set for heart rate were clearly inadmissible (values of 372, 7, 3); he changed those values to null values, and reclassified each of the cases as "un-classified" as he felt he then had insufficient information to make a classification. He dealt with a further instance in cell (B, E) similarly.
- **Extrapolated Data Points:** Several times the expert agreed that the actual information provided in an instance was not sufficient to make a decision, and agreed, for several of the missing values, he had looked at the corresponding values in the immediately preceding and following time-periods and had effectively used extrapolated values when making his decisions. In all cases he agreed that the instances should have their classifications changed to "unclassified". (This raises the issue of whether a further trending feature should be developed for INSIGHT and used with selected features.)
- **Significant values overlooked:** In many instances, e.g., cell (D, B), the expert agreed that the annotation should be changed as he had failed, when doing his initial classification, to note an important feature-value pair, in this case FiO_2 values of .55.
- **Deciding borderline values:** In handling many of the "adjacent" cells where the distance between them is just one (e.g., cells such as (A, B) (B, C) (C, B) etc); the expert in some circumstances reclassified the instances, and in others he modified the appropriate rules to achieve his desired classification for the instances.

This expert made 300[4] changes to annotations. Note some annotations might well have been changed several times: an instance originally annotated as an A, might initially be re-annotated as a "D", and finally following a rule change, might be re-annotated as a "C".

In summary, this expert during the process of this refinement modified 52.1% (300/576) of his annotations. 7 changes (1.2%) were to reclassify an instance as unclassifiable (due to missing information, which in some cases the expert had overcome by "extrapolation" as discussed above); 274 changes (47.6%) were to adjust instances which were on the borderline between two of the A-E categories; and the remaining 25 (4.3%) were due to the expert overlooking a piece of information in the patient record which he accepted was important when it had been brought to his attention (by INSIGHT).

5.1.1 Rule Refinements

To date we have observed two significant types of rules / rule-sets refinements, namely:

---

[4] 307 annotations were viewed, but 7 of these were left as the original annotation.

- Adding a new rule, e.g., clinician-1 in phase 2 added a new conjunctive rule to category "E": ADRENALINE (high) AND NORADRENALINE (high)
- Refining the conditions of a set of rules based on a common feature, say FiO_2. Note that all the values returned for FiO_2 are effectively integers; also note that all the ranges for FiO_2 are continuous. Below we give the values for FiO_2 for a number of categories, both before and after refinement:

Table 7. FiO_2 Values Before and After Refinement

Category	Before refinement	After Refinement
C	0.55 – 0.69	0.55 – 0.69
D	0.70 – 0.84	0.70 – 0.83
E	0.85 – 1.00	0.84 – 1.00

Below we give an overall summary of the actions taken during this analysis:

Table 8. Summary of Actions taken by Clinician-1 in Phase 1

Number of instances in the set	576
Number of instances / annotations viewed	307
Number of data values edited / removed	5
Number of annotations changed to unclassified	7
Number of annotations left as "inconsistencies"	7
Number of annotations changed to another A-E level (excluding "unclassified")	46
Number of annotations changed to be consistent with the rules (excluding unclassified)	242
Number of changes to the rule-set	10

5.2   Review of Study with Clinician-1 (Phase-2)

In this session, which lasted about 5 hours, we started with the rule-set which had been produced in this first session (when the expert had processed the data associated with Patient 705), and used that as the starting point to make the annotations of the remaining 9 patients (see table 5) consistent with this rule-set or a variant of this rule set. In this session the number of annotated instances to be dealt with was 2130 (ie 2706 – 576). It should be noted that as a result of the changes made earlier to INSIGHT the progress in this session was considerably faster.

At the start of this session, the rule-set produced in Phase 1 gave a 58.3% (1609 of 2760) agreement with the annotations created by the domain expert across all 10 patients; when the 135 unclassifiable instances are removed we get a 58.9% (1545 of 2625) agreement. By the end of the refinement session this agreement had increased to 96.4% (2663 of 2761), or when the 170 unclassifiable instances had been removed, to 100.0% (2591 of 2591). The expert initially chose to view the same parameters as he did at the end of the first session, but part way through he added Dobutamine. The strategy followed by the expert for refining these instances was very similar to that given above.

Again we conclude this section by providing a similar summary to the one given in the previous section, see Table 9.

Given that the number of instances considered here is nearly four times as large as considered in Phase 1, there are a relatively smaller number of changes, the exception being the number of instances which have been reclassified as "Unclassified". As noted before many instances are unclassified as "core" data elements are missing; clearly one is never going to capture all the data, but the expert noticed that data is often missing at critical points when patients experience a significant deterioration; this issue will be raised with nursing staff to see if the overall data collection rates can be improved. We also noted earlier that data tends to be sparse when patients first come to the ICU and just before they are discharged.

Table 9. Summary of Actions taken by Clinician-1 in Phase 2

Number of instances in the set	2130
Number of instances / annotations viewed	225[5]
Number of data values edited / removed	7
Number of annotations changed to unclassified	97
Number of annotations left as "inconsistencies"	16
Number of annotations changed to another A-E level (excluding "unclassified")	1
Number of annotations changed to be consistent with the rules (excluding unclassified)	104
Number of changes to the rule-set	6

5.3   Review of Study with Clinician-2

In this session with clinician-2 (JK), which lasted about 2 hours, we started with the rule-set which had been produced in the second session by clinician-1 as the result of reviewing all 10 patients (see table 5). Further this clinician, clinician-2, had annotated three patient data-sets, namely those of patients 708, 728 and 733, giving a total of 717 instances. (Clinician-1 annotated time points from 10 patients; the smaller number of 3 patients was chosen for subsequent clinicians to make the task more manageable.) This clinician also decided it was hard to review all the parameters reported for each time instance and chose, generally, to limit the ones he considered to Adrenaline, blood pump speed, CVP, Dobutamine, FiO_2, Gelofusin, Hartmanns, heart rate (HR), LiDCO Cl, MAP, Noradrenaline, PiCCO, Propofol, Sodium Chloride, SpO_2, temperature, urine output, and Vasopressin (18 parameters).

The strategy followed by the expert for refining these instances was very similar to that used by clinician-1. At the start of this session the final rule-set produced by clinician-1 gave a 40% agreement with the annotations created by this domain expert for patient 708, and by the end of this session the agreement had increased to 97.5%. These percentages are further improved, as one can see from Table 10 when the unclassifiable instances are removed. This table also gives results for patients 728 & 733 as well as for all three patients; in all cases results including & excluding unclassifiable instances, are

---

[5] This figure is approximate as there are several ways in which it could be calculated.

reported. The percentage agreement, after data set & rule refinements, for all these data sets is remarkably high: being ~97.5% when unclassified cases are included and ~99% when they are not. Note too that initially 5 unclassifiable instances had been detected, after the refinement process this number rose to 11.

Table 10. Summary of Clinician-2's Refinement

	P708 (before)	P708 (after)	P728 (before)	P728 (after)	P733 (before)	P733 (after)
All instances considered	40% 16/40	97.5% 39/40	10.7% 30/281	97.5% 274/281	8.1% 32/396	97.6% 387/396
Unclassifable instances excluded	40% 16/40	100% 39/39	10.8% 30/278	99.6% 274/275	8.1% 32/394	98.7% 387/392

	All 3 patients (before)	All 3 patients (after)
All instances considered	10.7% 77/717	97.6% 700/717
Unclassifable instances excluded	10.8% 77/712	99.6% 700/703

5.4 COMPARISON between Final Data-Sets and Rule-Sets for Clinician-1 and Clinician-2

The results in the diagonal cells of Table 11 are those for the individual clinicians and as such are reported at the end of sections 5.2 & 5.3 respectively. The figures in the off-diagonal cells give the agreements between the final rule-set & datasets of the 2 clinicians. As can be seen when unclassified instances are included in the analyses the results are 94.0% & 93.0% & when the unclassified items are removed from the analyses the agreement becomes: ~96% in both cases.

Table 11. Comparison between final data-sets & rule-sets for Clinician-1 & Clinician-2

	Clinician-1's final *data-set*	Clinician-2's final *data-set*
Clinician-1's final *rule-set*	96.4% (2663 of 2761) 100.0%* (2591 of 2591)	94.0% (674 of 717) 95.9%* (674 of 703)
Clinician-2's final *rule-set*	93.0% (2567 of 2761) 96.3%* (2495 of 2591)	97.6% (700 of 717) 99.6%* (700 of 703)

* These results correspond to analyses when the Unclassified instances are removed from the calculation.

These analyses suggest extremely high correlations between both the annotations & the rule-sets produced by these clinicians.

6 Contributions of This Work

- Produced a simple and useful tool to help experts appreciate how two perspectives on the same task is inconsistent and allows them to explore ways in which the two sources of knowledge can be made (more) consistent
- Confirmed the advantage, in some circumstance, of a simple information checking system as opposed to a more complex system which is able to (semi-) automatically extract the knowledge from a set of labelled instances.
- Challenged the accepted wisdom of Cognitive Science (Expertise Studies) that a domain expert's "active" knowledge is more reliable than his "passive" knowledge
- Confirmed the need, when acquiring knowledge from domain experts, to determine whether a particular category has sub-categories & if so to get the expert to articulate them.
- Confirmed the need for sizable numbers of instances for each of the (sub-) categories when Machine Learning algorithms are used to infer associations.
- Confirmed the need to have a domain expert critically review any rules (knowledge) produced by an automated system. More particularly, INSIGHT has shown the benefits of experts being able to see their knowledge *applied* on a series of relevant tasks, and being able to comment on the outcome.

7 Further Work

The following are some of the activities planned:

Plan to evaluate the ICU scoring system across several ICUs & with a larger number of experts. The central task which INSIGHT has been used to investigate, to date, is the development of a reliable patient scoring / classification system. So far, we have applied INSIGHT to data from only 10 patients from a single ICU, and this information has been evaluated by just two domain experts. Clearly, if the scoring system is to be used widely it will need to be evaluated with a larger and more disparate group of patients and with considerably more domain experts. This evaluation is currently being planned.

Excluding "unclassifiable" records from the analysis. Modify the rule-set such that all records which do not contain values for the core parameters will be excluded from the analysis. Before this can be implemented, a decision will have to be made about what constitutes this set of core parameters.

Use of INSIGHT with other domains. We plan to use INSIGHT with a range of other tasks including the classification of botanical species and other clinical diseases. In many situations, experts find it hard to articulate the actual distinctions between different categories; INSIGHT should help with this process.

Extend INSIGHT so that it could be used to achieve consistency between more that 2 knowledge sources.

Use of INSIGHT's mode to create rules from instances. We noted in section 4 that INSIGHT had such a mode, and that to date it had not been used by domain experts on a range of demanding real-world tasks. Clearly, we believe that this mode will be valuable for domain experts who will not then need to create a set of rules which correspond to each of the categories. We need to test this hypothesis with a number of domains and with a range of experts.

Develop a variant of INSIGHT to apply to planning / synthetic tasks. This will be more demanding than for classification tasks, but we believe it is possible, and moreover that it would be a useful additional tool in assessing Expertise.

Acknowledgements

- Sunil Sharma and Mark Winter implemented the earlier versions of REFINER.
- Andy Aiken & Laura Moss were both partially supported by EPSRC Studentships when they undertook this work.
- Consultants at the ICU at Glasgow Royal Infirmary for useful discussions & support on a range of related studies.

References

Buchanan, B.G., Shortliffe, E.H. (eds.): Rule-Based Expert Systems: The MYCIN Experiments of the Stanford Heuristic Programming Project. Addison-Wesley, San Francisco (1984)

Corsar, D.: KBS Development Through Ontology Mapping and Ontology Driven Acquisition. In: Sleeman, D., Barker, K. (eds.) K-CAP 2007: Proceedings of the 4th International Conference on Knowledge Capture, Whistler, BC, Canada, pp. 23–30. ACM, New York (2007)

Diaper, D.: Knowledge Elicitation: principles, techniques & applications. Ellis Horwood, London (1989)

Ericsson, K.A., Simon, H.A.: Protocol analysis; Verbal reports as data (revised edition). Bradfordbooks/MIT Press, Cambridge (1993)

Etzioni, O., Cafarella, M., Downey, D., Kok, S., Popescu, A., Shaked, T., Soderland, S., Weld, D., Yates, A.: Unsupervised named-entity extraction from the web: An experimental study. Artificial Intelligence 165(1), 91–134 (2005)

Johnson, P.E.: What kind of expert should a system be? Journal of Medicine and Philosophy 8, 77–97 (1983)

Knaus, W.A., Wagner, D.P., Draper, E.A., Zimmerman, J.E., Bergner, M., Bastos, P.G., Sirio, C.A., Murphy, D.J., Lotring, T., Damiano, A., et al.: The APACHE III prognostic system. Risk prediction of hospital mortality for critically ill hospitalized adults. Chest 100(6), 1619–1636 (1991)

Langley, P., Simon, H.A.: Applications of machine learning and rule induction. Communications of the ACM 38, 55–64 (1995)

Marcus, S., McDermott, J.: SALT: a Knowledge Acquisition Language for Propose-and-Revise Systems. Artificial Intelligence 39(1), 1–38 (1989)

Michalski, R., Chilausky, R.: Knowledge Acquisition by Encoding Expert Rules versus Computer Induction from Examples: A Case Study Involving Soybean Pathology. International Journal of Man-Machine Studies 12(1), 63–87 (1980)

Moss, L., Sleeman, D.: ACHE: An Architecture for Clinical Hypothesis Examination. In: Proceedings of 21st IEEE International Symposium on Computer-Based Medical Systems (CBMS 2008), Jyvaskyla, Finland, pp. 158–160 (2008)

Pearl, J.: Probabilistic Reasoning in Intelligent Systems. Morgan Kaufmann, San Francisco (1988)

Quinlan, J.R.: Induction of decision trees. Machine Learning 1, 81–106 (1986)

Sharma, S., Sleeman, D.: Refiner: A Case-based Differential Diagnosis Aide for Knowledge Acquisition and Knowledge Refinement. In: Sleeman, D. (ed.) Proceedings of EWSL 1988, pp. 201–210. Pitman, London (1988)

Singh, P., Lin, T., Mueller, E., Lim, G., Perkins, T., Zhu, W.L.: Open Mind Common Sense: Knowledge acquisition from the general public. In: Proceedings of the First Intern. Conf. on Ontologies, Data bases, and applications of Semantics for large Scale Information Systems. LNCS. Springer, Heidelberg (2002)

Sleeman, D.: Towards a Technology and a Science of Machine Learning. AI Communications 7(1), 29–38 (1994)

Sleeman, D., Barker, K., Corsar, D.: Report on the Fourth International Conference on Knowledge Capture (K-CAP 2007). AI Magazine (2007) (in press)

von Ahn, L.: Games with a Purpose. IEEE Computer Magazine, 96–98 (2006)

APPENDIX A: High Level Summary of Qualitative Assessments

Below we give outline descriptions for each of the 5 categories, where E corresponds to the most severely ill patients:

A.   Patient's cardiovascular system (CVS) normal, with no Adrenaline or Noradrenaline (ADR / NADR) and low levels of Oxygen; Urine production often essentially normal (or is well established on renal replacement therapy).

B.   Patient CVS nearly normal, probably needs low levels of ADR / NADR and Oxygen.

C.   Patient CVS system is effectively stable; probably on moderate dosages of ADR / NADR and Oxygen.

   Most parameters suggest the time-slot is in category A or B, but if any of the following conditions are met, then it should be assigned to category C:
   - Heart rate: Moderately Low OR Moderately High
   - MAP: Moderately Low OR Moderately High
   - Adrenaline: Moderate dose
   - Noradrenaline: Moderate dose
   - FiO_2: Moderate
   - SpO_2: Moderately Low

D.   Patient's CVS system is moderately unstable and / or on high doses of ADR / NADR/ fluid to retain stability.

   Most parameters suggest the time-slot is in category A or B, but if any of the following conditions are met, then it should be assigned to category D:

- Heart rate: Low OR High
- MAP: Low or High
- Adrenaline: High dose
- Noradrenaline: High dose
- FiO_2: High
- SpO_2: Low

E. Patient's CVS is very unstable (which is usually true in early phases of re-suscitation, or following a new event) with low BP and high HR or rapidly changing ADR / NADR dosage, and requires substantial fluid inputs.

Most parameters suggest the time-slot is in category A or B, but if any of the following conditions are met, then it should be assigned to category E:

- Heart rate: Extremely Low OR Extremely High
- MAP: Extremely Low OR Extremely High
- Adrenaline: Extremely High dose
- Noradrenaline: Extremely High dose
- FiO_2: Extremely High
- SpO_2: Extremely Low

The Dynamics of Multiagent Q-Learning in Commodity Market Resource Allocation

Eduardo R. Gomes and Ryszard Kowalczyk

Swinburne University of Technology, John Street, Hawthorn, VIC 3122, Australia
egomes@groupwise.swin.edu.au, rkowalczyk@groupwise.swin.edu.au

Abstract. The Commodity Market (CM) economic model offers a promising approach for the distributed resource allocation in large-scale distributed systems. Existing CM-based mechanisms apply the Economic Equilibrium concepts, assuming price-taking entities that will not engage in strategic behaviour. In this paper we address the above issue and investigate the dynamics of strategic learning agents in a specific type of CM-based mechanism called Iterative Price Adjustment. We investigate the scenario where agents use utility functions to describe preferences in the allocation and learn demand functions adapted to the market by Reinforcement Learning. The reward functions used during the learning process are based either on the individual utility of the agents, generating selfish learning agents, or the social welfare of the market, generating altruistic learning agents. Our experiments show that the market composed exclusively of selfish learning agents achieve results similar to the results obtained by the market composed of altruistic agents. Such an outcome is significant for a series of other domains where individual and social utility should be maximized but agents are not guaranteed to act cooperatively in order to achieve it or they do not want to reveal private preferences. We further investigate this outcome and present an analysis of the agents' behaviour from the perspective of the dynamic process generated by the learning algorithm employed by them. For this, we develop a theoretical model of Multiagent Q-learning with ε-greedy exploration and apply it in simplified version of the addressed scenario.

1 Introduction

This paper investigates the impacts of the introduction of participants with learning capabilities in market-based resource allocation mechanisms. More specifically, it studies the individual and the social dynamics of participants that apply the Q-learning with ε-greedy exploration algorithm to develop strategic behaviour in a commodity-market mechanism based on the Walrasian economic equilibrium principles.

Market-based mechanisms for resource allocation have been attracting increasingly research interest over the last years. Most of this interest results from the emergence of large-scale distributed systems, such as the Grid [13], peer-to-peer (P2P) [23] networks and service-oriented architectures (SOAs) [11], and their needs for efficient and flexible resource allocation, qualities usually endowed to market-based approaches [42]. There are several advantages in framing the distributed resource allocation problem into economic terms [42]. In the Grid context, for example, if users are willing to share or trade

J. Koronacki et al. (Eds.): Advances in Machine Learning II, SCI 263, pp. 315–349.
springerlink.com © Springer-Verlag Berlin Heidelberg 2010

access to their idle resources, they need to somehow account for the ratio between the benefits and costs of doing it. A similar need is clearly present in P2P and SOAs where sharing and using the resources require incentives and mutual benefits for the providers and consumers. Market-based mechanisms are suitable for these types of problem because they intrinsically incorporate the notion of relative worth, usually abstracting it into a *price* or other type of *exchange unit*. Additionally, most economic models are well-understood as they have been subject of study in the Economics disciplines along the years, making it possible to utilise this body of knowledge also to analyse and design market-based resource allocation (MBRA) mechanisms.

A promising approach for MBRA is based on the Commodity Market (CM) economic model [6, 8, 30, 32, 39, 40]. A CM offers a marketplace where sellers and buyers trade resources based on a common price known by all the participants. The price is defined by the market using a pricing mechanism usually driven by the *economic equilibrium* principles founded by the nineteenth-century French economist Walras [36] and its law of demand-and-supply. Walras proposed that markets are able to coordinate the allocation of resources by finding an *equilibrium price* where the total demand matches the total supply. When this price is found, the market enters into a state of *economic equilibrium* that, by the First Fundamental Theorem of Welfare Economics, is a Pareto-Optimal (PO) solution for the resource allocation problem. A solution is PO if there is no other solution that can improve one agents' outcome without deteriorating others'.

One of the conditions for the *Walrasian economic equilibrium* is the existence of a *Perfect Competition* market, that is, a market where no participant (buyer or seller) has the power to influence the price. To obtain perfect competition, participants are assumed to be price-taking entities that consider the price as a given aspect of their decision making, an aspect outside their control. In other words, the participants are regarded as passive entities that will not actively attempt to influence the mechanism in order to obtain higher profits. One participant's role is thus reduced to solving simple optimization problems, perhaps taking into account its budget and the costs and revenues derived from the allocation, but with no perceived dependence on the actions of other agents.

The assumption of price-taking participants is a convenient condition imposed by Walrasian-based approaches in order to ensure PO allocations. It is, however, hard to be satisfied in large-scale distributed systems. In such systems there is little control over the behaviour of the participants, making it impossible to guarantee that they will behave in an ordered manner. Buyers and sellers will typically have different objectives, preferences and demand/supply patterns which they will try to satisfy. Therefore, one cannot assume they will not attempt to exploit the system by engaging in strategic behaviour to obtain higher profits, possibly decreasing the overall quality of the allocation. Hence, to understand the impacts of these attempts and to develop mechanisms that are robust in the presence of strategic participants, being able to deliver optimal allocation also in this situation, are important aspects to enable the next generation of large-scale distributed systems. Moreover, most current CM-based mechanisms focus on the achievement of a PO allocation, but usually disregard how fair and how desirable the allocation is for both the system and the participants. Different PO outcomes generate different utility gains to the parties involved in the resource allocation process. Being able to find a fair

PO outcome, in which all the participants are equally or near-equally satisfied, is an important aspect of the problem.

In this paper we address the above issues by introducing participants with learning capabilities into the market. We approach the market-based resource allocation problem from the premise that the participants can behave like the entities composing real economies and will engage in strategic behaviour in order to satisfy their preferences. We present the IPA with RL, a market-based resource allocation mechanism that enhances the original Iterative Price Adjustment (IPA) [12, 41] mechanism. The IPA is a Walrasian-based mechanism in which the equilibrium price is calculated through an iterative process. The mechanism cyclically asks the agents for the amount of resources they would be willing to buy at a particular price and uses this information to update the price, which then is increased if the total demand requested by the agents is higher than the supply and decreased otherwise. This iterative process continues until the equilibrium between demand and supply is found, when the market is cleared and the resources are sold. In the IPA with RL the agents use utility functions to describe preferences over different resources attributes and develop strategic behaviour by learning demand functions adapted to the market through Reinforcement Learning (RL). We apply this approach to investigate the market in the presence of two types of agents: selfish learning agents, whose learning goal is to improve their individual utility; and altruistic learning agents, whose learning objective is to improve the social welfare of the market.

The paper is organised in two main parts. The first part, Section 2, presents results of experimental investigation in the IPA with RL. Our focus is to study the impacts of the introduction of the strategic learning agents on the individual and social performance of the IPA market. The second part, Section 3, presents the theoretical analysis of the results obtained in Section 2. We examine the dynamics generated by the learning algorithm and analyse how it affects the individual and the social results of the market. For this, we develop a theoretical model of Multiagent Q-learning with ε-greedy exploration and apply it in a simplified version of the scenario addressed in Section 2. The related works are presented in Section 4 and the conclusions and future directions in Section 5.

2   The IPA with RL

This section presents the experimental investigation in the IPA with RL. In the next sub-section we introduce the general scenario addressed in the paper. We review the IPA mechanism and the RL algorithm used in the experiments and present the modelling of the IPA as a RL problem. Sub-section 2.2 presents the experiments in the mechanism. We investigate the individual and the social performance of the market in two cases: when the market is composed of learning and non-learning agents and when it is composed exclusively of learning agents.

2.1   Scenario

We address the scenario in which a limited amount of resources has to be allocated to a set of agents in a commodity-market resource allocation system using the IPA mechanism. Agents use utility functions to describe preferences in the allocation and learn their demand functions from interaction with the market using RL (as illustrated in Figure 1).

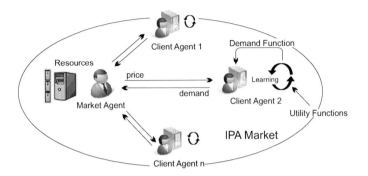

Fig. 1. The IPA mechanism with RL

2.1.1 Iterative Price Adjustment

The IPA decomposes the resource allocation optimization problem into smaller and easier sub-problems. Its behaviour mimics the law of demand and supply. The price is increased if the demand exceeds the supply and decreased otherwise. The mechanism process is a cycle that begins with a facilitator (the market) announcing the initial prices for the resources. Based on the initial price, agents decide on the amount of resources that maximize their private utilities (the sub-problems) and send these values to the facilitator. The facilitator adjusts the prices according to the total demand received and announces the new prices. The process continues until an equilibrium price is reached, when we say that the market is *cleared* and the resources are sold. In the equilibrium, the total demand equals the supply or the price of the excessive supply is zero. Under some circumstances, the equilibrium price may not exist [19], but that problem is out of the scope of this thesis.

The standard IPA method is formalized as follows. Let $C = \{C_1, \cdots, C_i, \cdots, C_m\}$ be the total supply of resources available, where C_i is the total supply of resource i; Let $P(t) = \{p_1(t), \cdots, p_i(t), \cdots, p_m(t)\}$ be the price vector for the resources C at time t, where p_i is the price for the resource i; there are n self-interested agents and each agent has a utility function $u_j(D_j)$, which is the utility over $\{d_{1,j}, d_{2,j}, \cdots, d_{m,j}\}$, where $d_{i,j}$ is the amount of resource i that the agent j requires. Let $D_{j(t)} = \{d_{1,j}(t), d_{2,j}(t), \cdots, d_{m,j}(t)\}$ be the total amount of resources agent j requires at time t. At time t, the objective of the agent j is to find $D_j(t) = \arg_{D_j(t)} \max(u_j(D_j(t)))$. At time $t+1$, the price can change, and the new price of resource i is given by:

$$p_i(t+1) = \max\{0, p_i(t) + \alpha(\sum_{j=1}^{n} d_{i,j}(t) - C_i)\}, \qquad 1 \le i \le m \qquad (1)$$

where α is a small positive constant.

It should be noted that the agent's utility maximization task in the IPA is actually the maximization of its instantaneous profit. As described by [41], an agent has a revenue function and a cost function over the resources and, at each time step, its task is to find the demand request that maximizes the difference between these two functions given the current price. The result is the existence of a demand function, in which all the points lying in the characteristic curve are equally preferred by the agent. This approach limits the power of the agents. As they cannot express preferences over different attributes of the allocation, in particular over the price and the amount of resources, they cannot develop a strategic behaviour to influence the mechanism.

In our approach to the IPA the agents use utility functions to describe their preferences in the allocation and learn demand functions optimized for the market by RL. The idea is that, given a particular market configuration, there is at least one demand function that maximizes the utility obtained from the aggregation of the agent's utility functions. Learning such a demand function leads to the development of the agent's strategic behaviour.

2.1.2 Reinforcement Learning

The task of a RL agent is to learn a mapping from environment states to actions so as to maximize a numerical reward signal [33]. The RL framework is formalized by a tuple (S,A,T,R), where S is a discrete set of environment states, A is a discrete set of actions, T is a state transition function $S \times A \times S \rightarrow [0,1]$, and R is a reward function $S \times A \rightarrow \mathbb{R}$. On each step the agent receives a signal from the environment indicating its state $s \in S$ and chooses an action $a \in A$. Once the action is performed, it changes the state of the environment, what generate a reinforcement signal $r \in R$. The agent uses this reinforcement signal to evaluate the quality of the decision. The task of the agent is then to maximize (or minimize) some long-run measure of the reinforcement signal.

Q-learning [37] is a commonly used algorithm for RL. The main attractive of Q-learning is that it assumes no knowledge about state transitions and reward functions, that is, the agent does not need to have a model of the environment, enabling the problem to be approached *on-line*. For this, the agent maintains a table of $Q(s, a)$-values that are updated as it gathers more experience in the environment. Q-values are estimations of $Q^*(s, a)$-values, which are the sum of the immediate reward r obtained by performing an action a in a state s and the total discounted expected future rewards obtained by following the optimal policy thereafter. By updating $Q(s, a)$, the agent eventually makes it converge to $Q^*(s, a)$. The optimal policy π^* is then followed by selecting the actions where the Q^*-values are maximum. The algorithm is to indefinitely perform the following 4 steps:

1. Observe the current state s, select an action a and execute it
2. Observe the new state s' and the reward $r(s,a)$
3. Update the Q-table according to:
   $$Q(s,a) := Q(s,a) + \alpha(r(s,a) + \gamma \max_{a'} Q(s',a') - Q(s,a))$$
4. Make the new state s' the current state s

where $\alpha \in \,]0,1[$ is the learning rate and $\gamma \in \,]0,1[$ is the discount rate. Considering that the probability of making transitions T and receiving specific reward signals R do not change over time, i.e. a stationary environment, if each action is executed in each state an infinite number of times and α is decayed appropriately, the Q-values will converge with probability 1 to the optimal ones [33].

An important component of Q-learning is the action selection mechanism. This mechanism is responsible for selecting the actions that the agent will perform during the learning process. Its purpose is to harmonize the trade-off between exploitation and exploration such that the agent can reinforce the evaluation of the actions it already knows to be good but also explore new actions.

In our experiments in the IPA with RL we have applied the ε-greedy action-selection mechanism. This mechanism selects a random action with probability ε and the best action, i.e. the one that has the highest Q-value at the moment, with probability 1-ε. As such, it can be seen as defining a probability vector over the action set of the agent for each state. If we let $\mathbf{x} = (x_1, x_2, ..., x_j)$ be one of these vectors, then the probability x_i of playing action i is given by:

$$x_i = \begin{cases} (1-\varepsilon) + (\varepsilon/n), & \text{if } Q(s,a) \text{ is currently the highest} \\ \varepsilon/n, & \text{otherwise} \end{cases} \tag{2}$$

where n is the number of actions in the set.

Multiagent Q-learning is a natural extension of single-agent Q-learning to multiagent scenarios. In this approach, the agents are equipped with a standard Q-learning algorithm each and learn independently without considering the presence of each other in the environment. The rewards and the state transitions, however, depend on the joint actions of all agents. The problem is formalized as a tuple $(n, S, A_{1 \cdots n}, T, R_{1 \cdots n})$, where n is the number of players, S is the set of states, A_i is the set of actions available to agent i with A being the joint action space $A_1 \times \cdots \times A_n$, T is the transition function $S \times A \times S \to [0,1]$, and R_i is the reward function for the ith player $S \times A \to \mathbb{R}$. Note that both T and R are defined over the joint action space.

2.1.3   The IPA as a RL Problem

The first task to apply the Q-learning algorithm is to define what are the states, the actions and the rewards. We use the current price of the resources as the environment states and the possible demand requests as the actions of the agents. In this case, Q-values represent estimations of how good it is for the agent to request demand d at price p, so $Q(p,d)$. The optimal policy π^* is then an optimal demand function and is followed by selecting the demand requests with highest Q^*-values.

The application of RL in the IPA changes the objective of the agents in the resource allocation. Instead of maximizing their private utility functions in an immediate fashion, they now have to maximize the accumulated reward:

$$\sum_{t=0}^{\infty} \gamma^t r_t \tag{3}$$

where r is the reward given by the reward function in use and $\gamma \in \,]0,1[$ is the discounting factor for infinite horizon problems.

During the experiments we evaluate the application of two different reward functions:

- **The Individual Reward Function (IRF)** is based on the private utility of the agents. Using only local information, the agent receives a positive reward equal to its utility when the market reaches an equilibrium state, and zero for all the other states:

$$r = \begin{cases} U & \text{if equilibrium found,} \\ 0 & \text{otherwise} \end{cases}$$

- **The Social Reward Function (SRF)** uses global information. The agent receives a reward equal to the SW of the market for the equilibrium states, and zero for the others:

$$r = \begin{cases} SW & \text{if equilibrium found,} \\ 0 & \text{otherwise} \end{cases}$$

The idea for the individual reward function is to develop strategic agents that present *selfish* behaviour. This function generates a competitive learning problem where the objective of one agent is to learn a demand function that maximizes its discounted expected individual utility. The idea for the social reward function, on the other hand, is to develop strategic agents with *altruistic* behaviour. The function generates a cooperative learning problem, where one agent's goal is to learn a demand function that maximizes the discounted expected SW of the market.

To calculate the market's SW we use the Nash Product (NP) function. The NP is given by the product of the individual utility of the agents:

$$SW = NP = \prod_{i=1}^{n} U_i \qquad (4)$$

where U_i is the utility of agent i. It is suitable for the resource allocation scenario because it emphasizes the improvement and the balance among the utility of the agents. In addition, this function is regarded as a good compromise between the Utilitarian Social Welfare (USW), obtained from the sum of the individual utilities, and the Egalitarian Social Welfare (ESW), given by the utility of the agent which is worst off [7].

2.2 Experimental Investigation

The general configuration for the experiments is as follows. The agents have preferences over price and amount of resources. They use an utility function for each attribute, $U_1(p)$ for price and $U_2(m)$ for amount of resource. The total utility of an agent is given by the product of these two utility functions, $U(p,m) = U_1(p) * U_2(m)$. The actual utility functions used by the agents in the experiments are shown in Figure 2.

During the experiments price and demand requests were bounded in $[0, 10]$. Q-learning formally relies on discrete sets, so both prices and demand requests were rounded to 1 decimal place. Therefore, the market has 101 possible states and each agent has 101 actions to choose from. In the IPA market, the only information available to the agents is the current price of the resources. Therefore, the agents do not know the actions taken and the rewards received by the other agents.

We set the market with 5 units of resources per agent. From the analysis of the utility functions, we can note that such an amount does not allow for all the agents to have a complete satisfaction in the allocation, but it permits the analysis of the market and the learning under a condition of limited supply.

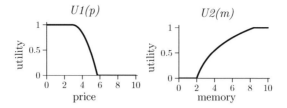

Fig. 2. Agents' utility functions

A series of preliminary experiments had been performed to identify a feasible configuration for the values of the parameters used in the learning algorithm. Based on these experiments we set $\alpha = 0.1$, $\gamma = 0.9$ and $\varepsilon = 0.4$. The price of the resources is adjusted by the IPA market using a constant parameter α set to 0.05.

We evaluate the agents' and market's performance using demand functions obtained at pre-defined intervals of learning episodes: 5000 for the experiments shown in Section 2.2.1 and 100 for Section 2.2.2. The evaluations are done with the trends of the actual demand functions learnt by the agents. One of the reasons for using the trends is that to implement the learning algorithm we had to transform prices (states) and demands (actions) into discrete sets. However, in the IPA market, such a discretization may lead to small losses of economical efficiency. Other reason is that by using the trends we avoid local instabilities present in the learnt demand functions. The trends were obtained by a process of curve fitting using the Sigmoidal-Boltzmann model. This monotonic model is described by the equation:

$$y = a + \frac{b - a}{1 + e^{-\frac{x-c}{d}}} \tag{5}$$

2.2.1 Learning with Static Agents

This section reports on the case of agents learning in the presence of other agents with static demand functions. The demand functions of the static agents, shown in Figure 3, were defined "by hand" and based on subjective criteria. They also use the utility functions $U_1(p)$ and $U_2(m)$ to evaluate the quality of the allocation.

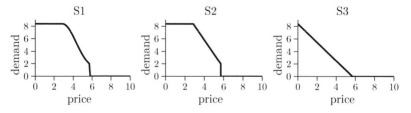

Fig. 3. Static Agents demand functions

We ran learning experiments with both reward functions using 3 agents at a time and iterating the number of learning agents per static agent. Each configuration was run 10 times of 500 000 learning episodes. The demand functions were extracted and evaluated in intervals of 5000 learning episodes. Next section presents the results.

Results

We first discuss the results for this scenario from the social perspective. Figure 4 shows the evolution of the average NP resulting from the use of the individual and the social reward functions over the learning episodes. The main point to note is that the NP achieves a level of relative stability quickly for both reward functions. This level is reached before 100 000 learning episodes in most of the experiments, suggesting that the learning process may be shorter than the 450 000 learning episodes we have originally applied in [16, 17].

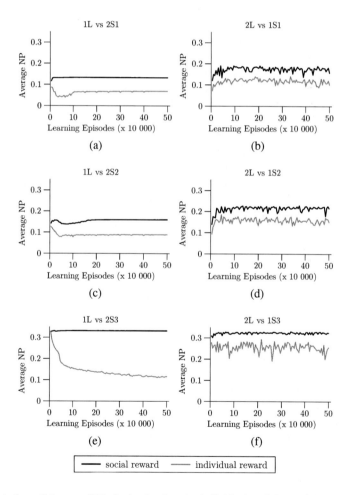

Fig. 4. Evolution of Average NP obtained using the individual and the social reward functions over the learning episodes

Figure 4 also shows that the stability of the average NP depends on the number of learning agents in the market. For experiments with 1 learning agent, a very stable level is achieved early and maintained along the learning episodes. Experiments with 2 learning agents presented only a relative stability. It evolves to a level, which is maintained, but fluctuates around it. We defer the discussion of the causes for this to the next section.

Still in Figure 4, the reward function based on social welfare improves the NP of the market in all cases. In addition, the average NP using the individual reward function increases with the number of learning agents. The application of the social reward function follows the same pattern, with the exception of S3, where a decrease was found. The analysis of S3's demand function (see Figure 3) and market configuration point that the equilibrium price is found at a lower level when only S3 agents are used, which improves the social welfare. When the social reward function is applied with these agents, the learners are able to follow that same strategy, but not as efficiently as S3's, generating the decrease. In the case of the individual reward function, the decrease is not seen because the learning agents develop a strategy to exploit S3's demand function, leaving to it a few resources. The more S3 agents are used in this case, the lower the social welfare.

The case of S3 suggests that a good strategy for the IPA would be to wait until the price is lower enough and then request the desired demand. However, this strategy only works when the agents have complete information, which is not realistic. In the case of agents with same demand functions this strategy is adopted implicitly. The implicit coordination contributes to the good social performance achieved when only learning agents are used in the market, shown in the next section.

Figure 4 e) presents an interesting point. It shows the average NP of the market decreasing over the learning episodes for the case of 1 learning and 2 S3 using the individual reward function. This decrease results from the fact that the learning agent develops a demand function which is able to exploit S3's demand function. This exploitation has been commented above and is clearly understood from Figure 5 e), which shows a large difference when the individual utilities received by those agents are compared. In that graph, it is also noticeable a decrease in the utility received by S3 along the learning episodes.

Figure 5 shows the evolution of the average utility obtained by learning and static agents using both reward functions over the learning episodes. The main point to note is the quick evolution of the individual utilities to a level of relative stability, which is achieved around or before 100 000 learning episodes for most of the agents.

In general, the application of the social reward function slightly decreases the individual utilities obtained by the learning agents and sharply increases the utility of the static clients. The application of the social reward causes a small movement of the demand function to the left. This small movement is able to lower the equilibrium price of the market. A lower equilibrium price means more resources to the static agents, increasing their utility in two components (price and amount of resource). In addition, the same small movement increases the learner utility in the price component, compensating a little bit the losses made in the resource component. Besides, the decrease noticed in the utilities received by the learning agents with the application of the social reward function is reduced with the addition of learning agents.

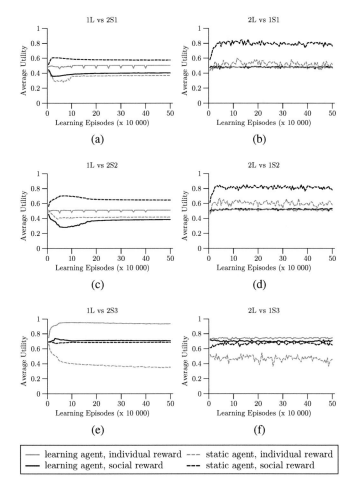

Fig. 5. Evolution of Average Utility obtained by the agents using the individual and the social reward functions over the learning episodes

It is important to comment that the application of the individual reward function does not imply in the learning agents beating the static ones by reaching higher utility. The goal for them is to get the most they can from the allocation taking into account the utility functions and static agents' demand functions. On the other hand, the application of the social reward function implies in the learning agents working to maximize the social welfare, even if it means to present lower individual performance. Therefore, the individual and social behaviours found are coherent.

2.2.2 Learning with Learning Agents

In this section we investigate the case where only learning agents are used in the market. We ran learning experiments with both reward functions using 2, 4, 6 and 8 agents. Each configuration was run 20 times of 10 000 learning episodes. The demand functions were extracted and evaluated in intervals of 100 learning episodes.

Results

Figure 6 shows the evolution of the market's average NP using the individual and the social reward functions. The first point to note is the quick evolution to a level of relative stability. It evolves to this level before 1000 learning episodes and fluctuates around it afterwards. The same type of instability was also seen in Figures 4 b), d) and f). They are caused by a non-appropriate decay rule for the learning parameter and by the application of Q-learning in a multiagent learning scenario. Q-learning is only proven to converge in stationary environments. With more than one learner, the environment becomes dynamic given the co-adaptation effect [28, 38]. There are RL algorithms that are more appropriate than Q-learning for the multiagent configuration [31]. They will be considered in our further investigations. Nevertheless, Q-learning has been applied with success in multiagent environments in the past [14, 20] and the results obtained in this research with this algorithm are satisfactory.

Results in the previous section have shown the social welfare improved with the application of the social reward function. This improvement is not clearly seen here. In fact, both reward functions presented similar results, approaching the optimal social welfare. It is not possible to identify which one performed better in general. The same type of result was found in the individual utilities received by the agents, shown in Figure 7.

The most important observation we can draw from the experiments is that, using both reward functions, there is a trend for the learning agents to divide the resources equally. While this strategy seems obvious for the case of the social reward function, since the optimal reward is received when the price is low enough and all the agents receive the same share of the allocation, it is not so intuitive for the individual reward function. Similar behaviour was observed in the results shown in the previous section. There, the agents using the individual reward function get the most utility they can from the allocation, but divide the resources more or less equally between them. So, the social welfare between the learning agents approaches the optimal value, given the demand functions of the static agents.

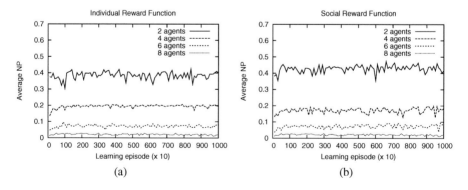

Fig. 6. Comparison of Average NP obtained by the market using the individual and the social reward functions over the learning episodes

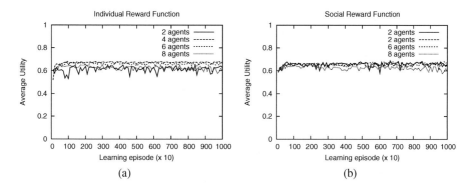

Fig. 7. Comparison of Average NP obtained by the market using the individual and the social reward functions over the learning episodes

An important comment to make is that a number of experiments in this scenario presented some problems in the curve-fitting in each checkpoint, generating demand functions unable to achieve equilibrium or not converging at all. These problems are due to the short learning process used here. Experiments made using the same amounts of learning used in the previous section, with checkpoints at each 100 000 learning episodes for a total of 500 000, presented less problems. Nevertheless, the experiments that have not presented problems, achieved good outcomes, as the graphs show.

3   Dynamic Analysis

In this section we present the analysis of agents' behaviour in the IPA with RL from the perspective of the dynamic process generated by the learning algorithm. Our focus is to gain some understanding on the reasons for the results shown in the previous section, which are particularly interesting since the learning problem generated by the social reward function is cooperative while the problem generated by the individual reward function is competitive. To understand the results, we first develop a model for the dynamics of Multiagent *Q*-learning with ε-greedy exploration. We then apply this model to theoretically analyse two games formulated with basis on the IPA with RL and that reproduce the application of the individual and the social reward functions.

3.1   A Model of Multiagent *Q*-Learning with ε-Greedy Exploration

The problem of modelling the dynamics of multiagent learning algorithms is challenging for a series of reasons. One of the difficulties is to cope with the very dynamic environment generated by multiple learners. There is also the co-adaptation effect, in which one agent adapts its strategy to the others', and vice-versa, in a cyclic fashion. In addition, the rewards that one agent receives depend on the actions of the other agents. All these features make it especially difficult to predict and to model the learning behaviour [24].

An important research in the area is the work of [34]. The authors studied the case of Q-learning agents with *Boltzmann* exploration. They developed a continuous time model for the learning process and have shown a link between the model and the Replicator Dynamics (RD) of Evolutionary Game Theory [18]. The main principle of the RD is that the growth in the probability of playing a given action is directly proportional to the performance of that action against the others. The ε-greedy mechanism, however, produces different dynamics. This mechanism defines a semi-uniform probability distribution in which the current best action is selected with probability $1 - \varepsilon$ and a random action with probability ε. Hence, that research cannot be directly applied in our case.

In another approach, [35] present a framework to track the error in one agent's decision during the multiagent learning process. The framework is generic enough to cover several different algorithms. However, it requires the tuning of some parameters that might not be known *a priori* or even impossible to obtain without extensive simulations.

As in our research, the above works share the property of being based on the analysis of differential or difference equations [26]. The topic has a long research tradition in the mathematical disciplines, a considerable theoretical framework and forms the standard approach to the study of dynamical systems. Other examples of the application of the approach to analyse multiagent reinforcement algorithms are the works of [3], who applied differential equations to study the dynamics of the *Weighted Policy Learner* algorithm [1], and [22], who studied the asymptotic behaviour of variants of the *Boltzmann*-based multiagent Q-learning. The approach has also been used for the analysis of single-agent reinforcement learning algorithms [4, 5].

As far as we are aware, none of the existing approaches has explored the specific case of Multiagent Q-learning with ε-greedy exploration. Apart from the necessity of explaining our results in the IPA with RL, the importance of obtaining such a model is justified through its large number of applications. For example [14] applies the algorithm to develop a decentralized resource allocation mechanism and [43] investigates the development of bidding strategies.

To develop the model we study how the ε-greedy mechanism and the presence of other agents affect the learning process of one agent. For this, we first show the derivation of a continuous time equation for the Q-learning rule. We then analyse the limits of this equation for the case of a single learner and show how they change dynamically when multiple learners are considered. Finally, we show how the ε-greedy mechanism affects the shape of the modelled function. The observations and results from this study are used to develop a system of difference equations to model the behaviour of the learners.

For simplicity of explanation, to develop the model we consider scenarios composed of 2 agents with 2 actions each[1]. The reward functions of the agents in this case can be described using payoff tables of the form:

$$A = \begin{bmatrix} a_{11} & a_{12} \\ a_{21} & a_{22} \end{bmatrix} \qquad B = \begin{bmatrix} b_{11} & b_{12} \\ b_{21} & b_{22} \end{bmatrix}$$

---

[1] This constraint is relaxed in our evaluation of the market-based resource allocation games (Section 3.2).

where A describes the rewards, or payoffs, for the first agent and B the rewards for the second. Given the existence of only one state, the Q-learning update rule can be simplified to

$$Q_{a_i} := Q_{a_i} + \alpha(r_{a_i} - Q_{a_i}) \qquad (6)$$

where Q_{a_i} is the Q-value of agent a for action i and r_{a_i} is reward that agent a receives for executing action i. Please note that this notation is slightly different from the notation applied in Section 2.

3.1.1 Analysis

We start the study by rewriting the update rule for the first agent as follows:

$$Q_{a_i}(k+1) - Q_{a_i}(k) = \alpha(r_{a_i}(k+1) - Q_{a_i}(k)) \qquad (7)$$

This difference equation describes the absolute growth in Q_{a_i} between times k and $k+1$. To obtain its continuous time version, consider $\Delta t \in [0,1]$ to be a small amount of time and

$$Q_{a_i}(k+\Delta t) - Q_{a_i}(k) \approx \Delta t \times \alpha(r_{a_i}(k+\Delta t) - Q_{a_i}(k))$$

to be the approximate growth in Q_{a_i} during Δt. Note that this equation becomes: an identity when $\Delta t = 0$; Equation 7 when $\Delta t = 1$; and a linear approximation when Δt is between 0 and 1. Dividing both sides of the equation by Δt,

$$\frac{Q_{a_i}(k+\Delta t) - Q_{a_i}(k)}{\Delta t} \approx \alpha(r_{a_i}(k+\Delta t) - Q_{a_i}(k))$$

and taking the limit for $\Delta t \to 0$,

$$\lim_{\Delta t \to 0} \frac{Q_{a_i}(k+\Delta t) - Q_{a_i}(k)}{\Delta t} \approx \alpha(r_{a_i}(k) - Q_{a_i}(k))$$

we obtain

$$\frac{dQ_{a_i}(k)}{dt} \approx \alpha(r_{a_i}(k) - Q_{a_i}(k)) \qquad (8)$$

which is an approximation for the continuous time version of Equation 7. This result is in line with [34].

The general solution for Equation 8 can be found by integration:

$$Q_{a_i}(k) = Ce^{-\alpha t} + r_{a_i} \qquad (9)$$

where C is the constant of integration. As e^{-x} is a monotonic function and $\lim_{x \to \infty} e^{-x} = 0$, it is easy to observe that the limit of Equation 9 when $t \to \infty$ is r_{a_i}:

$$\lim_{t \to \infty} Q_{a_i}(k) = \underbrace{\lim_{t \to \infty} Ce^{-\alpha t}}_{0} + \underbrace{\lim_{t \to \infty} r_{a_i}}_{r_{a_i}} = r_{a_i}$$

If we consider that only the first agent is learning and that the second is using a pure strategy, and assuming that the rewards are noise-free, playing a particular action will always generate the same reward for the first agent. In this case, the derivation above is enough to confirm that Q_{a_i} will monotonically increase or decrease towards r_{a_i}, for any initial value of Q_{a_i}. More specifically, the function is monotonically increasing if $Q_{a_i}(0) < r_{a_i}$ and monotonically decreasing if $Q_{a_i}(0) > r_{a_i}$. Examples of this behaviour are shown in Figure 8, which also shows the slope field associated with Equation 8. The figure plots the slope field obtained when $\alpha = 0.2$ and $r_{a_i} = 5$, and the sample paths for $Q_{a_i}(0)$ equal to 0, 2, 8 and 10. The line at $Q_{a_i} = 5$ is the equilibrium point for the slope field and the limit for the sample paths.

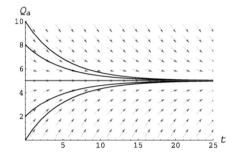

Fig. 8. Slope field associated with Equation 8 ($\alpha = 0.2$ and $r_{a_i} = 5$) and examples of specific solutions obtained when $Q_{a_i}(0) \in \{0, 2, 8, 10\}$

If the second agent is using a mixed strategy and the game is played repeatedly, then r_{a_i} can be replaced by

$$E[r_{a_i}] = \sum_j a_{ij} y_j \tag{10}$$

which is the expected payoff of the first agent given the mixed strategy **y** of the second. Note that a pure strategy is the specific case of a mixed strategy in which probability 1 is given to one of the actions. We then rewrite Equation 8 and 9 respectively as

$$\frac{dQ_{a_i}(k)}{dt} \approx \alpha(E[r_{a_i}(k)] - Q_{a_i}(k)) \tag{11}$$

$$Q_{a_i}(k) = Ce^{-\alpha t} + E[r_{a_i}] \tag{12}$$

Thus, if the adversary is not learning, Q_{a_i} will move in expectation towards $E[r_{a_i}]$ in a monotonic fashion. With a learning adversary, however, the situation is more complex. In this case, there is a possibility that the expected reward will change over time. A learning adversary can change its probability vector, which affects the expected reward. If we first look at Equation 11, changes in the expected reward will modify the associated direction field and, consequently, the equilibrium points of it. At this level, every time the expected reward changes, a new direction field is generated. If we now look at Equation 12, the changes will modify the limit and, possibly, the direction of Q_{a_i}. Hence, it is important to identify when they will occur.

The ε-greedy mechanism updates the probability vector whenever a new action becomes the one with the highest Q-value. Thus, we need to identify the intersection points in the functions of the adversary. It follows that the overall behaviour of the agent depends on these intersection points as they define which values Q_{a_i} will converge to.

From the analysis point of view, the fact that the expected rewards can change over time implies that Equation 11 cannot be solved in the same way we solved Equation 8. However, one can easily derive the paths given the initial Q-values.

Another important aspect to be considered in the model is the *speed* in which the Q-values are updated. During the learning process, the actions have different probabilities of being played. For example, if $\varepsilon = 0.2$, the Q-value of the current best action has a probability of 0.9 of being updated, while the other has a probability of 0.1 (considering a 2-actions game). It means that the Q-values are updated at different *speeds*. To simulate this behaviour, we define the growth in the Q-values as directly proportional to the probabilities. Then, Equation 11 becomes

$$\frac{dQ_{a_i}(k)}{dt} \approx x_i(k)\alpha(E[r_{a_i}(k)] - Q_{a_i}(k)) \tag{13}$$

where $x_i(k)$ is the probability of playing action i at time k.

It is important to emphasize that the *speed* of the updates affects the shape of the functions and, as a consequence, the points at which they will intersect each other. As such, this component plays a very significant role in the model. Roughly speaking, the expected reward indicates the values Q_{a_i} will converge to, the *speed* of the updates defines the paths that it will follow to get there and the presence of intersection points in the functions of the adversary determines if it is ever going to get there.

It should be clarified, however, that while the presence of intersection points in one agent's function does not affect the limits of its equations and the equilibrium points of the associated slope fields, it does affect the speed of the convergence and the slope field itself. To illustrate it, suppose that x_i and $E[r_{a_i}]$ are constants. Then, by integration we can find the general solution for Equation 13:

$$Q_{a_i}(k) = Ce^{-x_i\alpha t} + E[r_{a_i}] \tag{14}$$

Note that the only difference between this equation and Equation 12 is the exponential term. Because the limit of this term is 0 for $t \to \infty$, the limit of the equation remains $E[r_{a_i}]$, regardless of the value of x_i. On the other hand, different values of x_i generate different slope fields. This can be seen in Figure 9 where we plotted the slope fields obtained when $E[r_{a_i}] = 5$ and $\alpha = 0.2$ for $x_i \in \{0.1, 0.9\}$. For the sake of comparison, we have also plotted the sample paths for $Q_{a_i}(0)$ equal to 0, 2, 8 and 10.

In the next section we show how the observations above come together to model the behaviour of the Q-values during the multiagent learning process.

3.1.2 The Model

For the first and the second players, respectively, let A and B be the payoff matrices, $\mathbf{x}$ and $\mathbf{y}$ be the probability vectors, and Q_a and Q_b be the vectors of Q-values. Then, based

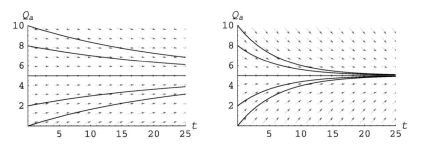

Fig. 9. Slope fields associated with Equation 13 ($\alpha = 0.2$ and $r_{a_i} = 5$) for $x_i = 0.1$ (Left) and $x_i = 0.9$ (Right), and examples of specific solutions obtained when $Q_{a_i}(0) \in \{0, 2, 8, 10\}$

on the analysis above, the expected behaviour for the Q-values can be modelled by the system of equations:

$$Q_{a_i}(k+1) = Q_{a_i}(k) + x_i(k)\alpha(\sum_j a_{ij} y_j(k) - Q_{a_i}(k))$$

$$Q_{b_i}(k+1) = Q_{b_i}(k) + y_i(k)\alpha(\sum_j b_{ij} x_j(k) - Q_{b_i}(k))$$

$$x_i(k) = \begin{cases} (1-\varepsilon) + (\varepsilon/n), & \text{if } Q_{a_i}(k) \text{ is currently the highest} \\ \varepsilon/n, & \text{otherwise} \end{cases}$$

$$y_i(k) = \begin{cases} (1-\varepsilon) + (\varepsilon/n), & \text{if } Q_{b_i}(k) \text{ is currently the highest} \\ \varepsilon/n, & \text{otherwise} \end{cases} \qquad (15)$$

Having the above model for the Q-values, the expected behaviour of the agents can be derived by tracking the actions with highest Q-value over the learning process of each agent.

3.2   Analysis of the Market-Based Resource Allocation Games

We now apply the model to analyse the dynamics of the learning agents in the IPA with RL. For the sake of simplicity, without losing the generality, we use a simplified version of the scenario addressed in the previous section. We consider agents with preferences over the amount of resources, using an increasing utility function to describe it, and a single IPA market with 4 units of indivisible resources, so only discrete units can be sold or requested. The price of the resources is not important at this stage as we are more interested in explaining why the agents reach the equilibrium that optimizes the SW. So, we consider the price to be constant.

Based on the simplified scenario we can define a two-player five-action stage game for each reward function. Tables 1 and 2 show the games. The actions represent the possible demand requests. Note that the joint-actions in the minor diagonal represent the possible equilibrium states for the market. Also note that the payoffs simulate the reward functions, generating a competitive game with symmetric payoff table for the

Table 1. Social Reward Game

	0	1	2	3	4
0	**0,0**	0,0	0,0	0,0	**0,0**
1	0,0	0,0	0,0	**3,3**	0,0
2	0,0	0,0	**4,4**	0,0	0,0
3	0,0	**3,3**	0,0	0, 0	0,0
4	**0,0**	0,0	0,0	0, 0	**0,0**

Table 2. Individual Reward Game

	0	1	2	3	4
0	0,0	0,0	0,0	0,0	*0,4*
1	0,0	0,0	0,0	*1,3*	0,0
2	0,0	0,0	*2,2*	0,0	0,0
3	0,0	*3,1*	0,0	0,0	0,0
4	*4,0*	0,0	0,0	0,0	**0,0**

individual reward function and a cooperative game for the social reward function. In the tables, *pure* Nash Equilibria (NEs) are presented in **bold** and Pareto-Optimal (PO) joint-actions in *italic*.

According to the results presented in the previous section, the agents learn to request 2 units of resources (joint-action $< 3,3 >$) in both games, which maximizes the SW. The analysis of the social reward game (Table 1) shows that this strategy is the only one that is PO and a NE. Furthermore, the strategy is the most profitable, so the convergence to this strategy is not surprising. The analysis of the individual reward game (Table 2), however, shows Pareto-Optimality and NE in all the strategies in the minor-diagonal. Supported by the model, in the next-subsections we investigate why and how the agents develop those strategies.

3.2.1 The Social Reward Game

We first focus on the analysis of the social reward game. This game has seven pure NEs: $< 1,5 >, < 2,4 >, < 3,3 >, < 4,2 >, < 5,1 >, < 5,5 >$ and $< 1,1 >$. And only one PO solution: $< 3,3 >$. The payoff matrices for the first and the second players are respectively:

$$A = \begin{bmatrix} 0 & 0 & 0 & 0 & 0 \\ 0 & 0 & 0 & 3 & 0 \\ 0 & 0 & 4 & 0 & 0 \\ 0 & 3 & 0 & 0 & 0 \\ 0 & 0 & 0 & 0 & 0 \end{bmatrix} \qquad B = \begin{bmatrix} 0 & 0 & 0 & 0 & 0 \\ 0 & 0 & 0 & 3 & 0 \\ 0 & 0 & 4 & 0 & 0 \\ 0 & 3 & 0 & 0 & 0 \\ 0 & 0 & 0 & 0 & 0 \end{bmatrix}$$

The results presented in the previous section were obtained with all the initial Q-values set to 0, so $Q_a = [0,0,0,0,0]$ and $Q_b = [0,0,0,0,0]$. This configuration generates a uniform probability distribution in which both agents play each action with probability $1/5$ in the first round of the learning process. Figure 10 presents the graphs obtained in this situation when $\alpha = 0.1$ and $\varepsilon = 0.45$. The graphs on the left-hand side of the figure show the theoretical evolution of the Q-values. The graphs on the center show the median Q-values aggregated from 100 learning experiments. The graphs on the right-hand side show the frequency in which each action has been adopted by the agents. By adopted action or strategy we mean the action or strategy with the highest Q-value at that particular time.

As seen in Figure 10, the dynamics of the agents' Q-values follow the dynamics obtained by the model relatively well. There is, however, a small difference in the shape of

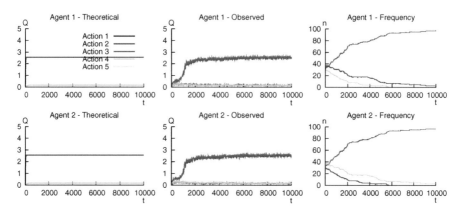

Fig. 10. Graphs for the Social Reward Game when the initial Q-values are $Q_a = Q_b = [0,0,0,0,0]$: the theoretical Q-values derived by the model (left); the median Q-values observed in the experiments (center); the observed frequency of the actions adopted by the agents (right) during the experiments

the curves of action 3 in the initial part of the learning process. This difference results from the uniform probability distribution applied by the agents in the first round, which has different impacts in the model and the experiments. For the model, the uniform distribution gives to action 3 the highest expected reward and, therefore, the highest Q-value from the second round on. Consequently, from the model point of view, the evolution of the Q-values is quite similar to the evolution obtained when strategy $< 3,3 >$ is the initial strategy of the agents, as shall be seen later. For the experiments, on the other hand, the uniform distribution gives to any strategy the same probability of being played. However, only strategies $< 2,4 >$, $< 3,3 >$ and $< 4,2 >$ are able to modify the Q-values as they are the only ones that return a payoff other than 0. Therefore, the first time any of these strategies is played, it will become the one with highest Q-values and the learning process from that point on will be similar to the process obtained when the initial strategy of the agents is set to $< 2,4 >$, $< 3,3 >$ or $< 4,2 >$, depending on which one is played first. Therefore, the impact that the initial uniform probability distribution has in the model is different from the impact that it has in the experiments. While in the model the strategy $< 3,3 >$ will be certainly adopted by the agents from the second round on, in the experiments strategies $< 2,4 >$, $< 3,3 >$ and $< 4,2 >$ have the same probability of being adopted. This difference generates the discrepancy between model and experiments in the initial part of the learning process. This behaviour is further illustrated by the graphs showing the frequencies of the actions, where it can be seen the equilibrium between actions 2, 3 and 4 during the initial part of the learning process.

Still in Figure 10, despite the different starting points between model and experiments, the observed Q-values end up stabilizing around the expected levels. In order to explain this behaviour, it is necessary to investigate the dynamics of the agents when they start the learning process playing specific strategies. For this, we modify the starting Q-value vectors in order to reflect the desired starting strategy. In particular we investigate the starting strategies $< 1,5 >$, $< 2,4 >$ and $< 3,3 >$ as they are the strategies that would be able to equilibrate demand and supply in the actual IPA with RL scenario. The analysis of initial strategies $< 5,1 >$ and $< 4,2 >$ are not shown because they are symmetric to strategies $< 1,5 >$ and $< 2,4 >$, respectively.

Figure 11 plots the graphs obtained when the agents start the process playing strategy $< 1,5 >$. The initial Q-values are set to $Q_a = [0.01,0,0,0,0]$ and $Q_b = [0,0,0,0,0.01]$, the learning parameters are set to $\alpha = 0.1$ and $\varepsilon = 0.45$. The results found for this case are quite similar to the results found in the previous case. The dynamics of the agents follow the dynamics obtained by the model with a small difference in the initial shapes of the action 3's curves. From the model perspective, as the payoff of actions 1 and 5 is 0, action 3 will have the highest expected reward and, consequently, the highest Q-value from the second round on, making the process similar to the one obtained in the first case. From the experimental perspective, as strategy $< 1,5 >$ returns 0, the first time any of the strategies $< 2,4 >$, $< 3,3 >$ and $< 4,2 >$ is played, it will be adopted by the agents from that point on. As the three strategies have the same probability of being played, then the process from the experimental perspective is also similar to the process obtained in the first case.

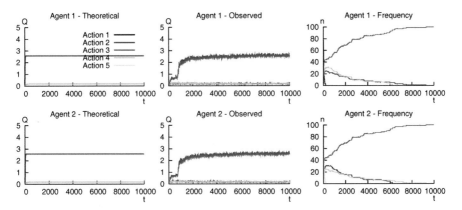

Fig. 11. Graphs for the Social Reward Game when the initial strategy is $< 1,5 >$: the theoretical Q-values derived by the model (left); the median Q-values observed in the experiments (center); the observed frequency of the actions adopted by the agents (right) during the experiments

Figure 12 plots the graphs obtained when the agents start the process playing strategy $< 2,4 >$. The initial Q-values are set to $Q_a = [0,0.01,0,0,0]$ and $Q_b = [0,0,0,0.01,0]$, the learning parameters are set to $\alpha = 0.1$ and $\varepsilon = 0.45$. The results obtained for this case are quite interesting. The dynamics of the agents in the initial part of the learning process is similar to the dynamics predicted by the model. After episode 2000, however, there are a sudden increase in the median Q-values of action 3 and a drecrease in the median Q-values of actions 2 and 4 of the first and second agents respectively. This strategy change, from $< 2,4 >$ to $< 3,3 >$, is not captured by the model, which predicts that strategy $< 2,4 >$ will be kept throughout the learning process. The graphs that show the frequency of the strategies further illustrate the fact that the agents eventually converge to strategy $< 3,3 >$ during the learning process. We will get back to this example later when we show the typical behaviour found in the analysis of the individual runs of the learning experiments.

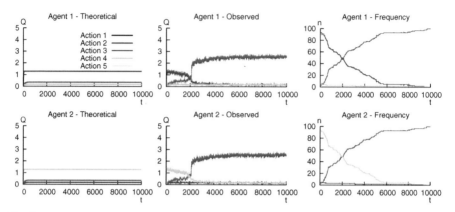

Fig. 12. Graphs for the Social Reward Game when the initial strategy is $< 2,4 >$: the theoretical Q-values derived by the model (left); the median Q-values observed in the experiments (center); the observed frequency of the actions adopted by the agent (right) during the experiments

In Figure 13 we plot the graphs obtained when the agents start playing strategy $< 3,3 >$. The initial Q-values in this case are set to $Q_a = Q_b = [0,0,0.01,0,0]$ and the learning parameters set to $\alpha = 0.1$ and $\varepsilon = 0.45$. The main point to note in this case is that the agents follow the dynamics described by the model remarkably well. The median Q-values of action 3 increases quite quickly and stabilizes around the expected levels. As commented above, the curves of theoretical Q-values are quite similar to the curves shown in Figure 10. Recall that in that case the agents apply a uniform probability distribution in the first round of the learning process, which gives to action 3 the highest expected reward and, therefore, the highest Q-value from the second round on. Consequently, both processes have similar starting points from the model perspective. From the experimental perspective, however, the starting points are significantly different. The uniform probability distribution of that case gives to strategies $< 2,4 >$, $< 3,3 >$ and $< 4,2 >$ the same probability of being adopted in the second round. The pre-defined distribution of this case, on the other hand, will always give to strategy $< 3,3 >$ the highest probability. Such a difference can be seen in the graphs that show the frequency of the agents' strategies. Note that, while in Figure 10 there is an initial equilibrium between the frequencies, which is generated by the uniform probability distribution, in Figure 13 the frequency of action 3 is always higher than the others and kept stable.

To conclude the discussion on the Social Reward Game, Figure 14 and 15 present the typical behaviours found during the learning experiments. The analysis of the individual runs revealed the presence of two types of typical behaviour. In the first type (Figure 14) the agents converge quite quickly to strategy $< 3,3 >$ and keep this strategy throughout

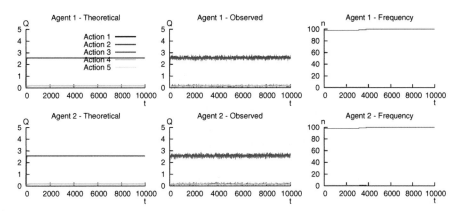

Fig. 13. Graphs for the Social Reward Game when the initial strategy is $< 3,3 >$: the theoretical Q-values derived by the model (left); the median Q-values observed in the experiments (center); the observed frequency of the actions adopted by the agents (right) during the experiments

the learning process. In the second type (Figure 15) the agents converge initially to strategies $< 2,4 >$ or $< 4,2 >$ and then change to strategy $< 3,3 >$, keeping it during the rest of the learning process. These behaviours were obtained from the analysis of experiments performed with initial Q-values set to 0, which is the configuration used in the actual IPA with RL. Nevertheless, they are illustrative of the typical behaviours found in the other configurations adopted in this section and can be used to further explain the results found for them. In particular, the second type of behaviour shows that when agents start playing strategy $< 2,4 >$ they will eventually converge to strategy $< 3,3 >$, which is the general behaviour presented in Figure 12. One point to note is that when the agents are playing strategy $< 3,3 >$, the behaviour of the observed Q-values is similar to the behaviour predicted by the model when the starting strategy of the agents is modified to $< 3,3 >$, shown in Figure 13. Likewise, the behaviour when the agents are playing strategy $< 2,4 >$ is similar to the behaviour predicted when the starting strategy is modified to $< 2,4 >$, shown in Figure 12.

3.2.2 The Individual Reward Game

We now focus on the analysis of the individual reward game. This game has six pure NEs: $< 1,5 >$, $< 2,4 >$, $< 3,3 >$, $< 4,2 >$, $< 5,1 >$ and $< 5,5 >$. And five PO solutions: $< 1,5 >$, $< 2,4 >$, $< 3,3 >$, $< 4,2 >$, $< 5,1 >$. The payoff matrices for the first and the second players are respectively:

$$A = \begin{bmatrix} 0 & 0 & 0 & 0 & 0 \\ 0 & 0 & 0 & 1 & 0 \\ 0 & 0 & 2 & 0 & 0 \\ 0 & 3 & 0 & 0 & 0 \\ 4 & 0 & 0 & 0 & 0 \end{bmatrix} \qquad B = \begin{bmatrix} 0 & 0 & 0 & 0 & 4 \\ 0 & 0 & 0 & 3 & 0 \\ 0 & 0 & 2 & 0 & 0 \\ 0 & 1 & 0 & 0 & 0 \\ 0 & 0 & 0 & 0 & 0 \end{bmatrix}$$

As for the social reward game, we first show the results obtained when $Q_a = [0,0,0,0,0]$ and $Q_b = [0,0,0,0,0]$, which has been the configuration applied in experiments in the

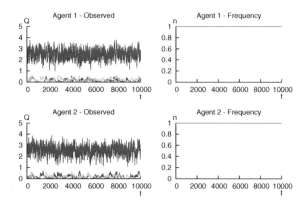

Fig. 14. Example of the typical behaviour found in the learning experiments with the Social Reward Game when the initial Q-values are $Q_a = Q_b = [0,0,0,0,0]$

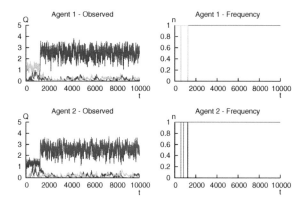

Fig. 15. Example of the typical behaviour found in the learning experiments with the Social Reward Game when the initial Q-values are $Q_a = Q_b = [0,0,0,0,0]$

IPA with RL. This configuration generates a uniform probability distribution in which both agents play each action with probability $1/5$ in the first round of the learning process. Figure 16 presents the graphs obtained in this situation when $\alpha = 0.1$ and $\varepsilon = 0.45$. The graphs on the left-hand side of the figure show the theoretical evolution of the Q-values. The graphs on the center show the median Q-values aggregated from 100 learning experiments. The graphs on the right-hand side show the frequency in which each action is adopted by the agents during the experiments. Again, by adopted action or strategy we mean the action or strategy that has the highest Q-value at that particular time.

As seen in Figure 16, the dynamics of the agents follow the dynamics predicted by the model relatively well in the beginning of the learning process. However, there is an increase in the median Q-value of action 3 around the learning episode 4000 that is not predicted. According to the model, the Q-values would converge constantly towards

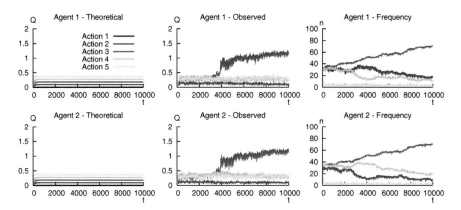

Fig. 16. Graphs for the Individual Reward Game when the initial Q-values are $Q_a = Q_b = [0,0,0,0,0]$: the theoretical Q-values derived by the model (left); the median Q-values observed in the experiments (center); the observed frequency of the actions adopted by the agents (right) during the experiments

$Q_a = Q_b = [0, 0.09, 0.18, 0.27, 0.36]$. These values are the expected rewards when the agents play strategy $< 5,5 >$. The development of this strategy is a consequence of the uniform probability distribution applied in the first round of the process, which gives to action 5 the highest expected reward and, therefore, the highest Q-value in the second round. As there are no intersection points in the curves of either agent, the Q-values would stabilize around those values. The experimental results show the median Q-values following this dynamic relatively well during the first learning episodes. Around episode 4000, however, there is a sudden increase in the median Q-value of action 3, which is not captured.

As commented above, the uniform distribution gives to action 5 the highest Q-value in the second round of the learning process according to the model. Therefore, it is necessary to investigate the dynamics of the agents when they start the learning process playing strategy $< 5,5 >$. The graphs for this case are presented in Figure 17. The initial Q-values are set to $Q_a = Q_b = [0,0,0,0,0.01]$ and the learning parameters set to $\alpha = 0.1$ and $\varepsilon = 0.45$. As it can be seen, the results obtained in this case are similar to the results obtained in the previous case. The main difference between them is in the observed frequency of the actions during the very beginning part of the learning process. While in the first case the frequency of action 5 starts at the same level as the frequency of actions 2, 3 and 4, which is a result of the uniform probability distribution, in this case it starts higher. In both situations, however, the frequency of action 5 decreases quite quickly and action 3 ends up dominating the learning process.

Figure 18 plots the graphs when the agents starting the learning process playing strategy $< 1,5 >$. The initial Q-values are set to $Q_a = [0.01,0,0,0,0]$ and $Q_b = [0,0,0,0,0.01]$ and the learning parameters set to $\alpha = 0.1$ and $\varepsilon = 0.45$. The results obtained in this case are also quite similar to the results obtained in the previous two cases. The median Q-values follow the theoretical Q-values relatively well in the first part of the learning process. In the second part there is a sudden increase in the Q-values of action 3, which

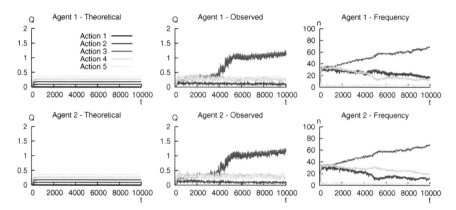

Fig. 17. Graphs for the Individual Reward Game when the initial strategy is $< 5,5 >$: the theoretical Q-values derived by the model (left); the median Q-values observed in the experiments (center); the observed frequency of the actions adopted by the agents (right) during the experiments

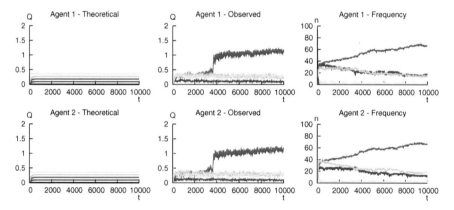

Fig. 18. Graphs for the Individual Reward Game when the initial strategy is $< 1,5 >$: the theoretical Q-values derivated by the model (left); the median Q-values observed in the experiments (center); the observed frequency of the actions adopted by the agents (right) during the experiments

is not predicted by the model. The frequency of the actions also reveals an interesting behaviour. For the first agent, the frequency of action 1 starts higher than the others but decrease rather quickly because the payoff of playing $< 1,5 >$ is 0. Since the other actions have the same probability ε/n of being played, their frequencies start at the same level. As the learning process progress, however, the frequency of action 5 decreases quickly while the frequency of actions 2, 3 and 4 are kept in equilibrium until action 3 dominates the process. For the second agent, the behaviour is similar to the behaviour found when the starting strategy is set to $< 5,5 >$: the frequency of action 5 starts higher than the others but decreases quickly and action 3 dominates the process.

The previous three examples have presented the similar behaviour in which the median Q-values follow the dynamics derived by the model relatively well during the first part of the learning process, when a sudden increase in the Q-values of action 3 is not captured. In the three cases the model predicts the convergence of the agents to strategy $< 5, 5 >$. When this strategy is adopted, however, the experiments show the frequency of action 5 decreasing quite quickly and the convergence to strategy $< 3, 3 >$. Additionally, the experiments show the frequency of actions 2, 3 and 4 in equilibrium for some time before the frequency of action 3 starts to rise. We will next present the analysis of the dynamics when the starting strategies are $< 2, 4 >$ and $< 3, 3 >$. Before that, however, we will present the analysis of the individual runs of the learning experiments as it will help in identifying why the agents always converge to strategy $< 3, 3 >$.

The analysis of the individual runs of the learning experiments has revealed the existence of two types of typical behaviour. In the first type the agents converge to joint-action $< 3, 3 >$ in the beginning of the learning process and keep this strategy quite stable during the rest of it. In the second type the agents converge initially to joint-actions $< 2, 4 >$ or $< 4, 2 >$, which are kept quite unstable, particularly by the agent that is playing action 2, until they eventually converge to strategy $< 3, 3 >$. In some cases the agents alternate between strategies $< 2, 4 >$ and $< 4, 2 >$ before converging to $< 3, 3 >$.

Figures 19 and 20 show examples of the typical behaviours. Figure 19 plots the example for the first type. Note the quick convergence to action 3 and the very few changes of strategy that occurs throughout the learning process. Figure 20 plots the example for the second type. Note the initial convergence to strategy $< 2, 4 >$ and then the convergence to strategy $< 3, 3 >$. Also note the high number of changes that occurs in the strategy of the first agent while it is playing action 2. From these observations we can infer that strategy $< 3, 3 >$ is more stable than strategies $< 2, 4 >$ and $< 4, 2 >$. We can then justify the rise in the Q-values of action 3: if the typical behaviour is to eventually converge to strategy $< 3, 3 >$ and if such a strategy is more stable than the others, then it is reasonable to say that it will dominate the learning process in the long run.

In Figure 21 we plot the graphs obtained when the agents start the process playing strategy $< 2, 4 >$. The initial Q-values are set to $Q_a = [0, 0.01, 0, 0, 0]$ and $Q_b = [0, 0, 0, 0.01, 0]$, the learning parameters are set to $\alpha = 0.1$ and $\varepsilon = 0.45$. The results for this case are quite interesting. According to the model the Q-values would converge to $Q_a = [0, 0.63, 0.18, 0.27, 0.36]$ and $Q_b = [0, 0.09, 0.18, 1.92, 0.36]$, so the agents would keep playing strategy $< 2, 4 >$. This dynamic is followed by the experiments during the first learning episodes, but then there is a decrease in the median Q-values of actions 2 and 4 for the first and second agent respectivelly. At the same time, there is an increase in the median Q-value of action 3. This behaviour is also noticed in the graphs for the frequency of the strategies, which show the frequency of actions 2 and 4 decreasing while the frequency of action 3 is increasing. These observations illustrate the instability of strategy $< 2, 4 >$. It is important to mention that the values found by the model are coherent with the values shown in Figure 20 for the periods where the agents are playing strategy $< 2, 4 >$. Furthermore, the typical behaviour found is similar to the

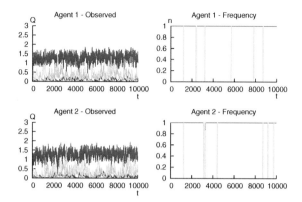

Fig. 19. Example of the typical behaviour found in the learning experiments with the Individual Reward Game when the initial Q-values are $Q_a = Q_b = [0,0,0,0,0]$

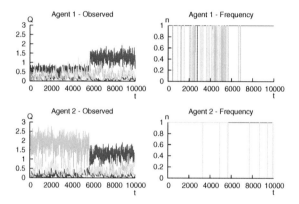

Fig. 20. Example of the typical behaviour found in the learning experiments with the Individual Reward Game when the initial Q-values are $Q_a = Q_b = [0,0,0,0,0]$

type of typical behaviour shown in Figure 20: the agents start playing strategy $< 2,4 >$ and eventually converge to $< 3,3 >$, which is then kept very stable.

In Figure 22 we plot the graphs obtained when the agents start the process playing strategy $< 3,3 >$. The initial Q-values are set to $Q_a = Q_b = [0,0,0.01,0,0]$ and the learning parameters set to $\alpha = 0.1$ and $\varepsilon = 0.45$. The first point to note in the graphs is that, in this case, the dynamics found in the experiments is captured very well by the model. It can also be seen that the frequency of strategy $< 3,3 >$ is kept constant throughout the learning process, which illustrates the stability of this strategy. Another point to note is that the theoretical Q-values described by the model is coherent with the values found in the individual runs of the learning experiments for the periods where the agents are playing strategy $< 3,3 >$ (see Figure 19). As in the previous example, the typical behaviour found in this case is similar to the first type of typical behaviour

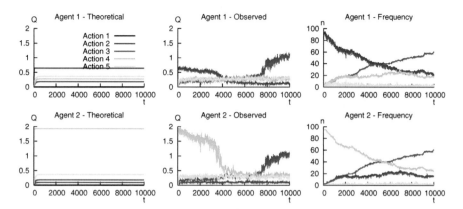

Fig. 21. Graphs for the Individual Reward Game when the initial Q-values strategy is $< 2, 4 >$: the theoretical Q-values derived by the model (left); the median Q-values observed in the experiments (center); the observed frequency of the actions adopted by the agents (right) during the experiments

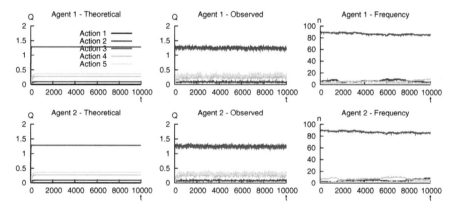

Fig. 22. Graphs for the Individual Reward Game when the initial strategy is $< 3, 3 >$: the theoretical Q-values derived by the model (left); the median Q-values observed in the experiments (center); the observed frequency of the actions adopted by the agents (right) during the experiments

shown in that Figure: the agents converge to strategy $< 3, 3 >$ in the beginning of the process and keep this strategy quite stable during the rest of it.

The experiments indicate the existence of certain relationships between the payoffs tables, the learning rate and the exploration rate of the ε-greedy mechanism. In the general case, these relationships may be responsible for inaccuracies of the model in the sense that the model is not able to capture the actual stochasticity present in the experiments. For instance, consider the case where the Q-values of two actions are quite close to each other (the theoretical Q-values shown in Figure 16 can serve as an example). With a high exploration, there is a high chance for the occurrence of intersections between these two curves during the real experimentation, even if the model does not predict it. The occurrence of non-predicted intersections can decrease the quality of the model as they can change the expected rewards, changing the curves in general. In our particular case of the individual reward game, the relationships are responsible for the instability presented by strategies $< 2,4 >$, $< 4,2 >$, $< 1,5 >$ and $< 5,1 >$, and contribute to the convergence of the system to strategy $< 3,3 >$, once this strategy is the most stable of them. To identify those relationships can improve the general quality of the model and constitute an important future work for this research.

4   Related Works

There is a significant amount of work on learning for market-based systems. While most of the approaches have not been developed specifically for resource allocation, they can be naturally extended for this end. One branch of research corresponds to the improvement of bidding and asking strategies in auction mechanisms. For example, [27] investigate the use of belief-based learning in simultaneous English Auctions. Agents build belief functions to model the valuations held by other participants and coordinate their bids accordingly in multiple auction houses. [15] apply a similar approach but for single Double Auctions. Another area of research is the use of learning on Mechanism Design [9, 25]. Mechanism design studies the problem of optimizing auction parameters in order to maximize an objective function (e.g. auctioneer revenue).

There is also some work on the automated pricing. [20] studied the use of Q-Learning in a scenario where two competitive "pricebots" have to set the price of a commodity. [21] investigated a similar problem for the scenario in which one agent has the power to enforce its strategy on the other.

[42] surveyed 35 different market-based mechanisms for resource allocation in large-scale distributed systems. From those, learning is applied only in Catnets [29]. Catnets' agents apply an evolutionary-like approach to learn negotiation strategies in a completely decentralized bargaining model. Most of work on learning for negotiation can be naturally transferred to the resource allocation field and, therefore, can be used to improve bargaining models.

The fundamental difference between the above works and our approach is the expression of the preferences of the agents. Previous works usually consider the utility of one agent as a function of the profit it makes in the market. Our agents, in contrast, describe their preferences by explicitly modelling utility functions for attributes of the allocation. It has the advantage of providing more power to the agents in developing strategic behaviours, which is particularly interesting in small markets where the possible gains from strategic attempts are larger, and achieving fair-optimal allocations.

Learning has also been explored to improve non-market-based resource allocation. In [14], agents learn which resource nodes to submit their jobs to, given that the nodes are managed by local schedulers. [2] proposed a gradient ascent learning algorithm and applied it in distributed task allocation. [10] modelled a resource allocation problem with precedence constraints as a Markov Decision Process and applied a distributed RL approach to solve it. Those works also follow a different approach for the preferences of the agents, which are typically modelled as functions of scheduling parameters, such as reducing the time interval between a job submission and its completion. Utility functions are not considered.

Finally, the work of [8] is also related to ours. The authors develop a pricing mechanism to maximize the aggregated utility for resource allocation in the Grid. Apart from the fact that learning is not applied, they also follow a different approach for agents' preferences. Their experiments use the maximization of resource usage as resource owners' preferences and functions of concrete parameters (e.g. price of the resources and time to complete a task) as users' preferences.

To the best of our knowledge, the approach presented herein is the first attempt to address the problem of learning demand functions based on the agents' preferences.

5   Conclusion

The commodity-market economic model offers a promising approach for resource allocation in large-scale distributed systems. Existing mechanisms based on this model usually employ the *Walrasian Economic Equilibrium* concepts, assuming that all the participants of the allocation process are price-taking entities that will not attempt to strategically influence the mechanism in order to improve their profits or welfare. Such a condition is hardly satisfied in large-scale distributed systems where there is little control over the behaviour of the agents, making it impossible to guarantee that they will behave in an ordered manner. To understand the impacts of these attempts and to develop mechanisms that are robust in the presence of strategic participants, being able to deliver optimal allocation also in this situation, are important aspects of the problem. Additionally, most mechanisms focus on the achievement of a Pareto-Optimal allocation but usually disregard how fair and how desirable the allocation is for both the system and the participants.

In this paper we have addressed the above issues and studied the dynamics of strategic learning agents in the IPA with RL. This mechanism enhances the original Iterative Price Adjustment (IPA) mechanism by introducing into the market participants with learning capabilities. These participants use utility functions to describe preferences over different resources attributes and develop strategic behaviour by learning demand functions adapted to the market through RL. The reward function used during the learning process are based either on the individual utility of the participants, generating selfish learning agents, or the social welfare of the market, generating altruistic learning agents.

The study has been conducted in two parts. In the first part we have experimentally investigated the impacts of the learning agents on the individual and social performance of the market. The results of this step have shown that the market composed exclusively of selfish agents achieved social performance similar to the performance obtained by the market composed exclusively of altruistic agents, both achieving near-optimal SW, measured by the NP social welfare function. Such an outcome is significant not only for the market-based resource allocation domain but also for a series of other domains where individual and social utility should be maximized but agents are not guaranteed to act cooperatively in order to achieve it or they do not want to reveal private preferences.

The second part of the study involved the analysis of the behaviour of the agents from the perspective of the dynamic process generated by the learning algorithm employed by them. The focus was to gain some understanding on the reasons for the results found in the experimental investigation, which are particularly interesting since the learning problem generated by the social reward function is cooperative while the problem generated by the individual reward function is competitive. For this, we first developed a model for the dynamics of Multiagent Q-learning with ε-greedy exploration. We then applied this model to theoretically analyse two games formulated with basis on the IPA with RL and that reproduce the application of the individual and the social reward functions.

The game-theoretical analysis of the games has shown Pareto-optimality and a Nash Equilibrium at the joint-strategies in which the agents request the same amount of resources, maximizing the social welfare. While in the social reward game this strategy returns the most profitable payoff, in the individual reward game there are strategies that are more profitable. The analysis supported by the model, however, has shown that such strategies are less stable, therefore, justifying the interesting results in which the selfish agents end up optimising the social welfare.

There are several opportunities for future works in this research. One area for extension involves the further reduction of the required learning episodes in the market, which is in general related to well-known scalability issues of RL. Moreover, it is necessary to evaluate the approach in extended scenarios, including agents with preferences described over multiple attributes, multiple markets, and the existence of resource provider agents. The later deals with a limitation of the IPA mechanism. In particular, the IPA does not model resource provider agents. Simply adding those agents to the model may change its theoretical implications as the resource supply becomes dynamic and, therefore, has to be carefully considered.

There are also several opportunities for future works in our model of Multiagent Q-learning with ε-greedy exploration. In particular, its application in the market-based resource allocation games has indicated the existence of some relationships between the payoffs, the exploration and the learning rate that may be responsible for inaccuracies of the model. To identify what are those relationships constitute an important future work. It is also important to extend the approach to multi-state scenarios. One of the ideas for this extension is to consider each state as a separated game and then analyse the basins of attraction in between the games/states. To enable it, however, we first need to explore ways of obtaining a more complete view of the expected joint-policy of the agents. Our current work in this area involves aggregating different starting Q-values with similar dynamics. From this aggregation we will generate a single graph of the joint policy space, which is discrete in our case, relating each starting point with its attracting basin in the asymptotic case. Another area for future work is to develop new techniques to allow the analysis of scenarios composed of multiple states and multiple agents. Currently, the dynamics of the algorithm is analysed in 2-dimensional graphs that show the development of the expected Q-values and strategies of the agents over the time, two graphs per agent. Such a technique is not scalable to the multi-state multi-agent scenario.

References

1. Abdallah, S., Lesser, V.: Learning the Task Allocation Game. In: Proceedings of the Fifth International Joint Conference on Autonomous Agents and Multi-Agent Systems (AAMAS 2006), Hakodate, Japan, pp. 850–857. ACM Press, New York (2006),
   http://mas.cs.umass.edu/paper/431
2. Abdallah, S., Lesser, V.: Learning the task allocation game. In: Proceedings of the Fifth International Joint Conference on Autonomous Agents and Multiagent Systems (AAMAS 2006), pp. 850–857. ACM Press, New York (2006),
   http://doi.acm.org/10.1145/1160633.1160786
3. Abdallah, S., Lesser, V.: Non-linear Dynamics in Multiagent Reinforcement Learning Algorithms. In: Proceedings of the Seventh International Conference on Autonomous Agents and Multiagent Systems (AAMAS 2008), Estoril, Portugal, pp. 1321–1324. IFAAMAS (2008),
   http://mas.cs.umass.edu/paper/450
4. Abounadi, J., Bertsekas, D.P., Borkar, V.: Stochastic approximation for nonexpansive maps: Application to q-learning algorithms. SIAM J. Control Optim. 41(1), 1–22 (2002)
5. Borkar, V.S., Meyn, S.P.: The o.d.e. method for convergence of stochastic approximation and reinforcement learning. SIAM J. Control Optim. 38, 447–469 (2000)
6. Buyya, R., Murshed, M., Abramson, D.: A deadline and budget constrained cost-time optimization algorithm for scheduling task farming applications on global grids. In: Proceedings of the 2002 International Conference on Parallel and Distributed Processing Techniques and Applications, PDPTA 2002 (2002)
7. Chevaleyre, Y., Dunne, P.E., Endriss, U., Lang, J., Lemaître, M., Maudet, N., Padget, J., Phelps, S., Rodríguez-Aguilar, J.A., Sousa, P.: Issues in multiagent resource allocation. Informatica 30, 3–31 (2006),
   http://www.illc.uva.nl/~ulle/MARA/mara-survey.pdf
8. Chunlin, L., Layuan, L.: Pricing and resource allocation in computational grid with utility functions. In: Proceedings of the International Conference on Information Technology: Coding and Computing (ITCC 2005), vol. II, pp. 175–180. IEEE Computer Society, Washington (2005)
9. Conitzer, V., Sandholm, T.: Self-interested automated mechanism design and implications for optimal combinatorial auctions. In: EC 2004: Proceedings of the 5th ACM conference on Electronic commerce, pp. 132–141. ACM, New York (2004),
   http://doi.acm.org/10.1145/988772.988793
10. Csáji, B.C., Monostori, L.: Adaptive algorithms in distributed resource allocation. In: Proceedings of the 6th international workshop on emergent synthesis, IWES 2006 (2006)
11. Erl, T.: Service-Oriented Architecture: Concepts, Technology, and Design. Prentice Hall PTR, Upper Saddle River (2005)
12. Everett, H.: Generalized lagrange multiplier method for solving problems of optimum allocation of resources. Operations Research 11(3), 399–417 (1963)
13. Foster, I., Kesselman, C. (eds.): The Grid: Blueprint for a Future Computing Infrastructure. Morgan Kaufmann, San Francisco (1999)
14. Galstyan, A., Czajkowski, K., Lerman, K.: Resource allocation in the grid using reinforcement learning. In: Proceedings of the Third International Joint Conference on Autonomous Agents and Multiagent Systems (AAMAS 2004), vol. 3, pp. 1314–1315. IEEE Computer Society, Washington (2004), http://dx.doi.org/10.1109/AAMAS.2004.232
15. Gjerstad, S., Dickhaut, J.: Price formation in double auctions. In: E-Commerce Agents, Marketplace Solutions, Security Issues, and Supply and Demand, London, UK, pp. 106–134. Springer, Heidelberg (2001)

16. Gomes, E.R., Kowalczyk, R.: Learning the ipa market with individual and social rewards. In: Proceedings of the International Conference on Intelligent Agent Technology (IAT 2007), pp. 328–334. IEEE Computer Society Press, Los Alamitos (2007)

17. Gomes, E.R., Kowalczyk, R.: Reinforcement learning with utility-aware agents for market-based resource allocation. In: Proceedings of the 7th International Joint Conference on Autonomous Agents and Multiagent Systems, AAMAS 2007 (2007)

18. Hofbauer, J., Sigmund, K.: Evolutionary Games and Population Dynamics. Cambridge University Press, Cambridge (1998)

19. Jennergren, P.: A price schedules decomposition algorithm for linear programming problems. Econometrica 41(5), 965–980 (1973)

20. Kephart, J.O., Tesauro, G.: Pseudo-convergent q-learning by competitive pricebots. In: Proceedings of the Seventeenth International Conference on Machine Learning (ICML 2000), pp. 463–470. Morgan Kaufmann Publishers Inc., San Francisco (2000)

21. Könönen, V.: Dynamic pricing based on asymmetric multiagent reinforcement learning: Research articles. Int. J. Intell. Syst. 21(1), 73–98 (2006),
http://dx.doi.org/10.1002/int.v21:1

22. Leslie, D.S., Collins, E.J.: Individual q-learning in normal form games. SIAM J. Control Optim. 44(2), 495–514 (2005),
http://dx.doi.org/10.1137/S0363012903437976

23. Noll, J.: A peer-to-peer architecture for workflow in virtual enterprises. In: Fifth International Conference on Quality Software (QSIC 2005), pp. 365–372 (2005), doi:10.1109/QSIC.2005.6

24. Panait, L., Luke, S.: Cooperative multi-agent learning: The state of the art. Autonomous Agents and Multi-Agent Systems 11(3), 387–434 (2005),
http://dx.doi.org/10.1007/s10458-005-2631-2

25. Pardoe, D., Stone, P., Saar-Tsechansky, M., Tomak, K.: Adaptive mechanism design: a metalearning approach. In: Proceedings of the 8th International Conference on Electronic Commerce, pp. 92–102. ACM Press, New York (2006),
http://doi.acm.org/10.1145/1151454.1151480

26. Perko, L.: Differential Equations and Dynamical Systems. Springer, New York (1996)

27. Preist, C., Byde, A., Bartolini, C.: Economic dynamics of agents in multiple auctions. In: AGENTS 2001: Proceedings of the fifth international conference on Autonomous agents, pp. 545–551. ACM Press, New York (2001),
http://doi.acm.org/10.1145/375735.376441

28. Sandholm, T.W., Crites, R.H.: On multiagent Q–learning in a semi–competitive domain. In: Weiß, G., Sen, S. (eds.) Adaptation and Learning in Multi–Agent Systems, pp. 191–205. Springer, Berlin (1996)

29. Schnizler, B., Neumann, D., Veit, D., Reinicke, M., Streitberger, W., Eymann, T., Freitag, F., Chao, I., Chacin, P.: Catnets - wp 1: Theoretical and computational basis (2005)

30. Sherwani, J., Ali, N., Lotia, N., Hayat, Z., Buyya, R.: Libra: a computational economy-based job scheduling system for clusters. Softw. Pract. Exper. 34(6), 573–590 (2004),
http://dx.doi.org/10.1002/spe.581

31. Shoham, Y., Powers, R., Grenager, T.: Multi-agent reinforcement learning: a critical survey (2003)

32. Subramoniam, K., Maheswaran, M., Toulouse, M.: Towards a micro-economic model for resource allocation in grid computing systems. In: IEEE Canadian Conference on Electrical and Computer Engineering, CCECE 2002, vol. 2, pp. 782–785 (2002)

33. Sutton, R.S., Barto, A.G.: Reinforcement Learning: An Introduction. MIT Press, Cambridge (1998)

34. Tuyls, K., Verbeeck, K., Lenaerts, T.: A selection-mutation model for q-learning in multi-agent systems. In: Proceedings of the Second International Joint Conference on Autonomous Agents and Multiagent Systems (AAMAS 2003), pp. 693–700. ACM, New York (2003), http://doi.acm.org/10.1145/860575.860687

35. Vidal, J.M., Durfee, E.H.: Predicting the expected behavior of agents that learn about agents: the CLRI framework. Autonomous Agents and Multi-Agent Systems 6(1), 77–107 (2003), http://jmvidal.cse.sc.edu/papers/clri.pdf

36. Walras, L.: Eleements d'Economie Politique Pure. Corbaz (1874)

37. Watkins, C.J.C.H.: Learning from delayed rewards. Ph.D. thesis, King's College, Cambridge, UK (1989)

38. Weinberg, M., Rosenschein, J.S.: Best-response multiagent learning in non-stationary environments. In: Proceedings of the International Joint Conference on Autonomous Agents and Multiagent Systems (AAMAS 2004), vol. 2, pp. 506–513. IEEE Computer Society, Los Alamitos (2004)

39. Wellman, M.P.: A market-oriented programming environment and its application to distributed multicommodity flow problems. Journal of Artificial Intelligence Research 1, 1–23 (1993)

40. Wolski, R., Plank, J.S., Brevik, J., Bryan, T.: Analyzing market-based resource allocation strategies for the computational grid. International Journal of High Performance Computing Applications 15(10), 258–281 (2001), http://hipersoft.rice.edu/grads/publications/gc-jour.pdf

41. Wu, T., Ye, N., Zhang, D.: Comparison of distributed methods for resource allocation. International Journal of Production Research 43(3), 515–536 (2005)

42. Yeo, C.S., Buyya, R.: A taxonomy of market-based resource management systems for utility-driven cluster computing. Softw. Pract. Exper. 36(13), 1381–1419 (2006), http://dx.doi.org/10.1002/spe.v36:13

43. Ziogos, N.P., Tellidou, A.C., Gountis, V.P., Bakirtzis, A.G.: A reinforcement learning algorithm for market participants in ftr auctions. In: 2007 IEEE POWERTECH, pp. 943–948. IEEE, Los Alamitos (2007), http://dx.doi.org/10.1109/PCT.2007.4538442

Simple Algorithms for Frequent Item Set Mining

Christian Borgelt

European Center for Soft Computing
c/ Gonzalo Gutiérrez Quirós s/n, 33600 Mieres, Asturias, Spain
christian.borgelt@softcomputing.es

Abstract. In this paper I introduce SaM, a split and merge algorithm for frequent item set mining. Its core advantages are its extremely simple data structure and processing scheme, which not only make it quite easy to implement, but also very convenient to execute on external storage, thus rendering it a highly useful method if the transaction database to mine cannot be loaded into main memory. Furthermore, I review RElim (an algorithm I proposed in an earlier paper and improved in the meantime) and discuss different optimization options for both SaM and RElim. Finally, I present experiments comparing SaM and RElim with classical frequent item set mining algorithms (like Apriori, Eclat and FP-growth).

1 Introduction

It is hardly an exaggeration to say that the popular research area of *data mining* was started by the tasks of frequent item set mining and association rule induction. At the very least, these tasks have a strong and long-standing tradition in data mining and *knowledge discovery in databases* and triggered an abundance of publications in data mining conferences and journals. The huge research efforts devoted to these tasks have led to a variety of sophisticated and efficient algorithms to find frequent item sets. Among the best known approaches are Apriori [1,2], Eclat [17] and FP-growth [11].

Nevertheless, there is still room for improvement: while Eclat, which is the simplest of the mentioned algorithms, can be relatively slow on some data sets (compared to other algorithms, see the experimental results reported in Section 5), FP-growth, which is usually the fastest algorithm, employs a sophisticated and rather complex data structure and thus requires to load the transaction database into main memory. Hence a simpler processing scheme, which maintains efficiency, is desirable. Other lines of improvement include filtering found frequent item sets and association rules (see, e.g., [22,23]), identifying temporal changes in discovered patterns (see, e.g., [4,5]), and discovering fault-tolerant or approximate frequent item sets (see, e.g., [9,14,21]).

In this paper I introduce SaM, a split and merge algorithm for frequent item set mining. Its core advantages are its extremely simple data structure and processing scheme, which not only make it very easy to implement, but also convenient to execute on external storage, thus rendering it a highly useful method if the transaction database to mine cannot be loaded into main memory. Furthermore, I review RElim, an also very simple algorithm, which I proposed in an earlier paper [8] and which can be seen as a precursor of SaM. In addition, I study different ways of optimizing RElim and SaM, while preserving, as far as possible, the simplicity of their basic processing schemes.

J. Koronacki et al. (Eds.): Advances in Machine Learning II, SCI 263, pp. 351–369.

The rest of this paper is structured as follows: Section 2 briefly reviews the task of frequent item set mining and especially the basic divide-and-conquer scheme underlying many frequent item set mining algorithms. In Section 3 I present my new SaM (Split and Merge) algorithm for frequent item set mining, while in Section 4 I review its precursor, the RElim algorithm, which I proposed in an earlier paper [8]. In Section 5 the basic versions of both SaM and RElim are compared experimentally to classical frequent item set mining algorithms like Apriori, Eclat, and FP-growth, and the results are analyzed. Based on this analysis, I suggest in Section 6 several optimization options for both SaM and RElim and then present the corresponding experimental results in Section 7. Finally, in Section 8, I draw conclusions from the discussion.

2   Frequent Item Set Mining

Frequent item set mining is a data analysis method, which was originally developed for market basket analysis and which aims at finding regularities in the shopping behavior of the customers of supermarkets, mail-order companies and online shops. In particular, it tries to identify sets of products that are frequently bought together. Once identified, such sets of associated products may be used to optimize the organization of the offered products on the shelves of a supermarket or the pages of a mail-order catalog or web shop, or may give hints which products may conveniently be bundled.

Formally, the task of frequent item set mining can be described as follows: we are given a set B of *items*, called the *item base*, and a database T of *transactions*. Each item represents a product, and the item base represents the set of all products offered by a store. The term *item set* refers to any subset of the item base B. Each transaction is an item set and represents a set of products that has been bought by an actual customer. Since two or even more customers may have bought the exact same set of products, the total of all transactions must be represented as a vector, a bag or a multiset, since in a simple set each transaction could occur at most once.[1] Note that the item base B is usually not given explicitly, but only implicitly as the union of all transactions.

The *support* $s_T(I)$ of an item set $I \subseteq B$ is the number of transactions in the database T, it is contained in. Given a user-specified *minimum support* $s_{\min} \in \mathbb{N}$, an item set I is called *frequent* in T iff $s_T(I) \geq s_{\min}$. The goal of frequent item set mining is to identify all item sets $I \subseteq B$ that are frequent in a given transaction database T. Note that the task of frequent item set mining may also be defined with a *relative* minimum support, which is the fraction of transactions in T that must contain an item set I in order to make I frequent. However, this alternative definition is obviously equivalent.

A standard approach to find all frequent item sets w.r.t. a given database T and support threshold $s_{\min}$, which is adopted by basically all frequent item set mining algorithms (except those of the Apriori family), is a *depth-first search* in the subset lattice of the item base B. Viewed properly, this approach can be interpreted as a simple *divide-and-conquer* scheme. For some chosen item i, the problem to find all frequent item sets is split into two subproblems: (1) find all frequent item sets containing the item i and (2) find all frequent item sets *not* containing the item i. Each subproblem is then further

---

[1] Alternatively, each transaction may be enhanced by a unique *transaction identifier*, and these enhanced transactions may then be combined in a simple set.

divided based on another item j: find all frequent item sets containing (1.1) both items i and j, (1.2) item i, but not j, (2.1) item j, but not i, (2.2) neither item i nor j, and so on.

All subproblems that occur in this divide-and-conquer recursion can be defined by a *conditional transaction database* and a *prefix*. The prefix is a set of items that has to be added to all frequent item sets that are discovered in the conditional database. Formally, all subproblems are tuples $S = (C, P)$, where C is a conditional database and $P \subseteq B$ is a prefix. The initial problem, with which the recursion is started, is $S = (T, \emptyset)$, where T is the transaction database to mine and the prefix is empty. A subproblem $S_0 = (C_0, P_0)$ is processed as follows: Choose an item $i \in B_0$, where B_0 is the set of items occurring in C_0. This choice is arbitrary, but usually follows some predefined order of the items. If $s_{C_0}(i) \geq s_{\min}$, then report the item set $P_0 \cup \{i\}$ as frequent with the support $s_{C_0}(i)$, and form the subproblem $S_1 = (C_1, P_1)$ with $P_1 = P_0 \cup \{i\}$. The conditional database C_1 comprises all transactions in C_0 that contain the item i, but with the item i removed. This also implies that transactions that contain no other item than i are entirely removed: no empty transactions are ever kept. If C_1 is not empty, process S_1 recursively. In any case (that is, regardless of whether $s_{C_0}(i) \geq s_{\min}$ or not), form the subproblem $S_2 = (C_2, P_2)$, where $P_2 = P_0$ and the conditional database C_2 comprises all transactions in C_0 (including those that do not contain the item i), but again with the item i removed. If C_2 is not empty, process S_2 recursively.

Eclat, FP-growth, RElim and several other frequent item set mining algorithms rely on this basic scheme, but differ in how they represent the conditional databases. The main approaches are horizontal and vertical representations. In a *horizontal representation*, the database is stored as a list (or array) of transactions, each of which is a list (or array) of the items contained in it. In a *vertical representation*, a database is represented by first referring with a list (or array) to the different items. For each item a list (or array) of identifiers is stored, which indicate the transactions that contain the item.

However, this distinction is not pure, since there are many algorithms that use a combination of the two forms of representing a database. For example, while Eclat uses a purely vertical representation, FP-growth combines in its FP-tree structure a vertical representation (links between branches) and a (compressed) horizontal representation (prefix tree of transactions). RElim uses basically a horizontal representation, but groups transactions w.r.t. their leading item, which is, at least partially, a vertical representation. The SaM algorithm presented in the next section is, to the best of my knowledge, the first frequent item set mining algorithm that is based on the general processing scheme outlined above and uses a purely horizontal representation.[2]

The basic processing scheme can easily be improved with so-called *perfect extension pruning*, which relies on the following simple idea: given an item set I, an item $i \notin I$ is called a *perfect extension* of I, iff I and $I \cup \{i\}$ have the same support, that is, if i is contained in all transactions containing I. Perfect extensions have the following properties: (1) if the item i is a perfect extension of an item set I, then it is also a perfect extension of any item set $J \supseteq I$ as long as $i \notin J$ and (2) if I is a frequent item set and K is the set of all perfect extensions of I, then all sets $I \cup J$ with $J \in 2^K$ (where 2^K denotes the power set of K) are also frequent and have the same support as I.

---

[2] Note that Apriori, which also uses a purely horizontal representation, relies on a different processing scheme, since it traverses the subset lattice level-wise rather than depth-first.

These properties can be exploited by collecting in the recursion not only prefix items, but also, in a third element of a subproblem description, perfect extension items. Once identified, perfect extension items are no longer processed in the recursion, but are only used to generate all supersets of the prefix that have the same support. Depending on the data set, this can lead to a considerable speed-up. It should be clear that this optimization can, in principle, be applied in all frequent item set mining algorithms.

3   A Split and Merge Algorithm

The SaM (Split and Merge) algorithm I introduce in this paper can be seen as a simplification of the already fairly simple RElim (Recursive Elimination) algorithm, which I proposed in [8]. While RElim represents a (conditional) database by storing one transaction list for each item (partially vertical representation), the split and merge algorithm presented here employs only a single transaction list (purely horizontal representation), stored as an array. This array is processed with a simple split and merge scheme, which computes a conditional database, processes this conditional database recursively, and finally eliminates the split item from the original (conditional) database.

SaM preprocesses a given transaction database in a way that is very similar to the preprocessing used by many other frequent item set mining algorithms. The steps are illustrated in Figure 1 for a simple example transaction database. Step 1 shows the transaction database in its original form. In step 2 the frequencies of individual items are determined from this input in order to be able to discard infrequent items immediately. If we assume a minimum support of three transactions for our example, there are no infrequent items, so all items are kept. In step 3 the (frequent) items in each transaction are sorted according to their frequency in the transaction database, since it is well known that processing the items in the order of increasing frequency usually leads to the shortest execution times. In step 4 the transactions are sorted lexicographically into descending order, with item comparisons again being decided by the item frequencies, although here the item with the lower frequency precedes the item with the higher frequency. In step 5 the data structure on which SaM operates is built by combining equal transactions and setting up an array, in which each element consists of two fields: an

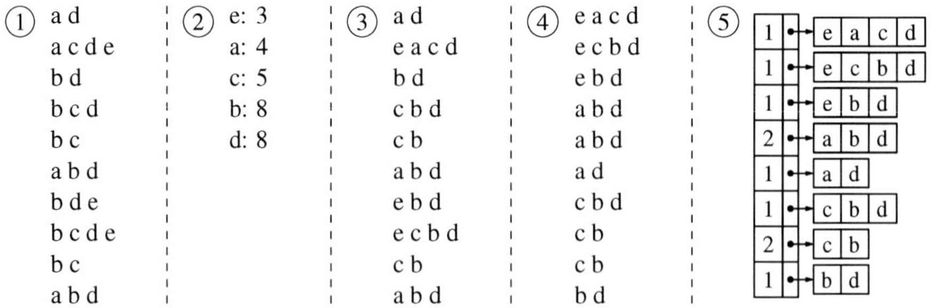

Fig. 1. The example database: original form (1), item frequencies (2), transactions with sorted items (3), lexicographically sorted transactions (4), and the data structure used by SaM (5)

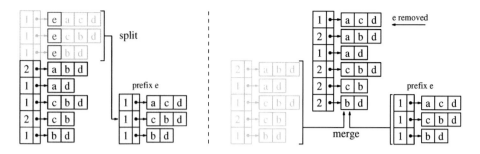

Fig. 2. The basic operations of the SaM algorithm: split (left) and merge (right)

occurrence counter and a pointer to the sorted transaction (array of contained items). This data structure is then processed recursively to find the frequent item sets.

The basic operations of the recursive processing, which follows the general depth-first/divide-and-conquer scheme reviewed in Section 2, are illustrated in Figure 2. In the *split step* (see the left part of Figure 2) the given array is split w.r.t. the leading item of the first transaction (item *e* in our example): all array elements referring to transactions starting with this item are transferred to a new array. In this process the pointer (in)to the transaction is advanced by one item, so that the common leading item is "removed" from all transactions. Obviously, this new array represents the conditional database of the first subproblem (see Section 2), which is then processed recursively to find all frequent items sets containing the split item (provided this item is frequent).

The conditional database for frequent item sets *not* containing this item (needed for the second subproblem, see Section 2) is obtained with a simple *merge step* (see the right part of Figure 2). The new array created in the split step and the rest of the original array (which refers to transactions starting with a different item) are combined with a procedure that is almost identical to one phase of the well-known *mergesort* algorithm. Since both arrays are obviously lexicographically sorted, a single traversal suffices to create a lexicographically sorted merged array. The only difference to a *mergesort* phase is that equal transactions (or transaction suffixes) are combined. That is, there is always just one instance of each transaction (suffix), while its number of occurrences is kept in the occurrence counter. In our example this results in the merged array having two elements less than the input arrays together: the transaction (suffixes) *cbd* and *bd*, which occur in both arrays, are combined and their occurrence counters are increased to 2.

Note that in both the split and the merge step only the array elements (that is, the occurrence counter and the (advanced) transaction pointer) are copied to a new array. There is no need to copy the transactions themselves (that is, the item arrays), since no changes are ever made to them. (In the split step the leading item is not actually removed, but only skipped by advancing the pointer (in)to the transaction.) Hence it suffices to have one global copy of all transactions, which is merely referred to in different ways from different arrays used in the recursive processing.

Note also that the merge result may be created in the array that represented the original (conditional) database, since its front elements have been cleared in the split step. In addition, the array for the split database can be reused after the recursion for the split

```
function SaM (a: array of transactions,      (* conditional database to process *)
              p: array of items,             (* prefix of the conditional database a *)
              s_min: int) : int              (* minimum support of an item set *)
var i: item;                                 (* buffer for the split item *)
    s: int;                                  (* support of the current split item *)
    n: int;                                  (* number of found frequent item sets *)
    b, c: array of transactions;             (* split item and merged database *)
begin                                        (* — split and merge recursion — *)
    n := 0;                                  (* initialize the number of found item sets *)
    while a is not empty do                  (* while conditional database is not empty *)
        i := a[0].items[0]; s := 0;          (* get leading item of the first transaction *)
        while a is not empty and a[0].items[0] = i do   (* split database w.r.t. this item *)
            s := s + a[0].wgt;               (* sum the occurrences (compute support) *)
            remove i from a[0].items;        (* remove the split item from the transaction *)
            if    a[0].items is not empty    (* if the transaction has not become empty *)
            then remove a[0] from a and append it to b;
            else  remove a[0] from a; end;   (* move it to the conditional database, *)
        end;                                 (* otherwise simply remove it *)
        c := empty;                          (* initialize the output array *)
        while a and b are both not empty do  (* merge split and rest of database *)
            if    a[0].items > b[0].items    (* copy lex. smaller transaction from a *)
            then  remove a[0] from a and append it to c;
            else if a[0].items < b[0].items  (* copy lex. smaller transaction from b *)
            then  remove b[0] from b and append it to c;
            else  b[0].wgt := b[0].wgt + a[0].wgt;  (* sum the occurrence counters/weights *)
                  remove b[0] from b and append it to c;
                  remove a[0] from a;        (* move combined transaction and *)
            end;                             (* delete the other, equal transaction: *)
        end;                                 (* keep only one instance per transaction *)
        while a is not empty do              (* copy the rest of the transactions in a *)
            remove a[0] from a and append it to c; end;
        while b is not empty do              (* copy the rest of the transactions in b *)
            remove b[0] from b and append it to c; end;
        a := c;                              (* second recursion is executed by the loop *)
        if s ≥ s_min then                    (* if the split item is frequent: *)
            append i to p;                   (* extend the prefix item set and *)
            report p with support s;         (* report the found frequent item set *)
            n := n + 1 + SaM(b, p, s_min);   (* process the created database recursively *)
            remove i from p;                 (* and sum the found frequent item sets, *)
        end;                                 (* then restore the original item set prefix *)
    end;
    return n;                                (* return the number of frequent item sets *)
end;   (* function SaM() *)
```

Fig. 3. Pseudo-code of the SaM algorithm. The actual C code is even shorter than this description, despite the fact that it contains additional functionality, because certain operations needed in this algorithm can be written very concisely in C (using pointer arithmetic to process arrays).

w.r.t. to the next item. As a consequence, each recursion step, which expands the prefix of the conditional database, only needs to allocate one new array, with a size that is limited to the size of the input array of that recursion step. This makes the algorithm not only simple in structure, but also very efficient in terms of memory consumption.

Finally, note that due to the fact that only a simple array is used as the data structure, the algorithm can fairly easily be implemented to work on external storage or a (relational) database system. There is, in principle, no need to load the transactions into main memory and even the array may easily be stored as a simple (relational) table. The split operation can then be implemented as an SQL select statement, the merge operation is very similar to a join, even though it may require a more sophisticated comparison of transactions (depending on how the transactions are actually stored).

Pseudo-code of the recursive procedure is shown in Figure 3. As can be seen, a single page of code is sufficient to describe the whole recursion in detail. The actual C code I developed is even shorter than this pseudo-code, despite the fact that the C code contains additional functionality (like, for example, perfect extension pruning, see Section 2), because certain operations needed in this algorithm can be written very concisely in C (especially when using pointer arithmetic to process arrays).

4   A Recursive Elimination Algorithm

The RElim (Recursive Elimination) algorithm [8] can be seen as a precursor of the SaM algorithm introduced in the preceding section. It also employs a basically horizontal transaction representation, but separates the transactions (or transaction suffixes) according to their leading item, thus introducing a vertical representation aspect.

The transaction database to mine is preprocessed in essentially the same way as for the SaM algorithm (cf. Figure 1). Only the final step (step 5), in which the data structure to work on is constructed, differs: instead of listing all transactions in one array, they are grouped according to their leading item (see Figure 4 on the right). In addition, the transactions are organized as lists (at least in my implementation), even though, in principle, using arrays would also be possible. These lists are sorted descendingly w.r.t. the frequency of their associated items in the transaction database: the first list is associated with the most frequent item, the last list with the least frequent item.

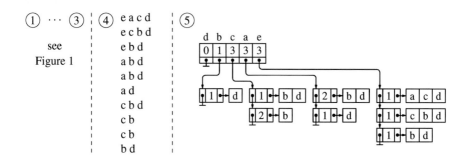

Fig. 4. Preprocessing the example database for the RElim algorithm and the initial data structure

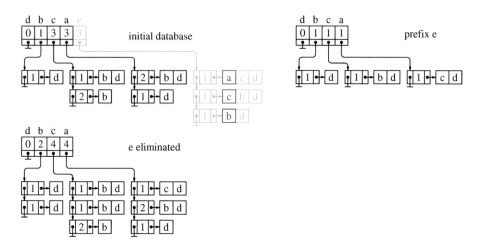

Fig. 5. The basic operations of the RElim algorithm. The rightmost list is traversed and reassigned: once to an initially empty list array (conditional database for the prefix e, see top right) and once to the original list array (eliminating item e, see bottom left). These two databases are then both processed recursively.

Note that each transaction list contains in its header a counter for the number of transactions. For the last (rightmost) list, this counter states the support of the associated item in the represented (conditional) database. For a preceding list, however, the value of this counter may be lower than the support of the item associated with the list, because this item may also occur in transactions starting with items following it in the item order (that is, with items that occur less frequently in the transaction database).

Note also that the leading item of each transaction has already been removed from all transactions, as it is implicitly represented by which list a transaction is contained in and thus need not be explicitly present. As a consequence, the counter associated with each transaction list need not be equal to the sum of the weights of the transactions in the list (even though this is the case in the example), because transactions (or transaction suffixes) that contain only the item (and thus become empty if the item is removed) are not explicitly represented by a list element, but only implicitly in the counter.

The basic operations of the RElim algorithm are illustrated in Figure 5. The next item to be processed is the one associated with the last (rightmost) list (in the example this is item e). If the counter associated with the list, which states the support of the item, exceeds the minimum support, the item set consisting of this item and the prefix of the conditional database (which is empty for the example) is reported as frequent. In addition, the list is traversed and its elements are copied to construct a new list array, which represents the conditional database of transactions containing the item. In this operation the leading item of each transaction (suffix) is used as an index into the list array to find the list it has to be added to. In addition, the leading item is removed (see Figure 5 on the right). The resulting conditional database is then processed recursively to find all frequent item sets containing the list item (first subproblem in Section 2).

```
function RElim (a: array of transaction lists,      (* conditional database to process *)
                p: array of items,                   (* prefix of the conditional database a *)
                s_min: int) : int                    (* minimum support of an item set *)
var i, k: int;                                       (* buffer for the current item *)
    s: int;                                          (* support of the current item *)
    n: int;                                          (* number of found frequent item sets *)
    b: array of transaction lists;                   (* conditional database for current item *)
    t, u: transaction list element;                  (* to traverse the transaction lists *)
begin                                                (* — recursive elimination — *)
  n := 0;                                            (* initialize the number of found item sets *)
  while a is not empty do                            (* while conditional database is not empty *)
    i := length(a) − 1; s := a[i].wgt;               (* get the next item to process *)
    if s ≥ s_min then                                (* if the current item is frequent: *)
      append item(i) to p;                           (* extend the prefix item set and *)
      report p with support s;                       (* report the found frequent item set *)
      b := array [0..i−1] of transaction lists;      (* create an empty list array *)
      t := a[i].head;                                (* get the list associated with the item *)
      while t ≠ nil do                               (* while not at the end of the list *)
        u := copy of t; t := t.succ;                 (* copy the transaction list element, *)
        k := u.items[0];                             (* go to the next list element, and *)
        remove k from u.items;                       (* remove the leading item from the copy *)
        if u.items is not empty                      (* add the copy to the conditional database *)
        then u.succ = b[k].head; b[k].head = u; end;
        b[k].wgt := b[k].wgt +u.wgt;                 (* sum the transaction weight *)
      end;                                           (* in the list weight/transaction counter *)
      n := n + 1 + RElim(b, p, s_min);               (* process the created database recursively *)
      remove item(i) from p;                         (* and sum the found frequent item sets, *)
    end;                                             (* then restore the original item set prefix *)
    t := a[i].head;                                  (* get the list associated with the item *)
    while t ≠ nil do                                 (* while not at the end of the list *)
      u := t; t := t.succ;                           (* note the current list element, *)
      k := u.items[0];                               (* go to the next list element, and *)
      remove k from u.items;                         (* remove the leading item from current *)
      if u.items is not empty                        (* reassign the noted list element *)
      then u.succ = a[k].head; a[k].head = u; end;
      a[k].wgt := a[k].wgt +u.wgt;                   (* sum the transaction weight *)
    end;                                             (* in the list weight/transaction counter *)
    remove a[i] from a;                              (* remove the processed list *)
  end;
  return n;                                          (* return the number of frequent item sets *)
end;   (* function RElim() *)
```

Fig. 6. Pseudo-code of the RElim algorithm. The function "item" yields the actual item coded with the integer number that is given as an argument. As for the SaM algorithm the removal of items from the transactions (or transaction suffixes) is realized by pointer arithmetic in the actual C implementation, thus avoiding copies of the item arrays. In addition, the same list array is always reused for the created conditional databases b.

In any case, the last list is traversed and its elements are reassigned to other lists in the original array, again using the leading item as an index into the list array and removing this item (see Figure 5 at the bottom left). This operation eliminates the item associated with the last list and thus yields the conditional database needed to find all frequent item sets not containing this item (second subproblem in Section 2).

Pseudo-code of the RElim algorithm can be found in Figure 6. It assumes that items are coded by consecutive integer numbers starting at 0 in the order of descending frequency in the transaction database to mine. The actual item can be retrieved by applying a function "item" to the code. As for the SaM algorithm the actual C implementation makes use of pointer arithmetic in order to avoid copying item arrays.

5   Experiments with the Basic Versions

In order to evaluate SaM and RElim in their basic forms, I ran them against my own implementations of Apriori [6], Eclat [6], and FP-growth [7], all of which rely on the same code to read the transaction database and to report found frequent item sets. Of course, using my own implementations has the disadvantage that not all of these implementations reach the speed of the fastest known implementations.[3] However, it has the important advantage that any differences in execution time can only be attributed to differences in the actual processing scheme, as all other parts of the programs are identical. Therefore I believe that the measured execution times are still reasonably expressive and allow me to compare the different approaches in an reliable manner.

I ran experiments on five data sets, which were also used in [6,7,8]. As they exhibit different characteristics, the advantages and disadvantages of the different algorithms can be observed well. These data sets are: census (a data set derived from an extract of the US census bureau data of 1994, which was preprocessed by discretizing numeric attributes), chess (a data set listing chess end game positions for king vs. king and rook), mushroom (a data set describing poisonous and edible mushrooms by different attributes), T10I4D100K (an artificial data set generated with IBM's data generator [24]), and BMS-Webview-1 (a web click stream from a leg-care company that no longer exists, which has been used in the KDD cup 2000 [12]). The first three data sets are available from the UCI machine learning repository [3]. The shell script used to discretize the numeric attributes of the census data set can be found at the URLs mentioned below. The first three data sets can be characterized as "dense", meaning that on average a rather high fraction of all items is present in a transaction (the number of different items divided by the average transaction length is 0.1, 0.5, and 0.2, respectively, for these data sets), while the last two are rather "sparse" (the number of different items divided by the average transaction length is 0.01 and 0.005, respectively).

For the experiments I used an IBM/Lenovo Thinkpad X60s laptop with an Intel Centrino Duo L2400 processor and 1 GB of main memory running openSuSE Linux 10.3 (32 bit) and gcc (Gnu C Compiler) version 4.2.1. The results for the five data sets mentioned above are shown in Figure 7. Each diagram refers to one data set and shows the decimal logarithm of the execution time in seconds (excluding the time to load the

---

[3] In particular, in [15] an FP-growth implementation was presented, which is highly optimized to how modern processor access their main memory [16].

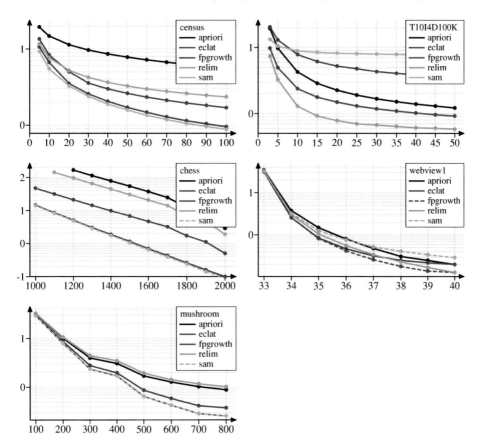

Fig. 7. Experimental results on five different data sets. Each diagram shows the minimum support (as the minimum number of transactions that contain an item set) on the horizontal axis and the decimal logarithm of the execution time in seconds on the vertical axis. The data sets underlying the diagrams on the left are rather dense, those underlying the diagrams on the right are rather sparse.

transaction database) over the minimum support (stated as the number of transactions that must contain an item set in order to render it frequent).

These results show a fairly clear picture: SaM performs extremely well on "dense" data sets. It is the fastest algorithm for the census data set and (though only by a very small margin) on the chess data set. On the mushroom data set it performs on par with FP-growth, while it is clearly faster than Eclat and Apriori. On "sparse" data sets, however, SaM struggles. On the artificial data set T10I4D100K it performs particularly badly and catches up with the performance of other algorithms only at the lowest support levels.[4] On BMS-Webview-1 it performs somewhat better, but again reaches the performance of other algorithms only for fairly low support values.

---

[4] It should be noted, though, that SaM's execution times on T10I4D100K are always around 8–10 seconds and thus not unbearable.

RElim, on the other hand, performs excellently on "sparse" datasets. On the artificial data set T10I4D100K it beats all other algorithms by a quite large margin, while on BMS-Webview-1 it is fairly close to the best performance (achieved by FP-growth). On "dense" data sets, however, RElim has serious problems. While on census it at least comes close to being competitive, at least for low support values, it performs very badly on chess and is the slowest algorithm on mushroom.

Even though their performance is uneven, these results give clear hints how to choose between these two algorithms: for "dense" data sets (that is, a high fraction of all items is present in a transaction) use SaM, for "sparse" data sets use RElim. This yields excellent performance, as each algorithm is applied to its "area of expertise".

6 Optimizations

Given SaM's processing scheme (cf. Section 3), the cause of the behavior observed in the preceding section is easily found: it is clearly the merge operation. Such a merge operation is most efficient if the two lists to merge do not differ too much in length. Because of this, the recursive procedure of the *mergesort* algorithm splits its input into two lists of roughly equal length. If, to consider an extreme case, it would always merge single elements with the (recursively sorted) rest of the list, its time complexity would deteriorate from $O(n \log n)$ to $O(n^2)$ (as it would actually execute a processing scheme that is equivalent to insertion sort). The same applies to SaM: in a "dense" data set it is more likely that the two transaction lists do not differ considerably in length, while in a "sparse" data set it can be expected that the list of transactions containing the split item will be rather short compared to the rest. As a consequence, SaM performs very well on "dense" data sets, but rather poorly on "sparse" ones.

The main reason for the merge operation in SaM is to keep the list sorted, so that (1) all transactions with the same leading item are grouped together and (2) equal transactions (or transaction suffixes) can be combined, thus reducing the number of objects to process. The obvious alternative to achieve (1), namely to set up a separate list (or array) for each item, is employed by the RElim algorithm, which, as these experiments show, performs considerably better on sparse data sets. On the other hand, the RElim algorithm does not combine equal transactions except in the initial database, since searching a list, to which an element is reassigned, for an equal entry would be too costly. As a consequence, (several) duplicate list elements may occur (see Figure 5 at the bottom left), which slow down RElim's operation on "dense" data sets.

This analysis immediately provides several ideas for optimizations. In the first place, RElim may be improved by removing duplicates from its transaction lists. Of course, this should not be done each time a new list element is added (as this would be too time consuming), but only when a transaction list is processed to form the conditional database for its associated item. To achieve this, a transaction list to be copied to a new list array is first sorted with a modified mergesort, which combines equal elements (similar to the merge phase used by SaM). In addition, one may use some heuristic in order to determine whether sorting the list leads to sufficient gains that outweigh the sorting costs. Such a simple heuristic is to sort the list only if the number of items occurring in the transactions is less than a user-specified threshold: the fewer items are left, the higher the chances that there are equal transactions (or transaction suffixes).

```
function merge (a, b: array of transactions) : array of transactions
var l, m, r: int;                          (* binary search variables *)
      c: array of transactions;            (* output transaction array *)
begin                                      (* — binary search based merge — *)
   c := empty;                             (* initialize the output array *)
   while a and b are both not empty do     (* merge projection and rest of database *)
      l := 0; r := length(a);              (* initialize the binary search range *)
      while l < r do                       (* while the search range is not empty *)
         m := ⌊l+r/2⌋;                      (* compute the middle index in the range *)
         if     a[m] < b[0]                (* compare the transaction to insert *)
         then l := m + 1; else r := m;     (* and adapt the binary search range *)
      end;                                 (* according to the comparison result *)
      while l > 0 do                       (* copy lexicographically larger transactions *)
         remove a[0] from a and append it to c; l := l − 1; end;
      remove b[0] from b and append it to c;    (* copy the transaction to insert and *)
      i := length(c) − 1;                  (* get its index in the output array *)
      if    a is not empty and a[0].items = c[i].items
      then c[i].wgt = c[i].wgt +a[0].wgt;  (* if there is a transaction in the rest *)
            remove a[0] from a;            (* that is equal to the one just appended, *)
      end;                                 (* then sum the transaction weights and *)
   end;                                    (* remove the transaction from the rest *)
   while a is not empty do                 (* copy the rest of the transactions in a *)
      remove a[0] from a and append it to c; end;
   while b is not empty do                 (* copy the rest of the transactions in b *)
      remove b[0] from b and append it to c; end;
   return c;                               (* return the resulting transaction array *)
end;   (* function merge() *)
```

Fig. 8. Pseudo-code of a binary search based merge procedure

For SaM, at least two possible improvements come to mind. The first is to check, before the merge operation, how unequal in length the two arrays are. If the lengths are considerably different, a modified merge operation, which employs a binary search to find the locations where the elements of the shorter array have to be inserted into the longer one, can be used. Pseudo-code of such a binary search based merge operation is shown in Figure 8. Its advantage is that it needs less comparisons between transactions if the length ratio of the arrays exceeds a certain threshold. Experiments with different thresholds revealed that best results are obtained if a binary search based merge is used if the length ratio of the arrays exceeds 16:1 and a standard merge otherwise.

A second approach to improve the SaM algorithm relies deviating from using only one transaction array. The idea is to maintain two "source" arrays, and always merging the split result to the shorter one, which increases the chances that the array lengths do not differ so much. Of course, such an approach adds complexity to the split step, because now the two source arrays have to be traversed to collect transactions with the same leading item, and these may even have to be merged. However, the costs for the merge operation may be considerably reduced, especially for sparse data sets, so that overall gains can be expected. Furthermore, if both source arrays have grown beyond a user-specified length threshold, they may be merged, in an additional step, into one, so that one source gets cleared. In this way, it becomes more likely that a short split result can be merged to an equally short source array. Experiments showed that using a length threshold of 8192 (that is, the source arrays are merged if both are longer than 8192 elements) yields good results (see the following section).

7   Experiments with the Optimized Versions

The different optimization options for SaM and RElim discussed in the preceding section were tested on the same data sets as in Section 5. The results are shown for RElim in Figure 9 and for SaM in Figures 10 and 11, while Figure 12 shows the results of the best optimization options in comparison with Apriori, Eclat and FP-growth.

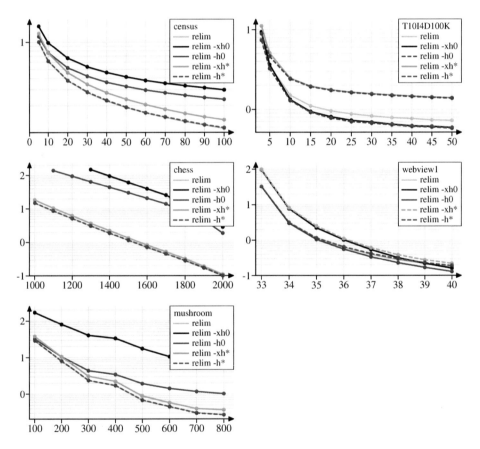

Fig. 9. Experimental results for different versions of the RElim algorithm. "-x" means that perfect extension pruning was switched off and "-h0" that no transactions lists, "-h*" that all transaction lists were sorted before processing. Without any "-h" option transaction lists were sorted if no more than 32 items were left in the conditional database.

Figure 9 shows the effects of sorting (or not sorting) a transaction list before it is processed, but also illustrates the effect of perfect extension pruning (see Section 2).[5] The option "-h" refers to the threshold for the number of items in a transaction list, below which the list is sorted before processing. Therefore "-h0" means that no transaction list is ever sorted, while "-h*" means that every transaction list is sorted, regardless of the number of items it contains. Without any "-h" option, a transaction list is sorted if it contains less than 32 items. The option "-x" refers to perfect extension pruning and indicates that it was disabled. The results show that perfect extension pruning is clearly advantageous in RElim, except for T10I4D100K, where the gains are negligible. Similarly, sorting a transaction list before processing is clearly beneficial on "dense" data sets and yields particularly high gains for census (though mainly for larger minimum support values) and chess. On "sparse" data sets, however, sorting causes higher costs than what can be gained by having fewer transactions to process.

Figure 10 show the effects of using, in SaM, a binary search based merge operation for arrays that differ considerably in length and also illustrates the gains that can result from pruning with perfect extensions. As before, the option "-x" refers to perfect extension pruning and indicates that it was disabled. The option "-y" switches off the use of a binary search based merge for arrays with a length ratio exceeding 16:1 (that is, with "-y" all merging is done with a standard merge operation). As can be seen from these diagrams, SaM also benefits from perfect extension pruning, though less strongly than RElim. Using an optional binary search based merge operation has only minor effects on "dense" data sets and even on BMS-Webview-1, but yields significant gains on T10I4D100K, almost cutting the execution time in half.

The effects of a double "source" buffering approach for SaM are shown in Figure 11. On all data sets except T10I4D100K the double source buffering approach performs completely on par with the standard algorithm, which is why the figure shows only the results on census (to illustrate the identical performance) and T10I4D100K. On the latter, clear gains result, which, with all option described in Section 6 activated, reduces the execution time by about 30% over those obtained with an optional binary search based merge. Of course, for the merge operations in the double source buffering approach, such an optional binary search based merge may also be used (option "-y"). Although it again improves performance, the gains are smaller, because the double buffering reduces the number of times it is exploited. The option "-h" disables the optional additional merging of the two source arrays if they both exceed 8192 elements, which was described in Section 6. Although the gains from this option are smaller than those resulting from the binary search based merge, they are not negligible.

Finally, Figure 12 compares the performance of the optimized versions of SaM and RElim to Apriori, Eclat and FP-growth. Clearly, RElim has become highly competitive on "dense" data sets without losing (much) of its excellent performance on "sparse" data sets. Optimized SaM, on the other hand, performs much better on "sparse" data sets, but is truly competitive only for (very) low support values. Its excellent behavior on "dense" data sets, however, is preserved.

---

[5] All results reported in Figure 7 were obtained *with* perfect extension pruning, because it is easy to add to all algorithms and causes no cost, but rather always improves performance.

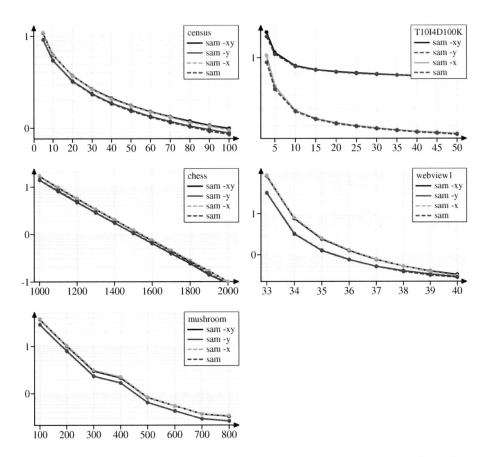

Fig. 10. Experimental results for different versions of the SaM algorithm. "-x" means that perfect extension pruning was disabled and "-y" that the binary search based merge procedure for highly unequal transaction arrays was switched off (that is, all merge operations use the standard merge procedure).

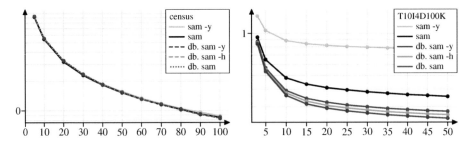

Fig. 11. Experimental results for the double source buffering version of the SaM algorithm

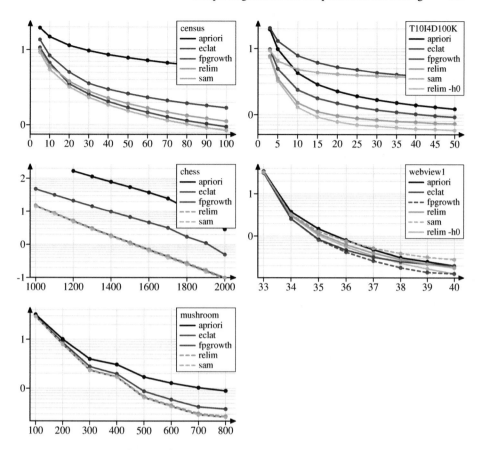

Fig. 12. Experimental results with the optimized versions. Each diagram shows the minimum support (as the minimum number of transactions that contain an item set) on the horizontal axis and the decimal logarithm of the execution time in seconds on the vertical axis. The data sets underlying the diagrams on the left are rather dense, those underlying the diagrams on the right are rather sparse.

8   Conclusions

In this paper I introduced the simple SaM (Split and Merge) algorithm and reviewed (an improved version) of the RElim algorithm, both of which distinguish themselves from other algorithms for frequent item set mining by their simple processing scheme and data structure. By comparing them to classical frequent item set mining algorithms like Apriori, Eclat and FP-growth the strength and weaknesses of these algorithms were analyzed. This led to several ideas for optimizations, which could improve the performance of both algorithms on those data sets on which they struggled in their basic form. The resulting optimized version are competitive with other frequent item set mining algorithms (with the exception of SaM on sparse data sets) and are only slightly more complex than the basic versions. In addition, it should be noted that SaM in particular is very well suited for an implementation that works on external storage, since it employs a simple array that can easily be represented as a table in a relational database system.

Software

Implementation of SaM and RElim in C can be found at:

```
http://www.borgelt.net/sam.html
http://www.borgelt.net/relim.html
```

References

1. Agrawal, R., Imielienski, T., Swami, A.: Mining Association Rules between Sets of Items in Large Databases. In: Proc. Conf. on Management of Data, pp. 207–216. ACM Press, New York (1993)
2. Agrawal, R., Mannila, H., Srikant, R., Toivonen, H., Verkamo, A.: Fast Discovery of Association Rules. In: [10], pp. 307–328
3. Blake, C.L., Merz, C.J.: UCI Repository of Machine Learning Databases. Dept. of Information and Computer Science, University of California at Irvine, CA, USA (1998), http://www.ics.uci.edu/~mlearn/MLRepository.html
4. Böttcher, M., Spott, M., Nauck, D.: Detecting Temporally Redundant Association Rules. In: Proc. 4th Int. Conf. on Machine Learning and Applications (ICMLA 2005), Los Angeles, CA, pp. 397–403. IEEE Press, Piscataway (2005)
5. Böttcher, M., Spott, M., Nauck, D.: A Framework for Discovering and Analyzing Changing Customer Segments. In: Perner, P. (ed.) ICDM 2007. LNCS (LNAI), vol. 4597, pp. 255–268. Springer, Heidelberg (2007)
6. Borgelt, C.: Efficient Implementations of Apriori and Eclat. In: Proc. Workshop Frequent Item Set Mining Implementations (FIMI 2003), Melbourne, FL, USA, Aachen, Germany. CEUR Workshop Proceedings, vol. 90 (2003)
7. Borgelt, C.: An Implementation of the FP-growth Algorithm. In: Proc. Workshop Open Software for Data Mining (OSDM 2005 at KDD 2005), Chicago, IL, pp. 1–5. ACM Press, New York (2005)
8. Borgelt, C.: Keeping Things Simple: Finding Frequent Item Sets by Recursive Elimination. In: Proc. Workshop Open Software for Data Mining (OSDM 2005 at KDD 2005), Chicago, IL, pp. 66–70. ACM Press, New York (2005)
9. Cheng, Y., Fayyad, U., Bradley, P.S.: Efficient Discovery of Error-Tolerant Frequent Itemsets in High Dimensions. In: Proc. 7th Int. Conf. on Knowledge Discovery and Data Mining (KDD 2001), San Francisco, CA, pp. 194–203. ACM Press, New York (2001)
10. Fayyad, U.M., Piatetsky-Shapiro, G., Smyth, P., Uthurusamy, R. (eds.): Advances in Knowledge Discovery and Data Mining. AAAI Press / MIT Press (1996)
11. Han, J., Pei, H., Yin, Y.: Mining Frequent Patterns without Candidate Generation. In: Proc. Conf. on the Management of Data (SIGMOD 2000), Dallas, TX, pp. 1–12. ACM Press, New York (2000)
12. Kohavi, R., Bradley, C.E., Frasca, B., Mason, L., Zheng, Z.: KDD-Cup 2000 Organizers' Report: Peeling the Onion. SIGKDD Exploration 2(2), 86–93 (2000)
13. Pei, J., Han, J., Mortazavi-Asl, B., Zhu, H.: Mining Access Patterns Efficiently from Web Logs. In: Terano, T., Chen, A.L.P. (eds.) PAKDD 2000. LNCS, vol. 1805, pp. 396–407. Springer, Heidelberg (2000)

14. Pei, J., Tung, A.K.H., Han, J.: Fault-Tolerant Frequent Pattern Mining: Problems and Challenges. In: Proc. ACM SIGMOD Workshop on Research Issues in Data Mining and Knowledge Discovery (DMK 2001), Santa Babara, CA, ACM Press, New York (2001)

15. Rász, B.: nonordfp: An FP-growth Variation without Rebuilding the FP-Tree. In: Proc. Workshop Frequent Item Set Mining Implementations (FIMI 2004), Brighton, UK, Aachen, Germany. CEUR Workshop Proceedings, vol. 126 (2004)

16. Rász, B., Bodon, F., Schmidt-Thieme, L.: On Benchmarking Frequent Itemset Mining Algorithms. In: Proc. Workshop Open Software for Data Mining (OSDM 2005 at KDD 2005), Chicago, IL, pp. 36–45. ACM Press, New York (2005)

17. Zaki, M., Parthasarathy, S., Ogihara, M., Li, W.: New Algorithms for Fast Discovery of Association Rules. In: Proc. 3rd Int. Conf. on Knowledge Discovery and Data Mining (KDD 1997), Newport Beach, CA, pp. 283–296. AAAI Press, Menlo Park (1997)

18. Mannila, H., Toivonen, H., Verkamo, A.I.: Discovery of Frequent Episodes in Event Sequences. Report C-1997-15, University of Helsinki, Finland (1997)

19. Kuok, C., Fu, A., Wong, M.: Mining Fuzzy Association Rules in Databases. SIGMOD Record 27(1), 41–46 (1998)

20. Moen, P.: Attribute, Event Sequence, and Event Type Similarity Notions for Data Mining. Ph.D. Thesis/Report A-2000-1, Department of Computer Science, University of Helsinki, Finland (2000)

21. Wang, X., Borgelt, C., Kruse, R.: Mining Fuzzy Frequent Item Sets. In: Proc. 11th Int. Fuzzy Systems Association World Congress (IFSA 2005), Beijing, China, pp. 528–533. Tsinghua University Press and Springer-Verlag (2005)

22. Webb, G.I., Zhang, S.: k-Optimal-Rule-Discovery. Data Mining and Knowledge Discovery 10(1), 39–79 (2005)

23. Webb, G.I.: Discovering Significant Patterns. Machine Learning 68(1), 1–33 (2007)

24. Synthetic Data Generation Code for Associations and Sequential Patterns. Intelligent Information Systems, IBM Almaden Research Center,
http://www.almaden.ibm.com/software/quest/Resources/
index.shtml

Monte Carlo Feature Selection and Interdependency Discovery in Supervised Classification

Michał Dramiński[1,*], Marcin Kierczak[2,*], Jacek Koronacki[1], and Jan Komorowski[2,3]

[1] Institute of Computer Science, Polish Acad. Sci., Ordona 21, Warsaw, Poland
Michal.Draminski@ipipan.waw.pl, Jacek.Koronacki@ipipan.waw.pl
[2] The Linnaeus Centre for Bioinformatics, Uppsala University
and The Swedish University of Agricultural Sciences,
Box 758, Uppsala, Sweden
Marcin.Kierczak@lcb.uu.se
[3] Interdisciplinary Centre for Mathematical and Computer Modelling,
Warsaw University, Poland
Jan.Komorowski@lcb.uu.se

Abstract. Applications of machine learning techniques in Life Sciences are the main applications forcing a paradigm shift in the way these techniques are used. Rather than obtaining the best possible supervised classifier, the Life Scientist needs to know which features contribute best to classifying observations into distinct classes and what are the interdependencies between the features. To this end we significantly extend our earlier work [Dramiński *et al.* (2008)] that introduced an effective and reliable method for ranking features according to their importance for classification. We begin with adding a method for finding a cut-off between informative and non-informative features and then continue with a development of a methodology and an implementation of a procedure for determining interdependencies between informative features. The reliability of our approach rests on multiple construction of tree classifiers. Essentially, each classifier is trained on a randomly chosen subset of the original data using only a fraction of all of the observed features. This approach is conceptually simple yet computer-intensive. The methodology is validated on a large and difficult task of modelling HIV-1 reverse transcriptase resistance to drugs which is a good example of the aforementioned paradigm shift. In this task, of the main interest is the identification of mutation points (i.e. features) and their combinations that model drug resistance.

1 Introduction

A major challenge in the analysis of many data sets, especially those presently generated by advanced biotechnologies, is their size: a very small number of records (samples, observations), of the order of tens, versus thousands of attributes or features per each record. Typical examples include microarray gene expression experiments (where the features are gene expression levels) or data coming from next generation DNA or RNA sequencing projects. Another obvious example are transactional data of commercial origin. In all these tasks supervised classification is quite different from a typical data mining problem in which every class has a large number of examples. In the latter

---

* These authors contributed equally.

J. Koronacki et al. (Eds.): Advances in Machine Learning II, SCI 263, pp. 371–385.
springerlink.com   © Springer-Verlag Berlin Heidelberg 2010

context, the main task is to propose a classifier of the highest possible quality of classification. In class prediction for typical gene expression data it is not the classifier *per se* that is crucial; rather, selection of informative (discriminative) features and the discovered interdependencies among them are to give the Life Scientists a much desired possibility of the interpretation of the classification results.

Given such data, all reasonable classifiers can be claimed to be capable of providing essentially similar results (if measured by error rate or the like criteria; cf. [Dudoit and Fridlyand (2003)]). However, since it is rather a rule than an exception that most features in the data are *not* informative, it is indeed of utmost interest to select the few ones that are informative and that may form the bases for class prediction. Equally interesting is a discovery of interdependencies between the informative features.

Generally speaking, feature selection may be performed either prior to building the classifier, or as an inherent part of this process. These two approaches are referred to as *filter* methods and *wrapper* methods, respectively. Currently, the wrapper methods are often divided into two subclasses: one retaining the name "wrapper methods" and the other, termed *embedded* methods. Within this finer taxonomy, the former refer to such classification methods in which feature selection is "wrapped" around the classifier construction and the latter one to those in which feature selection is directly built into the classifier construction.

A significant progress in these areas of research has been achieved in recent years; for a brief account, up to 2002, see [Dudoit and Fridlyand (2003)] and for an extensive survey and later developments see [Saeys *et al.* (2007)]. Regarding the wrapper and embedded approaches, an early successful method, not mentioned by [Saeys *et al.* (2007)], was developed by Tibshirani et al. (see [Tibshirani *et al.* (2002), Tibshirani *et al.* (2003)]) and is called nearest shrunken centroids. Most recently, a Bayesian technique of automatic relevance determination, the use of support vector machines, and the use of ensembles of classifiers, all these either alone or in combination, have proved particularly promising. For further details see [Li *et al.* (2002), Lu *et al.* (2007), Chrysostomou *et al.* (2008)] and the literature there. In the context of feature selection the last developments by the late Leo Breiman deserve special attention. In his Random Forests, he proposed to make use of the so-called variable (i.e. feature) importance for feature selection. Determination of the importance of the variable is not necessary for random forest construction, but it is a subroutine performed in parallel to building the forest; cf. [Breiman and Cutler (2008)]. Ranking features by variable importance can thus be considered to be a by-product of building the classifier. While ranking variables according to their importance is a natural basis for a filter, nothing prevents one from using such importances within, say, the embedded approach; cf., e.g., [Diaz-Uriarte and de Andres (2006)]. In any case, feature selection by measuring variable importance in random forests should be seen as a very promising method, albeit under one proviso. Namely, the problem with variable importance as originally defined is that it is biased towards variables with many categories and variables that are correlated; cf. [Strobl *et al.* (2007), Archer and Kimes (2008)]. Accordingly, proper debiasing is needed, in order to obtain true ranking of features; cf. [Strobl *et al.* (2008)].

One potential advantage of the filter approach is that it constructs a group of features that contribute the most to the classification task, and therefore are informative or "relatively important", to a given classification task *regardless* of the classifier that will be used. In other words, the filter approach should be seen as a way of providing an objective measure of relative importance of each feature for a particular classification task. Of course, for this to be the case, a filter method used for feature selection should be capable of incorporating interdependencies between the features. Indeed, the fact that a feature may prove informative only in conjunction with some other features, but not alone, should be taken into account. Clearly, the aforementioned algorithms for measuring variable importance in random forests possess the last capability.

Recently, a novel, effective and reliable filter method for ranking features according to their importance for a given supervised classification task has been introduced by [Dramiński *et al.* (2008)]. The method is capable of incorporating interdependencies between features. It bears some remote similarity to the Random Forest methodology, but differs entirely in the way features ranking is performed. Specifically, our method does not use bootstrapping and is conceptually simpler. A more important and new result is that it provides explicit information about interdependencies among features. Within our approach, discovering interdependencies builds on identifying features which "cooperate" in determining that samples belong to the same classes. It is worthwhile to emphasize that this is completely different from the usual approach which aims at finding features that are similar in some sense.

The procedure from [Dramiński *et al.* (2008)] for Monte Carlo feature selection is briefly recapitulated in Section 2. Since the original aim was only to rank features according to their classification ability, no distinction was made among informative and non-informative features. In that section we introduce an additional procedure to find a cut-off separating informative from non-informative features in the ranking list. In Section 3, a way to discover interdependencies between features is provided. In Section 4 application of the method is illustrated on the HIV-1 resistance to Didanosine. Interpretation of the obtained results is provided in Subsection 4.1. We close with concluding remarks in Section 5.

2 Monte Carlo Feature Selection

The Monte Carlo feature selection (MCFS) part of the algorithm is conceptually simple, albeit computer-intensive. We consider a particular feature to be important, or informative, if it is likely to take part in the process of classifying samples into classes "more often than not". This "readiness" of a feature to take part in the classification process, termed relative importance of a feature, is measured via intensive use of classification trees. The use of classification trees is motivated by the fact that they can be considered to be the most flexible classifiers within the family of all classification methods. In our method, however, the classifiers are used for measuring relative importance of features, not for classification *per se*.

In the main step of the procedure, we estimate relative importance of features by constructing thousands of trees for randomly selected subsets of the features. More precisely, out of all d features, we select s subsets of m features, m being fixed and $m << d$, and for each subset of features, t trees are constructed and their performance assessed. Each of the t trees in the inner loop is trained and evaluated on a different,

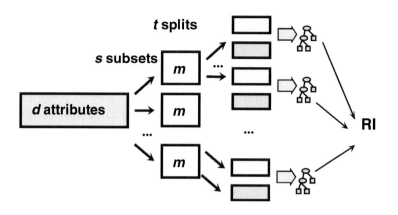

Fig. 1. Block diagram of the main step of the MCFS procedure

randomly selected training and test data sets. These sets come from a random split of the full data set into two subsets. Every time, about 66% out of all n samples, is used for training and the remaining samples are used for testing. The split is performed so that the proportions of classes in the original data set are preserved. See Fig. 1 for a block diagram of the procedure.

Eventually, $s \cdot t$ trees are constructed and evaluated in the main step of the procedure. Both s and t should be sufficiently large, so that each feature has a chance to appear in many different subsets of features and that randomness due to inherent variability in the data is properly accounted for. A crude measure of relative importance of a particular feature could be defined as the overall number of splits made on that feature in all nodes of all the $s \cdot t$ trees. However, it is clear that for any particular split, its contribution to the overall relative importance of the feature should be weighted by the information gain achieved by the split, the number of samples in the split node and by the classification ability of the whole tree.

In order to determine relative importance of a particular feature, we first recall weighted accuracy of a tree as a means to assess classification ability of the tree on a test set. For a classification problem with c classes, let n_{ij} denote the number of samples from class i classified as those from class j; clearly, $i, j, = 1, 2, \ldots, c$ and $\sum_{i,j} n_{ij} = n$, the number of all samples. Now, we define weighted accuracy as

$$wAcc = \frac{1}{c} \sum_{i=1}^{c} \frac{n_{ii}}{n_{i1} + n_{i2} + \cdots + n_{ic}}, \tag{1}$$

i.e., as the mean of c true positive rates.

Further, if a particular split is made on feature g_k, then the more informative this feature is, the greater is $wAcc$ for the whole tree. Similarly, both the information gain on the split and the number of samples in the split node are greater. Information gain can be measured, e.g., by Gini Index or Gain Ratio, and the relative importance of feature g_k, RI_{g_k}, can be defined as

$$\text{RI}_{g_k} = \sum_{\tau=1}^{st} (wAcc)^u \sum_{n_{g_k}(\tau)} \text{IG}(n_{g_k}(\tau)) \left(\frac{\text{no. in } n_{g_k}(\tau)}{\text{no. in } \tau} \right)^v, \quad (2)$$

where summation is over all the $s \cdot t$ trees and, within each τ-th tree, overall nodes $n_{g_k}(\tau)$ of the tree on which the split is made on feature g_k, $\text{IG}(n_{g_k}(\tau))$ stands for information gain for node $n_{g_k}(\tau)$, (no. in $n_{g_k}(\tau)$) denotes the number of samples in node $n_{g_k}(\tau)$, (no. in τ) denotes the number of samples in the root of the τ-th tree, and u and v are fixed positive reals.

Note that by taking, say, $u = 2$, trees with low $wAcc$ are penalized more severely than when taking $u = 1$. Similarly, the greater is v, the smaller is the influence of node $n_{g_k}(\tau)$ with a given ratio (no. in $n_{g_k}(\tau)$)/(no. in τ) on RI_{g_k}, unless $n_{g_k}(\tau)$ is the root of the tree. And, for any fixed positive v, the influence of any particular node on RI_{g_k} decreases monotonically with the number of samples in this node. In this way, and especially for low level nodes in a tree, the fact that information gains can be very high is taken into account, while only very small subsets of data are split.

There are five parameters, m, s, t, u and v to be set by an experimenter. A detailed discussion on how to set values of these parameters can be found in [Dramiński *et al.* (2008)]. Our experience suggests to use u and v set to 1 as the default value. The choice of subset size m of features selected for each series of t experiments should take into account the trade-off between the need to prevent informative features from being masked too severely by the relatively most important ones and the natural requirement that s be not too large. Indeed, the smaller m, the smaller the chance of masking the occurrence of a feature. However, a larger s is then needed, since all features should have a high chance of being selected into many subsets of the features. For classification problems of dimension d ranging from several thousands to tens of thousands, we have found that taking m equal to a few hundreds (say, $m = 300 - 500$) and t equal to maximum 20 (even $t = 5$ usually suffices) is a good choice in terms of reliability and overall computational cost of the procedure.

Now, for a given m, s can be made a running parameter of the procedure, and the procedure executed for $s = s_1, s_1 + 10, s_1 + 20, \ldots$ until the rankings of the top scoring $p\%$ features prove (almost) the same for successive values of s. Minimal number of subsets, s_1, should in fact be random and such that the ranking based on these subsets includes $p\%$ of all the features present in the full data sample. Note that after having used s subsets of m features, at most $s \cdot m$ features can be ranked, and the probability of achieving this upper bound practically equals zero. More precisely, a distance between two successive rankings is defined, and the procedure is run until the values of the distance stabilize at some acceptably low level, i.e., close to zero. The distance between the ranking obtained after s subsets of m features have been used in the procedure and the ranking reached after using $s - 10$ subsets is defined as follows:

$$\text{Dist}(s, s-10) = \frac{1}{d_p} \sum_{g_k} |\text{rank}(g_k, s) - \text{rank}(g_k, s-10)|, \quad (3)$$

where summation is over top $p\%$ features obtained after having used $s - 10$ subsets; $\text{rank}(g_k, r)$ is the rank of feature g_k after having used r subsets, and d_p is the normalizing constant equal to the number of features taken into account ($d_p = dp/100$).

Note that the number of features ranked accordingly to their relative importance, d_p, should not be too large. Indeed, if d_p is such that only, say, $d_p/10$ features are informative, convergence of the distance function given by (3) to some positive value as s increases will be slow and this distance, as a function of s, will show a persistently oscillatory behavior. A moment of thought suffices to realize, as was amply confirmed by simulations, that distance (3) converges to some positive value, however erratically, even if all features are non-informative, i.e. the vector of all features is independent of the decision or class attribute.) Summarizing, while d_p should be such that all informative features are included in the ranking, distance (3) should not only converge but it should approach its asymptote smoothly as s increases (the latter property can be considered as an indication that the convergence is possibly fast).

The above calls for a means to discern between informative and non-informative features, as measured by their relative importance. Only in this way we can learn that all informative features indeed have been found and ranked. We address this issue by comparing the ranking obtained for the original data with the one obtained for the data modified in such a way that the class attribute (label) becomes independent of the vector of all features. Such a data set is obtained via a random permutation of the values of the class attribute (i.e. of the class labels of the samples). Note that this modification does not change the multivariate distribution of feature vectors.

Let Π such permutations be made and thus Π modified data sets with no relationship between the features and the class attribute be obtained. For each π-th data set, $\pi = 1, 2, \ldots, \Pi$, rank the features according to their relative importance (2) and retain the maximal value of importance (2) among all the features. This value will be denoted $maxRI_\pi$. Obviously, these $maxRI_\pi$ values are random. It follows from our experiments performed on many different data sets that $maxRI_\pi$ may be considered normally distributed (if judged, e.g., by the Shapiro-Wilk test). In all experiments, it was sufficient to set Π to 30. If this is the case, then the way to find statistically significant, that is informative, features in the original data set is straightforward. Since $maxRI_\pi$ has, after proper normalization, Student's t-distribution with $\Pi - 1$ degrees of freedom, it suffices to provide a critical value for the one-sided Student's t-test at any given significance level. A feature g_k is declared informative (at a given significance level) if its relative importance RI_{g_k} in the original ranking (without any permutation) exceeds the desired critical value.

For the two data sets that we analyzed in [Dramiński *et al.* (2008)], namely the leukemia data of [Golub *et al.* (1999)] and the second one: lymphoma data of [Alizadeh *et al.* (2000)], we found that, at significance level of 0.05, there are 22 informative features in the first data set and 50 informative features in the second data set. Concluding this section we should note that the above significance analysis does not tell anything about the validity of the ranking itself. However, in [Dramiński *et al.* (2008)] we describe in detail additional steps of the MCFS procedure that allow appraising the statistical significance of the resulting ranking of features.

3   Discovering Feature Interdependencies

Ranking features by the MCFS procedure is an efficient method of selecting the set of informative (discriminative) features. The other natural issue to be raised concerns possible interdependencies among features. Likewise in experimental design and analysis of variance, these interdependencies are often modeled using interactions. Perhaps the most widely used approach to recognizing interdependencies is finding correlations between features or finding groups of features that behave in some sense similarly across samples. A typical bioinformatics example of this problem is finding co-regulated features, most often genes or, rather more precisely, their expression profiles. Searching groups of similar features is usually done with the help of various clustering techniques, frequently specially tailored to a task at hand. See [Smyth *et al.* (2003), Hastie *et al.* (2001), Saeys *et al.* (2007), Gyenesei, A. *et al.* (2007)] and the literature there.

Our approach to interdependency discovery (abbreviated MCFSID) is significantly different in that we focus on identifying features that "cooperate" in determining that a sample belongs to a particular class. To be more specific, assume that for a given training set of samples, a rule-based classifier, a decision or a classification tree, has been constructed. Now, for each class, a set of decision rules defining this class is provided. Each decision rule is in fact a conjunction of conditions imposed on particular separate features and, thus, points to some interdependencies between the features appearing in the conditions.

The given idea of looking at interdependencies among features seems to be rather plausible but we need to refine it so that it is not dependent on just one classifier. Given a single classifier, our trust in the decision rules that are learned, and thus in the discovered interdependencies, is naturally limited by the predictive ability of that classifier. Even more importantly, the classifier is trained on just one training set and the final set of rules is dependent on the classifier. Therefore, our conclusions are necessarily dependent on the classifier and are conditional upon the training set, since they follow from just one solution of the classification problem. In the case of decisions trees, the problem is aggravated by their high variance, i.e., their tendency to provide varying results even for slightly different training sets. It should now be obvious, however, that the way out of the trouble is through an aggregation of the information provided by all the $s \cdot t$ trees, which anyhow are built within the MCFS part of the MCFSID algorithm.

In each of the $s \cdot t$ trees the nodes represent features; i.e. a given node represents the feature on which the split is made. For each path in a tree, we define the distance between two nodes as the number of edges between these two nodes. For example, in Fig. 2 the distance between node n_1 and n_5 is equal to 2 and between n_1 and n_9 is equal to 3. In turn, one may define the strength of the interdependence between features g_i and g_j as

$$\mathrm{Dep}(g_i, g_j) = \sum_{\tau=1}^{st} \sum_{\xi_\tau} \sum_{n_{g_i}(\xi_\tau), n_{g_j}(\xi_\tau)} \frac{1}{dist(n_{g_i}(\xi_\tau), n_{g_j}(\xi_\tau))}, \qquad (4)$$

where summation is over all the $s \cdot t$ trees, within each τ-th tree over all paths ξ_τ and, within each path ξ_τ, over all pairs of nodes $(n_{g_i}(\xi_\tau), n_{g_j}(\xi_\tau))$ on which the splits are made, respectively, on feature g_i and feature g_j; $\mathrm{Dep}(g_i, g_j) = 0$ if in none of the trees there is a path along which there are two splits made, respectively, on g_i and g_j. The rationale behind this definition is obvious. $\mathrm{Dep}(g_i, g_j)$, calculated on the basis of thousands of trees provides an incomparably more stable and reliable information about the

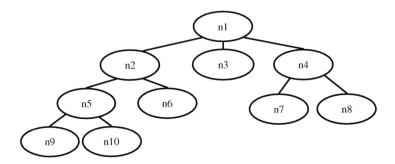

Fig. 2. An example of a tree graph

strength of interdependency between two features than that derived from just one single tree.

Analysis of all pairs of nodes for all decision trees is a time consuming task. To deal with this problem we introduce a parameter k that determines a maximum distance between two nodes to be taken into account in (4). By default, we set k to 3. Final results of the analysis may be presented in the form of a graph with nodes representing features and edges showing strength of dependence among the connected nodes as measured by $\text{Dep}(g_i, g_j)$. Only the nodes corresponding to the informative features should be included in the graph. If the graph proves hardly readable due to a large number of nodes and edges, it can be easily simplified by showing only these edges that represent a desired fraction of the strongest interdependencies. Typically, we show 20% of all the edges found for a selected set of features. Despite the obvious advantages of the presented approach, it should be kept in mind that it is a heuristic and, in particular, definition (4) is arbitrary. The real value of the presented method can be assessed by tests only. This can be achieved by examining a number of examples where the resulting graphs are scrutinized by domain experts who are able to verify and explain (or falsify) the claimed interdependencies.

4   Biological Validation Study

In one of our applications, we constructed classifiers for modeling drug resistance. We used publicly available HIV-1 drug resistance data from the Stanford HIV Resistance Database [HIV Resistance Database]. Using our domain knowledge, we evaluated the interdependencies discovered with the MCFSID methodology.

The data set consists of several protein sequences of the HIV-1 reverse transcriptase (RT) enzyme annotated with their drug resistance fold relative to the wild-type HIV-1 strain. Following [Rhee *et al.* (2006)], we have discretized these continuous resistance values into three classes: "susceptible", "intermediately resistant" and "resistant". These classes are commonly used by clinicians while deciding upon the treatment regimen. Every amino acid in the sequence is subsequently represented as a 7-vector of its biochemical properties (values relative to the wild-type virus). After [Rudnicki and Komorowski (2004)], we have used the following set of biochemical descriptors:

- D1 - Transfer free energy from octanol to water.
- D2 - Normalized van der Waals volume.
- D3 - Isoelectric point.
- D4 - Polarity.
- D5 - Normalized frequency of turn.
- D6 - Normalized frequency of alpha-helix.
- D7 - Free energy of solution in water.

In the sequel, we present and discuss the results that were obtained for Didanosine, a nucleoside analogue of adenosine, and a commonly administered anti-viral drug. Our results for the 6 other anti-varial drugs: Abacavir, Efavirenz, Nevirapine, Stavudine, Tenofovir and Zidovudine constitute a large problem that will appear in a separate paper.

Didanosine belongs to the so-called nucleoside RT inhibitors (NRTI) that mimic nucleosides that are natural substrates for the enzyme. Once inside the cell, the nucleoside analogs are phosphorylated to their triphosphate (TP) form and, as soon as they are incorporated into a newly synthetized HIV-1 DNA chain, they terminate its elongation and, thus, inhibit viral replication [Jonckheere, H. *et al.* (2000), Bauman, JD *et al.* (2008), Menédez-Arias, L. (2008)].

In this 3-class classification task, there were 706 samples, each described by 3920 features. We recall that they are coming from 560 amino acids in RT. Each aa is represented by 7 easy-to-interpret, low-correlated biochemical properties. The following parameters have been used in the MCFS part of the MCFSID algorithm: $s = 2500$, $t = 5$, $m = 0.05 \cdot 3920 = 196$, $u = 1$ and $v = 1$. The resulting graph of interdependencies is given in Fig. 3. The darker is the label of a graph node, the higher is the RI value of the corresponding feature. Following the default rule, only 20% of the strongest

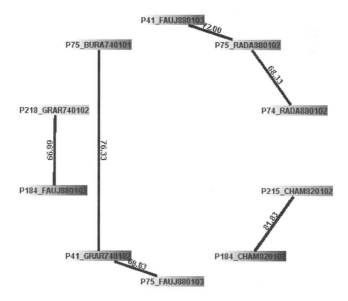

Fig. 3. Interdependencies graph for Didanosine important features

interdependencies are shown in the graph. This graph was created by a JAVA application called *graphViewer* that uses jgraph library [JGraph].

As it can be seen in the graph 3, there are 5 different biochemical properties were most frequently used by the MCFSID in constructing the trees.

4.1   Interpreation of the Interdependencies

Reverse Transcriptase is a viral enzyme responsible for transcribing single-stranded HIV-1 RNA genome into a double-stranded DNA provirus. In the course of infection, the provirus is incorporated into the host cell genome and may be replicated and transcribed by the native molecular machinery of the host. This results in new viral particles that are capable of infecting subsequent cells and thus complete the HIV life cycle [Jonckheere, H. *et al.* (2000), Menédez-Arias, L. (2008)]. The entire RT enzyme is a heterodimer consisting of two similar subunits: p51 (51 kDa) and p66 (66 kDa). The heavier p66 subunit contains two distinct domains: the N-terminal polymerase domain and the C-terminal RNase H domain. The lighter p51 is a product of the p66 proteolytic cleavage. It lacks RNase H domain and catalytic activity. Often, the ternary structure of the p66 domain is compared to the right hand and, accordingly, the palm, thumb and finger domains are distinguished (see Fig. 4). The palm domain contains three catalytically active re-si-du-es: Asp 110, Asp 185 and Asp 186 embedded in a hydrophobic region; cf. [Valverde-Garduño *et al.* (1998), Menédez-Arias, L. (2008),

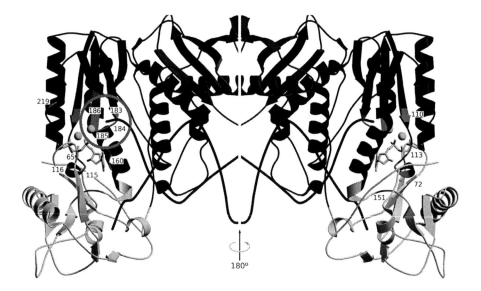

Fig. 4. Structure of the HIV-1 RT enzyme fragment (PDB structure 1RTD). Thumb domain (red), finger domains (yellow) and palm domain (green) constitute p66 subunit. Residues constituting dNTP binding pocket are shown in magenta, YXDD motif members are within the red circle. The incoming dNTP and two magnesium ions are shown in light green. (For clarity, the structure to the right is rotated 180°).

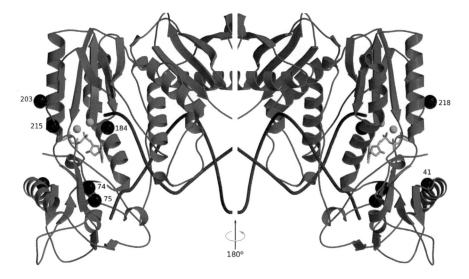

Fig. 5. Structure of the HIV-1 RT enzyme p66 subunit. Alpha-atoms of the residues selected by MCFSID as contributing to resistance to Didanosine represented as red spheres. The incoming dNTP and two magnesium ions shown in light green. (For clarity, the structure to the right is rotated 180°).

Ren, J. and Stammers, DK. (2008)]. The two latter residues, together with Tyr 183 and Met 184, constitute a highly conserved YXDD motif that is common to all reverse transcriptases ([Kaushik *et al.* (1996)]).

The fingers and both alpha-helices of the thumb are thought to form a clamp that holds the nucleic acid in place over the palm part that is required for polymerization. The template/primer interactions occur between the sugar-phosphate backbone of the DNA/RNA and p66 residues [Sarafianos, SG. *et al.* (1999)]. Apart from the catalytic triad, the palm domain contains a dNTP-binding pocket; cf. Fig. 4. As polymerization proceeds, new deoxyribonucleotide-triphosphates (dNTPs) come to the binding pocket and are incorporated into a newly synthesized provirus DNA strand. The catalytic triad plays crucial role in this process by interacting with magnesium cations that serve as activators of the enzymatic reaction. Mutagenesis studies described in [Valverde-Garduño *et al.* (1998), Harris, D. *et al.* (1998), Bauman, JD *et al.* (2008)] suggest that mutations surrounding the catalytic triad may lead to incorporation of non-specific nucleotides.

Our results for Didanosine are presented in Fig. 5. We show alpha-atoms of the residues selected as contributing to resistance to the drug.

The algorithm identified 5 biochemical properties that span over 6 aa residues. Both Fig. 6, graph 1 and Fig. 6, graph 4 contain site 184 and either of the 215 or the 218 sites. The two latter sites are both located on the side of the dNTP binding pocket opposite to residue 184. While the amino acids at positions 184 and 215 are known to be directly interacting with the incoming nucleotide, Asp 218 interacts directly with Lys 219 that is a part of the dNTP binding pocket; cf. [Harris, D. *et al.* (1998),

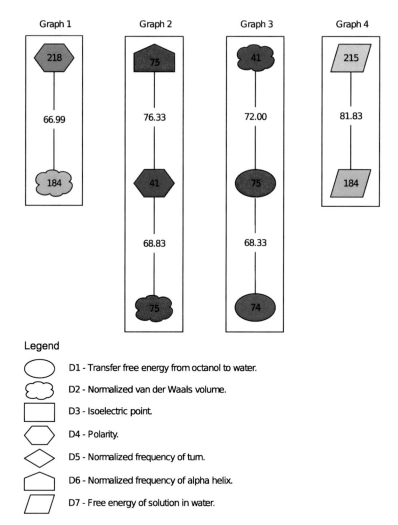

Fig. 6. Block diagram of the interdependencies between aminoacids descriptors of the didanosine. Shape corresponds to property, color to site number.

Menédez-Arias, L. (2008)]. "Polarity" has been selected as important at site 218, "free energy of solution in water" at site 215 and both "normalized van der Waals volume" and "free energy of solution in water" at site 184. While mutations at sites 184 and 215 may directly disturb properties of the dNTP binding pocket, mutations at site 218 may affect its geometry in an indirect way.

Pairs involving residues 41, 74 and 75 (Fig. 6, graphs: 2 and 3) appear to be good descriptors of the fingers domain properties. The fingers domain is known to be involved in the viral RNA template positioning; cf. [Sarafianos, SG. *et al.* (1999), Menédez-Arias, L. (2008)].

We conclude that two distinct resistance mechanisms characterize Didanosine: one is induced by mutations directly affecting the dNTP binding pocket and the other is caused by changes in the template positioning.

5 Concluding Remarks

The MCFSID algorithm provides an effective and reliable method for ranking features according to their importance in supervised classification. It achieves that by determining the set of informative features as contrasted with the remaining non-informative ones, and by discovering interdependencies between the informative features that jointly are instrumental in determining which samples belong to their particular classes and to what degree.

Reliability of the algorithm follows from the Monte Carlo approach; resampling is sufficiently numerous and extensive, and the tree classifier amply flexible. It also is a consequence of (i) the way in which we choose the number s of subsets of features while other parameters remain fixed (here we mean the requirement that the distance (3) between successive rankings stabilizes at some acceptably low level); and (ii) the way in which the strength of dependence among features is defined. It should be emphasized that the algorithm has been designed specifically to rank features with respect to their classification ability, not only to find those features that are important for classification.

Effectiveness and reliability of the algorithm has been confirmed on many data sets of biological and commercial origin. In addition to the HIV-1 case, we used the MCFSIF approach to identify interdependent features important to post-translational modifications of proteins. Our analyses also included transactional data from a major multinational FMCG company, geological data from oil wells operated by a major American company and data sets of samples with attributes being some functions of U.S. Census data. At the moment of completing this paper the MCFS part of the algorithm has been successfully used on some 15 data sets and the whole MCFSID algorithm on about 10.

In our opinion, the MCFSID approach appears to be quite promising in the area of systems biology where important features must be identified among hundreds if not thousands of them. In the case of HIV-1 we were able to automatically rediscover all the parameters that previously were identified by human experts as the important ones and thus rediscover known mechanisms of drug resistance. It is likely that applications of this approach for unknown problems will allow *in silico* discovery of new mechanisms that so far avoided human explanations.

References

[Alizadeh *et al.* (2000)] Alizadeh, A.A., et al.: Distinct Types of Diffuse Large B-cell Lymphoma Identified by Expression Profiling. Nature 403, 503–511 (2000)

[Archer and Kimes (2008)] Archer, K.J., Kimes, R.V.: Empirical Characterization of Random Forest Variable Importance Measures. Comp. Stat. & Data Anal. 52(4), 2249–2260 (2008)

[Bauman, JD *et al.* (2008)] Bauman, J.D., et al.: Crystal engineering of HIV-1 reverse transcriptase for structure-based drug design. Nucleic Acid Res. 36, 5083–5092 (2008)

[Breiman and Cutler (2008)] Breiman, L., Cutler, A.: Random Forests - Classification/ Clustering Manual (2008),
http://www.math.usu.edu/~adele/forests/cc_home.htm

[Chrysostomou *et al.* (2008)] Chrysostomou, K., et al.: Combining Multiple Classifiers for Wrapper Feature Selection. Int. J. Data Mining, Modelling and Management 1, 91–102 (2008)

[Diaz-Uriarte and de Andres (2006)] Diaz-Uriarte, R., de Andres, S.A.: Gene Selection and Classification of Microarray Data Using Random Forest. BMC Bioinformatics 7(3) (2006), doi:10.1186/1471-2105-7-3

[Dramiński *et al.* (2008)] Dramiński, M., et al.: Monte Carlo Feature Selection for Supervised Classification. Bioinformatics 24(1), 110–117 (2008)

[Dudoit and Fridlyand (2003)] Dudoit, S., Fridlyand, J.: Classification in Microarray Experiments. In: Speed, T. (ed.) Statistical Analysis of Gene Expression Microarray Data, pp. 93–158. Chapman & Hall/CRC, Boca Raton (2003)

[Golub *et al.* (1999)] Golub, T.R., et al.: Molecular Classification of Cancer: Class Discovery and Class Prediction by Gene Expression Monitoring. Science 286, 531–537 (1999)

[Gyenesei, A. *et al.* (2007)] Gyenesei, A., et al.: Mining Co-regulated Gene Profiles for the Detection of Functional Associations in Gene Expression data. Bioinformatics 23(15), 1927–1935 (2007)

[Harris, D. *et al.* (1998)] Harris, D., et al.: Functional analysis of amino acid residues constituting the dNTP binding pocket of HIV-1 reverse transcriptase. J. Biol. Chem. 273, 33624–33634 (1998)

[Hastie *et al.* (2001)] Hastie, T., et al.: Supervised Harvesting of Expression Trees. Genome Biology 2(1), research0003.1-0003.12 (2001)

[HIV Resistance Database] Stanford HIV Drug Resistance Database,
http://hivdb.stanford.edu

[JGraph] JGraph - The Java Open Source Graph Drawing Component,
http://www.jgraph.com/jgraph.html

[Jonckheere, H. *et al.* (2000)] Jonckheere, H., et al.: The HIV-1 reverse transcription (RT) process as target for RT inhibitors. Med. Res. Rev. 20, 129–154 (2000)

[Kaushik *et al.* (1996)] Kaushik, et al.: Biochemical analysis of catalytically crucial aspartate mutants of human immunodeficiency virus type 1 reverse transcriptase. Biochemistry 35(36), 11536–11546 (1996)

[Li *et al.* (2002)] Li, Y., et al.: Bayesian Automatic Relevance Determination Algorithms for Classifying Gene Expression data. Bioinformatics 18(10), 1332–1339 (2002)

[Lu *et al.* (2007)] Lu, C., et al.: Bagging Linear Sparse Bayesian Learning Models for Variable Selection in Cancer Diagnosis. IEEE Trans. Inf. Technol. Biomed. 11, 338–347 (2007)

[Menédez-Arias, L. (2008)] Menédez-Arias, L.: Mechanisms of resistance to nucleoside analogue inhibitors of HIV-1 reverse transcriptase. Virus Res. 134, 124–146 (2008)

[Ren, J. and Stammers, DK. (2008)] Ren, J., Stammers, D.K.: Structural basis for drug resistance mechanisms for non-nucleoside inhibitors of HIV reverse transcriptase. Virus Res. 134, 157–170 (2008)

[Rhee *et al.* (2006)] Rhee, S.Y., et al.: Genotypic predictors of human immunodeficiency virus type 1 drug resistance. Proc. Natl. Acad. Sci. USA 103, 17355–17360 (2006)

[Rudnicki and Komorowski (2004)] Rudnicki, W.R., Komorowski, J.: Feature synthesis and extraction for the construction of generalized properties of amino acids. In: Tsumoto, S., Słowiński, R., Komorowski, J., Grzymała-Busse, J.W. (eds.) RSCTC 2004. LNCS (LNAI), vol. 3066, pp. 786–791. Springer, Heidelberg (2004)

[Saeys *et al.* (2007)] Saeys, Y., et al.: A Review of Featrure Selection Techniques in Bioinformatics. Bioinformatics 23(19), 2507–2517 (2007)

[Sarafianos, SG. *et al.* (1999)] Sarafianos, S., et al.: Touching the heart of HIV-1 drug resistance: the fingers close down on the dNTP at the polymerase active site. Chem. & Biol. 6, R137–R146 (1999)

[Smyth *et al.* (2003)] Smyth, G.K.: Statistical Issues in cDNA Microarray Data Analysis. In: Brownstein, M.J., Khodursky, A.B. (eds.) Functional Genomics: Methods and Protocols. Methods in Molecular Biology, vol. 224, pp. 111–136. Humana Press (2003)

[Strobl *et al.* (2007)] Strobl, C., et al.: Bias in Random Forest Variable Importance Measures: Illustrations, Sources, and a Solution. BMC Bioinformatics 8(25) (2007), doi:10.1186/1471-2105-8-25

[Strobl *et al.* (2008)] Strobl, C., et al.: Conditional Variable Importance for Random Forests. BMC Bioinformatics 9(307) (2008), doi:10.1186/1471-2105-9-307

[Tibshirani *et al.* (2002)] Tibshirani, R., et al.: Diagnosis of Multiple Cancer Types by Nearest Shrunken Centroids of Gene Exressions. Proc. Natl. Acad. Sci. USA 99, 6567–6572 (2002)

[Tibshirani *et al.* (2003)] Tibshirani, R., et al.: Class Prediction by Nearest Shrunken Centroids, with Applications to DNA Microarrays. Statistical Science 18, 104–117 (2003)

[Valverde-Garduño *et al.* (1998)] Valverde-Garduño, et al.: Functional analysis of HIV-1 reverse transcriptase motif C: site-directed mutagenesis and metal cation interaction. J. Mol. Evol. 47, 73–80 (1998)

[Yousef *et al.* (2007)] Yousef, M., et al.: Recursive Cluster Elimination (RCE) for Classification and Feature Selection from Gene Expression Data. BMC Bioinformatics 8(144) (2007), doi:doi:10.1186/1471-2105-8-144

Machine Learning Methods in Automatic Image Annotation

Halina Kwaśnicka and Mariusz Paradowski

Wroclaw University of Technology, Wroclaw, Poland
halina.kwasnicka@pwr.wroc.pl, mariusz.paradowski@pwr.wroc.pl

Abstract. Machine learning methods are successfully applied in many branches of Computer Science. One of these branches is image analysis, and being more specific – *Automatic Image Annotation*. Automatic Image Annotation was found an important research domain several years ago. It grew from such research domains as image recognition and cross-lingual machine translation. Increase of computational, data storage and data transfer abilities of todays' computers has been one of key factors, making Automatic Image Annotation possible. Automatic Image Annotation methods, which have appeared during last several years, make a large use of many machine learning approaches. Clustering and classification methods are most frequently applied to annotate images. The chapter consists of three main parts. In the first, some general information concerning annotation methods is presented. In the second part, two original annotation methods are described. The last part presents experimental studies of the proposed methods.

Keywords: Automatic Image Annotation, Decision Trees.

1 Image Auto-annotation Methods – An Introduction

Automatic Image Annotation is a relatively new research topic. During last several years, various approaches and methods of automatic image annotation have been proposed. This research was made possible thanks to increase of computational power of todays computers. Data storage and transfer abilities have increased, making *Internet* a popular medium for transferring lots of visual data. A large part of the visual data are static images, often without any description or categorization. Such growth, both in number of images and in need of automatic captioning them, has made an automatic image annotation a very important and vital research area. Automatic image annotation methods try to answer the growing requirements for processing huge collections of image data available both in the *Internet* and large multimedia databases. Informally saying, *the task of automatic image annotation is to assign a subset of words from the given dictionary to a previously unseen image on the basis of weakly (incompletely, imprecisely, subjectively) annotated training set of images, without knowledge which parts of an image lead to which words.*

Machine learning methods are successfully applied in many branches of Computer Science, one of them is *Automatic Image Annotation*. In a certain way, the automatic image annotation tries to mimic the behavior of a human being, and thus usage of machine

J. Koronacki et al. (Eds.): Advances in Machine Learning II, SCI 263, pp. 387–411.
springerlink.com   © Springer-Verlag Berlin Heidelberg 2010

learning methods is almost natural. A host of methods developed within machine learning, especially classification and clustering methods, are now key components of automatic image annotation algorithms. These include nonparametric models [1, 2, 3, 4], neural networks [5, 6], Support Vector Machines [7, 8, 9], decision trees [10], ensembles of classifiers [11], K–Means clustering [12, 13], hierarchical clustering, dimensionality reduction [14, 15, 16], feature space discretization [17, 18], and others. Automatic annotation methods are often combinations of more than one of the mentioned machine learning methods.

Another important aspect of automatic image annotation is the availability vs. unavailability of semantic knowledge. Most of researchers, including the authors of this chapter, assume that the semantic knowledge is not available for the annotation process. However, one should notice that much research has been done on using semantic knowledge to advantage for automatic image annotation [19, 20, 21, 22, 23].

There are many similarities between *image recognition* and automatic image annotation. However, while many concepts and ideas are shared by these two research branches, there are also several important differences between them.

The main aim of image recognition (pattern recognition, pattern classification) methods is the identification of objects in images. It can thus be considered as *assigning a given image to one of the predefined classes* [24, 25, 26, 27]. The key differences between the stated goals of image recognition and automatic image annotation are the following. The first difference is that in the former, a class – which can be identified as a word – is assigned to an image, while in the latter, many words are assigned. The second difference is the weakly annotated training set ('weakly annotated' means that the descriptions in the training set may be incomplete, imprecise or subjective). The third difference is the lack of information which image parts cause which words in its description.

In order to formally define the problem of automatic image annotation, let us first define some useful basic terms.

Dictionary W is a set of n words $w_y, y = 1, \ldots, n$:

$$W = \{w_1, w_2, \ldots, w_n\}. \tag{1}$$

Description W_{I_x} of an image I_x is a subset of words from the dictionary W:

$$W_{I_x} \in W. \tag{2}$$

I denotes an annotated training set, which contains m pairs (I_x, W_{I_x}), where I_x represents image number x and $W_{I_x} \subseteq W$ its description, $x = 1, \ldots, m$:

$$I = \{(I_1, W_{I_1}), (I_2, W_{I_2}), \ldots, (I_m, W_{I_m})\}. \tag{3}$$

Selection of a subset of words from dictionary W may be formulated as ranking all words in dictionary W and selecting a subset of top ranked words. Automatic image annotation method A is then formulated in terms of ranking words from dictionary W:

$$A(I, J, W) = \begin{bmatrix} p_{w_1}^J \\ p_{w_2}^J \\ \vdots \\ p_{w_n}^J \end{bmatrix}, \tag{4}$$

where: J denotes a previously unseen image to be annotated, and $p_{w_y}^J$ is the value (score) generated by annotator A for word w_y.

The values generated by the automatic image annotator, according to which the ranking is done, may have various interpretations. They can be defined in terms of *probability* ($p_{w_y}^J$ can be seen as *a posteriori* probability of word w_y, given the image J) or may be less formally defined in terms of word support or score values. Automatic image annotation method definitions are usually approach specific, but generally they follow the presented idea.

In the next section, a taxonomy of automatic image annotation methods is presented, which takes into account methods being used for images annotation. In section 3, data preparation phase, which is of vital importance for the final result of automatic image annotation, is briefly described. Image segmentation as well as feature selection and extraction methods are presented. In section 4, the application of decision trees as a classification method for use in automatic image annotation is discussed. Also, two authors' annotation methods are presented: *Binary Machine Learning* (*BML*) and *Multi Class Machine Learning* (*MCML*); see [10]. The fifth section summarizes performed experiments, the proposed methods are compared with the selected reference method. In the last section, some concluding remarks are given and possible future research directions are pointed out.

2   Automatic Image Annotation Taxonomy

During last years many approaches have been proposed for automatic image annotation. Majority of them rely on machine learning methods, especially various types of clustering and classification. In this section some important approaches to automatic image annotation are shortly presented. Annotation methods are categorized according to the main mechanism, they are built with.

2.1   Approaches to Taxonomy

Several taxonomies are defined for automatic image annotation. We focus on two of them. The first one takes into account the data being processed by the methods. The second one focuses on methods being used to annotate images.

The first taxonomy is presented by Shah and it focuses on the data being processed by automatic image annotation methods [28]. On the highest level of taxonomy, automatic image annotation is split into two categories:

I. *Image-based methods*:
– Global feature-based methods (features extraction, annotation using single feature vector per image),
– Regional feature-based methods (segmentation process, features extraction applied for every segment, annotation using a set of feature vectors per image – single feature vector per each segment).
II. *Text-based methods*
The first category represents all methods in which the image is the most important source of information, the second ones are methods which use available text data presented in a context of the image.

We are dealing with *Image-based methods* with the regional feature-based approach. This category is much more prominent and it is the key issue in this chapter.

The second taxonomy is much less strict, and is loosely used by many automatic image annotation researchers. It focuses mainly on methods being used to describe (annotate) an image. Taking into account this criterion one can distinguish two main categories.

I. *Annotations in which distances between features play the main role*:
   – Clustering,
   – Feature space discretization.
II. *Annotations in which classification methods play the main role*:
   – Bayesian approaches,
   – Binary classification,
   – Multi-class classification.

These two approaches are the most interesting and they are shortly described in the next subsections.

2.2   Considering Distances between Features – Clustering and Feature Space Discretization

Clustering methods are used in many automatic annotation methods. Such methods usually split the feature space into similar regions. These feature space regions (clusters) are variously called by different researchers (*visterms, blobs, visual words*), however common ground is very similar. Instead of generating image description directly on the basis of feature values, various cluster properties are used.

From the historical point of view, one of the first methods, called *Mori*, is based on *K-means* clustering [12]. The concept of *K-means* usage is refined in later research in *Cross-Media Relevance Model* (*CMRM*) method [13]. Another idea of automatic image annotation combines clustering with machine translation methods [24, 29]. However, further research shows that hierarchical clustering increases results quality [30].

Feature space discretization constitutes other group of approaches based on distances between features. Both, fixed and information gain based discretization approaches are employed in *Dichotomic IMAge TEXt annotation* (*DIMATEX*) method [17]. We have developed *DIMATEX* into the method called *FastDiM* [18], by employing fixed, very dense feature space discretization. It makes make distance calculation much faster, but as similar as possible to the distance calculation on a continuous feature space.

Another interesting automatic image annotation method is based on *Self Organizing Map*, and is called *"Picture" and the self-organizing map* (*PicSOM*) [5, 6] . Neurons in SOM network represent image feature vectors and thus similar feature vectors are located near to each other in the network. The method is based on hierarchical *TS-SOM* algorithm, which allows to reduce the computational complexity of similar feature vectors search procedure.

Clustering based methods are usually very fast, because they do not have to search through the training set (lazy classification) but just refer to proper clusters. They also contain some level of generalization, which is given by the clustering process itself. However, it has been observed by researchers [1], that *a priori* unknown number of clusters (proper granularity of clustering) and other clustering parameters pose a large problem from the results quality point of view. Methods belonging to this category are usually very sensitive for clustering settings and these parameters have to be selected with great care.

2.3 Bayesian Approaches

Bayes probability framework is successfully applied in a number of automatic image annotation methods. These methods usually estimate *a posteriori* probability of single words given an input image. Conditional class densities are usually estimated by parametric or nonparametric models.

Continuous Relevance Model (CRM) [1] and its various modifications are very well known methods based on nonparametric model and *Gaussian kernel*. Nonparametric model is used to estimate conditional class densities for both features and word frequencies (this approach is called *doubly nonparametric*). *Normalized CRM* [2] assumes that all descriptions in the training set have equal length. Missing words are filled by empty words. Such approach is designed to nullify negative effects of description length differences, especially very short descriptions. *Modified CRM* [3] employs an additional distance measure between image segments, which is combined together with the Gaussian kernel. *Multiple Bernoulli Relevance Model (MBRM)* [4] employs a different word frequency distribution than the original *CRM* method. Authors of the original *CRM* method have observed that used word frequency distribution is not well suited to different annotated image collections.

Parametric models, especially *Gaussian mixture models (GMM)* are also used in automatic image annotation. *Mix-Hier (MH)* [31] and *Supervised Multiclass Labeling (SML)* [32, 33] methods introduce the concept of multi-class classification based on *GMM*. *Gaussian mixture model* parameters are estimated using the *Expectation Maximization* method. Additionally, both methods recognize the problem of formulating negative training examples in the training set, due to weak training set annotation. To reduce the problem and increase results quality, both methods are based on the concept of *multiple instance* learning.

2.4 Soft, Binary and Multi-class Classification

Binary and multi-class classification categories of automatic image annotation methods are based on various types of classification. In such approaches, a set of classifiers is used. Each classifier is usually responsible for learning a single word from the dictionary. After the classification process is completed, the results are gathered and combined together to form the final image annotation.

One of the most popular classifiers used by automatic image annotation researchers is *Support Vector Machine*. A method based on *multi-class Support Vector Machine* constructed according to *one-per-class* rule is presented by Cusano [7]. *CSD-Support Vector Machine (CSD-SVM)* [9] is also based on the multi-class approach. The method employs *MPEG–7 Colour Structure Descriptor* [34] which, according to the cited paper, is an efficient way to represent image features in automatic image annotation. *Content-Based Soft Annotation (CBSA)* [8] employs soft classification. Each classifier is responsible for classification to a single word. Two kinds of classifiers were tested: *Support Vector Machine* and *Bayes Point Machine*. According to the presented results *Bayes Point Machine* achieves better results quality than *Support Vector Machine*. Other method, *Confidence-based Dynamic Ensemble (CDE)* [11] employs ensemble of classifiers and *bagging* technique. Output of binary *Support Vector Machines* is combined into a multi-class classifier. The method is an adaptive one, it dynamically improves its work.

Classification based methods usually results in good quality of annotation results, especially if they are based on multi-class concept. On the other hand, they require very large training time, generated classifiers often require large amounts of memory, and additionally, extending the available image dataset poses a great difficulty.

3   Image Segmentation, Feature Selection and Extraction

Image segmentation and feature selection are very important steps in automatic image annotation task. They provide data for the whole annotation process, the selected features strongly influence the annotation results: the better data produce the better annotation results. It can be said that one of the goals of automatic image annotation studies is to verify the set of features defined for this task [24].

3.1   Image Segmentation

There are two main approaches to image segmentation in automatic image annotation methods. They represent the conceptual way of splitting objects in the image [35]: *region-based segmentation* and *block-based segmentation*.

Fig. 1. Exemplary input image for segmentation procedures

Region-based segmentation methods try to split into separate segments objects existing in the considered image. They represent the core idea of what image segmentation should be. Such approaches usually employ various machine learning methods and different kinds of similarity criteria. We have performed research on application of *K-means* clustering algorithm, enforced with such visual features like: pixel color, pixel neighborhood and pixel localization [18]. Such simple approach gives satisfying results, however generated regions have to be post-processed by such methods like segment merging or splitting of disjoint pixel groups. An exemplary image is presented in Fig. 1, and its region based segmentation in Fig. 2. A very popular method of image segmentation, used in several automatic annotation benchmark datasets is *Normalized Cuts* [36]. *Normalized Cuts* algorithm represents an image as a graph, segmentation is defined as graph split according to a given criterion. In the algorithm we use a criterion which measures differences between segmented areas.

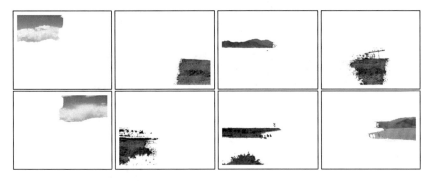

Fig. 2. Exemplary region-based segmentation. Image is split into eight visually and spatially similar segments.

Block-based segmentation is based on a simple concept of image segmentation. The image is split into a set of predefined areas, each image is segmented in the same way. Generated segments do not have any resemblance to objects existing in the image, they are just arbitrary defined pieces of the image. Such methods may seem awkward in the first glance, however their properties are very interesting. Resulting segments are completely insensitive to small changes in visual properties of the image. Such sensitivity is one of key weaknesses of region-based methods. It is enough to have the same object photographed in different lighting condition, scale or rotation and the segmentation results may vary very much. This behavior is very problematic from automatic image annotation point of view because feature vectors may differ very much for the similar objects in the image. Experiments performed by many researchers has shown that block-based segmentation yields in better quality of automatic image annotation. Fig. 3 presents an exemplary block-based segmentation for the image shown in Fig. 1.

3.2 Feature Selection and Extraction

Feature selection and extraction are image domain specific. For example, different features should be used for automatic annotation of landscape images and medical images. Most of researches in automatic image annotation field are performed for large collections of general topic images, containing among many others: landscapes, human

Fig. 3. Exemplary block-based segmentation. Image is split into 25 fixed rectangles.

beings, animals, wildlife, city scenes, plants, flowers, sea scenes or cars. These images are usually well described by such features as color, texture, localization and shape. In this section we present the most important approaches for feature extraction.

One of the most prominent features of an image segment is usually color information. Various color models are used for calculation of color features, among others: *RGB* [12, 14, 24], *Lab* [6, 24], *HSV* [7], *gray-scale* [37].

Textural information is usually an addition to color information. Using textural information allows to differentiate between objects with similar color, but with different structure, e.g. sky and sea. Various textural properties are used as features for automatic image annotation. Edge information [12] allows to differentiate between smoothly textured objects and objects with rough textures. Directional filters [24] provide information about directions of objects texture changes. There are many other approaches to extract textural information, such as: SIFT descriptor [38], Gabor filters [37, 38], texture correlation and energy [18]. Another frequently used approach is based on usage of coefficients extracted by the discrete cosine transform [31, 33, 39, 40]. Discrete cosine transform allows to extract very important image information in context of lossy image compression. Lossy image compression is usually designed to keep the essence of the image and maximally reduce the amount of necessary data. In these terms, such features operate on information which is important from the visual point of view.

Localization and shape are another examples of features used in automatic image annotation. These features are usually helpful to differentiate between objects which should have more or less fixed position and shape in the image. Exemplary features from this category are: normalized image region coordinates [24], shape convexity [24, 30] or localization descriptors [18].

Some automatic image annotation methods have inbuilt automated feature selection methods, however usually only basic approaches are used. Applied methods are usually treated by researchers more like feature space dimensionality reduction. We can mention *Singular Value Decomposition* [15] as an example of such approach. Also we can find research on feature verification and selection, e.g., a method that uses *Principal Component Analysis* and *Fisher Discriminant Analysis* [41].

4   Decision Trees in Automatic Image Annotation

This section presents our research in automatic image annotation area. We decided to use decision trees as binary and multi-class classifiers in the proposed annotation method. Decision trees are selected as classifiers for several reasons, namely:

– Decision trees offer a very fast processing phase with logarithmic computational complexity.
– Training phase is also relatively fast, comparing to other machine learning approaches.
– Additionally, training algorithms do not require randomization, which is not feasible in automatic image annotation case.
– Generated decision trees can be easily reformulated into a set of decision rules.

Our research is mainly focused on the quality of annotation, however great care is placed to computational complexity properties of the proposed solutions. We describe the methods which are constantly developed and improved in our research, they are: *Binary Machine Learning* (BML) and *Multi Class Machine Learning* (MCML). Both methods are improved and much more formally defined and deeper examined versions of those presented in [10].

4.1   General Concept

Although there are important differences between *Binary Machine Learning* and *Multi Class Machine Learning* methods, there are still a common ground and ideas. At the beginning, common architecture of both methods are presented and all aspects that these methods share.

Both methods take advantage of 'divide and conquer' approach: the annotation task is split into many smaller classification subtasks, and then the results are aggregated. Single subtask is a classification of feature vectors, extracted from the input image, performed for a single word in the dictionary. It must be done for every word in the dictionary. Given that there are n words in the dictionary W and there are q feature vectors extracted from the input image J, a number of classification tasks is equal to nq. As the result we receive specialized classifiers built for each word in the dictionary. It means that a number of specialized classifiers is equal to a size of the dictionary.

In both method we use *C4.5* decision trees. Decision trees are responsible for classification of specific words presence in the given image feature vectors. After classification problems are solved, the second part is executed, i.e., aggregation of results. Results from all classifications are combined together, normalized and presented as an answer of the automatic image annotation method.

4.1.1   Resulted Word Counts Optimization

Very important aspect of the automatic annotation process is a proper modification of output produced by the automatic annotation method. Its goal is to adjust the word frequency distribution produced by the annotator to this observed in the training set. We call this step the *optimization procedure* of an annotator. It is worth mentioning that optimization procedure strongly influences the annotators quality. This problem is not trivial and have been explored by many researchers [4, 14, 38, 42, 43, 44].

In this chapter we briefly present the optimization method called *Resulted Word Counts Optimization* (RWCO). This method is presented with more details in [45], it represents a general approach to conform a given annotator to the required word frequencies (according to the training data).

The idea of optimization procedure is to introduce a vector of coefficients V (one coefficient per each word), which correct the results generated by the 'raw' annotator:

$$OA(I,J,W,V) = \left(A(I,J,W)^T \begin{bmatrix} v_{w_1} & 0 & \cdots & 0 \\ 0 & v_{w_2} & \cdots & 0 \\ \vdots & \vdots & \ddots & \vdots \\ 0 & 0 & \cdots & v_{w_n} \end{bmatrix} \right)^T =$$

$$= \left[v_{w_1} p_{w_1}^J \ v_{w_2} p_{w_2}^J \cdots v_{w_n} p_{w_n}^J \right]^T = \begin{bmatrix} v_{w_1} p_{w_1}^J \\ v_{w_2} p_{w_2}^J \\ \vdots \\ v_{w_n} p_{w_n}^J \end{bmatrix}. \qquad (5)$$

where: OA denotes an optimized annotator, v_{w_i} – coefficient for i^{th} word from the dictionary, $A(I,J,W)$ and $p_{w_i}^J$ as in eq. 4.

A key factor in the annotator optimization is estimation of V vector. It is performed on the basis of the training set I. The training set is split into two disjoint parts. There are two iterations in the estimation procedure. In the first iteration, the first part of the training set is used as the training part, the second one plays the role of the test subset. In the second iteration an opposite setup is performed.

In each of above two iterations, vector of coefficients V is estimated by a specialized optimization method (a heuristic), which is designed to solve the following optimization task [45]:

$$V^* = \arg \min_{V \in R_+^{|W|}} \sum_{w \in W} \frac{1}{e_w} \max\left(e_w - r_w(V), 0\right), \qquad (6)$$

where: V^* – set of coefficients for optimized annotator OA, e_w – number of images from the test subset (part of the initial training set I) manually annotated by word w (ground truth), r_w – number of images from the test subset (part of the initial training set I) automatically annotated by word w using the automatic image annotator.

The results of two above iterations are aggregated giving the estimation of V^* vector of coefficients. Averaged vector is the output of the whole *Resulted Word Counts Optimization* method.

Both *Binary Machine Learning* and *Multi Class Machine Learning* methods employ the described optimization procedure. It is an important part of these methods and it results in large increase of quality of the annotations. Lots of automatic image annotation methods presented by various researchers contain specific routines for word frequency corrections. These routines usually assume some word frequency probability distributions. Parameters of these distributions are estimated during training phases of these methods. Such solution makes them method specific, it means that migration from one method to another is often conceptually and technically difficult, or even not possible. The proposed optimization method (RWCO) does not require any embedding, it is a wrapper like procedure, which can be applied to various image annotation methods.

4.1.2 Training Set Oversampling

During the research and development of both proposed methods, we have observed that training set oversampling can noticeably increase the quality of results. Various approaches to oversampling have been examined and, after this study, we have selected the approach presented below. In training set construction procedure we use an oversampling method similar to the one described in Machine Learning literature [46, 47].

The oversampling method is based on *entropy* calculation. The total value of entropy of a given training set is defined as follows:

$$H = \sum_{i=1}^{z} p(i) \log \frac{1}{p(i)} = -\sum_{i=1}^{z} p(i) \log p(i), \tag{7}$$

where: H – total training set entropy, z – number of possible random events, $p(i)$ – probability of i-th random event.

Let us assume that there are only two possible random events ($z = 2$):

 – description of a given image contains word w_y,
 – description of a given image does not contain word w_y.

Entropy is maximal when both of these random events have identical probabilities. In the such case, the entropy is equal to:

$$H = -\frac{1}{2} \log_2 \frac{1}{2} + -\frac{1}{2} \log_2 \frac{1}{2} = 1. \tag{8}$$

According to entropy calculation and the oversampling, training samples in the training set have to be duplicated. Parameters of the duplication routine are estimated on the basis of training set, using the following concept. In both *BML* and *MCML* methods we use the same parameters defined as follows. $p_{w_y}^+$ represents the total number of feature vectors for all images in training set I annotated by word w_y:

$$p_{w_y}^+ = \sum_{(I_x, W_{I_x}) \in I} |F_{I_x}| \left| W_{I_x} \cap \{w_y\} \right|, \tag{9}$$

where: F_{I_x} – feature vector set for image I_x.

$p_{w_y}^-$ represents the total number of feature vectors for all images in training set I *not* annotated by word w_y:

$$p_{w_y}^- = \sum_{(I_x, W_{I_x}) \in I} |F_{I_x}| \left(1 - \left| W_{I_x} \cap \{w_y\} \right| \right). \tag{10}$$

We denote by $t_{w_y}^+$ the oversampling factor (how many times a training example has to be duplicated) for all feature vectors annotated by word w_y:

$$
t_{w_y}^+ = \begin{cases} \left\lfloor \max\left(\frac{p_{w_y}^+ + p_{w_y}^-}{2p_{w_y}^+}, 1\right) + \frac{1}{2} \right\rfloor & p_{w_y}^+ > 0 \\ 0 & p_{w_y}^+ = 0 \end{cases} \tag{11}
$$

Analogously we denotes by $t_{w_y}^-$ the oversampling factor (how many times a training example has to be duplicated) for all feature vectors *not* annotated by word w_y:

$$
t_{w_y}^- = \begin{cases} \left\lfloor \max\left(\frac{p_{w_y}^+ + p_{w_y}^-}{2p_{w_y}^-}, 1\right) + \frac{1}{2} \right\rfloor & p_{w_y}^- > 0 \\ 0 & p_{w_y}^- = 0 \end{cases} \tag{12}
$$

Presented values are calculated for each word from dictionary W. After parameter values calculation is completed, a training set construction method is executed. Details of the training set construction are presented in a form of pseudo-code in Algorithms 1 and 2.

4.2   Binary Machine Learning

Binary Machine Learning (BML) automatic annotation method follows all concepts described in the previous section. It consists of a set of n binary *C4.5* decision trees and each tree is responsible for classification of a single word (n denotes a number of words in dictionary W). The training phase of the annotator requires training of all classifiers and the processing phase requires running all of them on the given feature vectors. The key idea of *Binary Machine Learning* is to classify only single word per each classifier. This concept of classification in automatic image annotation is known as *supervised one-versus-all* annotation [31].

Construction of the BML *C4.5* training set is performed according to Algorithm 1. The presented approach builds a training set for a single tree corresponding to word w_y. This tree produces the answer YES for presence or NO for absence of this word in the description of the analyzed image segment.

Processing phase applies the forest of constructed decision trees. Decision trees classify input feature vectors, their results are averaged. Formally, automatic image annotator *BML* is defined as follows:

$$
A_{BML}(I,J,W) = \begin{bmatrix} AS_{BML}(I,J,w_1) \\ AS_{BML}(I,J,w_2) \\ \vdots \\ AS_{BML}(I,J,w_n) \end{bmatrix}, \tag{13}
$$

where: $AS_{BML}(I,J,w_y)$ – automatic image annotator for a single word w_y.

$$
AS_{BML}(I,J,w_y) = \frac{1}{|F_J|} \sum_{F_J^r \in F_J} C45_{BML}^{w_y}(F_J^r), \tag{14}
$$

Algorithm 1. BML build decision tree training set

Require: Word w_y
Require: Annotated training set I
Ensure: Decision tree training set $T_{w_y}^{BML}$

1: Calculate $t_{w_y}^+$ on the basis of eq. 11
2: Calculate $t_{w_y}^-$ on the basis of eq. 12
3: $T_{w_y}^{BML} \Leftarrow \emptyset$
4: **for all** $(I_x, W_{I_x}) \in I$ **do**
5:    $F_{I_x} \Leftarrow$ feature vector set of I_x
6:    **for all** $F_{I_x}^q \in F_{I_x}$ **do**
7:      **if** $w_y \in W_{I_x}$ **then**
8:        **for** $t \Leftarrow 1$ **to** $t_{w_y}^+$ **do**
9:          $T_{w_y}^{BML} \Leftarrow T_{w_y}^{BML} \oplus (F_{I_x}^q, true)$ $\{\oplus - \text{add to the end}\}$
10:       **end for**
11:      **else**
12:        **for** $t \Leftarrow 1$ **to** $t_{w_y}^-$ **do**
13:          $T_{w_y}^{BML} \Leftarrow T_{w_y}^{BML} \oplus (F_{I_x}^q, false)$
14:       **end for**
15:      **end if**
16:    **end for**
17: **end for**
18: **return** $T_{w_y}^{BML}$

where: F_J – feature vector set extracted for image J; $C45_{BML}^{w_y}(F_J^r)$ – classification procedure for a single feature vector F_J^r, $r = 1, ..., q$, q – a number of segments.

$$C45_{BML}^{w_y}(F_J^r) = \begin{cases} 1 & if \quad F_J^r \quad belongs \quad to \quad class \quad w_y \\ 0 & otherwise \end{cases}. \qquad (15)$$

The answer of each decision tree is binary. Output value 1 represents the case in which decision tree classifies F_J^r as belonging to class w_y, value 0 represents the case it does not belong to the class.

4.3   Multi Class Machine Learning

Multi Class Machine Learning (MCML) method is an extension of *Binary Machine Learning* (BML) approach. The key idea of *MCML* method is to eliminate a serious weakness of *BML*. As it was mentioned earlier, automatic image annotation should effectively work on training sets which are weakly labeled. Weakly labeled images usually do not contain all words in their descriptions, which are present as objects in the image. This has a very important consequence: *a missing word in the description does not necessary mean that the word related object is not present in the image*, what results in large difficulties of forming *negative* training examples. The concept of problems with negative training examples is presented in [31] and it is suggested not to use negative training examples.

We propose to replace all negative examples from the training set with positive examples representing other words from the annotations. Such idea leads to reformulation of binary classification which now becomes *multi-class* classification, with as many classes as many words there are in the dictionary. Instead of being able to answer *YES* or *NO*, each decision tree, even if specialized for classification of a single word, can answer with any word from the dictionary. This concept is the key idea of *Multi Class Machine Learning* method and the essential difference to *Binary Machine Learning* method.

Algorithm 2. MCML build decision tree training set

Require: Word w_y
Require: Annotated training set I
Ensure: Decision tree training set $T_{w_y}^{MCML}$
  1: Calculate $t_{w_y}^+$ on the basis of eq. 11
  2: Calculate $t_{w_y}^-$ on the basis of eq. 12
  3: $T_{w_y}^{MCML} \Leftarrow \emptyset$
  4: **for all** $(I_x, W_{I_x}) \in I$ **do**
  5:     $F_{I_x} \Leftarrow$ feature vector set of I_x
  6:     **for all** $F_{I_x}^q \in F_{I_x}$ **do**
  7:         **if** $w_y \in W_{I_x}$ **then**
  8:             **for** $t \Leftarrow 1$ **to** $t_{w_y}^+$ **do**
  9:                 $T_{w_y}^{MCML} \Leftarrow T_{w_y}^{MCML} \oplus (F_{I_x}^q, w_y)$ $\{\oplus - \text{add to the end}\}$
 10:             **end for**
 11:         **else**
 12:             $w_z \Leftarrow$ select a word with index $(x+q+y) \mod |W_{I_x}|$ from W_{I_x}
 13:             **for** $t \Leftarrow 1$ **to** $t_{w_y}^-$ **do**
 14:                 $T_{w_y}^{MCML} \Leftarrow T_{w_y}^{MCML} \oplus (F_{I_x}^q, w_z)$
 15:             **end for**
 16:         **end if**
 17:     **end for**
 18: **end for**
 19: **return** $T_{w_y}^{MCML}$

A pseudo-code of the training phase is presented as Algorithm 2. There is large similarity to the one used in BML method, however there is an important difference. Generated training set has multi-class decisions instead of binary ones. To handle multi-class decisions, standard oversampling method has to be modified. The described oversampling method is fully deterministic and is based on *modulo* operator. A randomization element is removed from a practical reason. Training of decision trees for a single annotated dataset takes up to a week. Training of annotators requires training of many decision trees. Generated trees are usually large. Volume of these decision trees may reach several gigabytes of disk storage, using a compact representation. Usage of random version of the oversampling method and repeating the experiments to get proper average and standard deviation values is impractical on todays computers. The key aspect of automatic image annotation is to solve practical problems and thus the proposed approach may be seen as the proper one.

Processing phase of MCML annotation method uses *C4.5* decision trees constructed in the training phase. Automatic image annotator MCML is denoted by $A_{MCML}(I,J,W)$ and defined as follows:

$$A_{MCML}(I,J,W) = \begin{bmatrix} AS_{MCML}(I,J,w_1) \\ AS_{MCML}(I,J,w_2) \\ \vdots \\ AS_{MCML}(I,J,w_n) \end{bmatrix}, \tag{16}$$

where: $AS_{MCML}(I,J,w_y)$ – automatic image annotator for a single word w_y.

$$AS_{MCML}(I,J,w_y) = \frac{1}{|F_J|} \sum_{F_J^r \in F_J} AS_{MCML}^F(I,F_J^r,w_y), \tag{17}$$

where: F_J – feature vector set extracted from image J, $AS_{MCML}^F(I,F_J^r,w_y)$ – automatic image annotator for a single word and single feature vector r, extracted from image J.

$$AS_{MCML}^F(I,F_J^q,w_y) = \frac{1}{n} \sum_{w \in W} |\{C45_{MCML}^w(F_J^r)\} \cap \{w_y\}|, \tag{18}$$

where: $C45_{w_y}(F_J^r)$ – classification procedure of a single feature vector F_J^r done by decision tree trained for word w_y.

$$C45_{MCML}^{w_y}(F_J^r) = w_z \in W, \tag{19}$$

where: w_z – answer of the decision tree, it is a word from the dictionary W (we use multi-class classification in this process).

4.4   Computational Complexity Analysis

Usually automatic image annotation methods are used to process large collections of images, therefore processing of a single image should be as fast as possible. The most important factors in complexity analysis are: number of images in the training set – m, size of the dictionary – n. Both, number of words in the dictionary and number of examples in the training set are growing constantly.

We postulate that every annotation method which describes a single image in time smaller than $O(mn)$ can be classified as a fast image annotation method. Such complexity limitation is very important from a practical point of view. Many researchers are experimenting with datasets containing thousands of images.

Computational complexity analysis for annotation of a single image is presented in this section. Calculations are valid for both described annotation methods – *Binary Machine Learning* and *Multi Class Machine Learning*.

Computational complexity of a single classification process done by the decision tree is:

$$T_{tree}(d_{C4.5}) = O(d_{C4.5}), \tag{20}$$

where: $d_{C4.5}$ denotes the tree depth.

Each tree depth is much smaller than the total number of images in the training set multiplied by the average number of segments per image:

$$d_{C4.5} \ll m f_{avg},\qquad(21)$$

where: f_{avg} – average number of segments per image in the training set.

Aggregation of all classification results requires looping through all words in the dictionary and all segments of the image:

$$T_{agr}(n, f_{avg}) = O(n f_{avg}).\qquad(22)$$

Optimization of annotator results, done by the *Resulted Word Counts Optimizer* requires multiplying all aggregated results by the optimization coefficients:

$$T_{norm}(n) = O(n).\qquad(23)$$

Total complexity of a single annotation process of both MCML and BML methods is equal to:

$$T_{total}(n, f_{avg}, d_{C4.5}) = T_{tree}(d_{C4.5})T_{agr}(n, f_{avg}) + T_{norm}(n) =\qquad(24)$$
$$= O(d_{C4.5})O(n f_{avg}) + O(n) = O(d_{C4.5} n f_{avg}).$$

Let us now compare the calculated computational complexity with the assumed boundary computational complexity:

$$O(d_{C4.5} n f_{avg}) < O(m f_{avg} n f_{avg}) = O(mn).\qquad(25)$$

Computational complexities of both proposed methods are lower than the assumed boundary computational complexity, then both methods can be recognized as fast automatic image annotation methods.

5   Experimental Study of the Proposed Auto-annotation Methods

This section presents the experiments which the authors performed to test the quality of the proposed methods. Three annotators are compared – two authors' methods *Binary Machine Learning (BML)*, *Multi Class Machine Learning (MCML)* and *Continuous Relevance Model (CRM)* which is the reference method. In all experiments authors' implementations were used. Achieved results for the reference method, as well as for other methods described in literature, are almost identical (differences oscillate around 1%), what confirms that both the implementation and testing environment may be seen as identical to the ones presented in literature. Additionally, a list of results taken from the available research literature is presented with reference to the proposed methods.

5.1   The Aim of Experiments

The main goal of the experiments is to verify the quality of two proposed methods *Binary Machine Learning* and *Multi Class Machine Learning* against the reference method – *Continuous Relevance Model*. *Continuous Relevance Model* is chosen as the reference method because it is one of the best image annotation methods, very often cited by many researchers. Selection of this method should make quality comparison easier for the reader.

All experiments are performed on four annotated image sets: *MGV 2006, ICPR 2004, JMLR 2003* and *ECCV 2002*. They are large, automatic image annotation benchmark datasets. Datasets are both weakly and strongly annotated, they contain various numbers of images and various sizes of dictionaries. Feature vector generation and image segmentation are identical for *MGV 2006* and *ICPR 2004* datasets [18]. *ECCV 2002* and *JMLR 2004* use different set of feature vectors and *Normalized Cuts* method for image segmentation. Main characteristics of all tested datasets are presented in Table 1.

Table 1. The main characteristics of used datasets

Characteristic \ dataset	MGV	ICPR	ECCV	JMLR
Number of images	751	1109	5000	∼16000
Number of words in dictionary	74	407	374	∼160
Number of words used to annotate	74	407	260	∼160
Weakly annotated dataset	no	yes	yes	yes
Average annotation length	5.00	5.79	3.52	3.11
Maximum annotation length	9	23	5	5
Minimum annotation length	2	1	1	1
Length standard deviation	1.27	3.48	0.67	0.85
Segmentation method	rect. grid	rect. grid	N-Cuts	N-Cuts
Number of segments	25	25	3 – 10	3 – 10
Feature vector length	28	28	36	46
Validation dataset splits	3/4 - 1/4	3/4 - 1/4	1 split (4500 – 500)	10 splits

5.2 Annotation Quality

To evaluate our methods, we should be able to measure the quality of image annotation. Quality measures used for automatic image annotation reflect various important aspects of achieved results and have to be shown and discussed together. They are: *precision*, *recall*, and *accuracy*. *Precision* and *recall* quality measures are well known in cross-lingual machine translation applications. Accuracy is much more frequently used for classification problems. *Precision* and *recall* reflect the quality of annotation for single words, *accuracy* – for single images. Later on, to represent the whole dataset, their values are averaged. Let us recall the definitions of above measures.

Precision is defined by equation 26 [1, 4, 15, 20, 24, 30, 31, 42, 48, 49, 50, 51].

$$prec_{w_y} = \frac{c_{w_y}}{r_{w_y}}, \tag{26}$$

where: c_{w_y} – number of correctly annotated images by word w_y, r_{w_y} – number of images which are annotated by word w_y in the automatic annotation process.

Recall quality measure is defined by equation 27 [1, 4, 15, 20, 24, 30, 31, 42, 48, 49, 50, 51].

$$rec_{w_y} = \frac{c_{w_y}}{e_{w_y}}, \tag{27}$$

where: c_{w_y} – number of correctly annotated images by word w_y, e_{w_y} – number of images, which should be annotated by word w_y (ground truth).

Accuracy quality measure is defined by equation 28 [52].

$$acc_J = \frac{c_J}{l_J}, \tag{28}$$

where: c_J – number of correctly generated words for image J, l_J – length of image J annotation.

As mentioned before, all quality measures are averaged for the whole dataset. *Recall* and *precision* measures are averaged for all words in the dictionary or for a subset of these words. *Accuracy* is averaged for all images in the test set. All achieved results, shown in the next section, are presented using averaged values of these quality measures.

5.3   The Results

All experiments are grouped according to used datasets. For all datasets, the mentioned quality measures are calculated. Additionally, quality results reported in the literature are presented in Table 2. It makes easier the comparison the quality of the proposed methods with the results of other researchers.

Figure 4 presents the results achieved for *MGV* dataset. The results are very similar for all three methods, however both *Continuous Relevance Model* (CRM) and *Multi Class Machine Learning* (MCML) are slightly better than *Binary Machine Learning* (BML) method. The proposed optimization procedure *Resulted Word Counts Optimization* (RWCO) is used with two authors methods: BML and MCML. CRM has built-in specific optimization procedure, therefore using RWCO is not necessary. Large results similarity for various annotation methods indicates that, probably, the process of automatic image annotation has reached its maximum potentiality for this dataset. It seems that the results for this dataset can be further improved by introducing a new feature set or by using another segmentation routine.

Figure 5 presents the results achieved for *ICPR* dataset. *Binary Machine Learning* method achieves similar quality as *CRM* method. Quality of *Multi Class Machine Learning* annotation results are better than the other two methods.

Figure 6 presents the results achieved for *JMLR* dataset. The *Multi Class Machine Learning* annotation method outperforms all other tested annotation methods in all quality measures. The differences are large, because the dataset is weakly annotated. Automatic image annotator based on binary decision trees has the worst results, *CRM* is placed in the middle. One may note, that the conceptual change between binary decision trees and multi–class decision trees yielded in such large quality increase. Accuracy is increased from 28% to 41%, precision from 20% to 32% and recall from 20% to 33%.

Figure 7 presents the results achieved for *ECCV* dataset. The experiment with this dataset shows the highest differences between the quality of the used annotation methods. It should be stressed that it is definitely the hardest dataset to annotate. Results

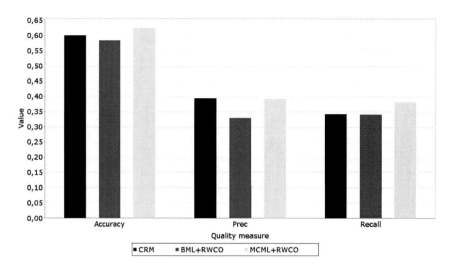

Fig. 4. The results quality for *MGV* dataset and three annotators: CRM, BML and MCML

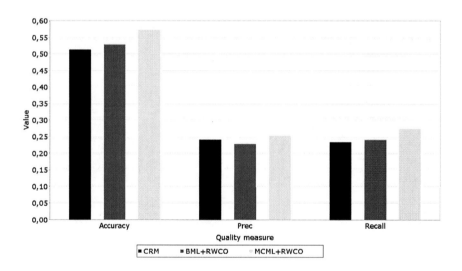

Fig. 5. The results quality for *ICPR* dataset and the three tested auto-annotators

achieved by the proposed *Multi Class Machine Learning* method combined with *Resulted Word Counts Optimizer* easily challenge the selected reference method. In comparison to *CRM*, achieved accuracy is increased from 39% to 48%, precision from 20% to 30%, and recall from 16% to 20%.

Table 2 presents values of *precision* and *recall* for various automatic image annotation methods, reported in the literature. All presented methods are tested on *COREL* image database. The *ECCV 2002* dataset is built using *COREL* images. However, not all researchers have used the same image segmentation methods and feature vectors. It

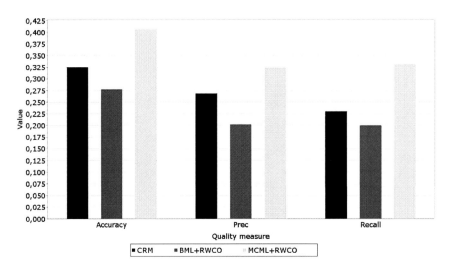

Fig. 6. Results quality for *JMLR* dataset

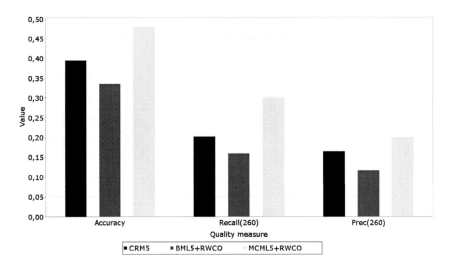

Fig. 7. Results quality for *ECCV* dataset

is worth mentioning that both, used segmentation method and selected set of features, strongly influence the quality of an annotation process. In general, using the original *ECCV 2002* dataset, the results are the worst. All modifications of the dataset preparation make the comparison very hard, or sometimes even impossible. These results are also presented in the table, however marked by an adequate comment on feature vector construction and image segmentation method. Fortunately, most researchers use the original *ECCV 2002* dataset, which makes the comparison reliable. Together with the method name, a literature reference is done where the cited results may be found.

Table 2. Annotation quality results of the *COREL* image dataset, taken from the available research literature

Automatic image annotation method	|precision|	recall
Dataset *ECCV 2002* based on *COREL* images		
Segmentation: *Normalized Cuts*, feature vector length: 36		
Mori Coocc. [31]	0.03	0.02
MT [4]	0.06	0.04
Binary Unigram [53]	0.08	0.11
Binary Unigram+Binary Bigram [53]	0.09	0.12
CMRM [1]	0.10	0.09
BA [54]	0.12	0.21
BA-T [54]	0.13	0.21
BA-TW [54]	0.14	0.22
SvdCos [55]	0.15	0.15
CRM [1]	0.16	0.19
EMD [49]	0.16	0.19
InfNet [56]	0.17	0.24
MCML+RWCO (presented method)	0.20	0.30
Images identical with *ECCV 2002*		
Segmentation: rect. grid, features as in *ECCV 2002*		
CRM-Rect [31]	0.22	0.23
MBRM [4]	0.24	0.25
Images identical with *ECCV 2002*		
Segmentation: rect. grid, feature values generated using *DCT*		
CRM-Rect-DCT [31]	0.21	0.22
Mix-Hier [31]	0.23	0.29
SML [33]	0.23	0.29
Images identical with *ECCV 2002*		
Segmentation: N.A., features values generated using *MPEG–7 CSD*		
CSD-Prop [55]	0.20	0.27
CSD-SVM [9]	0.25	0.28

Results achieved by the *Multi Class Machine Learning* method place it in the group of the best automatic image annotation methods. The method achieves the best results for all tested datasets, in all quality measures except of just one. *Precision, recall* and *accuracy* values for MCML are significantly higher than for the two other tested methods. Only *precision* value achieved for *MGV 2006* dataset is almost identical to the one calculated for *Continuous Relevance Model*. The concept of negative training examples removal helps to effectively process weakly annotated datasets. It is especially visible on the results achieved for the *ECCV 2002* dataset (Figure 7), which is one of the most difficult datasets for auto-annotation task due to existence of weak annotations.

Binary Machine Learning method is able to achieve satisfying results on easier datasets: *ICPR 2004* and *MGV 2006*. Only a small part of images is weakly annotated in these two datasets. The method performs only slightly worse than both *Multi Class Machine Learning* and *Continuous Relevance Model*. However on the most difficult dataset – *ECCV 2002*, the difference is much higher. Due to negative training examples usage, the method is not able to process weakly annotated images with satisfying quality.

It is important mentioning that one of the key components in both proposed annotation methods is the optimization routine, briefly described in section 4.1.1. *Resulted Word Counts Optimization* allows to construct good automatic image annotation methods relatively easy. It removes the problem of word frequencies correction from the designer of automatic image annotators, allowing to focus mostly on classification quality.

6 Concluding Remarks

Machine learning methods play an important role in automatic image annotation. Clustering and classification techniques are used in many approaches, proposed by various researchers. Recent researches indicate that classification based methods achieve better and better results, making it a potentially interesting direction for further work.

We have presented two annotation methods based on *C4.5* decision tree classifier. One of these methods, *Multi Class Machine Learning* can successfully compete with other state-of-the-art annotation methods. The concept of multi-class classification achieved by negative examples removal, allowed to increase results quality. *Multi Class Machine Learning* combined together with *Resulted Word Counts Optimization* proved to be able to annotate large and difficult datasets both in terms of high quality and low computational complexity. Experiments performed on several datasets show that the presented multi-class, decision tree based annotation concept is very promising.

Computational complexity analysis allowed to rank both presented solutions as *fast automatic image annotation* methods. Linear dependency on the training set size is avoided due to decision tree usage. *C4.5* decision trees introduce logarithmic computational complexity on the training set size. Such low computational complexity makes presented methods an interesting candidate for training and processing using very large image datasets.

Further research will be mostly focused on examination of different approaches to decision tree construction. Quality of annotation will be addressed especially in terms of training set construction, to include even more the features of weakly described image sets. Detection and retraining of decision trees with low classification quality will be introduced. Additionally, decision tree construction time has to be reduced to allow both efficient training and more effective research.

Acknowledgements. This work is partially financed from the Ministry of Science and Higher Education resources in 2007 - 2009 years as a research project N518 020 32/1454.

References

1. Lavrenko, V., Manmatha, R., Jeon, J.: A model for learning the semantics of pictures. In: Proceedings of NIPS. MIT Press, Cambridge (2003)
2. Lavrenko, V., Feng, S.L., Manmatha, R.: Statistical models for automatic video annotation and retrieval. In: Proceedings of IEEE International Conference on Acoustics, Speech, and Signal Processing (ICASSP 2004), vol. 3, pp. 1044–1047 (2004)

3. Jeon, J., Manmatha, R.: Automatic image annotation of news images with large vocabularies and low quality training data. MM 368, University of Massachusetts (2004)

4. Feng, S.L., Manmatha, R., Lavrenko, V.: Multiple bernoulli relevance models for image and video annotation. In: Proceedings of the IEEE Computer Society Conference on Computer Vision and Pattern Recognition (CVPR 2004), vol. 2, pp. 1002–1009 (2004)

5. Laaksonen, J., Koskela, M., Oja, E.: Picsom-self-organizing image retrieval with mpeg-7 content descriptors. IEEE Transactions on Neural Networks 13(4), 841–853 (2002)

6. Viitaniemi, V., Laaksonen, J.: Keyword-detection approach to automatic image annotation. In: Proceedings of 2nd European Workshop on the Integration of Knowledge, Semantic and Digital Media Technologies (EWIMT 2005), pp. 15–22 (2005)

7. Cusano, C., Ciocca, G., Schettini, R.: Image annotation using SVM. In: Proceedings of SPIE, Internet Imaging V, vol. 5304, pp. 330–338 (2003)

8. Chang, E., Goh, K., Sychay, G., Wu, G.: Cbsa: content-based soft annotation for multimodal image retrieval using bayes point machines. IEEE Transactions on Circuits and Systems for Video Technology 13(1), 26–38 (2003)

9. Tang, J., Lewis, P.H.: A study of quality issues for image auto-annotation with the corel data-set. IEEE Transactions on Circuits and Systems for Video Technology 17(3), 384–389 (2007)

10. Kwaśnicka, H., Paradowski, M.: Multiple class machine learning approach for image auto-annotation problem. In: Proceedings of The Sixth International Conference on Intelligent Systems Design and Applications (ISDA 2006), vol. 2, pp. 347–352 (2006)

11. Goh, K.-S., Chang, E.Y., Li, B.: Transactions on Knowledge and Data Engineering 17(10), 1333–1346 (2005)

12. Yasuhide, M., Hironobu, T., Ryuichi, O.: Image-to-word transformation based on dividing and vector quantizing images with words. In: Proceedings of the International Workshop on Multimedia Intelligent Storage and Retrieval Management (1999)

13. Jeon, J., Lavrenko, V., Manmatha, R.: Automatic image annotation and retrieval using cross-media relevance models. In: Proceedings of the ACM SIGIR Conference on Research and Development in Information Retrieval, pp. 119–126 (2003)

14. Monay, F., Gatica-Perez, D.: On image auto-annotation with latent space models. In: Proceedings of the eleventh ACM international conference on Multimedia, pp. 275–278 (2003)

15. Pan, J.-Y., Yang, H.-J., Duygulu, P., Faloutsos, C.: Automatic image captioning. In: Proceedings of the 2004 IEEE International Conference on Multimedia and Expo. (ICME 2004), vol. 3, pp. 1987–1990 (2004)

16. Liu, W., Tang, X.: Learning an image-word embedding for image auto-annotation on the nonlinear latent space. In: Proceedings of the 13th annual ACM international conference on Multimedia, pp. 451–454 (2005)

17. Glotin, H., Tollari, S.: Fast image auto-annotation with visual vector approximation clusters. In: Proceedings of Fourth International Workshop on Content-Based Multimedia Indexing, CBMI 2005 (2005)

18. Kwaśnicka, H., Paradowski, M.: Fast image auto-annotation with discretized feature distance measures. Machine Graphics and Vision 15(2), 123–140 (2006)

19. Hollink, L., Nguyen, G., Schreiber, G., Wielemaker, J., Wielinga, B., Worring, M.: Adding spatial semantics to image annotations. In: Proceedings of the 4th International Workshop on Knowledge Markup and Semantic Annotation at ISWC 2004 (2004)

20. Bashir, A., Khan, L.: A framework for image annotation using semantic web. In: Proceedings of ACM SIGKDD First International Workshop on Mining for and from the Semantic Web (MSW 2004) (2004)

21. Srikanth, M., Varner, J., Bowden, M., Moldovan, D.: Exploiting ontologies for automatic image annotation. In: Proceedings of the 28th annual international ACM SIGIR conference on Research and development in information retrieval, pp. 552–558 (2005)

22. Hollink, L., Little, S., Hunter, J.: Evaluating the application of semantic inferencing rules to image annotation. In: Proceedings of the 3rd international conference on Knowledge capture, pp. 91–98 (2005)
23. Hare, J.S., Sinclair, P.A.S., Lewis, P.H., Martinez, K., Enser, P.G.B., Sandom, C.J.: Bridging the semantic gap in multimedia information retrieval top-down and bottom-up approaches. In: Proceedings of Mastering the Gap: From Information Extraction to Semantic Representation / 3rd European Semantic Web Conference (2006)
24. Duygulu, P., Barnard, K., de Freitas, J.F.G., Forsyth, D.: Object recognition as machine translation: Learning a lexicon for a fixed image vocabulary. In: Heyden, A., Sparr, G., Nielsen, M., Johansen, P. (eds.) ECCV 2002. LNCS, vol. 2353, pp. 97–112. Springer, Heidelberg (2002)
25. Rangayyan, R.M.: Biomedical Image Analysis. Biomedical Engineering Series. CRC Press, Boca Raton (2004)
26. Kurzyñski, M.: Multistage diagnosis of myocardial infraction using a fuzzy relation. In: Proceesings of the 7th International Conference on Artificial Intelligence and Soft Computing, pp. 1014–1019 (2004)
27. Abe, S.: Support Vector Machines for Pattern Classification. Springer-Verlag London Limited, Heidelberg (2005)
28. Shah, B., Benton, R., Wu, Z., Raghavan, V.: Automatic and Semi-Automatic Techniques for Image Annotation, pp. 112–134 (2007)
29. Barnard, K., Duygulu, P., de Freitas, N., Forsyth, D.: Object recognition as machine translation - part 2: Exploiting image database clustering models (2002)
30. Barnard, K., Duygulu, P., Forsyth, D., de Freitas, N., Blei, D.M., Jordan, M.I.: Matching words and pictures. Journal of Machine Learning Research 3, 1107–1135 (2003)
31. Carneiro, G., Vasconcelos, N.: Formulating semantic image annotation as a supervised learning problem. In: Proceedings of the 2005 IEEE Computer Society Conference on Computer Vision and Pattern Recognition (CVPR 2005), vol. 2, pp. 163–168 (2005)
32. Chan, A.B., Moreno, P.J., Vasconcelos, N.: Using statistics to search and annotate pictures: an evaluation of semantic image annotation and retrieval on large databases. In: Joint Statistical Meetings (2006)
33. Carneiro, G., Chan, A.B., Moreno, P.J., Vasconcelos, N.: Supervised learning of semantic classes for image annotation and retrieval. IEEE Transactions on Pattern Analysis and Machine Intelligence 29(3), 394–410 (2007)
34. Buturovic, A.: Mpeg-7 color structure descriptor for visual information retrieval project vizir. In: Proceedings of the International Conference on Image Processing, vol. 1, pp. 670–673 (2001)
35. Tsai, C.-F., Hung, C.: Automatically annotating images with keywords a review of image annotation systems. Recent Patents on Computer Science (1), 55 (2008)
36. Shi, J., Malik, J.: Normalized cuts and image segmentation. IEEE Transactions on Pattern Analysis and Machine Intelligence 22(8), 888–905 (2000)
37. Cheng, P.-J., Chien, L.-F.: Effective image annotation for search using multi-level semantics. International Journal on Digital Libraries 4(4), 230 (2004)
38. Hare, J.S., Lewis, P.H.: Saliency-based models of image content and their application to auto-annotation by semantic propagation. In: Multimedia and the Semantic Web / European Semantic Web Conference (2005)
39. Vailaya, A., Jain, A., Zhang, H.J.: On image classification: city vs. landscape. Pattern Recognition 31(12), 1921–1935 (1998)
40. Westerveld, T., de Vries, A.P., van Ballegooij, A., de Jong, F., Hiemstra, D.: A probabilistic multimedia retrieval model and its evaluation. EURASIP Journal on Applied Signal Processing 2, 186–198 (2003)

41. Xu, F., Zhang, Y.-J.: A Novel Framework for Image Categorization and Automatic Annotation, pp. 90–111 (2007)
42. Kang, F., Jin, R., Chai, J.Y.: Regularizing translation models for better automatic image annotation. In: Proceedings of the 2004 ACM CIKM International Conference on Information and Knowledge Management, pp. 350–359 (2004)
43. Duygulu, P., Hauptmann, A.: What's news, what's not? associating news videos with words. In: Enser, P.G.B., Kompatsiaris, Y., O'Connor, N.E., Smeaton, A., Smeulders, A.W.M. (eds.) CIVR 2004. LNCS, vol. 3115, pp. 132–140. Springer, Heidelberg (2004)
44. Jin, R., Chai, J.Y., Si, L.: Effective automatic image annotation via a coherent language model and active learning. In: Proceedings of the 12th annual ACM international conference on Multimedia, pp. 892–899 (2004)
45. Kwaśnicka, H., Paradowski, M.: Resulted word counts optimization a new approach for better automatic image annotation. Pattern Recognition 41, 3562–3571 (2008)
46. Mitchell, T.M.: Machine Learning. McGraw-Hill Science Engineering (1997)
47. Japkowicz, N., Stephen, S.: The class imbalance problem: A systematic study. Intelligent Data Analysis 6(5), 429–449 (2002)
48. Wang, L., Khan, L.: Automatic image annotation and retrieval using weighted feature selection. Multimedia Tools and Applications 29(1), 55–71 (2006)
49. Yavlinsky, A., Schofield, E., Rüger, S.M.: Automated image annotation using global features and robust nonparametric density estimation. In: Leow, W.-K., Lew, M., Chua, T.-S., Ma, W.-Y., Chaisorn, L., Bakker, E.M. (eds.) CIVR 2005. LNCS, vol. 3568, pp. 507–517. Springer, Heidelberg (2005)
50. Smeulders, A.W.M., Gupta, A.: Content-based image retrieval at the end of the early years. Transactions on Pattern Analysis and Machine Intelligence 22(12), 1349–1380 (2000)
51. Kwaśnicka, H., Paradowski, M.: A discussion on evaluation of image auto-annotation methods. In: Proceedings of The Sixth International Conference on Intelligent Systems Design and Applications (ISDA 2006), vol. 2, pp. 353–358 (2006)
52. Pan, J.-Y., Yang, H.-J., Faloutsos, C., Duygulu, P.: Gcap: Graph-based automatic image captioning. In: Proceedings of the 4th International Workshop on Multimedia Data and Document Engineering (MDDE 2004), in conjunction with Computer Vision Pattern Recognition Conference, CVPR 2004 (2004)
53. Jeon, J., Manmatha, R.: Using maximum entropy for automatic image annotation. In: Enser, P.G.B., Kompatsiaris, Y., O'Connor, N.E., Smeaton, A., Smeulders, A.W.M. (eds.) CIVR 2004. LNCS, vol. 3115, pp. 24–32. Springer, Heidelberg (2004)
54. Ye, J., Zhou, X., Pei, J., Chen, L., Zhang, L.: A stratification-based approach to accurate and fast image annotation. In: Fan, W., Wu, Z., Yang, J. (eds.) WAIM 2005. LNCS, vol. 3739, pp. 284–296. Springer, Heidelberg (2005)
55. Tang, J., Lewis, P.H.: Image auto-annotation using 'easy' and 'more challenging' training sets. In: Proceedings of 7th International Workshop on Image Analysis for Multimedia Interactive Services, pp. 121–124 (2006)
56. Metzler, D., Manmatha, R.: An inference network approach to image retrieval. In: Enser, P.G.B., Kompatsiaris, Y., O'Connor, N.E., Smeaton, A., Smeulders, A.W.M. (eds.) CIVR 2004. LNCS, vol. 3115, pp. 42–50. Springer, Heidelberg (2004)

Part V
Neural Networks and Other Nature Inspired Approaches

Integrative Probabilistic Evolving Spiking Neural Networks Utilising Quantum Inspired Evolutionary Algorithm: A Computational Framework

Nikola Kasabov

Knowledge Engineering and Discovery Research Institute, KEDRI
Auckland University of Technology, Auckland, New Zealand
nkasabov@aut.ac.nz
http://www.kedri.info

Abstract. Integrative evolving connectionist systems (iECOS) integrate principles from different levels of information processing in the brain, including cognitive-, neuronal-, genetic- and quantum, in their dynamic interaction over time. The paper introduces a new framework of iECOS called integrative probabilistic evolving spiking neural networks (ipSNN) that incorporate probability learning parameters. ipSNN utilize a quantum inspired evolutionary optimization algorithm to optimize the probability parameters as these algorithms belong to the class of estimation of distribution algorithms (EDA). Both spikes and input features in ipESNN are represented as quantum bits being in a superposition of two states (1 and 0) defined by a probability density function. This representation allows for the state of an entire ipESNN at any time to be represented probabilistically in a quantum bit register and probabilistically optimised until convergence using quantum gate operators and a fitness function. The proposed ipESNN is a promising framework for both engineering applications and brain data modeling as it offers faster and more efficient feature selection and model optimization in a large dimensional space in addition to revealing new knowledge that is not possible to obtain using other models. Further development of ipESNN are the neuro-genetic models – ipESNG, that are introduced too, along with open research questions.

1 Introduction: Integrative Evolving Connectionist Systems (iECOS)

Many successful artificial neural network (ANN) models have been developed and applied to date [3,9,10,19,21,26,30,32], the most recent ones being Spiking Neural Networks (SNN) [14,15,23-25,33-37]. SNN have a great potential for brain data analysis [1,4,5,7,45] and data modelling [8,38,40,42,44,46-48]. However, despite some past work [2,14,35,36,41], current SNN models cannot model *probabilistically* data that are large, complex, noisy and dynamically changing in a way that reflects the stochastic nature of many real-world problems and brain processes [4,16,28,38].

J. Koronacki et al. (Eds.): Advances in Machine Learning II, SCI 263, pp. 415–425.
springerlink.com   © Springer-Verlag Berlin Heidelberg 2010

The brain is a dynamic information processing system that evolves its structure and functionality in time through information processing at different levels: cognitive-, ensemble of neurons-, single neuron-, molecular (genetic)-, quantum [26-29]. The information processes at each level are very complex and difficult to understand as they evolve in time, but much more difficult to understand is the interaction between them and how this interaction affects learning and cognition in the brain. These information processes are manifested at different time scales, e.g. cognitive processes happen in seconds, neuronal – in milliseconds, molecular- in minutes, and quantum - in nano-seconds. They also happen in different dimensional spaces, but they "work" together in the brain and contribute together to its intelligence.

Recently new information about neuronal- [1,25], genetic- [5,31,45] and quantum [6,22,43] levels of information processes in the brain has been obtained. For example, whether a neuron spikes or does not spike at any given time could depend not only on input signals but also on other factors such as gene and protein expression levels or physical properties [22,31,45]. The paradigm of *Integrative Evolving Connectionist Systems (iECOS)* [27-29, 39], previously proposed by the author, considers the *integrated optimisation* of all these factors represented as parameters and features (input variables) of an ANN model. This approach will be used here to develop a principally new framework - integrative probabilistic evolving SNN (ipESNN).

2   Evolving Spiking Neural Network Models

2.1 SNN – General Principles

SNN represent information as trains of spikes, rather than as single scalars, thus allowing the use of such features as frequency, phase, incremental accumulation of input signals, time of activation, etc. [3,5,14,23,47]. Neuronal dynamics of a spiking neuron are based on the increase in the inner potential of a neuron (post synaptic potential, PSP), after every input spike arrival. When a PSP reaches a certain threshold, the neuron emits a spike at its output (Fig. 1).

A wide range of models to simulate spiking neuronal activity have been proposed (for a review, see [25]). The Hodgkin- Huxley model is based on experimental study of the influence of conductance of three ion channels on the spike activity of the axon. The spike activity is modelled by an electric circuit, where the *chloride channel* is modelled with a parallel resistor-capacitor circuit, and the *sodium* and *potassium* channels are represented by voltage-dependent resistors.

In another model - the spike response model (SRM), a neuron i receives input spikes from pre-synaptic neurons j ∈ Γi, where Γi is a pool of all neurons pre-synaptic to neuron i. The state of the neuron i is described by the state variable $u_i(t)$ that can be interpreted as a total postsynaptic potential (PSP) at the membrane of soma – fig.1. When $u_i(t)$ reaches a firing threshold $\vartheta_i(t)$, neuron i fires, i.e. emits a spike. The value of the state variable $u_i(t)$ is the sum of all postsynaptic potentials, i.e.

$$u_i(t) = \sum_{j \in \Gamma_i} \sum_{t_j \in F_j} J_{ij} \varepsilon_{ij}(t - t_j - \Delta_{ij}^{ax})$$

(1)

where: the weight of synaptic connection from neuron j to neuron i is denoted by Jij, which takes positive (negative) values for excitatory (inhibitory) connections, respectively; depending on the sign of Jij, a pre-synaptic spike, generated at time t_j increases (or decreases) $u_i(t)$ by an amount of $\varepsilon_{ij}(t - t_j - \Delta_{ij}^{ax})$, where Δ_{ij}^{ax} is an axonal delay between neurons i and j which increases with Euclidean distance between neurons. The positive kernel $\varepsilon_{ij}(t - t_j - \Delta_{ij}^{ax}) = \varepsilon_{ij}(s)$ expresses an individual postsynaptic potential (PSP) evoked by a pre-synaptic neuron j on neuron i and can be expressed by a double exponential formula (2):

$$\varepsilon_{ij}^{synapse}(s) = A^{synapse}\left(\exp\left(-\frac{s}{\tau_{decay}^{synapse}}\right) - \exp\left(-\frac{s}{\tau_{rise}^{synapse}}\right)\right) \tag{2}$$

where: $\tau_{decay\,/\,rise}^{synapse}$ are time constants of the rise and fall of an individual PSP; A is the PSP's amplitude; the parameter *synapse* represents the type of the activity of the synapse from the neuron j to neuron i, that can be measured and modeled separately for a *fast_excitation, fast_inhibition, slow_excitation, and slow_inhibition.*

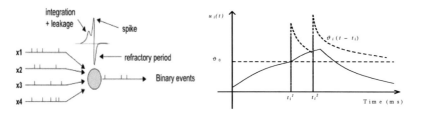

Fig. 1. A schematic representation of a spiking neuron model (from [5])

External inputs from the input layer of a SNN are added at each time step, thus incorporating the background noise and/or the background oscillations. Each external input has its own weight $J_{ik}^{ext\_input}$ and amount of signal $\varepsilon_k(t)$, such that:

$$u_i^{ext\_inpu}(t) = J_{ik}^{ext\_input}\varepsilon_{ik}(t) \tag{3}$$

2.2 Evolving Spiking Neural Networks (ESNN)

ESNN evolve/develop their structure and functionality in an incremental way from incoming data based on the following principles [26]:

(i) New spiking neurons are created to accommodate new data, e.g. new patterns belonging to a class or new output classes, such as faces in a face recognition system;

(ii) Spiking neurons are merged if they represent the same concept (class) and have similar connection weights (defined by a threshold of similarity).

In [40] an ESNN architecture is proposed where the change in a synaptic weight is achieved through a simple spike time dependent plasticity (STDP) learning rule:

$$\Delta w_{j,i} = \text{mod}^{\ order\ (j)} \tag{4}$$

where: $w_{j,i}$ is the weight between neuron j and neuron i, $mod \in (0,1)$ is the modulation factor, $order(j)$ is the order of arrival of a spike produced by neuron j to neuron i.

For each training sample, it is the *winner-takes-all* approach used, where only the neuron that has the highest *PSP* value has its weights updated. The postsynaptic threshold (PSP_{Th}) of a neuron is calculated as a proportion $c \in [0, 1]$ of the maximum postsynaptic potential, max(PSP), generated with the propagation of the training sample into the updated weights, such that:

$$PSP_{Th} = c \max(PSP) \tag{5}$$

Creating and merging neurons based on localised incoming information and on system's performance are main operations of the ESNN architecture that make it continuously *evolvable*. Successful applications of ESNN for taste recognition, face recognition and multimodal audio-visual information processing, have been previously reported [40,46,47].

2.3 Computational Neurogenetic Models as iECOS

A further extension of the SRM, that takes into account the ion channel activity (and thus brings the benefits of both Hodging-Huxley model and the SRM), that is also based on neurobiology, is called computational neuro-genetic model (CNGM) as proposed in [5,26]. Here different synaptic activities that are influencing the spiking activity of a neuron are represented as functions of different proteins (neuro-transmitters, neuro-receptors and ion channels) that affect the PSP value and the PSP threshold. Some proteins and genes known to be affecting the spiking activity of a neuron such as fast_excitation, fast_inhibition, slow_excitation, and slow_inhibition (see formula (2)) are summarized in Table 1. Besides the genes coding for the proteins mentioned above and directly affecting the spiking dynamics of a neuron, a CNGM may include other genes relevant to a problem in hand, e.g. modeling a brain

Table 1. Neuronal parameters and related proteins: PSP - postsynaptic potential, AMPAR - (amino- methylisoxazole- propionic acid) AMPA receptor, NMDR - (n-methyl-d-aspartate acid) NMDA receptor, GABRA - (gamma-aminobutyric acid) GABAA receptor, GABRB - GABAB receptor, SCN - sodium voltage-gated channel, KCN = kalium (potassium) voltage-gated channel, CLC = chloride channel (from [5])

Neuronal parameter	Protein
Fast excitation PSP	AMPAR
Slow excitation PSP	NMDAR
Fast inhibition PSP	GABRA
Slow inhibition PSP	GABRB
Firing threshold	SCN, KCN, CLC

function or a brain disease, for example: c-jun, mGLuR3, Jerky, BDNF, FGF-2, IGF-I, GALR1, NOS, S100beta [5,45]). CNGM are iECOS as they integrate principles from neuronal and molecular level of information processing in the brain.

However, it is also known that the spiking activity of the brain is stochastic [1,5,6,7]. And this is what is missing in the above SNN-, ESNN- and CNGM models that leave them not very suitable so far as large scale modeling techniques to model complex tasks. The problem is how to represent and process probabilities associated with spiking activity and how to build large ESNN probabilistic models.

3 Integrative Probabilistic Evolving SNN (ipESNN)

3.1 Biological Motivations

Some biological facts support the idea of ipESNN models [1,5,6,7]:

– For a neuron to spike or not to spike at a time t, is a "matter" of probability.
– Transmission of an electrical signal in a chemical synapse upon arrival of action potential into the terminal is probabilistic and depends on the probability of neurotransmitters to be released and ion channels to be open.
– Emission of a spike on the axon is also probabilistic.

The challenge is to develop a probabilistic neuronal model and to build ipESNN and ipESNG models for brain study and engineering applications. As the proposed below ipESNN model use quantum computation to deal with probabilities, we fist introduce some principles of quantum computing.

3.2 The Quantum Principle of *Superposition*

The smallest information unit in today's digital computers is one *bit*, existing as state '1' or '0' at any given time. The corresponding analogue in a quantum inspired representation is the quantum bit (*qbit*) [12,18,20]. Similar to classical bits a *qbit* may be in '1'or '0' states, but also in a *superposition* of both states. A qbit state $|\Psi\rangle$ can be described as:

$$|\Psi\rangle = \alpha|0\rangle + \beta|1\rangle \tag{6}$$

where α and β are complex numbers that are used to define the probability of which of the corresponding states is likely to appear when a *qbit* is read (measured, collapsed). $|\alpha|^2$ and $|\beta|^2$ give the probability of a qbit being found in state '0' or '1' respectively. Normalization of the states to unity guarantees:

$$|\alpha|^2 + |\beta|^2 = 1 \tag{7}$$

at any time. The *qbit* is not a single value entity, but is a function of parameters which values are complex numbers. In order to modify the probability amplitudes, *quantum gate operators* can be applied to the states of a *qbit or a qbit* vector. A quantum gate is represented by a square matrix, operating on the amplitudes $|\alpha|$ and $|\beta|$ in a Hilbert space, with the only condition that the operation is reversible. Such gates are: NOT-gate, rotation gate, Hadamard gate, and others [18, 20].

Another quantum principle is *entanglement* - two or more particles, regardless of their location, can be viewed as "correlated", undistinguishable, "synchronized", coherent. If one particle is "measured" and "collapsed", it causes for all other entangled particles to "collapse" too.

The main motivations for the development of the ipSNN that utilize quantum computation are: (1) The biological facts about stochastic behavior of spiking neurons and SNN; (2) The properties of a quantum representation of probabilities; (3) The recent manifestation that quantum inspired evolutionary algorithms (QiEA) are probability estimation of distribution algorithms (EDA) [11].

3.3 QiEA belong to the Class of EDA Algorithms

It was proven in [11] that QiEA, such as the vQEA, belong to the class of EDA. Using a quantum gate operator over consecutive iterations would lead to a change of the global state of a system described by a qbit chromosome(s) as shown in fig. 2. At the beginning (fig.2a) all states are equally probable and in the end – (fig2.c) the system converges to a local minimum according to a chosen fitness function.

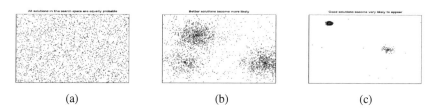

(a)   (b)   (c)

Fig. 2. An example of state convergence to local minima for a system described by a qbit register (chromosome) over 2 applications of a rotation quantum gate operator. The darker points represent states that have a higher probability of occurrence (the figure is produced by Stefan Schliebs).

3.4 The ipESNN Framework

The proposed here ipESNN framework is based on the following principles:

(a) A quantum probabilistic representation of a spike: A spike, at any time *t*, is both present (1) and not present (0), which is represented as a *qbit* defined by a probability density amplitude. When the spike is evaluated, it is either present or not present. To modify the probability amplitudes, a quantum gate operator is used, for example *the rotation gate:*

$$\begin{bmatrix} \alpha_i^j(t+1) \\ \beta_i^j(t+1) \end{bmatrix} = \begin{bmatrix} \cos(\Delta\theta) & -\sin(\Delta\theta) \\ \sin(\Delta\theta) & \cos(\Delta\theta) \end{bmatrix} \begin{bmatrix} \alpha_i^j(t) \\ \beta_i^j(t) \end{bmatrix} \qquad (8)$$

More precisely, a spike arriving at a moment t at each synapse Sij connecting a neuron Ni to a pre-synaptic neuron Nj, is represented as a qbit $Qij(t)$ with a probability to be in state "1" $\beta_{ij}(t)$ (probability for state "0" is $\alpha_{ij}(t)$). From the SNN architecture perspective this is equivalent to the existence (non-existence) of a connection Cij between neurons Nj and Ni.

(b) A quantum probabilistic model of a spiking neuron for ipSNN: A neuron *Ni* is represented as a qbit vector, representing all m synaptic connections to this neuron:

$$\begin{bmatrix} \alpha_1 & | & \alpha_2 & | & \cdots & | & \alpha_m \\ \beta_1 & | & \beta_2 & | & \cdots & | & \beta_m \end{bmatrix} \tag{9}$$

At a time t each synaptic qbit represents the probability for a spike to arrive at the neuron. The post-synaptic inputs to the neuron are collapsed into spikes (or no spikes) and the cumulative input $u_i(t)$ to the neuron Ni is calculated as per. eq.(1).

Based on the above principles two architectures of a feed-forward ipESNN and a recurrent ipESNN are developed as illustrated in fig.3(a) and 3(b) respectively. All input features (x1, x2,...,xn), the ESNN parameters (q1,q2,.., qs), the connections between the inputs and the neurons, including recurrent connections (C1, C1, ..., Ck) and the probability of the neurons to spike (p1,p2,...,pm) at time (t) are represented in an integrated *qbit* register that is operated upon as a whole [27,29] (fig.3c).

This framework goes beyond the traditional "wrapper" mode for feature selection and modelling [16]. The ipESNN single qbit vector (a chromosome) is optimised through vQEA, therefore an ipESNN is a probabilistic EDA model. It was demonstrated that the vQEA is efficient for integrated feature and SNN parameter optimisation in a large dimensional space and also useful for extracting unique information from the modelled data [11,39]. All probability amplitudes together define a probability density ψ of the state of the ipESNN in a Hilbert space. This density will change if a quantum gate operator is applied according to an objective criterion (fitness function). This representation can be used for both tracing the learning process in an ipESNN system or the reaction of the system to an input vector.

(c) ipSNN learning rules: As the ipESNN model is an ESNN, in addition to the ESNN learning rules (formulas 4,5) there are rules to change the probability density amplitudes of spiking activity of a neuron. The probability $\beta ij(t)$ of a spike to arrive from neuron Nj to neuron Ni (the connection between the two be present) will change according to STDP rule, which is implemented using the quantum rotation gate. In a more detailed model, $\beta ij(t)$ will depend on the strength and the frequency of the spikes, on the distance Dij between neurons Nj and Ni, and on many other physical and chemical parameters that are ignored in this model but can be added if necessary.

(d) The principle of feature superposition representation [27,29]: A vector of n qbits represents the probability of using each input variable $x1,x2,...,xn$ in the model at a time t. When the model computes, all features are "collapsed", where "0" represents that a variable is not used, and "1" – the variable is used.

The above principles can also be used to develop more sophisticated ipESNN models, as the one presented below – the ipESNG model.

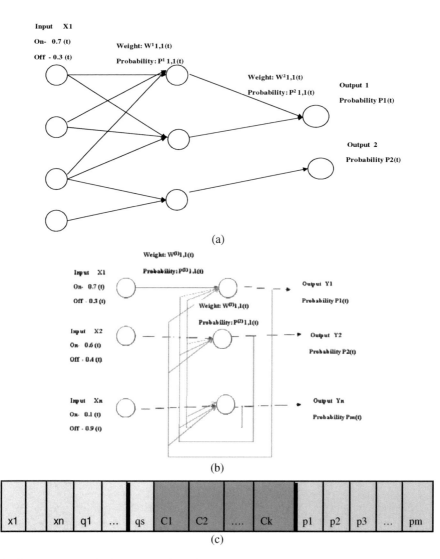

Fig. 3. (a) A schematic diagram of a feed-forward ipESNN; (b) and a recurrent ipESNN. (c)All input features (x1, x2,…,xn), the ESNN parameters (q1,q2,.., qs), the connections between neurons (C1, C2,., Ck) and the probability of the neurons to spike (p1,p2,…,pm) at time (t) are represented as an integrated qbit register that is operated as a whole [27,29].

4   Integrative Probabilistic Evolving Spiking Neuro-Genetic Models (ipESNG)

A schematic diagram of an ipESNG model is given in fig.4a. The framework combines ipESNN and a gene regulatory network (GRN) similar to the CNGM [5] from 2.3. The qbit vector for optimization through the QEA is given in fig.4b. In addition to the ipESNN parameters from fig.3c, the ipESNG model has gene expression parameters g1,g2,... gl, each of them also represented as a qbit with two states (state "1" – gene is expressed; state "0" – gene is not expressed"0). Each link Li (i=1,2,...,r) between two genes in the GRN is represented as a quantum bit with 3 states ("1" positive connection; "0" – no connection; "-1" – negative connection).

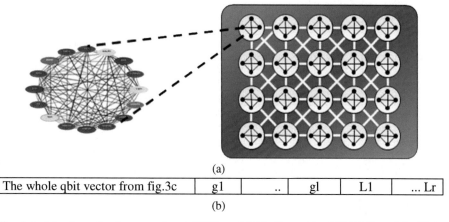

(a)

The whole qbit vector from fig.3c	g1		..	gl	L1	... Lr

(b)

Fig. 4. (a)A schematic diagram of an ipSNG model (similar to the CNGM from [5]; (b) In addition to the parameters shown in fig.3c, an ipESNG represents in a qbit register also gene expression levels (g1,g2,...,gl) and the connections between the genes (L1,...,Lr) from a GRN

5 Open Questions and Future Research

A significant originality of the proposed ipSNN models is that for the first time elements from the mathematical apparatus of quantum computation are used to develop a probabilistic ESNN model. Both input features, SNN parameters and output neuronal activities are represented as an integrated qbit vector defined by a probability wave function and used to calculate *probabilistically* the state of the whole system at any given time. The learning process of an ipSNN involves not only changing the connection weights by using learning rules, but also changing the probability density function of the qbit vector by the application of a quantum-inspired evolutionary algorithm [11]. The proposed ipSNN framework will be used for the development of a feed-forward, and a recurrent models as shown in fig.3a,b,c. The feed-forward ipSNN will be tested on a large-scale multimodal audiovisual information processing task [47]. Our hypothesis is that the ipSNN model will be faster and more accurate than other techniques, revealing useful information such as new features. The recurrent ipSNN model will be used to create and test an associative memory that is anticipated to be capable of storing and retrieving much larger number of patterns when compared to traditional associative memories [13,21]. Other QEA will be tested for probability estimation, such as quantum inspired particle swarm optimization. These models will also be used for a single neuron modeling on data from a *2-photon Laser Laboratory* [31] with the hypothesis that they are applicable to modeling processes in a neuron. We expect that new information about learning processes in the brain can be discovered.

Acknowledgement. Stefan Schliebs and Peter Hwang for technical support, Prof. H. Kojima, Dr Liam McDaid, Prof. A.Villa for fruitful discussions.

References

1. Abbott, L.F., Sacha, B.: Synaptic plasticity: taming the beast. Nature Neuroscience 3, 1178–1183 (2000)
2. Ackley, D.H., Hinton, G.E., Sejnowski, T.J.: A learning algorithm for Boltzmann machines. Cognitive Science 9, 147–169 (1985)
3. Arbib, M. (ed.): The Handbook of Brain Theory and Neural Networks. MIT Press, Cambridge (2003)
4. Belatreche, A., Maguire, L.P., McGinnity, M.: Advances in Design and Application of Spiking Neural Networks. Soft Comput. 11(3), 239–248 (2006)
5. Benuskova, L., Kasabov, N.: Comput. Neurogenetic Modelling. Springer, NY (2007)
6. Bershadskii, A., et al.: Brain neurons as quantum computers: in vivo support of background physics. Reports of the Bar-Ilan University, Israel 1-12 (2003)
7. Brette, R., et al.: Simulation of networks of spiking neurons: A review of tools and strategies. Journal of Computational Neuroscience 23(3), 349–398 (2007)
8. Castellani, M.: ANNE - A New Algorithm for Evolution of ANN Classifier Systems. In: IEEE Congress on Evolutionary Computation, CEC 2006, pp. 3294–3301 (2006)
9. Dayan, P., Hinton, G.E.: Varieties of Helmholtz machines. Neural Networks 9, 1385–1403 (1996)
10. Dayan, P., Hinton, G.E., Neal, R., Zemel, R.S.: The Helmholtz machine. Neural Computation 7, 1022–1037 (1995)
11. Defoin-Platel, M., Schliebs, S., Kasabov, N.: Quantum-inspired Evolutionary Algorithm: A multi-model EDA. IEEE Trans. Evolutionary Computation (in print, 2009)
12. Deutsch, D.: Quantum computational networks. Proceedings of the Royal Society of London A(425), 73–90 (1989)
13. Ezhov, A., Ventura, D.: Quantum neural networks. In: Kasabov, N. (ed.) Future Directions for Intelligent Systems and Information Sciences. Springer, Heidelberg (2000)
14. Gerstner, W., Kistler, W.M.: Spiking Neuron Models. Cambridge Univ. Press, Cambridge (2002)
15. Gerstner, W.: What's different with spiking neurons? In: Mastebroek, H., Vos, H. (eds.) Plausible Neural Networks for Biological Modelling, pp. 23–48. Kl. Ac. Publ., Dorchet (2001)
16. Guyon, I., et al. (eds.): Feature Extraction, Foundations and Applications. Springer, Heidelberg (2006)
17. Han, K.-H., Kim, J.-H.: Quantum-inspired evolutionary algorithm for a class of combinatorial optimization. IEEE Trans. on Evolutionary Computation, 580–593 (2005)
18. Hey, T.: Quantum computing: an introduction. Comp. & Control Eng. J. 10(6) (1999)
19. Hinton, G.E., Dayan, P., Frey, B.J., Neal, R.: The wake-sleep algorithm for unsupervised neural networks. Science 268, 1158–1161 (1995)
20. Hirvensalo, M.: Quantum computing. Springer, Heidelberg (2004)
21. Hopfield, J.J.: Neural networks and physical systems with emergent collective computational abilities. Proc. Natl. Acad. Sci. USA 79, 2554–2558 (1982)
22. Huguenard, J.R.: Reliability of axonal propagation: The spike doesn't stop here. PNAS 97(17), 9349–9350 (2000)
23. Izhikevich, E., Desai, N.: Relating STDP to BCM. Neural Comp. 15, 1511–1523 (2003)
24. Izhikevich, E.: Simple model of spiking neurons. IEEE Tr. NN 14(6), 1569–1572 (2003)
25. Izhikevich, E.: Which model to use for cortical spiking neurons? IEEE TrNN 15(5), 1063–1070 (2004)
26. Kasabov, N.: Evolving Connectionist Systems: The Knowl. Eng. Appr. Springer, Heidelberg (2007)
27. Kasabov, N.: Integrative Connectionist Learning Systems Inspired by Nature: Current Models, Future Trends and Challenges. Natural Computing (January 2008)

28. Kasabov, N.: Brain-, Gene-, and Quantum Inspired Computational Intelligence: Challenges and Opportunities. In: Duch, W., Manzduk, J. (eds.) Challenges in Computational Intelligence, pp. 193–219. Springer, Heidelberg (2007)
29. Kasabov, N.: Evolving Intelligence in Humans and Machines: Integrative Evolving Connectionist Systems Approach. IEEE Computational Intelligence Magazine 3(3), 23–37 (2008)
30. Kasabov, N.: Found. of neural networks, fuzzy systems and knowl. eng. MIT Press, Cambridge (1996)
31. Katsumata, S., Sakai, K., Toujoh, S., Miyamoto, A., Nakai, J., Tsukada, M., Kojima, H.: Analysis of synaptic transmission and its plasticity by glutamate receptor channel kinetics models and 2-photon laser photolysis. In: Köppen, M., Kasabov, N., Coghill, G. (eds.) ICONIP 2008, Part I. LNCS, vol. 5506. Springer, Heidelberg (2009)
32. Kohonen, T.: Self-Organizing Maps, 2nd edn. Springer, Heidelberg (1997)
33. Kistler, G., Gerstner, W.: Spiking Neuron Models - Single Neurons, Populations, Plasticity. Cambridge Univ. Press, Cambridge (2002)
34. Maass, W., Bishop, C. (eds.): Pulsed Neural Networks. MIT Press, Cambridge (1999)
35. Pavlidis, N.G., Tasoulis, O.K., Plagianakos, V.P., Nikiforidis, G., Vrahatis, M.N.: Spiking neural network training using evolutionary algorithms. In: Proceedings IEEE International Joint Conference on neural networks, vol. 4, pp. 2190–2194 (2005)
36. Pfister, J.P., Barber, D., Gerstner, W.: Optimal Hebbian Learning: a Probabilistic Point of View. In: Kaynak, O., Alpaydın, E., Oja, E., Xu, L. (eds.) ICANN 2003 and ICONIP 2003. LNCS, vol. 2714, pp. 92–98. Springer, Heidelberg (2003)
37. Sander, M., Bohte, H.A., La Poutré, J.N.: Error-Backpropagation in Temporally Encoded Networks of Spiking Neurons. Neurocomputing 48(1-4), 17–37 (2002)
38. Sander, M., Bohte, J.N.: Applications of spiking neural networks. Information Processing Letters 95(6), 519–520 (2005)
39. Schliebs, S., Defoin-Platel, M., Kasabov, N.: Integrated Feature and Parameter Optimization for an Evolving Spiking Neural Network. In: Köppen, M., Kasabov, N., Coghill, G. (eds.) ICONIP 2008, Part I. LNCS, vol. 5506. Springer, Heidelberg (2009)
40. Soltic, Wysoski, S., Kasabov, N.: Evolving spiking neural networks for taste recognition. In: Proc. WCCI 2008, Hong Kong. IEEE Press, Los Alamitos (2008)
41. Specht, D.F.: Enhancements to probabilistic neural networks. In: Proc. Int. Joint Conference on Neural Networks, June 1992, vol. 1, pp. 761–768 (1992)
42. Tuffy, F., McDaid, L., Wong Kwan, V., Alderman, J., McGinnity, T.M., Kelly, P., Santos, J.: Spiking Neuron Cell Based on Charge Coupled Synapses. In: Proc. IJCNN, Vancouver (2006)
43. Ventura, D., Martinez, T.: Quantum associative memory. Information Sciences 124(1-4), 273–296 (2000)
44. Verstraeten, D., Schrauwen, B., Stroobandt, D., Van Campenhout, J.: Isolated word recog. with the Liquid State Machine: a case study. Inf. Proc. Letters 95(6), 521–528 (2005)
45. Villa, A.E.P., et al.: Cross-channel coupling of neuronal activity in parvalbumin-deficient mice susceptible to epileptic seizures. Epilepsia 46(suppl. 6), 359 (2005)
46. Wysoski, S., Benuskova, L., Kasabov, N.: On-line learning with structural adaptation in a network of spiking neurons for visual pattern recognition. In: Kollias, S.D., Stafylopatis, A., Duch, W., Oja, E. (eds.) ICANN 2006. LNCS, vol. 4131, pp. 61–70. Springer, Heidelberg (2006)
47. Wysoski, S., Benuskova, L., Kasabov, N.: Brain-like Evolving Spiking Neural Networks for Multimodal Information. In: Proc. ICONIP 2007, Kitakyushu. LNCS. Springer, Heidelberg (2007)
48. Yadav, A., Mishra, D., Yadav, R.N., Ray, S., Kalra, P.K.: Time-series prediction with single integrate-and-fire neuron. Applied Soft Computing 7(3), 739–745 (2007)

Machine Learning in Vector Models of Neural Networks

Boris Kryzhanovsky, Vladimir Kryzhanovsky, and Leonid Litinskii

Center of Optical Neural Technologies of Scientific Research Institute for
System Analysis of Russian Academy of Sciences, Moscow

Abstract. We present the review of our works related to the theory of vector
neural networks. The interconnection matrix always is constructed according to
the generalized Hebb's rule, which is well-known in the Machine Learning. We
accentuate the main principles and ideas. Analytical calculations are based on
the probability approach. The obtained theoretical results are verified with the
aid of computer simulations.

1 Introduction

The Hopfield Model (HM) is a well-known version of a binary auto-associative neural network. With the aid of this approach one can retrieve binary N-dimensional
initial patterns when their distorted copies are given. As is well known, the storage
capacity of HM is rather small[1]. The statistical physics approach gives the estimate
$M_{HM} \sim 0.14 \cdot N$, where M_{HM} is a number of randomized patterns, which can be restored by HM [1]. At the same time, elementary calculations show that if the number
of initial patterns is less than N, the most effective approach is to store patterns in the
computer directly and to associate the input vector with that pattern, the distance to
which is the smallest one. Thus, small storage capacity of HM makes it useless for
practical applications.

In the end of 80-th attempts were made to improve recognizing characteristics of
auto-associative memory by considering q-ary patterns, whose coordinates can take q
different values, where $q > 2$ ([2]-[5]). These models were designed for processing of
color images. The number q in this case is the number of different colors and in order
of magnitude $q \sim 10^2 - 10^3$. All these models (with one exception) had the storage
capacity even smaller than HM. The exception was the Potts-glass neural network
(PG), for which the statistical physics approach gave the estimate
$M_{PG} \sim \dfrac{q(q-1)}{2} M_{HM}$ [3]. Since in this case the storage capacity is rather large,
$M_{PG} \sim 10^4 \cdot N$, the recognizing characteristics of Potts-glass neural network are of
interest for practical applications. Unfortunately, the reasons of large storage capacity
of Potts-glass neural network were not understood. The statistical physics approach
does not provide an explanation for this result.

---

[1] All over this article by the storage capacity we mean the number of stored patterns, which can
be retrieved from distortions.

J. Koronacki et al. (Eds.): Advances in Machine Learning II, SCI 263, pp. 427–443.
springerlink.com   © Springer-Verlag Berlin Heidelberg 2010

On the other hand, in [6], [7] the optical model of auto-associative memory was examined. Such a network is capable to hold and handle information that is encoded in the form of the phase-frequency modulation. In the network the signals propagate along interconnections in the form of quasi-monochromatic pulses at q different frequencies. The model is based on a *parametrical neuron* that is a cubic nonlinear element capable to transform and generate pulses in four-wave mixing processes [8]. If the interconnections have a generalized Hebbian form, the storage capacity of such a network also exceeds the storage capacity of Hopfield model in $\dfrac{q(q-1)}{2}$ times. The authors of [6], [7] called their model the *parametrical neural network* (PNN).

Further analysis of PNN showed that it can be described adequately in the framework of the vector formalism, when q different states of neurons are described by basis vectors of q-dimensional space. Everything else in this model (interconnections, a dynamical rule) is a direct generalization of HM.

The vector formalism allowed us to determine the identity of basic principles of Potts-glass neural network and PNN. We succeeded in the understanding of the mechanism of suppression of internal noise, which guarantees high recognizing properties for both architectures. Some variants of auto-associative PNN-architectures were suggested, including the one that has the record storage capacity for today (PNN-2). Moreover, we managed to build up: *i*) *hetero-associative* variant of a q-ary neural network that in orders of magnitude exceeds the auto-associative PNN with respect to the speed of operations; *ii*) *decorrelating PNN* that is an architecture aimed at processing of binary patterns; it allows one to retrieve a polynomially many patterns, $\sim N^R$, where $R \approx 2-4$; *iii*) binarized variant of PNN, when one cuts elements of the connection matrix by signs. After that the matrix elements are significantly simpler, however the storage capacity of the binarized network is remained rather large.

In the present review we summarize the results obtained in [10]-[28] by a group of authors from Center of Optical Neural Technologies Scientific Research Institute for System Analysis of Russian Academy of Sciences. This chapter is an extended version of the paper originally published in [20].

2   Vector Formalism for q-Ary Networks

To describe different states of neurons we will use unit vectors $\mathbf{e}_l$ of the space $\mathbf{R}^q$, $q \geq 1$:

$$\mathbf{e}_l = \begin{pmatrix} 0 \\ \vdots \\ 1 \\ \vdots \\ 0 \end{pmatrix}, \; l = 1, \ldots, q. \tag{1}$$

The state of the ith neuron is given by the column-vector $\mathbf{x}_i$,

$$\mathbf{x}_i = x_i \mathbf{e}_{l_i}, \; x_i \in \mathbf{R}^1, \; \mathbf{e}_{l_i} \in \mathbf{R}^q, \; 1 \leq l_i \leq q, \; i = 1, \ldots, N. \tag{2}$$

Always the real scalar x_i in the expression (2) is equal 1 in modulus: $|x_i| = 1$. Models in which the scalar x_i can be equal to ± 1, are called *phase PNN*. If x_i is equal to 1 identically, $x_i \equiv 1$, the models are called *phaseless PNN*. These names of PNN are related to the fact that if we compare the vector formalism with the original optical model [6], [7], the unit vectors $\mathbf{e}_l$ are modeling q different frequencies ω_l of quasi-monochromatic pulses, while a binary variable x_i is modeling the presence of a phase $\phi = \{0/\pi\}$ of a pulse. In fact, for the phase PNN we use the product of the spin variable $x_i = \pm 1$ and the vector variable $\mathbf{e}_l$ to describe the neuron states. Formally, here the number of different states of neurons is equal to $2q$. The presence of the spin variable leads to substantial differences in properties of the phase and phaseless networks. Below we discuss these differences. Now we continue to describe general principles that are the same for all the PNN-architectures.

An N-dimensional image with q-ary coordinates is given by a set of N q-dimensional vectors $\mathbf{x}_i$: $X = (\mathbf{x}_1,...,\mathbf{x}_N)$. *Initial patterns* are M given in advance sets of q-dimensional vectors:

$$X^{(\mu)} = \left(\mathbf{x}_1^{(\mu)},...,\mathbf{x}_N^{(\mu)}\right), \ \mathbf{x}_i^{(\mu)} = x_i^{(\mu)}\mathbf{e}_{l_i^{(\mu)}}, \ \mu = 1,...,M, \ i = 1,...,N, \ l_i^{(\mu)} \in [1,q]; \quad (3)$$

We will call *input vectors* the distorted copies these patterns. The network has to retrieve the initial patterns from their distorted copies.

The local field at the ith neuron has the usual form

$$\mathbf{h}_i = \frac{1}{N}\sum_{j=1}^{N}\mathbf{T}_{ij}\mathbf{x}_j, \quad (4)$$

where $(q \times q)$-matrix $\mathbf{T}_{ij}$ defines interconnection between ith and jth vector-neurons. The interconnections are chosen in the generalized Hebbian form:

$$\mathbf{T}_{ij} = \left(1 - \delta_{ij}\right)\sum_{\mu=1}^{M}\mathbf{x}_i^{(\mu)}\mathbf{x}_j^{(\mu)+}, \ i, j = 1,...,N. \quad (5)$$

Here $\mathbf{x}^+$ denotes q-dimensional vector-row, and $\mathbf{xy}^+$ denotes the product of vector-column $\mathbf{x}$ and vector-row $\mathbf{y}^+$ carried out according the matrix product rule. In other words, $\mathbf{xy}^+$ is a tensor product of two q-dimensional vectors. Analogously to the standard Hebb's rule the interconnection $\mathbf{T}_{ij}$ is determined by superposition of the states of ith and jth neurons over all patterns only. This allows us to interpret rule (5) as the Hebbian-like.

In the expression (4) for the local field $\mathbf{h}_i$ the matrices $\mathbf{T}_{ij}$ act on the vectors $\mathbf{x}_j \in \mathbf{R}^q$. After summation over all j, the local field at the ith neuron will be a linear combination of unit vectors $\mathbf{e}_l$. The dynamic rule generalizing the standard asynchronous dynamics of the HM is defined as follows: at the moment $t+1$ the ith neuron is oriented in the direction that is the nearest to the direction of the local field $\mathbf{h}_i$ at the moment t. In other words, if at the moment t

$$\mathbf{h}_i(t) = \sum_{l=1}^{q} A_l^{(i)} \mathbf{e}_l, \quad \text{where} \quad A_l^{(i)} \sim \sum_{j(\neq i)}^{N} \sum_{\mu=1}^{M} \left(\mathbf{e}_l \mathbf{x}_i^{(\mu)} \right) \left(\mathbf{x}_j^{(\mu)} \mathbf{x}_j(t) \right), \tag{6}$$

and $A_k^{(i)}$ is the largest in modulus amplitude, $\left| A_k^{(i)} \right| = \max_{1 \leq l \leq q} \left| A_l^{(i)} \right|$, then

$$\mathbf{x}_i(t+1) = \text{sign}\left(A_k^{(i)} \right) \mathbf{e}_k. \tag{7}$$

The evolution of the system consists in successive orientation of vector-neurons according this rule. It is not difficult to show that during the process of evolution the energy of the state,

$$E(t) \sim -\sum_{i=1}^{N} \mathbf{x}_i^+(t) \mathbf{h}_i(t),$$

decreases monotonically. Finally the system falls down to the energy minimum that is a fixed point of the network.

Here the description of the vector formalism relating to all PNN-architectures can be finished. In many respects the vector formalism resembles the approach that was suggested previously in the paper [9]. We find out this paper after working out our approach. The authors of [9] used not a proper dynamic rule that differs from Eq. (7). However, it seems that authors of [9] were the first who formulated clearly the vector approach for describing q-ary neurons.

3 Basic PNN-Architectures and Their Recognizing Characteristics

The estimates of recognizing characteristics of PNN can be obtained with the aid of the standard probability-theoretical approach [1]. The local field $\mathbf{h}_i$ (6) is divided into two parts, which are *the useful signal* s and *the internal noise* η, and their mean values and dispersions are calculated. As a rule, it is possible to do that when $N \to \infty$ the dispersion of the signal $\sigma^2(s)$ as well as the mean value of the internal noise $\overline{\eta}$ tend to zero. In this approach an important role plays the "ratio signal/noise" that is the ratio of the mean value of the signal $\overline{s}$ to the standard deviation of the noise $\sigma(\eta)$, $\gamma = \overline{s} / \sigma(\eta)$. The greater γ, the greater is the probability to recognize the initial pattern. In contrast to the standard situation described in [1], here partial noise components are not independent random variables, but uncorrelated variables only. This does not allow us to use the Central Limit Theorem to estimate the characteristics of internal noise. However, one can use the Chebyshev-Chernov statistical method [24], which allows us to obtain the exponential part of the estimated value in the case of uncorrelated random variables also.

3.1 Phase PNN

PNN-2. For a phase PNN, just the presence of spin variables $x_i = \pm 1$ leads to uncorrelated partial noise components. The corresponding neural network is called PNN-2.

Firstly it was described in [10], [11]. All calculations were performed accurately in [12], [13]. It was found that comparing with HM the dispersion of the internal noise is q^2 times less: $\sigma_{PNN-2}(\eta) \sim \sigma_{HM}(\eta)/q$. This is because overwhelming majority of the summands $\mathbf{x}_i^{(\mu)}\mathbf{x}_j^{(\mu)+}\mathbf{x}_j$, arising when calculating partial noisy components, vanish. At that the dispersion of the internal noise decreases. While for HM, the analogous terms $x_i^{(\mu)}x_j^{(\mu)}x_j$ can be "+1" or "-1" only. Consequently, each of these summands contributes to increase of the dispersion of the internal noise.

Let us right down the final expression for the probability of error of recognizing the initial pattern from its distorted copy. Let a be the probability of distortion of co-ordinates of the pattern in spin variable x_i, and let b be the probability of distortion of its coordinates in one of possible directions $\mathbf{e}_l$. (In terms of the original optical model [6], [7] a is the probability of a distortion of the quasi-monochromatic pulse in its phase, and b is the probability of a distortion in its frequency.) We suggested that the initial patterns (3) are randomized, e.g. their coordinates are independently and equiprobable distributed. Then, we obtain the estimate for the retrieval error probability of the pattern in one-step approximation (when the output pattern is evaluated from the input vector after one synchronous parallel calculation of all neurons):

$$\Pr_{err} \sim \frac{N(2q-1)}{\sqrt{2\pi}\gamma}\exp\left(-\frac{\gamma^2}{2}\right), \quad \text{where} \quad \gamma = \sqrt{\frac{N}{M}} \cdot q(1-2a)(1-b). \tag{8}$$

When $N, M \to \infty$ this probability tends to zero *for all patterns*, if the number of patterns M does not exceeds the critical value

$$M_{PNN-2} = \frac{N(1-2a)^2}{4\ln\left(Nq^{3/2}\right)} \cdot q^2(1-b)^2. \tag{9}$$

The last quantity can be considered as asymptotically obtainable storage capacity of PNN-2.

When $q = 1$ expressions (8), (9) turn into the known results for HM (in this case there is no frequency noise and it follows that we have to set $b = 0$). When q increases the retrieval error probability falls down exponentially: The noise immunity of the network increases significantly. In the same time the storage capacity (9) increases proportionally q^2. If we have in mind a processing of colored images, the neurons are pixels of the screen, and q is the number of different colors. It can be considered that $q \sim 10^2 - 10^3$. With such value of q the storage capacity of PNN-2 by 4-6 orders of magnitude greater than the storage capacity of HM.

PNN-2 allows one to store the number of patterns that is many times exceeds the number of neurons N. For example, let us set a constant value $\Pr_{err} = 0.01$. In the Hopfield model, with this error probability we can store $M = N/10$ patterns only, each of which is less then 30% noisy. In the same time, PNN-2 with $q = 64$ allows us to retrieve any of $M = 5N$ patterns with 90% noise, or any of $M = 50N$ patterns with 65% noise. Computer simulations confirm these estimates.

3.2 Phaseless PNNs

PNN-3. From the point of view of realization of PNN in the form of electronic device it is of interest a *phaseless variant* of PNN, when there are lacking of phases of quasi-monochromatic pulses. If we use the language of the vector formalism, this means that to describe different states of neurons one use unit vectors (1) only, and the spin variable x_i is absent:

$$\mathbf{x}_i = \mathbf{e}_{l_i}, \; \mathbf{e}_{l_i} \in \mathbf{R}^q, \; 1 \le l_i \le q, i = 1,\ldots,N. \tag{10}$$

Note, now neurons can be found in one of q different states, while for PNN-2 the number of different states of neurons is equal to $2q$ (due to the spin variable x_i - see Eq. (2)).

This model was analyzed in [13]-[15]. It was found that if as before the interconnection matrices $\mathbf{T}_{ij}$ are chosen in the form (5), the dispersion of internal noise increases disastrous. The way allowing us to overcome this problem is analogous to the one that is used in *the sparse coding* method [29], [30]: when calculating the matrices $\mathbf{T}_{ij}$, it is necessary to subtract the value of the averaged neuron activity from the vector coordinates $\mathbf{x}_i^{(\mu)}$. For randomized patterns the averaged neuron activity is equal to $\mathbf{e}/q$, where $\mathbf{e}$ is the sum of all the unit vectors $\mathbf{e}_l$: $\mathbf{e} = \sum_1^q \mathbf{e}_l$. So, in place of Eq. (5) the interconnection matrices have the form

$$\mathbf{T}_{ij} = \left(1 - \delta_{ij}\right) \sum_{\mu=1}^M (\mathbf{x}_i^{(\mu)} - \mathbf{e}/q)(\mathbf{x}_j^{(\mu)} - \mathbf{e}/q)^+, \; i, j = 1,\ldots,N. \tag{11}$$

The neural architecture that is determined by Eqs. (10), (11), is called PNN-3. The analogues of the expressions (8), (9) here are

$$\Pr_{err} \sim \frac{N(q-1)}{\sqrt{2\pi}\gamma} \exp\left(-\frac{\gamma^2}{2}\right), \text{ where } \gamma = \sqrt{\frac{N}{M}} \cdot \sqrt{\frac{q(q-1)}{2}} \left(1 - \tilde{b}\right), \tag{12}$$

and

$$M_{PNN-3} = \frac{N\left(1 - \tilde{b}\right)^2}{4\ln\left(N(q-1)^{3/2}\right)} \cdot \frac{q(q-1)}{2}, \quad \tilde{b} = \frac{q}{q-1} b. \tag{13}$$

When $q = 2$, the expressions (12), (13) turn into the known results for the HM.

The storage capacity of PNN-3 is two times less comparing with PNN-2. The superiority of PNN-2 is related to the presence of the spin variables $x_i = \pm 1$. Let us discuss this in details. Generally speaking, to double the storage capacity of PNN-3 it is necessary to increase the number of neuron states in $\sqrt{2}$ times. Indeed, if in Eq. (13) we replace q by $Q = \sqrt{2}q$, the approximately doubled storage capacity is obtained. However, in the same time the number of elements of the matrix $\mathbf{T}_{ij}$ is doubled too:

$Q^2 = 2q^2$. Consequently, we need the computer memory (RAM) that is two times greater. On the contrary, when for PNN-2 we use the spin variable, we avoid the doubling of RAM. In other words, the spin variable x_i is very effective for increasing the storage capacity.

Note 1. The detailed analysis shows, that PNN-3 is a refined version of the Potts-glass neural network that was examined for the first time in [3]. For this publication poorly chosen notations and its brief style built up a reputation of a difficult for understanding text (see for example [9]). In reality, the Potts-glass neural network is almost coincides with PNN-3. In both models interconnection matrices $\mathbf{T}_{ij}$ are defined by the same Eq. (11), with the aid of *the Potts vectors* $\mathbf{d}_l = \mathbf{e}_l - \mathbf{e}/q$. The only difference between these models is that in the Potts-glass model the vectors $\mathbf{d}_l$ are used to describe q different states of neurons, however in PNN-3 for this purpose one use unit vectors $\mathbf{e}_l$. Recognizing characteristics of PNN-3 and Potts-glass neural network are identical. In the same time the computer realization of PNN- 3 is q times quicker than the Potts-glass network.

PNN-4. One other phaseless neural network that is defined by Eq.(5) and Eq.(10) is of interest. The connection matrices of the type (5) are built with the aid of the unit vectors $\mathbf{e}_l$ (1) without the subtraction of the averaged neuron activity as it was done in Eq.(11). In fact we have *a modified* Potts-glass neural network. It was called PNN-4. Its matrix elements are very simple, and they are not negative always.

For PNN-4 expressions that are analogous to the expressions (8), (9) are:

$$\Pr_{err} \sim \frac{Nq}{\sqrt{2\pi\gamma}} \exp\left(-\frac{\gamma^2}{2}\right), \text{ where } \gamma = \sqrt{\frac{N}{M}} \cdot \frac{q(1-b)}{\sqrt{1+N/q}}. \tag{14}$$

and

$$M_{PNN-4} = \frac{N(1-b)^2}{4(1+N/q)\ln\left(\dfrac{Nq^{3/2}}{1+N/q}\right)} \cdot q^2 \tag{15}$$

It is easy to see that when N increases, the increase of the storage capacity of PNN-4 is only up to a critical value N_0. Its value is defined from an transcendental equation

$$1 + N_0/q = \ln(qN_0).$$

For $N > N_0$ the storage capacity is slowly (logarithmically) decreases. Such a behavior of the storage capacity is not interesting from the point of view of practical applications. It was found that PNN-4 is very effective when we deal with binarization of matrix elements (see below). Mainly it is due to the simplicity of matrix elements of PNN-4.

In conclusion of this item we note again that recognizing characteristics of PNN-architectures improve when the number of states of neurons q increases. We actively use this property in the next item.

4   Special Neural Architectures Based on PNN

Outstanding recognizing characteristics of PNN can be used to develop other architectures, directed onto solutions of one or other special problems. *The decorrelating
PNN* allows one to store a polynomial large number of binary patterns, and, this is
most important, even if these patterns are strongly correlated. *The q-ary identifier* allows one to accelerate the recognizing process by orders of magnitude, and in the
same time to decrease the working memory (RAM). *Binarization* of matrix elements
(*clipping*) allows one to construct a neural-architecture that is the most suitable for realization in the form of optoelectronic device.

4.1   The Decorrelating PNN

It is known that the memory capacity of HM falls down drastically if there are correlations between patterns. The way out is so named sparse coding [29], [30]. The
decorrelating PNN (DPNN) is an alternative to this approach [16]-[20].

   The main idea of DPNN is as follows. The initial binary patterns are one-to-one
mapped into internal representation using vector-neurons of large dimension q. Then,
one of above-mentioned PNN-architectures is being constructed on the basis of
the obtained vector-neuron patterns. The mapping has the following properties: First,
correlations between vector-neuron patterns become negligible, and, second, the
dimension q of vector-neurons increases exponentially as a function of the mapping
parameter. The algorithm of binary patterns recognition consists of three stages. At
first, the input binary vector is mapped into the q-ary representation. Then with the
aid of PNN its recognition occurs: it is identified with one of the q-ary patterns. At
last, the inverse mapping of the q-ary pattern into its binary representation takes place.
We can expect substantial increase of the storage capacity of the network, because the
dimension q of vector-neurons increases exponentially, and we know: the larger the
dimension of vector-neurons, the better recognizing properties of PNN.

   This scheme can be realized for any of above-mentioned PNNs. We explain the
approach for the case of PNN-2. The algorithm of binary patterns mapping into vector-neuron ones is very simple. Let $Y = \left(y_1, y_2, \ldots, y_N \right)$ be N-dimensional binary vector, $y_i = \pm 1$. We divide it into n fragments of $k+1$ elements each, $N = n(k+1)$. With
each fragment we associate a signed integer number $y \cdot l$ according the following
rules:

1) the first element of the binary fragment defines the sign $y = \pm 1$ of the number;
2) the other k elements of the fragment determine the absolute value l,

$$l = 1 + \sum_{i=2}^{k+1} (y_i + 1) \cdot 2^{k-i}, \quad 1 \le l \le 2^k.$$

In fact, we interpret the last k binary coordinates of the fragment as the binary notation of the integer l.

   After that we associate each fragment $(y_1, y_2, \ldots, y_{k+1})$ with a vector $\mathbf{x} = y_1 \mathbf{e}_l$,
where $\mathbf{e}_l$ is the lth unit vector in the space $\mathbf{R}^q$ of dimensionality $q = 2^k$, and the sign

of the vector is given by the first element y_1 of the fragment. So, the mapping of binary fragment into 2^k-dimensional unit vector occurs as follows:

$$\left(y_1, y_2, \ldots, y_{k+1}\right) \rightarrow y_1 \cdot l \rightarrow \mathbf{x} = y_1 \mathbf{e}_l \in \mathbf{R}^q, \quad q = 2^k. \tag{16}$$

According this scheme any binary vector $Y \in \mathbf{R}^N$ can be one-to-one mapped into a set of n q-dimensional unit vectors

$$Y = \left(y_1, \ldots, y_{k+1}, y_{k+2}, \ldots, y_{2k+2}, \ldots, y_{N-k}, \ldots, y_N\right) \rightarrow X = \left(\mathbf{x}_1, \mathbf{x}_2, \ldots, \mathbf{x}_n\right).$$

The result of this mapping is the vector-neuron image X. The number k is called *the mapping parameter.*

When the mapping for the given set of binary patterns $\{Y^{(\mu)}\}_1^M$ is done, we obtain vector-neuron images $\{X^{(\mu)}\}_1^M$. Note, even if binary patterns $Y^{(\mu)}$ are correlated, their vector-neuron images are practically not correlated. Such is the property of the mapping (16): It eliminates the correlations between patterns. Indeed, for two binary fragments it is sufficient to differ even in one coordinate, and the fragments are mapped into two absolutely different orthogonal unit vectors of the space $\mathbf{R}^q$.

Now, making use of the vector-neuron images $\{X^{(\mu)}\}_1^M$ we construct PNN-2 – see formula (5). Since the images $X^{(\mu)}$ can be considered randomized, we use estimates (8), (9) obtained previously. If a is the probability of distortion of binary coordinates (the level of binary noise), for constructed PNN-2 we obtain:

$$M_{DPNN} \approx \frac{N\left(1-2a\right)^2}{4\ln N} \cdot \frac{(2(1-a))^{2k}}{(k+1)(1+k/\ln N)}. \tag{17}$$

The first factor in the right-hand side is the storage capacity of HM. The second factor provides exponential increase of the storage capacity when the mapping parameter k increases (since always the inequality $2(1-a) > 1$ is fulfilled).

However, it is necessary to be careful: the parameter k cannot be an arbitrary large. First, the number of vector-neurons n must be sufficiently large to make it possible to use probability-theoretic estimates. Experiments show that as far as the dimension of vector-neuron patterns is greater than 100, $n \geq 100$, asymptotical estimates (8), (9) are fulfilled to a high accuracy. This leads to the first restriction on the value of k:

$$k + 1 \leq \frac{N}{100}. \tag{18}$$

Second, among the vector-coordinates of the distorted vector-neuron image at least one coordinate must be undistorted. Its contribution to the local field $\mathbf{h}_i$ allows one to retrieve the correct value of the ith vector-coordinate. From these argumentations it follows that correct recognition can be guaranteed only if the number of undistorted vector-coordinates is not less than two. It is easy to estimate the average number of undistorted vector-coordinates of the image X: $n(1-a)^{k+1}$. So, we obtain the second restriction on the value of k:

$$n(1-a)^{k+1} \geq 2 . \tag{19}$$

Thus, maximal possible value k_c must fulfill the inequalities (18), (19).

When k changes from 0 to k_c, the storage capacity of the network increases rapidly. When k becomes greater than k_c, the network no longer retrieves the patterns. However, even before k reaches the limiting value k_c, the storage capacity of the network can be done sufficiently large.

Let us write down the storage capacity of DPNN (17) for the critical value k_c in the form

$$M_{DPNN}(k_c) \sim N^R . \tag{20}$$

In Fig.1 the dependences of k_c (the top panel) and the exponent R in the expression (20) (the bottom panel) on the level of the binary noise a are shown for the dimension of the binary patterns $N=1000$. We see (the top panel) that for the most part of the variation interval of the noise a, the main restriction for k_c is inequality (18). This is because the dimension of binary patterns is comparatively small. Apparently, the

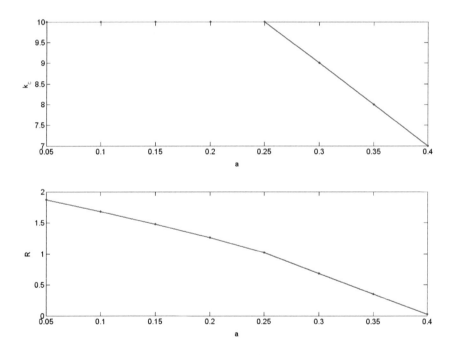

Fig. 1. For $N=1000$ the dependence of k_c (the top panel) and exponent R in the expression (20) (the bottom panel) on the level of the binary noise $a \in [0.05, 0.4]$ (abscissa axis)

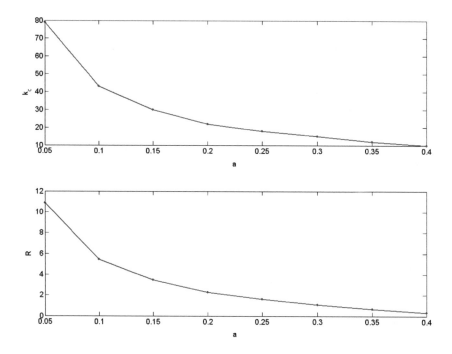

Fig. 2. The same as in Fig.1 for the dimension N=10000 of binary patterns

reasonable level of the noise is $a \le 0.25$. The bottom panel shows that in this region DPNN allows us to store N^R patterns with $R \in [1, 2]$. The Hopfield Model allows us to store in this region less than N patterns only.

The greater a binary dimension N, the large is the superiority of DPPN onto HM. In Fig.2 the analogous graphs are shown for the dimension of binary patterns N=10000. Now the main restriction for k_c is the inequality (19). From the graph on the bottom panel it follows that for this dimension N and the level of the noise a=0.1 the network can store $\sim N^{5.7}$ of binary patterns, and for a=0.15 we obtain $M_{DPNN} \sim N^{3.7}$. Note these results are valid for any level of correlations between the binary patterns.

4.2 q-Ary Identifier

The main idea of the given architecture is based on the following result: when PNN is working far from the limits of its possibility (for example, when the number of patterns is less than the critical value M_c), the retrieval of the correct value of each co-ordinate occurs at once, during the first processing of this coordinate by the network. This is because the retrieval error probability falls down exponentially when the number of patterns decreases (see expressions (8) and (12)). In other words, to retrieve the pattern correctly the network runs over all the coordinates only once. Then the following approach can be realized [21]-[23].

Suppose we have M initial patterns each of which is described by a set of N q-ary coordinates. As it is done for PNN-3, we represent these coordinates by unit vectors of q-dimensional space $\mathbf{R}^q$ (see the expressions (10)). We would like to emphasize, for us substantial meaning of the coordinates are of no importance. It is important only that coordinates can take q different values, and in the framework of the vector formalism they are described by unit vectors $\mathbf{e}_l \in \mathbf{R}^q$:

$$X^{(\mu)} = \left(\mathbf{x}_1^{(\mu)}, \mathbf{x}_2^{(\mu)}, \ldots, \mathbf{x}_N^{(\mu)}\right), \quad \mathbf{x}_i^{(\mu)} \in \{\mathbf{e}_l\}_1^q, \quad \mu = 1, \ldots, M. \tag{21}$$

We number all the patterns and write down their numbers in q-ary number system. To do this we need $n = \log_q M$ q-ary positions. Let us treat these n q-ary numbers as additional - *enumerated* – coordinates of patterns. Then we add them to the description of each initial pattern as first n coordinates. The enumerated coordinate takes one of q possible values, and from this point of view it does not differ from the true coordinates of the patterns. When we pass to the vector formalism, we also represent the values of these enumerated coordinates by q-dimensional unit vectors $\mathbf{y}_j \in \{\mathbf{e}_l\}_1^q$. If previously to represent a pattern we needed N q-ary coordinates, now it is represented by $n+N$ such coordinates. In place of the expression (21) we use expanded description of the patterns,

$$\hat{X}^{(\mu)} = \left(\underbrace{\mathbf{y}_1^{(\mu)}, \ldots, \mathbf{y}_n^{(\mu)}}_{\text{enumerated coordinates}}, \underbrace{\mathbf{x}_1^{(\mu)}, \ldots, \mathbf{x}_N^{(\mu)}}_{\text{true coordinates}} \right), \quad \mathbf{x}_i^{(\mu)}, \mathbf{y}_j^{(\mu)} \in \{\mathbf{e}_l\}_1^q, \mu = 1, \ldots, M.$$

We use these M patterns $\hat{X}^{(\mu)}$ of the dimensionality $n+N$ to construct PNN-like architecture, whose interconnection matrix, generally speaking, consists of $(n+N)^2$ $(q \times q)$-matrices. However, we suppose that only interconnections $\hat{\mathbf{T}}_{ij}$ *from true* coordinates *to enumerated* coordinates are nonzero. Other interconnections we forcedly replace by zero (we break these interconnections).

Assuming PNN-3 as the basis architecture, we use formula (11) for $i \in [1, n]$ and $j \in [n+1, N+n]$ when calculating the nonzero interconnection matrices,

$$\hat{\mathbf{T}}_{ij} = \begin{cases} \mathbf{T}_{ij} \text{ from Eq.(7),} & i \le n, \ j > n, \\ \mathbf{0}, & \text{in the rest cases,} \end{cases} \quad i, j \in [1, N+n], \tag{22}$$

where

$$\mathbf{T}_{ij} = \sum_{\mu=1}^{M} (\mathbf{y}_i^{(\mu)} - \mathbf{e}/q)(\mathbf{x}_j^{(\mu)} - \mathbf{e}/q)^+ .$$

The principle of work of such a network is as follows. Suppose we need to recognize distortion of what an initial pattern $X^{(\mu)}$ (21) is N-dimensional q-ary input image X. We transform N-dimensional image X into $(n+N)$-dimensional $\hat{X}$, using some *arbitrary* integers from the interval $[1,q]$ in place of the first n enumerated coordinates (in what follows we show that it is not significant what n integers namely are used). In terms of the vector formalism this means that from the image $X = (\mathbf{x}_1, \mathbf{x}_2, ..., \mathbf{x}_N)$ we pass to $\hat{X} = (\mathbf{y}_1, \mathbf{y}_2, ..., \mathbf{y}_n, \mathbf{x}_1, \mathbf{x}_2, ..., \mathbf{x}_N)$.

Just the image $\hat{X}$ of the extended dimensionality $n+N$ is presented for recognition at the input of PNN-like architecture (22). From the construction of the network it is seen that when working the network retrieves the enumerated coordinates only. Consequently, if we work far from the limits of PNN-3 possibilities, the correct value of each enumerated coordinate is obtained at once during the first processing (see above). No additional runs over all the coordinates of the pattern will be necessary. So, the number of the initial pattern, whose distorted copy is the input image X, will be retrieve at once during one run over n enumerated coordinates. But if the number of the pattern is known, it is not necessary to retrieve correct values of its N true coordinates $\mathbf{x}_i^{(\mu)}$. We simply extract this pattern from the database!

Here the main questions are: How much is the number n of the enumerated coordinates? How large is the additional distortion when we pass from the N-dimensional input image X to the $(n+N)$-dimensional image $\hat{X}$? It turns out that the number n of the enumerated coordinates is rather small. We remind that $n = \log_q M$. If as M we take the storage capacity of PNN-3 given by the expression (13), it is easy to obtain the asymptotic estimate for n:

$$n \sim 2 + \frac{\ln N}{\ln q}, \quad N \gg 1.$$

For example, if the number of pixels of the screen is $N \sim 10^6$ and the number of the colors is $q \sim 10^2$, we obtain: $n \approx 5$. In other words, the number of the enumerated coordinates n is negligible small comparing with the number of the true coordinates N. Since there are no connections between enumerated coordinates (see Eq.(22)), the usage of an arbitrary integers as first n enumerated coordinates does not lead to additional distortions.

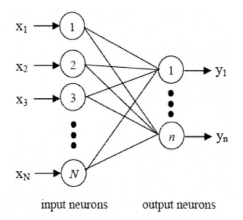

Fig. 3. The scheme of the vector perceptron

In this case the network ceases to be auto-associative one. It works like a q-ary perceptron whose interconnections are arranged according the generalized Hebb rule (11) – see Fig.3. The procedure of recognition of the pattern itself (the retrieval of the correct values of its true coordinates) is absent. Instead of it identification of the number of the pattern takes place. That is why this architecture was called q-ary identifier.

The advantages of the q-ary identifier are evident. First, the process of recognition speeds up significantly – the network has to retrieve correct values of $n \sim 5$ enumerated coordinates only, instead of doing the same for $N \sim 10^6$ coordinates. Second, the computer memory necessary to store interconnections $\mathbf{T}_{ij}$ decreases substantially: now we need only Nn $(q \times q)$-matrices, but not $(n + N)^2$.

4.3 Binarization of PNN

The *binarization* of matrix elements (*the clipping*) consists in the following procedure: positive elements are replaced by "+1", negative elements are replaced by "-1", and zero matrix elements do not change. The binarized matrices possess evident advantages: RAM is economized more than an order of magnitude and the calculations speed up by the same order of magnitude. The recognizing properties of the binarized networks are examined for the Hopfield model only [1]. The binarization of the Hebb matrix decreases the storage capacity by a factor of $\pi/2$. If this result takes place also for PNN, then with the aid of the binarization we should simplify the calculation scheme and in the same time keep rather large storage capacity. However, it turns out that the result of binarization differs for different PNN-architectures [25]-[28].

PNN-2. For this architecture the above-mentioned result for HM is true. Namely, for sufficiently large dimensionality of the problem, when $N > 5\ln(Nq)$, the storage capacity of the binarized PNN-2 is $\pi/2$ times less than the storage capacity of the original PNN-2. Note, that this not significant deterioration of recognizing characteristics is accompanied by significant simplification of the calculations. Firstly, RAM that we need to store the binarized matrix is an order of magnitude less. Secondly, the speed of calculations becomes at least an order of magnitude faster. It is easy to imagine a situation, when such an economy of computational resources is rather important.

It was found out that for networks of small dimensions, when $N < 5\ln(Nq)$, the binarized PNN-2 works better than its nonbinarized prototype. In this case for the binarized network the ratio "signal/noise" exceeds the analogous characteristic of PNN-2 by a factor of $\sqrt{2}$. In other words, the binarized network is capable to retrieve patterns from larger distortions than the standard PNN-2. This theoretical result is confirmed by computer simulations.

PNN-3 (the Potts-glass network). Computer experiments show that binarization of PNN-3 matrix destroys the capability of this network to be an associative memory. Our analysis shows that for PNN-3 the distribution of matrix elements crucially differs from the normal distribution. It is strongly asymmetric and it has several peaks. Evidently, too much information is lost when the true matrix elements are replaced by their simplified values +/-1.

PNN-4. Since in this case all matrix elements are positive, binarization of the connection matrix decreases RAM by a factor of 30, and the speed of calculation becomes more than q times larger. Computer simulations show that the storage capacity of PNN-4 *always* increases as a result of binarization. This rather unexpected result is confirmed not only by experiments, but also by the theoretical analysis.

For the binarized variant of PNN-4 one cannot use the standard probability-theoretical method based on the derivation of an explicit expression for the ratio "signal/noise". This is due to the fact that in this case the distribution of local fields differs from the normal distribution significantly. One needs to use combinatorial calculus to estimate the probability of pattern recognition. This way leads to a complicated nonlinear equation that has to be solved numerically. In Fig.4 for q=8 we show the dependence of the storage capacity M on the dimensionality of the problem N for PNN-4 (the lower graph) and its binarized variant (the upper graph). The dots show the results of the computer simulations, the solid lines correspond to the theoretical curves.

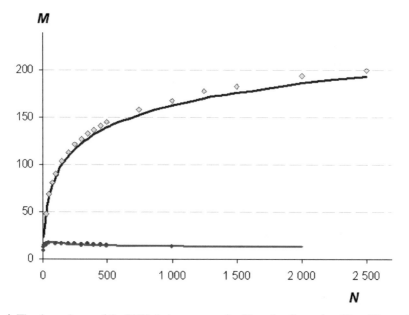

Fig. 4. The dependence of the PNN-4 storage capacity M on the dimensionality of the problem N for q=8. The lower plot corresponds to PNN-4, the upper to its binarized variant. The dots are the results of a computer experiments, the solid lines are theoretical curves.

For both networks we see a good agreement of our experiments with the theory. On the lower graph we clearly see the slow (logarithmical) *decrease* of the storage capacity when N increases (we mentioned that in the previous Section when describing PNN-4 characteristics – see Eq.(14) and Eq.(15)). On the contrary, the storage capacity of the binarized PNN-4 *increases* slowly but steadily. For large values of N this curve can be roughly approximated by the expression $q^2 \ln(Nq)$. Let β be the ratio of the storage capacity of binarized PNN-4 to the storage capacity of nonbinarized network: $\beta = M_{PNN4}^{bin} / M_{PNN4}$. When N increases, the value of β is steadily increasing approximately as $\ln^2(Nq)$. In our experiments for different q and N it reaches the value $\beta = 30$. These properties make the binarized PNN-4 very attractive.

5 Conclusions

In the previous Section we described some architectures each of which allows us to intensify different characteristics of neural network: to increase the binary storage capacity and noise immunity, to increase the speed of operation, to decrease RAM and/or to simplify computer realization of a neural network and so on. Note that one can combine these architectures in order to produce new, may be more effective, neural networks. For example, we can combine in one calculation scheme the decorrelating PNN with the q-ary identifier. Then we can apply the binarization procedure in order to estimate the resulting efficiency. In our plans we have the development of some other new architectures, which would extend the spectrum of neural instruments.

In conclusion it is necessary to say that in fact, up to now associative neural networks were not used for solving complicated practical problems. Seemingly, it is because of the poor recognition characteristics of HM. From this point of view PNNs are very fruitful and promising architectures. We hope that the development of PNNs and other architectures based on PNNs, will lead to increase of interest to the theory of associative neural networks, as well as their practical application.

Acknowledgement. The work was supported by the grant of President of Russian Federation #356.2008.9 and in part by Russian Basic Research Foundation (grant #09-07-00159).

References

[1] Hertz, J., Krogh, A., Palmer, R.: Introduction to the Theory of Neural Computation. Addison-Wesley, NY (1991)
[2] Noest, J.: Discrete-state phasor neural networks. Phys. Rev. A 38, 2196–2199 (1988)
[3] Kanter, I.: Potts-glass models of neural networks. Phys. Rev. A 37, 2739–2742 (1988)
[4] Cook, J.: The mean-field theory of a Q-state neural network model. J. Phys. A 22, 2000–2012 (1989)
[5] Vogt, H., Zippelius, A.: Invariant Recognition in Potts Glass Neural Networks. J. Phys. A 25, 2209–2226 (1992)
[6] Kryzhanovsky, B., Mikaelian, A.: On the Recognition Ability of a Neural Network on Neurons with Parametric Transformation of Frequencies. Doklady Mathematics 65(2), 286–288 (2002)
[7] Fonarev, A., Kryzhanovsky, B., et al.: Parametric dynamic neural network recognition power. Optical Memory & Neural Networks 10(4), 31–48 (2001)
[8] Bloembergen, N.: Nonlinear optics, Benjamin, NY (1965)
[9] Nakamura, Y., Torii, K., Munaka, T.: Neural-network model composed of multidimensional spin neurons. Phys. Rev. B 51, 1538–1546 (1995)
[10] Kryzhanovsky, B., Litinskii, L.: Vector models of associative memory. In: Lectures on Neuroinformatics on V Russian Conference Neuroinformatics – 2003, vol. 1, pp. 72–85. Moscow Engineering Physical Institute Press, Moscow (2003) (in Russian)
[11] Kryzhanovsky, B., Litinskii, L., Fonarev, A.: Parametrical Neural Network Based on the Four-Wave Mixing Process. Nuclear Instruments and Methods in Physics Research Section A 502, 517–519 (2003)

[12] Kryzhanovsky, B., Litinskii, L.: Vector models of associative memory. Automaton and Remote Control 64(11), 1782–1793 (2003)

[13] Kryzhanovsky, B., Litinskii, L., Mikaelian, A.: Parametrical Neural Network. Optical Memory & Neural Networks 12(3), 138–156 (2003)

[14] Kryzhanovsky, B., Litinskii, L., Mikaelian, A.: Vector neuron representation of q-valued neural networks. In: Prokhorov, D. (ed.) Proceedings of IEEE International Joint Conference on Neural Networks – 2004, vol. 1, pp. 909–915. IEEE Press, Budapest (2004)

[15] Kryzhanovsky, B., Litinskii, L., Mikaelian, A.: Vector Neurons Model of Associative Memory. Information technologies and computer systems 1, 68–81 (2004) (in Russian)

[16] Kryzhanovsky, B., Mikaelian, A.: An associative memory capable of recognizing strongly correlated patterns. Doklady Mathematics 67(3), 455–459 (2003)

[17] Kryzhanovsky, B., Litinskii, L., Fonarev, A.: An effective associative memory for pattern recognition. In: Berthold, M.R., Lenz, H.-J., Bradley, E., Kruse, R., Borgelt, C. (eds.) IDA 2003. LNCS, vol. 2810, pp. 179–186. Springer, Heidelberg (2003)

[18] Kryzhanovsky, B., Kryzhanovsky, V., Fonarev, A.: Decorrelating parametrical neural network. In: Proceedings of IEEE International Joint Conference on Neural Networks–2005, vol. 1, pp. 153–157. IEEE Press, Montreal (2005)

[19] Kryzhanovsky, B., Mikaelian, A., Fonarev, A.: Vector Neural Net Identifying Many Strongly Distorted and Correlated Patterns. In: International Conference on Information Optics and Photonics Technology, Photonics Asia-2004, vol. 5642, pp. 124–133. SPIE Press, Beijing (2004)

[20] Litinskii, L.: Parametrical Neural Networks and Some Other Similar Architectures. Optical Memory & Neural Networks 15(1), 11–19 (2006)

[21] Kryzhanovsky, B., Kryzhanovsky, V., Magomedov, B., et al.: Vector Perceptron as Fast Search Algorithm. Optical Memory & Neural Networks 13(2), 103–108 (2004)

[22] Kryzhanovsky, B.V., Kryzhanovsky, V.M.: Distinguishing Features of a Small Hopfield Model with Clipping of Synapses. Optical Memory & Neural Networks 17(3), 193–200 (2008)

[23] Alieva, D., Kryzhanovsky, B., Kryzhanovsky, V., et al.: Q-valued neural network as a system of fast identification and pattern recognition. Pattern Recognition and Image Analysis 15(1), 30–33 (2005)

[24] Kryzhanovsky, B.V., Mikaelian, A.L., Koshelev, V.N., et al.: On recognition error bound for associative Hopfield memory. Optical Memory & Neural Networks 9(4), 267–276 (2000)

[25] Kryzhanovsky, V.M., Simkina, D.I.: Properties of clipped phase-vector of associative memory. Bulletin of Computer and Information Technologies 11, 20–25 (2007) (in Russian)

[26] Kryzhanovsky, V.M.: Modified q-state Potts Model with Binarized Synaptic Coefficients. In: Kůrková, V., Neruda, R., Koutník, J. (eds.) ICANN 2008,, Part II. LNCS, vol. 5164, pp. 72–80. Springer, Heidelberg (2008)

[27] Kryzhanovsky, V.M., Kryzhanovsky, B., Fonarev, A.: Application of Potts-model Perceptron for Binary Patterns Identification. In: Kůrková, V., Neruda, R., Koutník, J. (eds.) ICANN 2008, Part I. LNCS, vol. 5163, pp. 553–561. Springer, Heidelberg (2008)

[28] Kryzhanovsky, V.M., Kryzhanovsky, B.V.: In: Lectures on Neuroinformatics on XI Russian Conference, Neuroinformatics – 2009. Moscow Engineering Physical Institute Press, Moscow (in press, 2009) (in Russian)

[29] Perez-Vicente, C.J., Amit, D.J.: Optimized network for sparsely coded patterns. J. Phys. A 22, 559–569 (1989)

[30] Palm, G., Sommer, F.T.: Information capacity in recurrent McCulloch-Pitts networks with sparsely coded memory states. Network 3, 1–10 (1992)

Nature Inspired Multi-Swarm Heuristics for Multi-Knowledge Extraction

Hongbo Liu[1,3], Ajith Abraham[2], and Benxian Yue[3]

[1] School of Information Science and Technology, Dalian Maritime University,
116026 Dalian, China
lhb@dlut.edu.cn
http://hongboliu.torrent.googlepages.com
[2] Centre for Quantifiable Quality of Service in Communication Systems,
Norwegian University of Science and Technology, Trondheim, Norway
ajith.abraham@ieee.org
http://www.softcomputing.net
[3] School of Electronic and Information Engineering, Dalian University of Technology,
Dalian 116023, China

Abstract. Multi-knowledge extraction is significant for many real-world applications. The nature inspired population-based reduction approaches are attractive to find multiple reducts in the decision systems, which could be applied to generate multi-knowledge and to improve decision accuracy. In this Chapter, we introduce two nature inspired population-based computational optimization techniques namely Particle Swarm Optimization (PSO) and Genetic Algorithm (GA) for rough set reduction and multi-knowledge extraction. A Multi-Swarm Synergetic Optimization (MSSO) algorithm is presented for rough set reduction and multi-knowledge extraction. In the MSSO approach, different individuals encodes different reducts. The proposed approach discovers the best feature combinations in an efficient way to observe the change of positive region as the particles proceed throughout the search space. We also attempt to theoretically prove that the multi-swarm synergetic optimization algorithm converges with a probability of 1 towards the global optimal. The performance of the proposed approach is evaluated and compared with Standard Particle Swarm Optimization (SPSO) and Genetic Algorithms (GA). Empirical results illustrate that the approach can be applied for multiple reduct problems and multi-knowledge extraction very effectively.

1 Introduction

Rough set theory [1,2,3] provides a mathematical tool that can be used for both feature selection and knowledge discovery. It helps us to find out the minimal attribute sets called '*reducts*' to classify objects without deterioration of classification quality and induce minimal length decision rules inherent in a given information system. The idea of reducts has encouraged many researchers in studying the effectiveness of rough set theory in a number of real world domains, including medicine, pharmacology, control systems, fault-diagnosis, text

J. Koronacki et al. (Eds.): Advances in Machine Learning II, SCI 263, pp. 445–466.
springerlink.com   © Springer-Verlag Berlin Heidelberg 2010

categorization, social sciences, switching circuits, economic/financial prediction, image processing, and so on [4,5,6,7,8,9,10].

The reduct of an information system is not unique. There may be many subsets of attributes, which preserve the equivalence class structure (i.e., the knowledge) expressed in the information system. Although several variants of reduct algorithms are reported in the literature, at the moment, there is no accredited best heuristic reduct algorithm. So far, it is still an open research area in rough sets theory.

Particle swarm algorithm is inspired by social behavior patterns of organisms that live and interact within large groups. In particular, it incorporates swarming behaviors observed in flocks of birds, schools of fish, or swarms of bees, and even human social behavior, from which the Swarm Intelligence (SI) paradigm has emerged [11]. The swarm intelligent model helps to find optimal regions of complex search spaces through interaction of individuals in a population of particles [12,13,14]. As an algorithm, its main strength is its fast convergence, which compares favorably with many other global optimization algorithms [15,16]. It has exhibited good performance across a wide range of applications [17,18,19,20,21]. The particle swarm algorithm is particularly attractive for feature selection as there seems to be no heuristic that can guide search to the optimal minimal feature subset. Additionally, it can be the case that particles discover the best feature combinations as they proceed throughout the search space.

The main focus of this Chapter is to investigate Multi-Swarm Synergetic Optimization (MSSO) algorithm and its application in finding multiple reducts for difficult problems. The rest of the Chapter is organized as follows. Some related terms and theorems on rough set theory are explained briefly in Section 3. Particle swarm model is presented and the effects on the change of the neighborhoods of particles are analyzed in Section 4. The proposed approach based on particle swarm algorithm is presented in Section 5. In this Section, we describe the MSSO model in detail and theoretically prove the properties related to the convergence of the proposed algorithm. Experiment settings, results and discussions are given in Section 6 and finally conclusions are given in Section 7.

2   Related Research Works

Usually real world objects are the corresponding tuple in some decision tables. They store a huge quantity of data, which is hard to manage from a computational point of view. Finding reducts in a large information system is still an NP-hard problem [22]. The high complexity of this problem has motivated investigators to apply various approximation techniques to find near-optimal solutions. Many approaches have been proposed for finding reducts, e.g., discernibility matrices, dynamic reducts, and others [23,24]. The heuristic algorithm is a better choice. Hu et al. [25] proposed a heuristic algorithm using discernibility matrix. The approach provided a weighting mechanism to rank attributes. Zhong and Dong [26] presented a wrapper approach using rough sets theory with greedy heuristics for feature subset selection. The aim of feature subset selection is to find out a minimum set of relevant attributes that describe the dataset as well as the original all attributes do. So finding reduct is similar to feature selection. Zhong's algorithm employed the number of consistent instances as heuristics. Banerjee et al. [27] presented various attempts of using Genetic Algorithms in order to obtain reducts.

Conventional approaches for knowledge discovery always try to find a good reduct or to select a set of features [28]. In the knowledge discovery applications, only the good reduct can be applied to represent knowledge, which is called a single body of knowledge. In fact, many information systems in the real world have multiple reducts, and each reduct can be applied to generate a single body of knowledge. Therefore, multi-knowledge based on multiple reducts has the potential to improve knowledge representation and decision accuracy [29]. However, it would be exceedingly time-consuming to find multiple reducts in an instance information system with larger numbers of attributes and instances. In most of strategies, different reducts are obtained by changing the order of condition attributes and calculating the significance of different condition attribute combinations against decision attribute(s). It is a complex multi-restart processing about condition attribute increasing or decreasing in quantity. Population-based search approaches are of great benefits in the multiple reduction problems, because different individual trends to be encoded to different reduct. So it is attractive to find multiple reducts in the decision systems.

3   Rough Set Reduction

The basic concepts of rough set theory and its philosophy are presented and illustrated with examples in [1,2,3,26,28,30,31]. Here, we illustrate only the relevant basic ideas of rough sets that are relevant to the present work.

In rough set theory, an information system is denoted in 4-tuple by $S = (U, A, V, f)$, where U is the universe of discourse, a non-empty finite set of N objects $\{x_1, x_2, \cdots, x_N\}$. A is a non-empty finite set of attributes such that $a : U \rightarrow V_a$ for every $a \in A$ (V_a is the value set of the attribute a).

$$V = \bigcup_{a \in A} V_a$$

$f : U \times A \rightarrow V$ is the total decision function (also called the information function) such that $f(x, a) \in V_a$ for every $a \in A$, $x \in U$. The information system can also be defined as a decision table by $S = (U, C, D, V, f)$. For the decision table, C and D are two subsets of attributes. $A = \{C \cup D\}$, $C \cap D = \emptyset$, where C is the set of input features and D is the set of class indices. They are also called condition and decision attributes, respectively.

Let $a \in C \cup D$, $P \subseteq C \cup D$. A binary relation $IND(P)$, called an equivalence (indiscernibility) relation, is defined as follows:

$$IND(P) = \{(x, y) \in U \times U \quad | \quad \forall a \in P, f(x, a) = f(y, a)\} \qquad (1)$$

The equivalence relation $IND(P)$ partitions the set U into disjoint subsets. Let $U/IND(P)$ denote the family of all equivalence classes of the relation $IND(P)$. For simplicity of notation, U/P will be written instead of $U/IND(P)$. Such a partition of the universe is denoted by $U/P = \{P_1, P_2, \cdots, P_i, \cdots\}$, where P_i is an equivalence class of P, which is denoted $[x_i]_P$. Equivalence classes U/C and U/D will be called condition and decision classes, respectively.

Lower Approximation: Given a decision table $T = (U, C, D, V, f)$. Let $R \subseteq C \cup D$, $X \subseteq U$ and $U/R = \{R_1, R_2, \cdots, R_i, \cdots\}$. The R-lower approximation set of X is the set of all elements of U which can be with certainty classified as elements of X, assuming knowledge R. It can be presented formally as

$$APR_{\overline{R}}(X) = \bigcup \{R_i \mid R_i \in U/R, R_i \subseteq X\} \tag{2}$$

Positive Region: Given a decision table $T = (U, C, D, V, f)$. Let $B \subseteq C$, $U/D = \{D_1, D_2, \cdots, D_i, \cdots\}$ and $U/B = \{B_1, B_2, \cdots, B_i, \cdots\}$. The B-positive region of D is the set of all objects from the universe U which can be classified with certainty to classes of U/D employing features from B, i.e.,

$$POS_B(D) = \bigcup_{D_i \in U/D} APR_{\overline{B}}(D_i) \tag{3}$$

Positive Region: Given a decision table $T = (U, C, D, V, f)$. Let $B \subseteq C$, $U/D = \{D_1, D_2, \cdots, D_i, \cdots\}$ and $U/B = \{B_1, B_2, \cdots, B_i, \cdots\}$. The B-positive region of D is the set of all objects from the universe U which can be classified with certainty to classes of U/D employing features from B, i.e.,

$$POS_B(D) = \bigcup_{D_i \in U/D} B_-(D_i) \tag{4}$$

Reduct: Given a decision table $T = (U, C, D, V, f)$. The attribute $a \in B \subseteq C$ is $D - dispensable$ in B, if $POS_B(D) = POS_{(B-\{a\})}(D)$; otherwise the attribute a is $D - indispensable$ in B. If all attributes $a \in B$ are $D - indispensable$ in B, then B will be called $D - independent$. A subset of attributes $B \subseteq C$ is a $D - reduct$ of C, iff $POS_B(D) = POS_C(D)$ and B is $D - independent$. It means that a reduct is the minimal subset of attributes that enables the same classification of elements of the universe as the whole set of attributes. In other words, attributes that do not belong to a reduct are superfluous with regard to classification of elements of the universe. Usually, there are many reducts in an instance information system. Let $2^{|A|}$ represent all possible attribute subsets $\{\{a_1\}, \cdots, \{a_{|A|}\}, \{a_1, a_2\}, \cdots, \{a_1, \cdots, a_{|A|}\}\}$. Let RED represent the set of reducts, i.e.,

$$RED = \{B \mid POS_B(D) = POS_C(D), POS_{(B-\{a\})}(D) < POS_B(D)\} \tag{5}$$

Multi-knowledge: Given a decision table $T = (U, C, D, V, f)$. Let RED represent the set of reducts, Let φ is a mapping from the condition space to the decision space. Then multi-knowledge can be defined as follows:

$$\Psi = \{\varphi_B \mid B \in RED\} \tag{6}$$

Reduced Positive Universe and *Reduced Positive Region*: Given a decision table $T = (U, C, D, V, f)$. Let $U/C = \{[u_1']_C, [u_2']_C, \cdots, [u_m']_C\}$, Reduced Positive Universe U' can be written as:

$$U' = \{u_1', u_2', \cdots, u_m'\}. \tag{7}$$

and

$$POS_C(D) = [u_{i_1}']_C \cup [u_{i_2}']_C \cup \cdots \cup [u_{i_t}']_C. \tag{8}$$

Where $\forall u_{i_s}' \in U'$ and $|[u_{i_s}']_C/D| = 1 (s = 1, 2, \cdots, t)$. Reduced positive universe can be written as:

$$U_{pos}' = \{u_{i_1}', u_{i_2}', \cdots, u_{i_t}'\}. \tag{9}$$

and $\forall B \subseteq C$, reduced positive region

$$POS_B'(D) = \bigcup_{X \in U'/B \wedge X \subseteq U_{pos}' \wedge |X/D| = 1} X \tag{10}$$

where $|X/D|$ represents the cardinality of the set X/D. $\forall B \subseteq C$, $POS_B(D) = POS_C(D)$ if $POS_B' = U_{pos}'$ [31]. It is to be noted that U' is the reduced universe, which usually would reduce significantly the scale of datasets. It provides a more efficient method to observe the change of positive region when we search the reducts. We didn't have to calculate U/C, U/D, U/B, $POS_C(D)$, $POS_B(D)$ and then compare $POS_B(D)$ with $POS_C(D)$ to determine whether they are equal to each other or not. We only calculate U/C, U', U_{pos}', POS_B' and then compare POS_B' with U_{pos}'.

4   Particle Swarm Optimization Algorithm

The classical particle swarm model consists of a swarm of particles, which are initialized with a population of random candidate solutions. They move iteratively through the d-dimension problem space to search the new solutions, where the fitness f can be measured by calculating the number of condition attributes in the potential reduction solution. Each particle has a position represented by a position-vector $\boldsymbol{p}_i$ (i is the index of the particle), and a velocity represented by a velocity-vector $\boldsymbol{v}_i$. Each particle remembers its own best position so far in a vector $\boldsymbol{p}_i^\#$, and its j-th dimensional value is $p_{ij}^\#$. The best position-vector among the swarm so far is then stored in a vector $\boldsymbol{p}^*$, and its j-th dimensional value is p_j^*. When the particle moves in a state space restricted to zero and one on each dimension, the change of probability with time steps is defined as follows:

$$P(p_{ij}(t) = 1) = f(p_{ij}(t-1), v_{ij}(t-1), p_{ij}^\#(t-1), p_j^*(t-1)). \tag{11}$$

where the probability function is

$$sig(v_{ij}(t)) = \frac{1}{1 + e^{-v_{ij}(t)}}. \tag{12}$$

At each time step, each particle updates its velocity and moves to a new position according to Eqs.(13) and (14):

$$v_{ij}(t) = wv_{ij}(t-1) + c_1 r_1 (p_{ij}^{\#}(t-1) - p_{ij}(t-1))$$
$$+ c_2 r_2 (p_j^*(t-1) - p_{ij}(t-1)) \tag{13}$$

$$p_{ij}(t) = \begin{cases} 1 & \text{if } \rho < sig(v_{ij}(t)); \\ 0 & \text{otherwise.} \end{cases} \tag{14}$$

Where c_1 is a positive constant, called as coefficient of the self-recognition component, c_2 is a positive constant, called as coefficient of the social component. r_1 and r_2 are the random numbers in the interval $[0,1]$. The variable w is called as the inertia factor, which value is typically setup to vary linearly from 1 to near 0 during the iterated processing. ρ is random number in the closed interval $[0,1]$. From Eq. (13), a particle decides where to move next, considering its current state, its own experience, which is the memory of its best past position, and the experience of its most successful particle in the swarm. Figure 1 illustrates how the position is reacted on by its velocity. The pseudo-code for particle swarm optimization algorithm is illustrated in Algorithm 1.

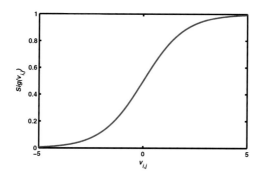

Fig. 1. Sigmoid function for PSO

The particle swarm algorithm can be described generally as a population of vectors whose trajectories oscillate around a region which is defined by each individual's previous best success and the success of some other particle. Some previous studies have discussed the trajectory of particles and its convergence [14,12,32,33]. Bergh and Engelbrecht [33] overviewed the theoretical studies, and extended these studies to investigate particle trajectories for general swarms to include the influence of the inertia term. They also provided a formal proof that each particle converges to a stable point. It has been shown that the trajectories of the particles oscillate as different sinusoidal waves and converge quickly, sometimes prematurely. Various methods have been used to identify some other particle to influence the individual.

Algorithm 1. Particle Swarm Optimization Algorithm

---

01. Initialize the size of the particle swarm n, and other parameters.
02. Initialize the positions and the velocities for all the particles randomly.
03. While (the end criterion is not met) do
04. $t = 1$;
05.   Calculate the fitness value of each particle;
06. $p^* = argmin_{i=1}^{n}(f(p^*(t-1)), f(p_1(t)), f(p_2(t)), \cdots, f(p_i(t)), \cdots, f(p_n(t)))$;
07.   For i= 1 to n
08. $p_i^{\#}(t) = argmin_{i=1}^{n}(f(p_i^{\#}(t-1)), f(p_i(t)))$;
09.   For $j = 1$ to d
10.   Update the j-th dimension value of p_i and v_i
10.   according to Eqs.(13),(14),(12);
12.   Next j
13.   Next i
13. $t + +$
14. End While.

Eberhart and Kennedy called the two basic methods as "$g$best model" and "$l$best model" [11]. In the $g$best model, the trajectory for each particle's search is influenced by the best point found by any member of the entire population. The best particle acts as an attractor, pulling all the particles towards it. Eventually all particles will converge to this position. In the $l$best model, particles have information only of their own and their nearest array neighbors' best ($l$best), rather than that of the whole swarm. Namely, in Eq. (13), $g$best is replaced by $l$best in the model. The $l$best model allows each individual to be influenced by some smaller number of adjacent members of the population array. The particles selected to be in one subset of the swarm have no direct relationship to the other particles in the other neighborhood. Typically $l$best neighborhoods comprise exactly two neighbors. When the number of neighbors increases to all but itself in the $l$best model, the case is equivalent to the $g$best model. Unfortunately there is a large computational cost to explore the neighborhood relation in each iteration when the number of neighbors is too little. Some previous studies has been shown that $g$best model converges quickly on problem solutions but has a weakness for becoming trapped in local optima, while $l$best model converges slowly on problem solutions but is able to "flow around" local optima, as the individuals explore different regions [36]. Some related research and development during the recent years are also reported in [21,34,35,37].

5   Rough Set Reduction Algorithm Based on Swarms

Blackwell and Branke [38] investigated a multi-swarm optimization specifically designed to work well in dynamic environments. The main idea is to split the population of particles into a set of interacting swarms. These swarms interact locally by an exclusion parameter and globally through a new anti-convergence operator. The results illustrated that the multiswarm optimizer significantly outperformed

the other considered approaches. Niu et al. [39] presented a multi-swarm cooperative particle swarm optimizer, inspired by the phenomenon of symbiosis in
natural ecosystems. The approach is based on a master - slave model, in which
a population consists of one master swarm and several slave swarms. The slave
swarms execute a single PSO or its variants independently to maintain the diversity of particles, while the master swarm evolves based on its own knowledge
and also the knowledge of the slave swarms. In the simulation studies, several
benchmark functions are performed, and the performances of their algorithms are
compared with the standard PSO (SPSO) to demonstrate the superiority. The
multi swarm approaches let several swarms of particles cooperate to find good
solutions. Usually they have to be designed for specific problems. In this Section,
we design a multi-swarm synergetic optimization algorithm for rough set reduction and multi-knowledge extraction. The sub-swarms are encoded with different
reducts, which is suitable for searching multiple reducts in decision systems.

5.1   Coding and Evaluation

Given a decision table $T = (U, C, D, V, f)$, the set of condition attributes, C,
consists of m attributes. We set up a search space of m dimensions for the reduction problem. Accordingly, each particle's position is represented as a binary
bit string of length m. Each dimension of the particle's position maps one condition attribute. The domain for each dimension is limited to 0 or 1. The value
'1' means the corresponding attribute is selected while '0' not selected. Each
position can be "decoded" to a potential reduction solution, a subset of C. The
particle's position is a series of priority levels of the attributes. The sequence
of the attribute will not be changed during the iteration. But after updating
the velocity and position of the particles, the particle's position may appear
real values such as 0.4, etc. It is meaningless for the reduction. Therefore, we
introduce a discrete particle swarm optimization technique for this reduction
problem. The particles updates its velocity according to Eq. (13), considering
its current state, its own experience, and the experience of its successful particle
in its neighborhood swarm. Each dimension of the particles' position would be
explored between 0 and 1 through Eqs. (12) and (14).

During the search procedure, each individual is evaluated using the fitness.
According to the definition of rough set reduct, the reduction solution must
ensure the decision ability is the same as the primary decision table and the
number of attributes in the feasible solution is kept as low as possible. In our
algorithm, we first evaluate whether the potential reduction solution satisfies
$POS'_E = U'_{pos}$ or not (E is the subset of attributes represented by the potential
reduction solution). If it is a feasible solution, we calculate the number of '1'
in it. The solution with the lowest number of '1' would be selected. For the
particle swarm, the lower number of '1' in its position, the better the fitness of
the individual is. $POS'_E = U'_{pos}$ is used as the criterion of the solution validity.

In the proposed encoding representations, we consider particle's position encoding as the binary representation of an integer. The step size is equal to 1, that
is, the dimension of the search space is then 1. In practice, when the binary string

is too long for a large scale attribute reduction problem, it is difficult to use it as an integer. It is time-consuming for each iteration. So we split it into a small number (say H) of shorter binary strings, each one is seen as an integer. Then the dimension of the problem is not anymore 1, but H. The swarm algorithm with two strategies is called as Bi-metrics Binary PSO (Figures 2 and 3).

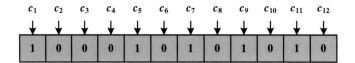

Fig. 2. Bi-metrics Binary Representation 1

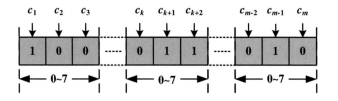

Fig. 3. Bi-metrics Binary Representation 2

5.2 Multi-Swarm Synergetic Model

To employ a multi-swarm, the solution vector is split amongst the different populations according to some rule in such a way that the simplest of the schemes does not allow any overlap between the spaces covered by different populations. To find a solution to the original problem, representatives from all the populations are combined to form the potential solution vector, which, in turn, is passed on the error function. This adds a new dimension to the survival game: cooperation between different populations [33,40].

As mentioned above, one of the most important applications is to solve multiple reduct problems and multi-knowledge extraction. Those reducts usually share some common classification characteristics in the information systems. They are apt to cluster into different groups. Sometime they are also the members of several groups at the same time [41]. To match the classification characteristics, we introduce a multi-swarm search algorithm for them. In the algorithm, all particles are clustered spontaneously into different sub-swarms of the whole swarm. Every particle can connect to more than one sub-swarm, and a crossover neighborhood topology is constructed between different sub-swarms. The particles in the same sub-swarm would carry some similar functions as possible and search their optimal. Each sub-swarm would approach its appropriate position (solution), which would be helpful for the whole swarm to keep in a good balance state. Figure 4 illustrates a multi-swarm topology. In the swarm system, a swarm

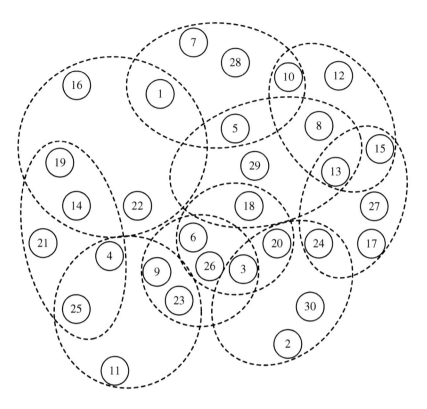

Fig. 4. A multi-swarm topology

with 30 particles is organized into 10 sub-swarms, each sub-swarm consisting of 5 particles. Particles 3 and 13 have the maximum membership level, 3. During the iteration process, the particle updates its velocity followed by the location of the best fitness achieved so far by the particle itself and by the location of the best fitness achieved so far across all its neighbors in all sub-swarms it belongs to. The process makes an important influence on the particles' ergodic and synergetic performance. The multi-swarm algorithm for the reduction problem is illustrated as follows:

Step 1. Calculate U', U'_{pos} using Eqs. (7) and (9).
Step 2. Initialize the size of the particle swarm n, and other parameters. Initialize the positions and the velocities for all the particles randomly.
Step 3. Multiple sub-swarms n are organized into a crossover neighborhood topology. A particle can join more than one sub-swarm. Each particle has the maximum membership level l, and each sub-swarm accommodates default number of particles m.
Step 4. Decode the positions and evaluate the fitness for each particles, if $POS'_E \neq U'_{pos}$, the fitness is punished as the total number of the condition attributes, else the fitness is the number of '1' in the position.

Step 5. Find the best particle in the swarm, and find the best one in each sub-swarms. If the "global best" of the swarm is improved, $noimprove = 0$, otherwise, $noimprove = 1$. Update velocity and position for each particle at the iteration t.

5.01 For $m = 1$ to $subs$

5.02 $p^* = argmin_{i=1}^{subs_m}(f(p^*(t-1)), f(p_1(t)),$

5.02 $f(p_2(t)), \cdots, f(p_i(t)), \cdots, f(p_{subs_m}(t)));$

5.03   For $ss = 1$ to $subs_m$

5.04 $p_i^{\#}(t) = argmin(f(p_i^{\#}(t-1)), f(p_i(t)));$

5.05   For $d = 1$ to D

5.06   Update the d-th dimension value of p_i and v_i

5.06   according to Eqs.(13), (12), and (14);

5.07   Next d

5.08   Next ss

5.09 Next m

Step 6. If $noimprove = 1$, goto Step 3, the topology is re-organized. If the end criterion is not met, goto Step 4. Otherwise, provide the best solution (output), the fitness.

5.3   Algorithm Analysis

For analyzing the convergence of the multi-swarm algorithm, we first introduce the definitions and lemmas [42,43,44], and then theoretically prove that the algorithm converges with a probability 1 or strongly towards the global optimal.

Xu et $al.$ [45] analyzed the search capability of an algebraic crossover through classifying the individual space of genetic algorithms, which is helpful to comprehend the search of genetic algorithms such that premature convergence and deceptive problems [46] could be avoided. In this Subsection, we also attempt to theoretically analyze the performance of the multi-swarm algorithm with crossover neighborhood topology. For the sake of convenience, let crossover operator $|_c$ denote the wheeling-round-the-best-particles process.

Consider the problem (P) as

$$(P) = min\{f(\boldsymbol{x}) : \boldsymbol{x} \in D\} \tag{15}$$

where $\boldsymbol{x} = (x_1, x_2, \cdots, x_n)^T$, $f(\boldsymbol{x}) : D \to R$ is the objective function and D is a compact Hausdorff space. Applying our algorithm the problem (P) may be transformed to P' as

$$(P') = \begin{cases} min f(\boldsymbol{x}) \\ \boldsymbol{x} \in \Omega = [-s, s]^n \end{cases} \tag{16}$$

where Ω is the set of feasible solutions of the problem. A swarm is a set, which consists of some feasible solutions of the problem. Assume S as the encoding space of D. A neighborhood function is a mapping $\mathcal{N} : \Omega \to 2^{\Omega}$, which defines for each solution $S \in \Omega$ a subset $\mathcal{N}(S)$ of Ω, called a neighborhood. Each solution in $\mathcal{N}(S)$ is a neighbor of S. A local search algorithm starts off with an initial solution and then continually tries to find better solutions by searching neighborhoods [47]. Most generally said, in swarm algorithms the encoding types S of particles in the search space D are often represented as strings of a fixed-length L over an alphabet. Without loss of generality, S can be described as

$$S = \underbrace{z_m \times \cdots \times z_m}_{L} \tag{17}$$

where z_m is a finite field about integer number mod m. Most often, it is the binary alphabet, *i.e.* $m = 2$.

Proposition 1. *If k alleles are '0's in the nontrivial ideal Ω, i.e. $L - k$ alleles are uncertain, then θ_Ω partitions Ω into 2^k disjoint subsets as equivalence classes corresponding to Holland's schema theorem [48,49], i.e., each equivalence class consists of some '1's which k alleles in Ω with '0' are replaced by '1's. Let $A \in S/\theta_\Omega$, then there is an minimal element m of A under partial order $(S, \vee, \wedge, \neg)$, such that $A = \{m \vee x \mid x \in \Omega\}$.*

Theorem 1. *Let A, B, C are three equivalence classes on θ_Ω, where θ_Ω is the congruence relation about Ω. $\exists\, x \in A$, $y \in B$, and $x \mid_c y \in C$, then $C = \{x \mid_c y \mid x \in A, y \in B\}$.*

Proof. Firstly, we verify that for any $d_1, d_2 \in \Omega$, if $x \mid_c y \in C$, then $(x \vee d_1) \mid_c (y \vee d_2) \in C$. In fact,

$$\begin{aligned}(x \vee d_1) \mid_c (y \vee d_2) &= (x \vee d_1)c \vee (y \vee d_2)\bar{c} \\ & (xc \vee y\bar{c}) \vee (d_1 c \vee d_2 \bar{c}) \tag{18} \\ & (x \mid_c y) \vee (d_1 c \vee d_2 \bar{c})\end{aligned}$$

Obviously, $(d_1 c \vee d_2 \bar{c}) \in \Omega$, so $(x \vee d_1) \mid_c (y \vee d_2) \equiv (x \mid_c y)(\bmod \theta_\Omega)$, *i.e.* $(x \vee d_1) \mid_c (y \vee d_2) \in \Omega$.

Secondly, from Proposition 1, $\exists m, n, d_3, d_4 \in \Omega$ of A, B, such that $x = m \vee d_3$, $y = n \vee d_4$. As a result of analysis in Eq.(18), $x \mid_c y \equiv (m \mid_c n)(\bmod \theta_\Omega)$, *i.e.*, $m \mid_c n \in C$.

Finally, we verify that $m \mid_c n$ is a minimal element of C and $(m \mid_c n) \vee d = (m \vee d) \mid_c (n \vee d)$. As a result of analysis in Eq.(18), if $d_1 = d_2 = d$, then $m \mid_c n \vee d = (m \vee d) \mid_c (n \vee d)$. Therefore $m \mid_c n$ is a minimal element of C.

To conclude, $C = \{(m \mid_c n) \vee d \mid d \in \Omega\} = \{x \mid_c y \mid x \in A, y \in B\}$. The theorem is proven.

Proposition 2. *Let A, B are two equivalence classes on θ_Ω, and there exist $x \in A$, $y \in B$, such that $x \mid_c y \in C$, then, $x \mid_c y$ makes ergodic search C while x and y make ergodic search A and B, respectively.*

Definition 1 (Convergence in terms of probability). *Let ξ_n a sequence of random variables, and ξ a random variable, and all of them are defined on the same probability space. The sequence ξ_n converges with a probability of ξ if*

$$\lim_{n \to \infty} P(|\xi_n - \xi| < \varepsilon) = 1 \tag{19}$$

for every $\varepsilon > 0$.

Definition 2 (Convergence with a probability of 1). *Let ξ_n a sequence of random variables, and ξ a random variable, and all of them are defined on the same probability space. The sequence ξ_n converges almost surely or almost everywhere or with probability of 1 or strongly towards ξ if*

$$P\left(\lim_{n\to\infty}\xi_n=\xi\right)=1;\tag{20}$$

or

$$P\left(\bigcap_{n=1}^{\infty}\bigcup_{k\geq n}[|\xi_n-\xi|\geq\varepsilon]\right)=0\tag{21}$$

for every $\varepsilon>0$.

Theorem 2. *Let $\boldsymbol{x}^*$ is the global optimal solution to the problem (P'), and $f^*=f(\boldsymbol{x}^*)$. Assume that the clubs-based multi-swarm algorithm provides position series $\boldsymbol{x}_i(t)$ $(i=1,2,\cdots,n)$ at time t by the iterated procedure. $\boldsymbol{p}^*$ is the best position among all the swarms explored so far, i.e.*

$$\boldsymbol{p}^*(t)=arg\min_{1\leq i\leq n}\left(f(\boldsymbol{p}^*(t-1)),f(\boldsymbol{p}_i(t))\right)\tag{22}$$

Then,

$$P\left(\lim_{t\to\infty}f(\boldsymbol{p}^*(t))=f^*\right)=1\tag{23}$$

Proof. Let

$$D_0=\{\boldsymbol{x}\in\Omega|f(\boldsymbol{x})-f^*<\varepsilon\}\tag{24}$$
$$D_1=\Omega\setminus D_0$$

for every $\varepsilon>0$.

While the different swarm searches their feasible solutions by themselves, assume Δp is the difference of the particle's position among different club swarms at the iteration time t. Therefore $-s\leq\Delta p\leq s$. $Rand(-1,1)$ is a normal distributed random number within the interval [-1,1]. According to the update of the velocity and position by Eqs.(13)~(14), Δp belongs to the normal distribution, i.e. $\Delta p\sim[-s,s]$. During the iterated procedure from the time t to $t+1$, let q_{ij} denote that $\boldsymbol{x}(t)\in D_i$ and $\boldsymbol{x}(t+1)\in D_j$. Accordingly the particles' positions in the swarm could be classified into four states: q_{00}, q_{01}, q_{10} and q_{01}. Obviously $q_{00}+q_{01}=1$, $q_{10}+q_{11}=1$. According to Borel-Cantelli Lemma and Particle State Transference [21], proving by the same methods, $q_{01}=0$; $q_{00}=1$; $q_{11}\leq c\in(0,1)$ and $q_{10}\geq1-c\in(0,1)$.

For $\forall\varepsilon>0$, let $p_k=P\{|f(\boldsymbol{p}^*(k))-f^*|\geq\varepsilon\}$, then

$$p_k=\begin{cases}0 & \text{if }\exists T\in\{1,2,\cdots,k\},\,\boldsymbol{p}^*(T)\in D_0\\\bar{p}_k & \text{if }\boldsymbol{p}^*(t)\notin D_0,\,t=1,2,\cdots,k\end{cases}\tag{25}$$

According to Particle State Transference Lemma,

$$\bar{p}_k = P\{\boldsymbol{p}^*(t) \notin D_0, t = 1, 2, \cdots, k\} = q_{11}^k \leq c^k. \tag{26}$$

Hence,

$$\sum_{k=1}^{\infty} p_k \leq \sum_{k=1}^{\infty} c^k = \frac{c}{1-c} < \infty. \tag{27}$$

According to Borel-Cantelli Lemma,

$$P\left(\bigcap_{t=1}^{\infty} \bigcup_{k \geq t} |f(\boldsymbol{p}^*(k)) - f^*| \geq \varepsilon\right) = 0 \tag{28}$$

As defined in Definition 2, the sequence $f(\boldsymbol{p}^*(t))$ converges almost surely or almost everywhere or with probability 1 or strongly towards f^*. The theorem is proven.

6   Experiment Results and Discussions

The algorithms used for performance comparison were the Standard Particle Swarm Optimization (SPSO) ([11]) and a Genetic Algorithm (GA) ([50,51]). These algorithms share many similarities. GA is powerful stochastic global search and optimization methods, which are also inspired from the nature like the PSO. Genetic algorithms mimic an evolutionary natural selection process. Generations of solutions are evaluated according to a fitness value and only those candidates with high fitness values are used to create further solutions via crossover and mutation procedures. Both methods are valid and efficient methods in numeric programming and have been employed in various fields due to their strong convergence properties. Specific parameter settings for the algorithms are described in Table 1, where D is the dimension of the position, i.e., the number of condition attributes. Besides the first small scale rough set reduction problem shown in Table 2, the maximum number of iterations is set as

Table 1. Parameter settings for the algorithms

Algorithm	Parameter name	Value
	Size of the population	$(even)(int)(10 + 2 * sqrt(D))$
GA	Probability of crossover	0.8
	Probability of mutation	0.01
	Swarm size	$(even)(int)(10 + 2 * sqrt(D))$
	Self coefficient c_1	$0.5 + log(2)$
PSO(s)	Social coefficient c_2	$0.5 + log(2)$
	Inertia weight w	0.91
	Clamping Coefficient ρ	0.5

$(int)(0.1 * recnum + 10 * (nfields - 1))$ for each trial, where $recnum$ is the number of records/rows and $nfields - 1$ is the number of condition attributes. Each experiment (for each algorithm) was repeated 10 times with different random seeds. If the standard deviation is larger than 20%, the number of trials were increased to 20.

To analyze the effectiveness and performance of the considered algorithms, first we tested a small scale rough set reduction problem shown in Table 2. In the experiments, the maximum number of iterations was fixed as 10. Each experiment were repeated 10 times with different random seeds. The results (the number of reduced attributes) for 10 GA runs were all 2. The results of 10 PSO runs were also all 2. The optimal result is supposed to be 2. But the reduction result for 10 GA runs is $\{2, 3\}$ while the reduction result for 10 PSO runs are $\{1, 4\}$ and $\{2, 3\}$. Table 3 depicts the reducts for Table 2 (Please also see Figure 5). For the small scale rough set reduction problem, GA has a same result than PSO. GA only provides one reduct, while PSOs provide one more reduct. There seems a conflict between the instances 13 and 15. It depends on conflict analysis and how to explain the knowledge, which will be tackled in future publications.

Table 2. A decision table

Instance	c_1	c_2	c_3	c_4	d
x_1	1	1	1	1	0
x_2	2	2	2	1	1
x_3	1	1	1	1	0
x_4	2	3	2	3	0
x_5	2	2	2	1	1
x_6	3	1	2	1	0
x_7	1	2	3	2	2
x_8	2	3	1	2	3
x_9	3	1	2	1	1
x_{10}	1	2	3	2	2
x_{11}	3	1	2	1	1
x_{12}	2	3	1	2	3
x_{13}	4	3	4	2	1
x_{14}	1	2	3	2	3
x_{15}	4	3	4	2	2

Further we considered the datasets in Table 4 from AFS[1], AiLab[2] and UCI[3]. Figures 6, 7 and 8 illustrate the performance of the algorithms for lung-cancer, lymphography and mofn-3-7-10 datasets, respectively. For lung-cancer dataset, the results (the number of reduced attributes) for 10 GA runs were 10: { 1, 4, 8, 13, 18, 34, 38, 40, 50, 55 } (The number before the colon is the number of condition attributes, the numbers in brackets are attribute index, which represents

---

[1] http://sra.itc.it/research/afs/
[2] http://www.ailab.si/orange/datasets.asp
[3] http://www.datalab.uci.edu/data/mldb-sgi/data/

Table 3. A reduction of the data in Table 2

Reduct	Instance	c_1	c_2	c_3	c_4	d
{1,4}						
	x_1	1			1	0
	x_2	2			1	1
	x_4	2			3	0
	x_6	3			1	0
	x_7	1			2	2
	x_8	2			2	3
	x_9	3			1	1
	x_{13}	4			2	1
	x_{14}	1			2	3
	x_{15}	4			2	2
{2,3}						
	x_1		1	1		0
	x_2		2	2		1
	x_4		3	2		0
	x_6		1	2		0
	x_7		2	3		2
	x_8		3	1		3
	x_9		1	2		1
	x_{13}		3	4		1
	x_{14}		2	3		3
	x_{15}		3	4		2

Table 4. Datasets used in the experiments

Dataset	Size	$|C|$	Class	GA		PSO		MSSO	
				L	R	L	R	L	R
lung-cancer	27	56	3	10	1	6	3	6	3
zoo	101	16	7	5	1	5	2	5	3
corral	128	6	2	4	1	4	1	4	1
lymphography	148	18	4	7	1	6	2	6	1
hayes-roth	160	4	3	3	1	3	1	3	1
shuttle-landing-control	253	6	2	6	-	6	-	6	-
soybean-large-test	296	35	15	12	1	10	3	8	3
monks	432	6	2	3	1	3	1	3	1
xd6-test	512	9	2	9	-	9	-	9	-
balance-scale	625	4	3	4	-	4	-	4	-
breast-cancer-wisconsin	683	9	2	4	1	4	2	4	3
mofn-3-7-10	1024	10	2	7	1	7	1	7	1
parity5+5	1024	10	2	5	1	5	1	5	1

a reduction solution); the results of 10 PSO runs were PSO 7: { 1, 6, 12, 27, 29, 35, 41 }, 6: { 2, 3, 12, 22, 25, 56 }, 7: { 2, 3, 8, 12, 22, 31, 49 }; the results of 10 MSSO runs were 6: { 4, 6, 14, 31, 49, 53 }, 6: { 4, 6, 9, 23, 27, 54 }, 6: { 3, 10,

20, 32, 34, 56 }. For lymphography dataset, the results of 10 GA runs all were 7: { 2, 11, 12, 13, 14, 16, 18 }; the results of 10 PSO runs were PSO 7: { 3, 8, 11, 12, 13, 14, 15 }, 6: { 2, 13, 14, 15, 16, 18 }, 6: { 2, 13, 14, 15, 16, 18 }; the results of 10 MSSO runs were 6:{ 2, 13, 14, 15, 16, 18 }. For soybean-large-test dataset, the results of 10 GA runs all were 12: { 1, 3, 4, 5, 6, 7, 13, 15, 16, 22, 32, 35 }; the results of 10 PSO runs were 10:{ 1, 3, 5, 6, 7, 12, 15, 18, 22, 31 }, 10: { 1, 3, 5, 6, 7, 15, 23, 26, 28, 30 }, 10: { 1, 2, 3, 6, 7, 9, 15, 21, 22, 30 }; the results of 10 MSSO runs were 9: { 1, 3, 5, 6, 7, 15, 22, 30, 34 }, 8: { 1, 3, 4, 6, 7, 10, 15, 22 }, 9: { 1, 3, 5, 6, 7, 13, 22, 25, 31 }. Other results are shown in Table 4, in which L is the minimum length and R is the number of the obtained reducts. "-" means that all features *cannot* be reduced. MSSO usually obtained better results than GA and PSO, specially for the large scale problems. Although the three algorithms achieved the same-length results for some datasets, MSSO usually can provide more reducts for multi-knowledge extraction. It indicates that MSSO has a better performance than other two algorithms for the larger scale rough set reduction problem. It is to be noted that PSO usually can obtain more candidate solutions for the reduction problems.

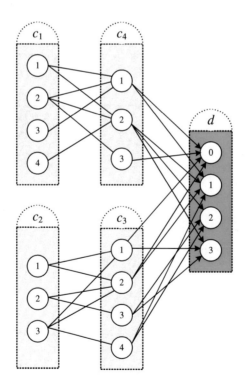

Fig. 5. Rule networks based on Table 3

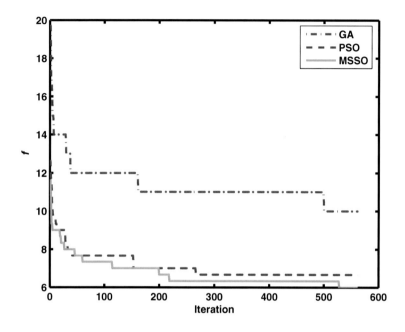

Fig. 6. Performance of rough set reduction for lung-cancer dataset

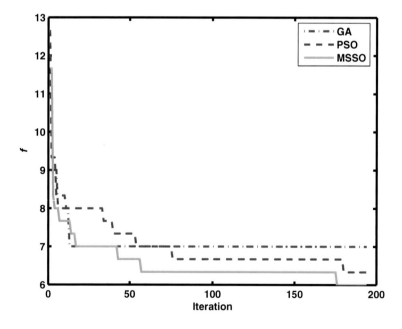

Fig. 7. Performance of rough set reduction for lymphography dataset

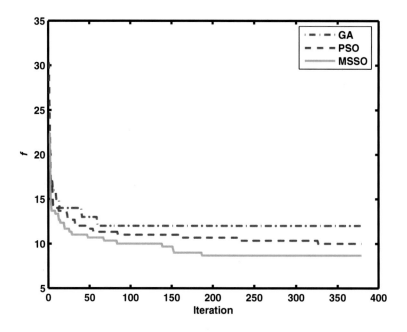

Fig. 8. Performance of rough set reduction for soybean-large-test dataset

7   Conclusions

In this Chapter, we investigated multi-knowledge extraction using particle swarm optimization and genetic algorithm techniques. The considered approaches discovered the good feature combinations in an efficient way to observe the change of positive region as the particles explored the search space. The multi-swarm search approach offer great benefits for multiple reduction problems, because different individuals encode different reducts. Empirical results indicate that the proposed approach usually obtained better results than GA and standard PSO, specially for large scale problems, although its stability need to be improved in further research. MSSO has better convergence than GA for the larger scale rough set reduction problem, although MSSO is worst for some small scale rough set reduction problems. MSSO also can obtain more candidate solutions for the reduction problems. Empirical results illustrated that the multi-swarm search approach was an effective approach to solve multi-knowledge extraction.

Acknowledgements

This work was partly supported by NSFC (60873054) and DLMU (DLMU-ZL-200709).

References

1. Pawlak, Z.: Rough Sets. International Journal of Computer and Information Sciences 11, 341–356 (1982)
2. Pawlak, Z.: Rough Sets: Present State and The Future. Foundations of Computing and Decision Sciences 18, 157–166 (1993)
3. Pawlak, Z.: Rough Sets and Intelligent Data Analysis. Information Sciences 147, 1–12 (2002)
4. Kusiak, A.: Rough Set Theory: A Data Mining Tool for Semiconductor Manufacturing. IEEE Transactions on Electronics Packaging Manufacturing 24, 44–50 (2001)
5. Shang, C., Shen, Q.: Rough Feature Selection for Neural Network Based Image Classification. International Journal of Image and Graphics 2, 541–555 (2002)
6. Tay, F.E.H.: Economic And Financial Prediction Using Rough Sets Model. European Journal of Operational Research 141, 641–659 (2002)
7. Świniarski, R.W., Skowron, A.: Rough Set Methods in Feature Selection and Recognition. Pattern Recognition Letters 24, 833–849 (2003)
8. Beaubouef, T., Ladner, R., Petry, F.: Rough Set Spatial Data Modeling for Data Mining. International Journal of Intelligent Systems 19, 567–584 (2004)
9. Shen, L., Tay, F.E.H.: Tay Applying Rough Sets to Market Timing Decisions. Decision Support Systems 37, 583–597 (2004)
10. Gupta, K.M., Moore, P.G., Aha, D.W., Pal, S.K.: Rough Set Feature Selection Methods for Case-Based Categorization of Text Documents. In: Pal, S.K., Bandyopadhyay, S., Biswas, S. (eds.) PReMI 2005. LNCS, vol. 3776, pp. 792–798. Springer, Heidelberg (2005)
11. Kennedy, J., Eberhart, R.: Swarm Intelligence. Morgan Kaufmann Publishers, San Francisco (2001)
12. Clerc, M., Kennedy, J.: The Particle Swarm-Explosion, Stability, and Convergence in a Multidimensional Complex Space. IEEE Transactions on Evolutionary Computation 6, 58–73 (2002)
13. Clerc, M.: Particle Swarm Optimization. ISTE Publishing Company, London (2006)
14. Liu, H., Abraham, A., Clerc, M.: Chaotic Dynamic Characteristics in Swarm Intelligence. Applied Soft Computing Journal 7, 1019–1026 (2007)
15. Parsopoulos, K.E., Vrahatis, M.N.: Recent approaches to global optimization problems through particle swarm optimization. Natural Computing 1, 235–306 (2002)
16. Abraham, A., Guo, H., Liu, H.: Swarm Intelligence: Foundations, Perspectives and Applications. In: Nedjah, N., Mourelle, L. (eds.) Swarm Intelligent Systems. Studies in Computational Intelligence, pp. 3–25. Springer, Germany (2006)
17. Salman, A., Ahmad, I., Al-Madani, S.: Particle Swarm Optimization for Task Assignment Problem. Microprocessors and Microsystems 26, 363–371 (2002)
18. Sousa, T., Silva, A., Neves, A.: Particle Swarm Based Data Mining Algorithms for Classification Tasks. Parallel Computing 30, 767–783 (2004)
19. Liu, B., Wang, L., Jin, Y., Tang, F., Huang, D.: Improved Particle Swarm Optimization Combined With Chaos. Chaos, Solitons and Fractals 25, 1261–1271 (2005)
20. Schute, J.F., Groenwold, A.A.: A Study of Global Optimization Using Particle Swarms. Journal of Global Optimization 3, 103–108 (2005)
21. Liu, H., Abraham, A.: An Hybrid Fuzzy Variable Neighborhood Particle Swarm Optimization Algorithm for Solving Quadratic Assignment Problems. Journal of Universal Computer Science 13(7), 1032–1054 (2007)

22. Boussouf, M.: A Hybrid Approach to Feature Selection. In: Żytkow, J.M. (ed.) PKDD 1998. LNCS, vol. 1510, pp. 231–238. Springer, Heidelberg (1998)

23. Skowron, A., Rauszer, C.: The Discernibility Matrices and Functions in Information Systems. In: Świniarski, R.W. (ed.) Handbook of Applications and Advances of the Rough Set Theory, pp. 331–362. Kluwer Academic Publishers, Dordrecht (1992)

24. Zhang, J., Wang, J., Li, D., He, H., Sun, J.: A New Heuristic Reduct Algorithm Base on Rough Sets Theory. In: Dong, G., Tang, C., Wang, W. (eds.) WAIM 2003. LNCS, vol. 2762, pp. 247–253. Springer, Heidelberg (2003)

25. Hu, K., Diao, L., Lu, Y.-c., Shi, C.-Y.: A Heuristic Optimal Reduct Algorithm. In: Leung, K.-S., Chan, L., Meng, H. (eds.) IDEAL 2000. LNCS, vol. 1983, pp. 139–144. Springer, Heidelberg (2000)

26. Zhong, N., Dong, J.: Using Rough Sets with Heuristics for Feature Selection. Journal of Intelligent Information Systems 16, 199–214 (2001)

27. Banerjee, M., Mitra, S., Anand, A.: Feature Selection Using Rough Sets. Studies in Computational Intelligence, vol. 16, pp. 3–20. Springer, Heidelberg (2006)

28. Wu, Q., Bell, D.: Multi-knowledge Extraction and Application. In: Wang, G., Liu, Q., Yao, Y., Skowron, A. (eds.) RSFDGrC 2003. LNCS (LNAI), vol. 2639, pp. 574–575. Springer, Heidelberg (2003)

29. Wu, Q.: Multiknowledge and Computing Network Model for Decision Making and Localisation of Robots. Thesis, University of Ulster (2005)

30. Wang, G.: Rough Reduction in Algebra View and Information View. International Journal of Intelligent Systems 18, 679–688 (2003)

31. Xu, Z., Liu, Z., Yang, B., Song, W.: A Quick Attibute Reduction Algorithm with Complexity of $Max(O(|C||U|), O(|C|^2|U/C|))$. Chinese Journal of Computers 29, 391–399 (2006)

32. Cristian, T.I.: The particle swarm optimization algorithm: convergence analysis and parameter selection. Information Processing Letters 85(6), 317–325 (2003)

33. van den Bergh, F., Engelbrecht, A.P.: A Study of Particle Swarm Optimization Particle Trajectories. Information Sciences 176(8), 937–971 (2006)

34. Grosan, C., Abraham, A., Nicoara, M.: Search Optimization Using Hybrid Particle Sub-swarms and Evolutionary Algorithms. International Journal of Simulation Systems, Science & Technology 6(10), 60–79 (2005)

35. Jiang, C., Etorre, B.: A hybrid Method of Chaotic Particle Swarm Optimization and Linear Interior for Reactive Power Optimisation. Mathematics and Computers in Simulation 68, 57–65 (2005)

36. Liu, H., Li, B., Ji, Y., Tong, S.: Particle Swarm Optimisation from lbest to gbest. In: Abraham, A., Baets, B.D., Koppen, M. (eds.) Applied Soft Computing Technologies: The Challenge of Complexity, pp. 537–545. Springer, Heidelberg (2006)

37. Liang, J.J., Qin, A.K., Suganthan, P.N., Baskar, S.: Comprehensive Learning Particle Swarm Optimizer for Global Optimization of Multimodal Functions. IEEE Transactions on Evolutionary Computation 10(3), 281–295 (2006)

38. Blackwell, T., Branke, J.: Multiswarms, exclusion, and anti-convergence in dynamic environments. IEEE Transactions on Evolutionary Computation 10(4), 459–472 (2006)

39. Niu, B., Zhu, Y., He, X., Wu, H.: MCPSO: A Multi-Swarm Cooperative Particle Swarm Optimizer. Applied Mathematics and Computation 185(2), 1050–1062 (2007)

40. Settles, M., Soule, T.: Breeding swarms: a GA/PSO hybrid. In: Proceedings of the Genetic and Evolutionary Computation Conference (GECCO), pp. 161–168 (2005)

41. Elshamy, W., Emara, H.M., Bahgat, A.: Clubs-based Particle Swarm Optimization. In: Proceedings of the IEEE International Conference on Swarm Intelligence Symposium, vol. 1, pp. 289–296 (2007)
42. Guo, C., Tang, H.: Global Convergence Properties of Evolution Stragtegies. Mathematica Numerica Sinica 23(1), 105–110 (2001)
43. He, R., Wang, Y., Wang, Q., Zhou, J., Hu, C.: An Improved Particle Swarm Optimization Based on Self-adaptive Escape Velocity. Journal of Software 16(12), 2036–2044 (2005)
44. Weisstein, E.W.: Borel-Cantelli Lemma, From MathWorld – A Wolfram Web Resource (2007), http://mathworld.wolfram.com/Borel-CantelliLemma.html
45. Xu, Z., Cheng, G., Liang, Y.: Search Capability for An Algebraic Crossover. Journal of Xi'an Jiaotong University 33(10), 88–99 (1999)
46. Whitley, L.D.: Fundamental Principles of Deception in Genetic Search. Foundation of Genetic Algorithms, pp. 221–241. Morgan Kaufmann Publishers, California (1991)
47. Mastrolilli, M., Gambardella, L.M.: Effective Neighborhood Functions for the Flexible Job Shop Problem. Journal of Scheduling 3(1), 3–20 (2002)
48. Holland, J.H.: Adaptation in Natural and Artificial Systems. University of Michigan Press, Ann Arbor (1975)
49. Goldberg, D.E.: Genetic Algorithms in search, optimization, and machine learning. Addison-Wesley Publishing Corporation, Inc., Reading (1989)
50. Cantú-Paz, E.: Efficient and Accurate Parallel Genetic Algorithms. Kluwer Academic Publishers, Netherland (2000)
51. Abraham, A.: Evolutionary Computation. In: Sydenham, P., Thorn, R. (eds.) Handbook for Measurement Systems Design, pp. 920–931. John Wiley and Sons Ltd., London (2005)

Discovering Data Structures Using Meta-learning, Visualization and Constructive Neural Networks

Tomasz Maszczyk, Marek Grochowski, and Włodzisław Duch

Department of Informatics, Nicolaus Copernicus University, Toruń, Poland
`tmaszczyk@is.umk.pl`, `grochu@is.umk.pl`, `wduch@is.umk.pl`

Abstract. Several visualization methods have been used to reveal hidden data structures, facilitating discovery of simplest data models. Insights gained in this way are used to create constructive neural networks implementing appropriate transformations that provide simplest models of data. This is an efficient approach to meta-learning, guiding the search for best models in the space of all data transformations. It can solve problems with complex inherent logical structure that are very difficult for traditional machine learning algorithms.

Keywords: Meta-learning, machine learning, constructive neural networks, projection pursuit, visualization.

1 Introduction

Ryszard Michalski has always been interested in discovering comprehensible structures in the data, and has developed his own multistrategy learning approach [10]. In this paper we shall look at the problem of finding the best set of transformations to solve classification and regression problems in the spirit of multistrategy learning. Searching for transformations that will reveal the inherent structure in the data is very difficult. Instead of trying to solve all problems with the same universal tool, such as one of the neural networks, or support vector machine (SVM) algorithms [21], a meta-learning approach [8] builds the final data model from components that are heterogeneous [7].

Each data model depends on some specific assumptions about the data distribution in the input space, and is successfully applicable only to some types of problems. For example SVM and many other statistical learning methods [21] rely on the assumption of uniform resolution, local similarity between data samples, and may completely fail in case of high-dimensional functions that are not sufficiently smooth [2]. In such case accurate solution may require an extremely large number of training samples that will be used as reference vectors, leading to high cost of computations and creating complex models that do not generalize well.

The type of solution offered by a given data model may not be appropriate for the particular data. Each data model defines a hypotheses space, that is a set of functions that this model may easily learn, providing a particular bias for the model. Although many basis set expansion networks, including the multilayer perceptron neural networks (MLPs), are universal approximators (given sufficient number of functions they may approximate arbitrary data distributions), models created by such networks are not necessarily optimal from the complexity point of view. Linear discrimination models are

J. Koronacki et al. (Eds.): Advances in Machine Learning II, SCI 263, pp. 467–484.
springerlink.com   © Springer-Verlag Berlin Heidelberg 2010

obviously not suitable for spherical distributions of data, requiring $O(N^2)$ parameters to approximately cover each spherical distribution in N dimensions, where an expansion in radial functions requires only $O(N)$ parameters. On the other hand many spherical functions are needed to approximate a hyperplane. Some problems, such as the multidimensional parity problem, cannot be easily approximated neither by radial nor by hyperplane functions [6]. An optimal solution may only be found if a model based on suitable transformations is defined.

In general each supervised learning system may be represented by an operator $\mathcal{T}$ that transforms a given vector $\mathbf{X}$ into some output vector $\mathbf{Y}$

$$\mathcal{T}\mathbf{X} = \mathbf{Y}$$

The goal is to find an operator $\mathcal{T}$ that not only gives correct answers on the training data but also provides a model that has low Kolmogorov complexity, facilitating easy interpretation. Operator $\mathcal{T}$ may in general be created as a sequence of transformations (for simplicity recurrent processes are not considered here):

$$\mathcal{T} = \mathcal{T}_1 \mathcal{T}_2 \ldots \mathcal{T}_k$$

For example, initial transformations may define data preprocessing based on linear scaling (standardization or normalization), projection on a low-dimensional space defined by principal components or by other criteria, or just select a subset of features. Data preparation may have a crucial influence on algorithms applicable for further training. Some computationally expensive algorithms require dimensionality reduction to deal with large datasets. Therefore model selection should not focus only on searching for suitable classifier or regression tool, selection of supportive transformations is also very important. Each layer of the MLP network can be seen as a single mapping $\mathcal{T}_i$, although taken separately these transformations do not have a well-defined goal. The final transformation should provide desired output; for classification tasks this output Y takes the form of discrete values of class labels, or estimations of probabilities of class labels. Composition of transformations covers also combinations of many learning algorithms (classifier committees, boosting etc).

With no *a priori* knowledge about a given problem finding an optimal sequence of transformations is a great challenge. Many meta-learning techniques have recently been developed to deal with the problem of model selection [22,3]. Most of them search for optimal model characterizing a given problem by some meta-features (e.g. statistical properties, landmarking, model-based characterization), and by referring to some meta-knowledge gained earlier. For a given data one can use the classifier that gave the best result on a similar dataset in the StatLog Project [18]. However, choosing good meta-features is not a trivial issue as most features do not characterize the complexity of data distributions. In addition the space of possible solutions generated by this approach is bounded to already known types of algorithms. General meta-learning algorithm should browse through all interesting models, searching for the best composition of transformations. The meta-learning problem may thus be seen as a search in the space of all possible models composed of transformations. These class of admissible transformations may be restricted, as for example it has been done in the similarity-based approach [8]. General search in model space requires a very sophisticated intelligent system and search strategies. Some ideas for building such systems and for controlling the process of composing transformations based on monitoring of model complexity have recently been proposed in [13,14].

Fig. 1. After initial transformation and visualization the final transformation is selected

Judicious transformation performed at the beginning of learning does not only lead to simplification of further learning, but also provides useful information guiding researcher through the set of possible models. Based on structures emerging in low-dimensional visualizations of a given problem an experienced researcher is able to construct the best learning strategies to learn from this data. Such approach to meta-learning is introduced and tested in this paper. In the next section a few dimensionality reduction algorithms suitable for visualization are presented, and applied in the third section to the real and artificial data. Upon visual inspection it becomes quite clear which type of transformation should be applied to the data to create it simplest model. Numerical tests confirm the superiority of models selected in this way.

2   Visualization Algorithms

Visualization methods are discussed in details in many books, for example [20,23]. Many visualization methods may be useful in our approach to meta-learning, but here only 5 methods are used. Below a short description of these algorithms is given: first three well known methods – principal component analysis (PCA), Fisher discriminant analysis (FDA), and multidimensional scaling (MDS), followed by a description of two new approaches recently introduced by us – SVM projections [17] and Quality of Projected Clusters (QPC) projections [16]. Only PCA and MDS are unsupervised methods, and only MDS is non-linear.

Principal Component Analysis (PCA) is a linear projection method that finds orthogonal combinations of input features $\mathbf{X} = \{x_1, x_2, ..., x_N\}$, with each new direction accounting for the largest remaining variation in the data. Since no information about class labels is used PCA transformation may be performed on all available data. Principal components $\mathbf{P}_i$ that result from diagonalization of data covariance matrix guarantee minimal loss of information when position of points are recreated from their low-dimensional projections. Taking 1, 2 or 3 principal components and projecting the data into the space defined by these components $y_{ij} = \mathbf{P}_i \cdot \mathbf{X}_j$ provides for each input vector its representative $(y_{1j}, y_{2j}, ... y_{kj})$ in the target space.

Fisher Discriminant Analysis (FDA) is a supervised method that uses information about the classes to find projections separating samples from these classes. This popular algorithm maximizes the ratio of between-class to within-class scatter, seeking a direction $\mathbf{W}$ such that

$$\max_{\mathbf{W}} J_{\mathbf{W}} = \frac{\mathbf{W}^T \mathbf{S}_B \mathbf{W}}{\mathbf{W}^T \mathbf{S}_I \mathbf{W}} \tag{1}$$

where the scatter matrices $\mathbf{S}_B$ and $\mathbf{S}_I$ are defined by

$$\mathbf{S}_B = \sum_{i=1}^{C} \frac{n_i}{n}(\mathbf{m}_i - \mathbf{m})(\mathbf{m}_i - \mathbf{m})^T; \qquad S_I = \sum_{i=1}^{C} \frac{n_i}{n}\hat{\Sigma}_i \qquad (2)$$

where $\mathbf{m}_i$ and $\hat{\Sigma}_i$ are the means and covariance matrices for each class and $\mathbf{m}$ is the total mean vector [23]. FDA is frequently used for classification projecting data on a single line. For visualization generating the second FDA vector in a two-class problem is not so trivial. This is due to the fact that the rank of the $\mathbf{S}_B$ matrix for the C-class problems is $C - 1$. Cheng *et al.* [4] proposed several solutions to this problem:

- stabilize the $\mathbf{S}_I$ matrix by adding a small perturbation matrix;
- use pseudoinverse, replacing S_I^{-1} by the pseudoinverse matrix $S_I^{\dagger}$;
- use rank decomposition method.

In our implementation pseudoinverse matrix has been used to generate higher FDA directions.

Multidimensional scaling (MDS) is a non-linear technique used for proximity visualization [5]. The main idea is to decrease dimensionality of data while preserving original distances between data points as defined in the high-dimensional space. MDS methods need only similarities between objects, so explicit vector representation of objects is not necessary. In metric scaling quantitative evaluation of similarity based on numerical distance measures (Euclidean, cosine, or any other measure) is used, while for non-metric scaling qualitative information about the pairwise similarities is sufficient. MDS methods also differ by their cost functions, optimization algorithms, the number of similarity matrices used, and the use of feature weighting. There are many measures of topographical distortions due to the reduction of dimensionality, most of them variants of the stress function:

$$S_T(\mathbf{d}) = \sum_{i>j}^{n} (D_{ij} - d_{ij})^2 \qquad (3)$$

where d_{ij} are distances (dissimilarities) in the target (low-dimensional) space, and D_{ij} are distances in the input space, pre-processed or calculated directly using some metric functions. These measures are minimized over positions of all target points, with large distances dominating in the $S_T(\mathbf{d})$. In the k-dimensional target space there are kn parameters for minimization. For visualization purposes the dimension of the target space is $k = 1 - 3$. The sum runs over all pairs of vectors and has $O(n^2)$ terms. The number of vectors n may be quite large, making the approximation to the minimization process necessary [19]. MDS cost functions are not easy to minimize, with multiple local minima representing different mappings. Initial configuration is either selected randomly or based on projection of data to the space spanned by principal components. Orientation of axes in the MDS mapping is arbitrary, and the values of coordinates do not have any simple interpretation, as only relative distances are important. However, if the data has clear clusters in the input space MDS may show it.

Linear SVM algorithm creates a hyperplane that provides a large margin of classification, using regularization term and quadratic programming. Non-linear versions are based on a kernel trick [21] that allows for implicit mapping of data vectors to a high-dimensional feature space where the best separating hyperplane (the maximum margin hyperplane) is constructed. Linear discriminant function is defined by:

$$g_{\mathbf{W}}(\mathbf{X}) = \mathbf{W}^T \cdot \mathbf{X} + w_0 \tag{4}$$

The best discriminating hyperplane should maximize the distance between decision hyperplane defined by $g_{\mathbf{W}}(\mathbf{X}) = 0$ and the vectors that are nearest to it, $\max_{\mathbf{W}} D(\mathbf{W}, \mathbf{X}^{(i)})$. The largest classification margin is obtained from minimization of the norm $\|\mathbf{W}\|^2$ with constraints:

$$Y^{(i)} g_{\mathbf{W}}(\mathbf{X}^{(i)}) \geq 1 \tag{5}$$

for all training vectors $\mathbf{X}^{(i)}$ that belong to class $Y^{(i)}$. Vector $\mathbf{W}$, orthogonal to the discriminant hyperplane, defines direction on which the data vectors are projected, and thus may be used to create one-dimensional projection. The same may be done using non-linear SVM based on kernel discriminant:

$$g_{\mathbf{W}}(\mathbf{X}) = \sum_{i=1}^{N_{sv}} \alpha_i K(\mathbf{X}^{(i)}, \mathbf{X}) + w_0 \tag{6}$$

where the summation is over support vectors $\mathbf{X}^{(i)}$ that are selected from the training set. The $x = g_{\mathbf{W}}(\mathbf{X})$ values for different classes may be smoothed and displayed as a histogram, estimating either the $p(x|C)$ class-conditionals or posterior probabilities $p(C|x) = p(x|C)p(C)/p(x)$.

SVM visualization in more than one dimension requires generation of additional discriminating directions. The first projection on $\mathbf{W}_1$ line should give $g_{\mathbf{W}_1}(\mathbf{X}) < 0$ for vectors from the first class, and $g_{\mathbf{W}_1}(\mathbf{X}) > 0$ for the second class. This is obviously possible only for linearly separable data. More directions may be found in the space orthogonalized to the first direction. An alternative approach is to select a subset $\mathscr{D}(\mathbf{W}_1)$ of all vectors that have projections in the $[a(\mathbf{W}_1), b(\mathbf{W}_1)]$ interval containing the zero point is selected. This interval should include all vectors for which $p(x|C_i)$ class-conditionals estimated along the $x = g_{\mathbf{W}_1}(\mathbf{X})$ direction overlap. The second best direction may then be obtained by repeating SVM calculations in the space orthogonalized to the already obtained $\mathbf{W}_1$ direction, using only the subset of $\mathscr{D}(\mathbf{W}_1)$ vectors, as the remaining vectors are already well separated in the first dimension. SVM training in its final phase is using anyway mainly support vectors that belong to this subset. However, vectors in the $[a(\mathbf{W}_1), b(\mathbf{W}_1)]$ interval should not include outliers that are far from the decision border, and therefore may generate a significantly different direction. This process may be repeated to obtain more dimensions. Each additional dimension should help to decrease errors, and the optimal dimensionality is obtained when new dimensions stop decreasing the number of errors in crossvalidation tests.

In the case of non-linear kernel, $g_{\mathbf{W}}(\mathbf{X})$ provides the first direction, while the second direction may be generated in several ways. The simplest approach is to repeat training on $\mathscr{D}(\mathbf{W})$ subset of vectors that are close to the hyperplane in the extended space using another kernel.

The Quality of Projected Clusters (QPC) is a supervised projection pursuit method that finds most interesting and informative linear projection maximizing the following index [16]:

$$QPC(w) = \sum_x \left(A^+ \sum_{x_k \in \mathscr{C}_x} G\left(w^T(x - x_k)\right) - A^- \sum_{x_k \notin \mathscr{C}_x} G\left(w^T(x - x_k)\right) \right) \qquad (7)$$

where $G(x)$ is a function with localized support and maximum for $x = 0$ (e.g. a Gaussian function), and $\mathscr{C}_x$ denotes the set of all vectors that have the same label as x. This index achieves maximum value for projections on the direction w that group vectors belonging to the same class into a compact and well separated clusters. It does not enforce linear separability and is suitable for multi-modal data. Parameters A^+, A^- control influence of each term in Eq. (7). For large value of A^- strong separation between classes is enforced, while increasing A^+ impacts mostly compactness and purity of clusters. The shape and width of the $G(x)$ function used in E.q. (7) has influence on convergence. For continuous $G(X)$ functions gradient-based methods may be used to maximize QPC index.

Visualizations of data presented in the next section have been obtained by maximization of QPC index based on an inverse quartic function

$$G(x) = \frac{1}{1 + (bx)^4} \qquad (8)$$

but any bell-shaped function (e.g. Gaussian or bicentral function) that achieves maximum value for $x = 0$ and vanish for $x \to \pm\infty$ is suitable here. The QPC index provides a leave-one-out estimator that measures quality of clusters projected on w direction. Direct calculation of the QPC index (7), as in the case of all nearest neighbor methods, requires $O(n^2)$ operations. The greatest advantage of using this index is that it is able to discover non-local structures and multimodal class distributions (e.g. k-separable datasets with $k > 2$ [6]). The QPC may be used also (in the same way as the SVM approach described above) as a base for creation of feature ranking and feature selection methods. Projection coefficients w_i indicate then significance of the i-th feature. For noisy and non-informative variables values of associated weights should decrease to zero during QPC optimization.

Not only global, but also local extrema of the QPC index are of interest, as they may also provide useful insight into data structures and may be used in a committee-based approach that combines different views on the same data. For complex problems usually more than one projection is required, therefore one should search for more linear projections either in the orthogonalized space, or using additional penalty term:

$$QPC(w; w_1) = QPC(w) - \lambda f(w, w_1) . \qquad (9)$$

This term should provide punishment for solutions that are too similar to those already found, therefore the value of $f(w, w_1)$ should become large for directions w close to the previous direction w_1. Powers of the scalar product between these two directions may be used for this purpose, $f(w, w_1) = (w_1^T \cdot w)^2$. Repeating this procedure leads to creation of a sequence of unique interesting projections [16].

3   Illustrative Examples

Visualization methods described above will be used to determine what kind of transformations should be used to discover structures hidden in the multidimensional data. The usefulness of this meta-learning approach has been tested on several datasets: one artificial binary dataset, four real datasets downloaded from the UCI Machine Learning Repository [1], and a microarray gene expression data [11]. A summary of these datasets is presented in Tab. 1; their short description follows:

1. **Parity_8**, 8-bit parity dataset, with 8 binary features and 256 vectors.
2. **Heart** disease dataset consisting of 270 samples, each described by 13 attributes, 150 cases labeled as "absence", and 120 as "presence" of heart disease.
3. **Wisconsin** breast cancer dataset [24] contains samples describing results of biopsies on 699 patients, with 458 biopsies labeled as "benign", and 241 as "malignant". Feature 6 has 16 missing values, removing corresponding vectors leaves 683 examples.
4. **Leukemia**, microarray gene expressions for two types of leukemia (ALL and AML), with a total of 47 ALL and 25 AML samples measured with 7129 probes [11]. Visualization and evaluations of this data is based here on pre-selected 100 best features, done by simple feature ranking using FDA index.
5. **Monks_1** dataset contains 124 cases, where 62 samples belong to the first class, and the remaining 62 to the second. Each sample is described by 6 attributes. Logical function has been used to created class labels.

Table 1. Summary of used datasets

Title	#Features	#Samples	#Samples per class		Source
Parity_8	8	256	128 C_0	128 C_1	artificial
Heart	13	270	150 "absence"	120 "presence"	[1]
Wisconsin	10	683	444 "benign"	239 "malignant"	[24]
Leukemia	100	72	47 "ALL"	25 "AML"	[11]
Monks_1	6	124	62 C_0	62 C_1	[1]

For each dataset one and two-dimensional mappings have been created using MDS, PCA, FDA, SVM and QPC algorithms (Figs. 2-6). On the basis of these visualizations classifier that should be most appropriate for particular data distribution has been chosen from the following list:

1. Naive Bayesian Classifier (NBC)
2. k-Nearest Neighbours (kNN)
3. Separability Split Value Tree (SSV) [12]
4. Support Vector Machines with Linear Kernel (SVML)
5. Support Vector Machines with Gaussian Kernel (SVMG)

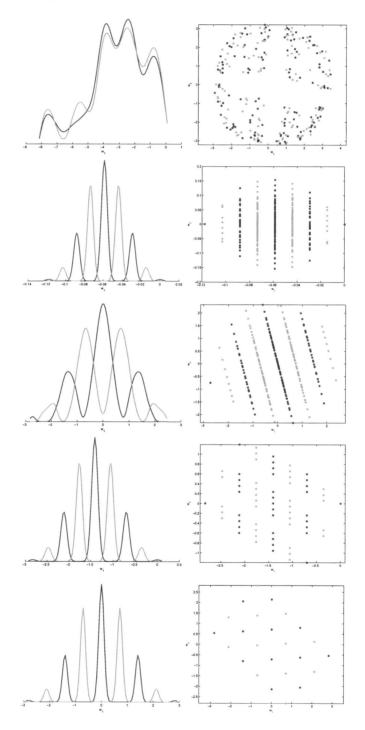

Fig. 2. 8-bit parity dataset, from top to bottom: MDS, PCA, FDA, SVM and QPC

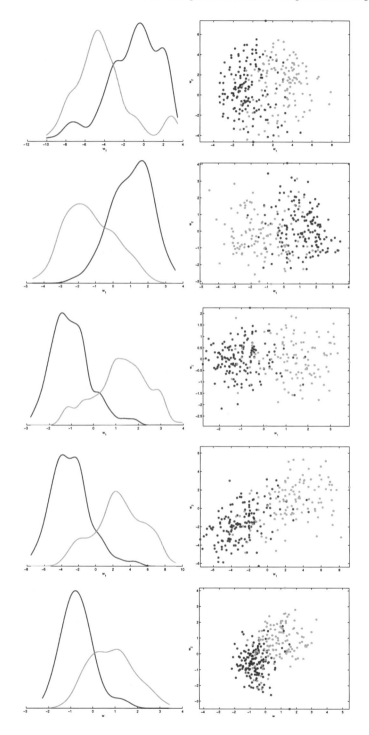

Fig. 3. Heart data set, from top to bottom: MDS, PCA, FDA, SVM and QPC

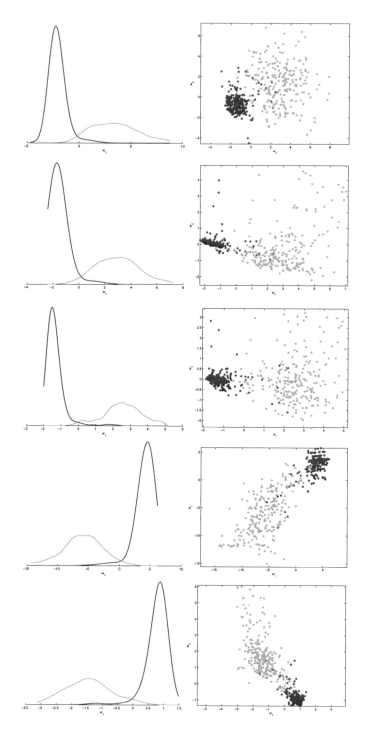

Fig. 4. Wisconsin data set, from top to bottom: MDS, PCA, FDA, SVM and QPC

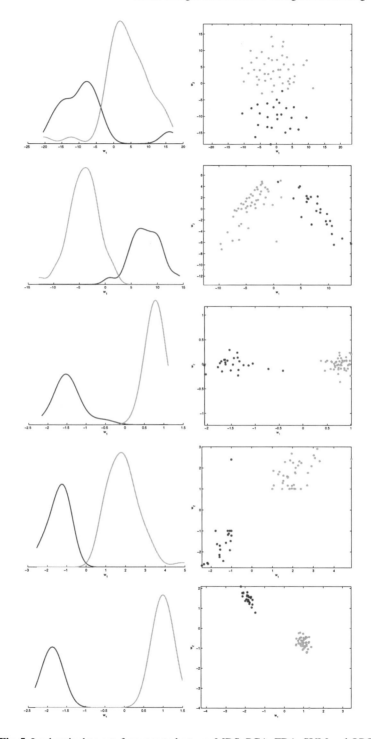

Fig. 5. Leukemia data set, from top to bottom: MDS, PCA, FDA, SVM and QPC

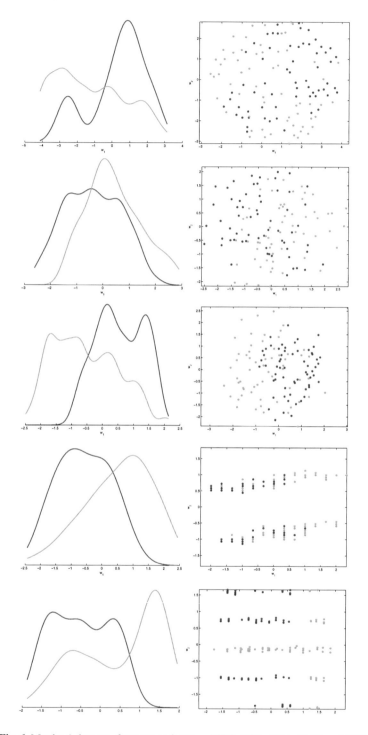

Fig. 6. Monks_1 data set, from top to bottom: MDS, PCA, FDA, SVM and QPC

Some methods require standardization of data before classification. Analyzing the one-dimensional probability distributions and two-dimensional scatterograms, Figs. 2 – 6 it is not difficult to determine the particular classifier bias that should lead to the simplest model of a given dataset. To check if an optimal choice has been made comparison with classification accuracies for each dataset using all classifiers listed above has been done, in the original as well as reduced one and two-dimensional spaces. 10-fold crossvalidation tests results are collected in Tables 2-6, with accuracies and standard deviations given for each dataset. For the kNN classifier the number of nearest neighbours has been automatically selected from the $1 - 10$ range using crossvalidation estimation. Also the SVM parameters C and σ have been optimized in an automatic way using crossvalidation estimations. All calculations have been performed using the Ghostminer package developed in our group [9].

For all datasets visualization helped to estimate reliability of predictions for individual data samples, showing how far each case is from the decision border and how strong are the overlaps between different classes in this region. Estimation of confidence of predictions is very important in many applications. It may be particularly useful in problems when decision borders are quite complex, and as a result simple rule-based systems will not be able to provide good approximation. The only way to understand such data will then be to use flexible classifiers with visualization of results, making the solution more comprehensible.

High-dimensional parity problem is very difficult for most classification methods. Many papers have been published on special neural network models for parity functions, and the reason is quite obvious, as Fig. 2 illustrates: linear separation cannot be achieved by simple transformations because this is a k-separable problem that for n bit strings may be easily separated only into $n + 1$ intervals [15,6]. MDS is completely lost and does not show any interesting structure, as all vectors from one class have their nearest neighbors from the opposite class. PCA and SVM find a very useful projection direction $[1, 1..1]$, but the second direction does not help at all. FDA shows significant overlaps for projection on the first direction.

Only the QPC index finds both directions that are quite useful. Points that are in small clusters projected on the first direction belong to a large cluster projected on the second direction, giving much better chance for correct classification. This is a very interesting example showing that visualization may help to solve a difficult problem in a perfect way even when almost all classifiers fail. Looking at these pictures one can notice that because data are not linearly separable, probably the best classifier to solve this problem should be:

- any decision tree, after transformation to one or two-dimensions by PCA, SVM or QPC;
- NBC, in one or two-dimensions, combining directions for the most robust solution, provided that it will use density estimation based on a sum of Gaussian functions or similar localized kernels;
- kNN on the 1D data reduced by PCA, SVM or QPC, with k=1, although it will make a small error for the two extreme points.

This choice agrees with the results from tables 2-6, where the highest accuracy (99.6 ± 1.2) is obtained by the SSV classifier on the 2D data transformed by SVM or QPC

Table 2. NBC 10-fold crossvalidation accuracy for datasets with reduced features

	# Features	Parity_8	Heart	Wisconsin	Leukemia	Monks1
PCA	1	99.21±1.65	80.74±6.24	97.36±2.27	98.57±4.51	56.98±14.12
PCA	2	99.23±1.62	78.88±10.91	96.18±2.95	98.57±4.51	54.67±13.93
MDS	1	38.35±7.00	75.55±6.80	96.63±1.95	92.85±7.52	67.94±11.24
MDS	2	30.49±13.79	80.74±9.36	95.16±1.70	98.57±4.51	63.52±16.02
FDA	1	75.84±10.63	85.18±9.07	97.07±0.97	100±0.00	72.05±12.03
FDA	2	74.56±10.69	84.07±8.01	95.46±1.89	100±0.00	64.48±17.54
SVM	1	99.23±1.62	85.92±6.93	95.90±1.64	100±0.00	70.38±10.73
SVM	2	99.21±1.65	83.70±5.57	97.21±1.89	100±0.00	71.79±8.78
QPC	1	99.20±2.52	81.48±6.53	96.33±3.12	100±0.00	72.56±9.70
QPC	2	98.41±2.04	84.44±7.96	97.21±2.44	100±0.00	100±0
	ALL	23.38±6.74	72.22±4.70	95.46±2.77	78.28±13.55	69.35±16.54

Table 3. kNN 10-fold crossvalidation accuracy for datasets with reduced features (optimal k value in parenthesis)

	# Features	Parity_8	Heart	Wisconsin	Leukemia	Monks1
PCA	1	99.20±1.68 (1)	75.92±9.44 (10)	96.92±1.61 (7)	98.57±4.51 (2)	53.97±15.61 (8)
PCA	2	99.21±1.65 (1)	80.74±8.51 (9)	96.34±2.69 (7)	98.57±4.51 (3)	61.28±17.07 (9)
MDS	1	43.73±7.44 (4)	72.96±7.62 (8)	95.60±1.84 (7)	91.78±7.10 (4)	69.48±10.83 (8)
MDS	2	48.46±7.77 (1)	80.37±8.19 (6)	96.48±2.60 (3)	97.32±5.66 (8)	67.75±16.51 (9)
FDA	1	76.60±7.37 (10)	84.81±5.64 (8)	97.35±1.93 (5)	100±0.00 (1)	69.35±8.72 (7)
FDA	2	99.23±1.62 (1)	82.96±6.34 (10)	96.77±1.51 (9)	100±0.00 (1)	69.29±13.70 (9)
SVM	1	99.61±1.21 (1)	82.59±7.81 (9)	97.22±1.98 (9)	100±0.00 (1)	70.12±8.55 (9)
SVM	2	99.61±1.21 (1)	82.96±7.44 (10)	97.36±3.51 (10)	100±0.00 (1)	69.29±10.93 (9)
QPC	1	99.21±1.65 (1)	81.85±8.80 (10)	97.22±1.74 (7)	100±0.00 (1)	81.34±12.49 (3)
QPC	2	98.44±2.70 (1)	85.55±4.76 (10)	96.62±1.84 (7)	100±0.00 (1)	100±0 (1)
	ALL	1.16±1.88 (10)	79.62±11.61 (9)	96.34±2.52 (7)	98.57±4.51 (2)	71.15±12.68(10)

method. NBC and kNN results are not worse from the statistical point of view (results are within one standard deviation). kNN results on the original data with k≤ 10 are always wrong, as all 8 closest neighbors belong to the opposite class. After dimensionality reduction k=1 is sufficient.

It is also interesting to note the complexity of other models: SVM takes all 256 vectors as support vectors, achieving results around the base rate of 50%, with the exception of SVM with Gaussian kernel that does quite well on the one and two-dimensional data reduced by SVM and QPC projections. SSV creates moderately complex tree with 7 leaves and a total of 13 nodes.

Visualization in Fig. 2 also suggest that using 2D QPC projected data the nearest neighbor rule may be easily modified: instead of a fixed number of neighbors for vector **X**, take its projections y_1, y_2 on the two dimensions, and count the number of neighbors $k_i(\varepsilon_i)$ in the largest interval $y_i \pm \varepsilon_i$ around y_i that contain vectors from a single class

Table 4. SSV 10-fold crossvalidation accuracy for datasets with reduced features (total number of nodes/leaves in parenthesis)

	# Features	Parity_8	Heart	Wisconsin	Leukemia	Monks1
PCA	1	99.21±1.65 (13/7)	79.25±10.64 (3/2)	97.07±1.68 (3/2)	95.71±6.90 (7/4)	57.94±11.00 (3/2)
PCA	2	99.23±1.62 (13/7)	79.62±7.03 (15/8)	97.36±1.92 (3/2)	95.81±5.16 (7/4)	61.34±11.82 (11/6)
MDS	1	47.66±4.69 (1/1)	77.40±6.16 (3/2)	97.07±1.83 (3/2)	91.78±14.87 (3/2)	68.58±10.44 (3/2)
MDS	2	49.20±1.03 (1/1)	81.11±4.76 (3/2)	96.19±2.51 (9/5)	95.71±6.90 (7/4)	66.98±12.21 (35/18)
FDA	1	73.83±6.97 (17/9)	84.07±6.77 (3/2)	96.92±2.34 (3/2)	100±0.00 (3/2)	67.82±9.10 (3/2)
FDA	2	96.87±3.54 (35/18)	83.70±6.34 (3/2)	96.93±1.86 (11/6)	100±0.00 (3/2)	68.65±14.74 (3/2)
SVM	1	99.23±1.62 (13/7)	83.33±7.25 (3/2)	97.22±1.99 (3/2)	100±0.00 (3/2)	70.32±16.06 (3/2)
SVM	2	99.61±1.21 (13/7)	84.81±6.63 (3/2)	97.22±1.73 (3/2)	100±0.00 (3/2)	69.35±9.80 (3/2)
QPC	1	99.20±2.52 (13/7)	82.22±5.46 (9/5)	96.91±2.01 (3/2)	100±0.00 (3/2)	82.43±12.22 (47/24)
QPC	2	99.61±1.21 (13/7)	83.33±7.25 (13/7)	96.33±2.32 (3/2)	100±0.00 (3/2)	98.46±3.24 (7/4)
	ALL	49.2±1.03 (1/1)	81.48±4.61 (7/4)	95.60±3.30 (7/4)	90.00±9.64 (5/3)	83.26±14.13 (35/18)

Table 5. SVML 10-fold crossvalidation accuracy for datasets with reduced features (number of support vectors in parenthesis)

	# Features	Parity_8	Heart	Wisconsin	Leukemia	Monks1
PCA	1	39.15±13.47 (256)	81.11±8.08 (118)	96.78±2.46 (52)	98.57±4.51 (4)	63.71±10.68 (98)
PCA	2	43.36±7.02 (256)	82.96±7.02 (113)	96.92±2.33 (53)	97.14±6.02 (4)	63.71±10.05 (95)
MDS	1	42.98±5.84 (256)	77.03±7.15 (170)	95.60±2.59 (54)	91.78±9.78 (28)	69.61±11.77 (88)
MDS	2	43.83±8.72 (256)	82.96±6.09 (112)	96.92±2.43 (52)	97.32±5.66 (5)	64.74±16.52 (103)
FDA	1	45.73±6.83 (256)	85.18±4.61 (92)	97.21±1.88 (52)	100±0.00 (2)	69.93±11.32 (80)
FDA	2	44.16±5.67 (256)	84.81±5.36 (92)	96.77±2.65 (51)	100±0.00 (3)	69.23±10.57 (80)
SVM	1	54.61±6.36 (256)	85.55±5.36 (92)	97.22±1.26 (46)	100±0.00 (2)	71.98±13.14 (78)
SVM	2	50.29±9.28 (256)	85.55±7.69 (92)	96.92±2.88 (47)	100±0.00 (5)	72.75±10.80 (80)
QPC	1	41.46±9.57 (256)	82.59±8.73 (118)	96.34±2.78 (62)	100±0.00 (2)	67.50±13.54 (82)
QPC	2	43.01±8.21 (256)	85.92±5.46 (103)	96.62±1.40 (54)	100±0.00 (2)	66.92±16.68 (83)
	ALL	31.61±8.31 (256)	84.44±5.17 (99)	96.63±2.68 (50)	98.57±4.51 (16)	65.38±10.75 (83)

Table 6. SVMG 10-fold crossvalidation accuracy for datasets with reduced features (number of support vectors in parenthesis)

	# Features	Parity_8	Heart	Wisconsin	Leukemia	Monks1
PCA	1	99.20±1.68 (256)	80.00±9.43 (128)	97.36±2.15 (76)	98.57±4.51 (20)	58.84±12.08 (102)
PCA	2	98.83±1.88 (256)	80.00±9.99 (125)	97.22±2.22 (79)	97.14±6.02 (22)	67.17±17.05 (99)
MDS	1	44.10±8.50 (256)	73.70±8.27 (171)	95.74±2.45 (86)	91.78±7.10 (36)	64.67±10.88 (92)
MDS	2	43.04±8.91 (256)	82.59±7.20 (121)	96.63±2.58 (78)	98.75±3.95 (27)	62.17±15.47 (104)
FDA	1	77.76±7.89 (256)	85.18±4.61 (106)	97.65±1.86 (70)	100±0.00 (12)	72.37±9.29 (85)
FDA	2	98.84±1.85 (256)	84.81±6.16 (110)	97.07±2.06 (74)	100±0.00 (15)	70.96±10.63 (85)
SVM	1	99.61±1.21 (9)	85.18±4.61 (107)	96.93±1.73 (69)	100±0.00 (14)	72.82±10.20 (77)
SVM	2	99.61±1.21 (43)	84.07±7.20 (131)	96.92±3.28 (86)	100±0.00 (21)	68.65±13.99 (93)
QPC	1	99.21±1.65 (256)	82.59±10.33 (130)	97.07±1.82 (84)	100±0.00 (10)	67.43±17.05 (84)
QPC	2	98.44±2.70 (24)	85.18±4.93 (132)	96.33±1.87 (107)	100±0.00 (12)	99.16±2.63 (45)
	ALL	16.80±22.76 (256)	82.22±5.17 (162)	96.63±2.59 (93)	98.57±4.51 (72)	78.20±8.65 (87)

only, summing results from both dimensions $k_1(\varepsilon_1) + k_2(\varepsilon_2)$. This is an interesting new version of the kNN method, but it will not be explored here further.

Cleveland Heart data Fig. 3 is rather typical for biomedical data. The information contained in the test data is not really sufficient to make a perfect diagnosis. Almost all projections show comparable separation of a significant portion of the data, although looking at probability distributions in one dimension SVM and FDA seem to have a bit of an advantage. In such case strong regularization is advised to improve generalization. For kNN this means rather large number of neighbors (in most cases 10, the maximum allowed here, was optimal), for decision tree strong pruning (SSV after FDA has only a root node and two leaves), while for SVM rather large value of C parameter and (for Gaussian kernels) large dispersions, that will lead to a significant number of support vectors.

The best recommendation for this dataset is to apply the simplest classifier – SSV or linear SVM on FDA projected data. Comparing this recommendation with calculations presented in tables 2-6 confirms that this is the best choice.

Wisconsin breast cancer dataset is similar to Cleveland Heart data. It shows much stronger separation (Fig. 4) of the cases that belong to the two classes for all types of visualization. It is quite likely that this data contains several outliers. All methods give comparable results, although reduction of dimensionality to two dimensions helps quite a bit to decrease the complexity of the data models, except for the SVM that achieves essentially the same accuracy requiring similar number of support vectors for the original and for the reduced data.

Again, the simplest classifier is quite sufficient here, SSV on FDA or QPC projections with a single threshold (a tree with just two leaves), or more complex (about 50 support vectors) SVM model with linear kernel on 2D data reduced by linear projection. One should not expect that more information can be extracted from this data.

Leukemia data showed a remarkable separation using both one and two-dimensional QPC, SVM and FDA projections (Fig. 5), providing more interesting solution than MDS or PCA methods. Choosing one of the three linear transformations (for example the QPC), and projecting original data to the one-dimensional space, SSV decision tree, kNN, NBC and SVM classifiers, give 100% accuracy in the 10CV tests (Table 5). All these models are very simple, with k=1 for kNN, or decision trees with 3 nodes, or only 2 support vectors for linear SVM. Results on the whole data are much worse than on the projected features.

In this case dimensionality reduction is the most important factor, combining the activity of many genes into a single profile. As the projection coefficients are linear the importance of each gene in this profile may be easily evaluated.

The last dataset used in this section is Monks_1. This data provides a very interesting example how visualization can help to choose which classificator should be used. Almost all visualization methods (MDS, PCA, FDA and SVM, Fig. 6) for one and two-dimensional projections do not show interesting structure. However, the 2D scatterplot of the QPC projection shows quite clear structure that can easily be separated using decision tree as classifier (SSV) on the reduced data. Moreover, a very simple tree with 7 nodes and 4 leaves is created. Comparing this with the results in Table 4 one can confirm that this is really the best possible classification method for this data, giving in most crossvalidations 100% accuracy. Moreover, analysis of the QPC projection coefficients helps to convert the obtained solution to a set of logical rules in the original space.

4   Conclusions

The holy grail of computational intelligence is to create algorithms that will automatically discover the best model for any data. There is no hope that a single method will be always the best and therefore multistrategy approaches [10] should be developed. Most machine learning methods developed so far may be presented as sequences of transformations. Searching in the space of all possible transformations may be done in an automatic way if a restricted framework for building models is provided, such as the similarity based framework [8]. Even in such restricted frameworks the space of possible transformations is huge, and thus the search for good models is very difficult. Linear separability is the most common, but not the best goal of learning. Initial transformation may show non-linear structures in the data that – if noticed – may be easy to handle with specific transformations.

Visualization may help to notice what type of algorithms are the most promising. Several linear and nonlinear visualization methods presented here proved to be useful in dimensionality reduction, evaluation of the reliability of classification for individual cases, but also discovering whether simple linear classifier, nearest neighbor approach, radial basis function expansion, naive Bayes, or a decision tree will provide simplest data model. In particular the QPC index recently introduced [16] proved to be quite helpful, showing structures in the Monk _1 problem that other methods were not able to reveal. Combining visualization and transformation-based systems should bring us significantly closer to practical systems that use meta-learning to create automatically the best data models.

References

1. Asuncion, A., Newman, D.: UCI repository of machine learning databases (2007),
   http://www.ics.uci.edu/~mlearn/MLRepository.html
2. Bengio, Y., Delalleau, O., Roux, N.L.: The curse of dimensionality for local kernel machines. Tech. Rep. Technical Report 1258, Dṕartement d'informatique et recherche opérationnelle, Université de Montréal (2005)
3. Brazdil, P., Giraud-Carrier, C., Soares, C., Vilalta, R.: Metalearning: Applications to Data Mining. In: Cognitive Technologies. Springer, Heidelberg (2009),
   http://www.liaad.up.pt/pub/2009/BGSV09
4. Cheng, Y.Q., Zhuang, Y.M., Yang, J.Y.: Optimal Fisher discriminant analysis using the rank decomposition. Pattern Recognition 25(1), 101–111 (1992)
5. Cox, T., Cox, M.: Multidimensional Scaling, 2nd edn. Chapman and Hall, Boca Raton (2001)
6. Duch, W.: k-separability. In: Kollias, S.D., Stafylopatis, A., Duch, W., Oja, E. (eds.) ICANN 2006. LNCS, vol. 4131, pp. 188–197. Springer, Heidelberg (2006)
7. Duch, W., Grąbczewski, K.: Heterogeneous adaptive systems. In: IEEE World Congress on Computational Intelligence, pp. 524–529. IEEE Press, Honolulu (2002)
8. Duch, W., Grudziński, K.: Meta-learning via search combined with parameter optimization. In: Rutkowski, L., Kacprzyk, J. (eds.) Advances in Soft Computing, pp. 13–22. Physica Verlag, Springer, New York (2002)
9. Duch, W., Jankowski, N., Grąbczewski, K., Naud, A., Adamczak, R.: Ghostminer data mining software. Tech. rep (2000-2005), http://www.fqspl.com.pl/ghostminer/
10. Michalski, R. (ed.): Multistrategy Learning. Kluwer Academic Publishers, Dordrecht (1993)

11. Golub, T.: Molecular classification of cancer: Class discovery and class prediction by gene expression monitoring. Science 286, 531–537 (1999)
12. Grąbczewski, K., Duch, W.: The separability of split value criterion. In: Proceedings of the 5th Conf. on Neural Networks and Soft Computing, pp. 201–208. Polish Neural Network Society, Zakopane (2000)
13. Grabczewski, K., Jankowski, N.: Versatile and efficient meta-learning architecture: Knowledge representation and management in computational intelligence. In: CIDM, pp. 51–58. IEEE, Los Alamitos (2007)
14. Grabczewski, K., Jankowski, N.: Meta-learning with machine generators and complexity controlled exploration. In: Rutkowski, L., Tadeusiewicz, R., Zadeh, L.A., Zurada, J.M. (eds.) ICAISC 2008. LNCS (LNAI), vol. 5097, pp. 545–555. Springer, Heidelberg (2008)
15. Grochowski, M., Duch, W.: Learning highly non-separable Boolean functions using Constructive Feedforward Neural Network. In: de Sá, J.M., Alexandre, L.A., Duch, W., Mandic, D.P. (eds.) ICANN 2007. LNCS, vol. 4668, pp. 180–189. Springer, Heidelberg (2007)
16. Grochowski, M., Duch, W.: Projection Pursuit Constructive Neural Networks Based on Quality of Projected Clusters. In: Kůrková, V., Neruda, R., Koutník, J. (eds.) ICANN 2008, Part II. LNCS, vol. 5164, pp. 754–762. Springer, Heidelberg (2008)
17. Maszczyk, T., Duch, W.: Support vector machines for visualization and dimensionality reduction. In: Kůrková, V., Neruda, R., Koutník, J. (eds.) ICANN 2008, Part I. LNCS, vol. 5163, pp. 346–356. Springer, Heidelberg (2008)
18. Michie, D., Spiegelhalter, D.J., Taylor, C.C.: Machine learning, neural and statistical classification. Elis Horwood, London (1994)
19. Naud, A.: An Accurate MDS-Based Algorithm for the Visualization of Large Multidimensional Datasets. In: Rutkowski, L., Tadeusiewicz, R., Zadeh, L.A., Żurada, J.M. (eds.) ICAISC 2006. LNCS (LNAI), vol. 4029, pp. 643–652. Springer, Heidelberg (2006)
20. Pękalska, E., Duin, R.: The dissimilarity representation for pattern recognition: foundations and applications. World Scientific, Singapore (2005)
21. Schölkopf, B., Smola, A.: Learning with Kernels. Support Vector Machines, Regularization, Optimization, and Beyond. MIT Press, Cambridge (2001)
22. Vilalta, R., Giraud-Carrier, C.G., Brazdil, P., Soares, C.: Using meta-learning to support data mining. IJCSA 1(1), 31–45 (2004),
    http://www.tmrfindia.org/ijcsa/V1I13.pdf
23. Webb, A.: Statistical Pattern Recognition. J. Wiley & Sons, Chichester (2002)
24. Wolberg, W.H., Mangasarian, O.: Multisurface method of pattern separation for medical diagnosis applied to breast cytology. Proceedings of the National Academy of Sciences 87, 9193–9196 (1990)

Neural Network and Artificial Immune Systems for Malware and Network Intrusion Detection

Vladimir Golovko, Sergei Bezobrazov, Pavel Kachurka, and Leanid Vaitsekhovich

Laboratory of Artificial Neural Networks, Brest State Technical University,
Moskovskaja str. 267, 224017 Brest, Belarus

Abstract. Neural network techniques and artificial immune systems (AIS) have been successfully applied to many problems in the area of anomaly activity detection and recognition. The existing solutions use mostly static approaches, which are based on collection viruses or intrusion signatures. Therefore the major problem of traditional techniques is detection and recognition of new viruses or attacks. This chapter discusses the use of neural networks and artificial immune systems for intrusion and virus detection. We studied the performance of different intelligent techniques, namely integration of neural networks and AIS for virus and intrusion detection as well as combination of various kinds of neural networks in modular neural system for intrusion detection. This approach has good potential to recognize novel viruses and attacks.

1 Introduction

At present one of the forms of world space globalization is cyber space globalization, because of increasing of a number of computers connected to the Internet. The rapid expansion of network-based computer systems has changed the computing world in the last years. As a result the number of attacks and criminals concerning computer networks are increasing. Due to the increasing computer incidents because of cyber-crime, construction effective protecting systems are important for computer systems security. There are many different techniques to build computer security systems [1,2,3]. The traditional approaches use as a rule static models, which are based mostly on signature analysis [4]. It consists of collecting and analyzing of viruses or intrusion signatures. The main problem of signature approach is inability to detect new viruses and attacks. Besides, this approach demands time for signature database updating. The methods of heuristic analysis [5], which were developed for disadvantages removal of traditional approach for malware detection, are still a long way off perfection. The heuristic analyzers are frequently finding malicious code where it absent and vice versa.

To overcome these limitations, the AIS and neural networks can be effectively used to build computer security systems. In order to achieve maximal performance we study different intelligent techniques, namely artificial neural networks and artificial immune systems. In comparison with conventional approaches such technique has ability to detect novel viruses and attacks in real time. Besides, this allows getting more accurate results.

J. Koronacki et al. (Eds.): Advances in Machine Learning II, SCI 263, pp. 485–513.
springerlink.com   © Springer-Verlag Berlin Heidelberg 2010

The rest of the chapter is organized as follows. Section 2 presents overview of artificial immune systems (AIS), as well as integration of AIS and neural networks for malicious code detection. Section 3 tackles different neural network techniques for intrusion detection and recognition.

2   Integration of Artificial Immune Systems and Neural Network Techniques for Malicious Code Detection

The actual researches in the information security field are directed to creation on such new methods that will be capable to detect unknown malicious code. The biologically-inspired and ready-built on basic principals of Biological Immune Systems (BIS) [1], Artificial Immune Systems method, thanks to distributed computational power [2, 3], is allow to detect not only known but unknown malware. Combining of two methods of artificial intelligence (Artificial Neural Networks method [6, 7] and Artificial Immune Systems method) let us developed a new technique of detection of malicious code. This technique allows to avoid the main weaknesses of signature analyzers and to detect unknown malware.

2.1   The Biological Immune System Overview

The biological immune system is unique protective mechanism which defends organism from invaders: harmful bacteria and viruses. The BIS capable to detect foreign cells and destroy them, and based on synthesis of special proteins – antibodies, which capable to bind with foreign material. Every day BIS face with a dozens invaders and successfully struggle against them.

The biological immune system is based on capability of antibodies to distinguish between self (cells of own body) and nonself (antigens, foreign substance) [1]. For complete and successful detection of wide variety of antigens the BIS must generate a large variety of detectors (B-lymphocytes and T-lymphocytes). Lymphocytes are formed from the bone marrow stem cells and initially incapable of antigens detect. In order to acquire immunological ability they have to go through maturation process. T-lymphocytes are mature in thymus and B- lymphocytes are mature in lymph nodes. During the maturation process only fittest lymphocytes are survived. Mature lymphocytes have on the own surface specific detectors which able to react on specific antigens. Lymphocytes circulate through the body and perform the function of antigens detection [8,9].

When lymphocyte detects an antigen the process called clonal selection is occurred [10]. The clonal selection process consists in proliferation (cycle of cell division) those lymphocytes who detected a virus. As a result the large population of identical detectors is formed. These generated lymphocytes are reacting on the same antigen and allow to BIS timely eliminating manifestation of disease.

Another important process in the BIS is immune memory [9]. Immune memory keeps information about previous infection and owing to this information defense body against repeated infection. Immune memory consists of detectors which in the past detected antigens. These detectors circulate in the body at long time and form the immune memory. By repeated infection antigens can be detected quickly since the BIS already have lymphocytes which react on this infection. Described processes showed in Figure 1.

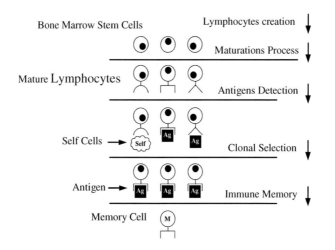

Fig. 1. Basic principles of the biological immune system: stage of lymphocytes evolution

2.2   The Artificial Immune Systems Overview

The AIS is founded on the same processes as BIS: detectors generation, detectors maturation, detection process, detectors cloning and mutation, immune memory creation. Let's view in detail each process (Fig. 2 shows processes as flow block).

Process of detectors creation in computer system represents a random generation of immune detectors population. Each of them can be, for example, as binary string of fixed size [11]. At this stage generated immune detectors have analogy with immature lymphocytes.

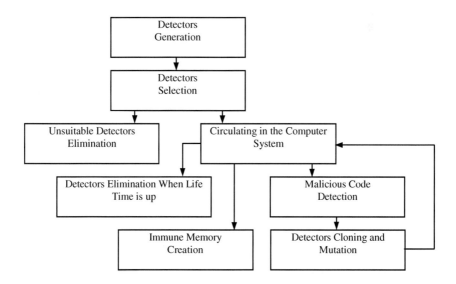

Fig. 2. Block-diagram model of artificial immune system: AIS interprocess communication

After generation detectors undergo a selection process. During the selection process detectors are received training in self – nonself recognition. But not all of immune detectors can get ability to correct pattern recognition. Even after training process some of them detect self as nonself and vice versa. These detectors named unsuitable detectors and should be eliminating. As a result of the selection process unsuitable detectors are eliminated and survive only those which able to distinguish between self and nonself. S. Forest at al. [12] proposed negative selection algorithm based on the principles of self – nonself discrimination in the BIS. According to negative selection algorithm immune detectors are compared with set of self pattern. If detector is similar to self pattern, it is reputed as negative and destroyed. Only those detectors survive which are structurally different from self data. For matching between detectors and pattern can be applied different rules: bit-by-bit comparison, r-contiguous matching [13] and r-chunk matching [14]. Mature detectors structurally different from self pattern therefore react only against nonself pattern.

Mature detectors circulate in computer systems. For maintenance of wide variety of structurally different detectors, each immune detector has a lifecycle [1,3]. The lifecycle is a time during immune detector can be found in the computer system. When the life time ends the detector is destroyed but if the detector detected anomaly (in our case it is malicious code) then lifecycle is prolong. Lifecycle mechanism allows the AIS to unload from weak detectors and permanently provide a space for new various immune detectors.

When malware enters into the computer system it often infects a large quantity of files. For quick reacting and eliminating virus manifestation we need a great number of similar detectors, which react on the same malicious code. For large quantity of similar detectors generations in AIS a cloning process is exists. During cloning process immune detector which found malicious code undergoes a cloning mechanism. Cloning means a large quantity of similar detectors creation. This mechanism allows the AIS infection elimination in a short space of time. Along with cloning a mutation mechanism is used [15]. Mutation process means introduce small random changes in detectors structure (for example, inverting of several bits in binary string) thereby as much as possible similar structure to finding virus acquires.

When the malicious code is eliminated then most of cloning detectors die because of lifecycle. However the fittest of them are kept as memory detectors. A set of such detectors are formed an immune memory. The immune memory keeps information about all malicious code which a computer system infects. The same as BIS the immune memory allows the AIS to quickly react on repeated infection and to fight against it.

2.3   Application of Neural Networks in Artificial Immune System to Malicious Code Detection

A quality of malicious code detections depends on the structure of immune detectors. We considered the immune detector as a binary string. This structure is comfortable, as it corresponds with data presentation in computer systems, and allows implement simple matching rules. However binary structure applies some restrictions. As it is well known bit-by-bit comparison is one of the slowest operations and needs heavy computational power. We propose the artificial neural networks (ANN) applying for the immune detectors formation. This approach for the detectors generation should remove weaknesses of the binary string structure and should increase a rate of the malicious code detection.

The ANN for vector quantization was proposed by T. Kohonen in 1982 and named as learning vector quantization (LVQ) [16]. The LVQ uses for classification and image segmentation problems. The LVQ is a feedforward artificial neural network with an input layer, a single hidden competitive Kohonen layer and an output layer (see Fig.3).

The output layer has as many elements as there are classes. Processing elements of the hidden (Kohonen) layer are grouped for each of these classes. Each class can be represents as a number of cells of the input space of samples. The centre of each cell corresponds to a codebook vector. One codebook vector represents one class only. The main idea of vector quantization is to cover the input space of samples with codebook vectors. A codebook vector can be seen as a hidden (Kohonen) neuron or a weight vector of the weights between all input neurons and the regarded Kohonen neuron respectively [17].

The Learning consists in modifying weights in accordance with adapting rules and, therefore, changing the position of a code vector in the input space. Many methods of training of the LVQ are exists [18]. We used the competitive training with one winner.

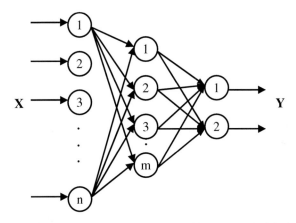

Fig. 3. LVQ one hidden competitive layer of neurons fully connected with the input layer, and the linear output layer consists of a number of neurons equal of a number of classes

Let's examine the process of detectors generation based on the LVQ. First an initial population of detectors is created. Each detector represents one LVQ. Further we will determine a set of self files consisting of different utilities of operating system, various software files etc, and one or a few malicious code (or signature of malicious code). Both self files and malicious virus will be used for LVQ learning. It is necessary to be sure that files from the set of self's are noninfected (without malicious code). Presence of malicious code or its signature in a learning sample allows a mature detector to tell the difference between self and nonself. Of course the more there are diverse files in the learning sample the more structurally different detectors are got. It is desirable to have all kind of malicious cod (worms, Trojans, file infectors etc.) in the learning sample. However it is not compulsory condition. As stated above there are differences between malicious software and noninfected files, which influence on the decision of a mature detector. By learning we denote to the LVQ where data from noninfected files and where data from malicious code (learning by instruction). A set of mature LVQ form a population of detectors which circulate into the computer system. In process of checking of a file the LVQ identifies unknown pattern and determines its proximity to one or another sample vector. Depending on this the LVQ takes a decision about the nature of files – self or malicious code.

General algorithm of neural immune detectors activity can be represents as next iterations:

1. Neuronet immune detectors generations. Each immune detector represents one neural network.
2. Detectors learning. The training set of self and nonself files is formed.
3. Unsuitable detectors eliminations.
4. Circulation of neuronet immune detectors in the computer system. On this stage detectors during scanning different files perform the function of malicious code detection.
5. Neuronet immune detectors eliminations by lifecycle.
6. Detection of malicious code.
7. Detectors cloning and mutation. On this iteration the AIS is formed a large quantity of similar detectors which react on the same malicious code.
8. Immune memory creation. Detectors of immune memory keep information about previous infections.

2.4   Description of Experimental Model of the AIS Security System

We used next structure of the LVQ for detectors formation – 128 neurons of the input layer, 10 neurons of the hidden layer and 2 neurons of the output layer (such detector is illustrated in Fig. 3, where $n = 128$, $m = 10$). A learning sample for one detector is formed as follows:

- four noninfected files from self's and one malicious code are selected randomly;
- from each selected file in fives fragments (binary string with length equal 128 bits) are randomly chosen. Then these fragments step by step will be inputted to the LVQ.

Competitive learning with one winner is used for the LVQ training. It is learning by instruction that is we indicate during training to the neural network where data from noninfected files is and where data from malicious code is. As a result of learning we get 10 code vectors in the hidden layer and they correspond with two output classes. The first class consists from 8 code vectors (noninfected files). The second class consists from 2 code vectors (malicious code).

As a result we will have a set of structurally different mature detectors since random process for files selecting is used for detectors learning. These detectors will be used for file identifications and decision making – is it self file or malicious virus? Experimental results in next section are described.

An immature detector compares any input pattern (independently of malicious code or noninfected file) to the first class (noninfected files) with probability 80% and to the second class (malicious code) with probability 20% since we divide the input space of samples in proportions 8 to 2 (see above). A mature detector (after the LVQ learning) will correlate an input pattern from a noninfected file with the first class with an expectancy of hitting more then 80%. Accordingly, the mature detector will correlate an input pattern from malicious code with the second class with expectancy of hitting more then 20%. The detector divides the under test file into pieces of 128 bytes apiece, examines them for malicious code in series and calculates total expectancy of hitting in one or another class:

$$P = \frac{X}{N} * 100\% , \qquad (1)$$

where X is a number of pieces running in one of a class, N is a total number of pieces of an under test file.

Let's review an example:

The file diskcopy.com (utility of operation system): file size is 7168 byte – 56 pieces of 128 bytes. A detector correlated 49 pieces with the first class (self) that was $P_S = \frac{49}{56} \cdot 100\% = 87,5\%$ expectancy of hitting. Accordingly an expectancy of hitting in the second class (malicious code) was $P_M = \frac{7}{56} \cdot 100\% = 12,5\%$. Detector's decision was noninfected file.

Experimental model of AIS for malicious code detection showed in Figure 4.

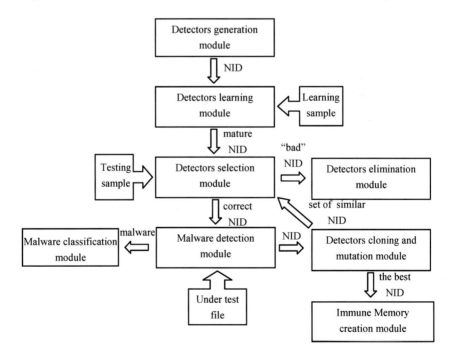

Fig. 4. Model of AIS for malicious code detection: NID – neuronet immune detector

2.5 Experimental Results

For our experiments we choose "wild" malwares, which were in the top 10 of the most prevalence at January and February in many countries. The latest malware using new algorithms and methods are chosen for test in order to observe the ability of neuronet immune detectors to find unknown malware.

In the first experiment we train detectors using well-known, not new different malware. They were owned to various classes: network worms (*Email-Worm.Win32. Eyeveg.m*, *Net-Worm.Win32.Bozori.a*), Trojans (*Trojan-Downloader.Win32.Adload.a*), classic viruses (*Virus.Win32.Hidrag.d*). Malware are classed according to Kaspersky classification [19]. Table 1 shows the results of malware detection. In all tables we used next parameters: P_S is expectancy that the under test file is noninfected (self) and P_M is expectancy that the under test file is malicious code. If $P_S > 0.8$ then detector takes under test file as self. If $P_S < 0.8$ then detector takes under test file as malware.

Table 1. The results of malware detection

Malware	Detector 1 P_S / P_M	Detector 2 P_S / P_M	Detector 3 P_S / P_M	Detector 4 P_S / P_M
Backdoor.Agobot	0.79/**0.21**	0.72/**0.28**	0.72/**0.28**	**0.85**/0.15
E-Worm.Bagle	0.69/**0.31**	0.51/**0.49**	**0.99**/0.01	0.74/**0.26**
E-Worm.Brontok	0.74/**0.26**	0.60/**0.40**	**0.98**/0.02	0.78/**0.22**
E-Worm.LovGate	0.72/**0.28**	0.53/**0.47**	**0.99**/0.01	0.74/**0.26**
E-WormMydoom	0.74/**0.26**	0.66/**0.34**	**0.81**/0.19	**0.83**/0.17
E-Worm.NetSky	0.77/**0.23**	0.70/**0.30**	0.77/**0.23**	0.75/**0.25**
E-Worm.Nyxem	**0.81**/0.19	0.76/**0.24**	0.72/**0.28**	**0.87**/0.13
E-Worm.Rays	**0.93**/0.07	**0.86**/0.14	0.79/**0.21**	**0.88**/0.12
E-Worm.Scano	**1.00**/0.00	**1.00**/0.00	**1.00**/0.00	**1.00**/0.00
Net-Worm.Mytob	0.69/**0.31**	0.54/**0.46**	**0.97**/0.03	0.75/**0.25**
Trojan.KillWin	**1.00**/0.00	**1.00**/0.00	**0.95**/0.05	**1.00**/0.00
Trojan.Dialer	**0.82**/0.18	0.76/**0.24**	0.80/0.20	**0.87**/0.13
Trojan.VB	**0.91**/0.09	**0.86**/0.14	0.69/**0.31**	**0.91**/0.09
Trojan-D.Small	0.79/**0.21**	0.75/**0.25**	0.68/**0.32**	**0.84**/0.16
Trojan-D.Zlob	**0.87**/0.13	0.74/**0.26**	**0.90**/0.10	0.80/0.20

The detector 1 is trained on *Email-Worm.Win32.Eyeveg.m* and able to detect 53% of all amount malware. The detector 2 is trained on *Net-Worm.Win32.Bozori.a* and able to detect 73% of all amount malware. The detector 3 is trained on *Trojan-Downloader.Win32.Adload.a* and able to detect % of all amount malware. The detector 4 is trained on *Virus.Win32.Hidrag.d* and able to detect 33% of all amount malware. In the result four detectors cover almost whole space of malware with the exception *Email-Worm.Win32.Scano.gen* and *Trojan.BAT.KillWin.c*.

In the second experiment we had subset all collection of the newest malware in to their classes, then we chosen a typical sample of every class and trained detectors on selected malware. The goal of this experiment is to research how different neuronet immune detectors react to unknown malware. Table 2 shows the results of the second experiment.

The detector 1 is trained on *Email-Worm.Win32.NetSky.c*, the detector 2 is trained on *Email-Worm.Win32.Nyxem.e*, the detector 3 is trained on *Net-Worm.Win32.Mytob.w* and the detector 4 is trained on *Trojan-Downloader.Win32.Zlob.jd*. As follows from results the first, second and third detectors find email worms and net worms very well as representatives of this class were included in learning sample for these detectors. Detection of relating to another class malware (in our case they are Trojans) is already not so well. The picture of malware detection by the firth detector is directly opposite. The detector 4 finds Trojans very well and net worms not so well. As a result all four detectors cover the whole space of malware (except *Email-Worm.Win32.Scano.gen* and *Trojan.BAT.KillWin.c*).

Table 2. The results of malware detection

Malware	Detector 1 P_S / P_M	Detector 2 P_S / P_M	Detector 3 P_S / P_M	Detector 4 P_S / P_M
Backdoor.Agobot	0.72/**0.28**	0.68/**0.32**	0.79/**0.21**	**0.87**/0.13
E-Worm.Bagle	0.69/**0.31**	0.73/**0.27**	0.61/**0.39**	0.60/**0.40**
E-Worm.Brontok	0.73/**0.27**	0.77/**0.23**	0.68/**0.32**	0.66/**0.34**
E-Worm.LovGate	0.70/**0.30**	0.76/**0.24**	0.63/**0.37**	0.60/**0.40**
E-WormMydoom	0.70/**0.30**	0.64/**0.36**	0.74/**0.26**	**0.82**/0.18
E-Worm.NetSky	0.71/**0.29**	0.66/**0.34**	0.77/**0.23**	**0.84**/0.16
E-Worm.Nyxem	0.75/**0.25**	0.70/**0.30**	**0.82**/0.18	**0.89**/0.11
E-Worm.Rays	**0.90**/0.10	**0.93**/0.07	**0.91**/0.09	0.79/**0.21**
E-Worm.Scano	**1.00**/0.00	**1.00**/0.00	**1.00**/0.00	**1.00**/0.00
Net-Worm.Mytob	0.68/**0.32**	0.71/**0.29**	0.63/**0.37**	0.63/**0.37**
Trojan.KillWin	**1.00**/0.00	**1.00**/0.00	**1.00**/0.00	**1.00**/0.00
Trojan.Dialer	0.77/**0.23**	0.74/**0.26**	**0.81**/0.19	0.73/**0.27**
Trojan.VB	**0.82**/0.18	0.79/**0.21**	**0.91**/0.09	0.75/**0.25**
Trojan-D.Small	0.75/**0.25**	0.72/**0.28**	0.79/**0.21**	0.70/**0.30**
Trojan-D.Zlob	**0.87**/0.13	**0.93**/0.07	**0.85**/0.15	0.71/**0.29**

In the third experiment we compare malware detection results by heuristic analyzer of ESET NOD32 antivirus software [20] and by our system. The results of experiment are displayed in the table 3.

Both *Trojan.BAT.KillWin.c* and *Email-Worm.Win32.Scano.gen* stay undetectable for NOD32 and AIS (we consider reason of this above). In addition NOD32 misses two malware (*Net-Worm.Win32.Mytob.q* and *Trojan.Win32.VB.at*), while AIS detects them.

Table 3. The comparative analysis of malware detection results

Malware	NOD32	AIS
Backdoor.Win32.Agobot.gen	Virus	Virus
Email-Worm.Win32.Bagle.gen	Virus	Virus
Email-Worm.Win32.Brontok.q	Virus	Virus
Email-Worm.Win32.LovGate.w	Virus	Virus
Email-Worm.Win32.Mydoom.l	Virus	Virus
Email-Worm.Win32.NetSky.aa	Virus	Virus
Email-Worm.Win32.Nyxem.e	Virus	Virus
Email-Worm.Win32.Rays	Virus	Virus
Email-Worm.Win32.Scano.gen	Ok	Ok
Net-Worm.Win32.Mytob.q	Ok	Virus
Trojan.BAT.KillWin.c	Ok	Ok
Trojan.Win32.Dialer.z	Ok	Virus
Trojan.Win32.VB.at	Ok	Virus
Trojan-Downloader.Win32.Small.to	Virus	Virus
Trojan-Downloader.Win32.Zlob.jd	Virus	Virus

Thus, as was shown the AIS for malicious code detection is able to discern between noninfected files of operation system and malicious code. The feature of the AIS consists in capability for unknown malicious code detection. Application of the ANN for detectors generation allows us to create the powerful detectors. Undesirable detectors are destroyed during the selection process which allows avoiding false detection appearance. Uniqueness of detectors consists in capability to detect several malicious viruses. That is detector can detect viruses analogous with that malicious code on which training are realized. In that way we significant increased probability of unknown malicious code detection. As experiments show it is necessary to large population of detectors creation. Presence of random probability by detectors generation enables to create different detectors. However it is significant that detectors ability depends on files on which they are trained. It is desirable for training process a various noninfected files and all types of malicious code to have. If your computer system with outdated antivirus bases can be unprotected in the face of new malicious code attack then the AIS gives you a high probability detect it. Applying of the AIS for malicious code detection will expand the potentialities of existing antivirus software and will increase level of computer systems security.

3   Neural Network Techniques for Intrusion Detection

The goal of Intrusion Detection Systems (IDS) is to protect computer networks from attacks. An IDS has been widely studied in recent years. There exist two main intrusion detection methods: misuse detection and anomaly detection. Misuse detection is based on the known signatures of intrusions or vulnerabilities. The main disadvantage of this approach is that it cannot detect novel or unknown attacks that were not previously defined. There are examples of misuse detection models: IDIOT [22], STAT [23] and Snort [24]. Anomaly detection defines normal behavior and assumes that an intrusion is any unacceptable deviation from normal behavior. The main advantage of anomaly detection model is the ability to detect unknown attacks. There are examples of anomaly detection models: IDES [25] and EMERALD [26].

Different defense approaches exist in order to protect the computer networks, namely, neural networks, data mining, statistical approach.

The principal component classifier is examined in [27, 28]. The data mining techniques were presented in [29, 30]. The other authors proposed a geometric framework for unsupervised anomaly detection and three algorithms: cluster, k-Nearest Neighbor (k-NN) and Support Vector Machine (SVM) [31, 32]. The different neural networks can be used for intrusion detection [33, 34]: Self Organizing Maps (SOM), MLP, Radial Basis Function (RBF) network.

The major problem of existing models is recognition of new attacks, low accuracy, detection time and system adaptability. The current anomaly detection systems are not adequate for real-time effective intrusion prevention [32]. Therefore processing a large amount of audit data in real time is very important for practical implementation IDS.

We use the KDD-99 data set [35] for training and testing of our approach. The data set contains approximately 5 000 000 connection records. Each record in the data set is a network connection pattern, which is defined as a sequence of TCP packets starting and ending at some well defined times, between which data flows to and from a source IP address to a target IP address under some well defined protocol.

Every record is described by 41 features and labeled either as an attack or non-attack. Every connection record consists of about 100 bytes. Among these features, 34 are numeric and 7 are symbolic. For instance, the first one is the duration of connection time, the second is protocol type, and the third is service name, and so on.

The goal of IDS is to detect and recognize attacks. There are 22 types of attacks in KDD-99 data set. All the attacks fall into four main classes: DoS – denial of service attack. This attack leads to overloading or crashing of networks; U2R – unauthorized access to local super user privileges; R2L – unauthorized access from remote user; Probe – scanning and probing for getting confidential data.

Every class consists of different attack types (Smurf, Neptune, Buffer Overflow, etc.)

3.1   Intrusion Detection Based on Recirculation Neural Networks

In the following sections, the recirculation neural network (RNN) based detectors to construct intrusion detection systems are discussed. The fusion classifier built up of these detectors is introduced to perform detection and recognition of network attacks.

3.1.1   RNN Based Detectors

The Anomaly Detector
Recirculation neural networks (Figure 5) differ from others ANNs that on the input information in the same kind is reconstructed on an output. They are applied to compression and restoration of the information (direct and return distribution of the information in the networks «with a narrow throat») [36], for definition of outliers on a background of the general file of entrance data [37].

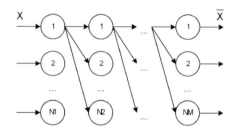

Fig. 5. *M* layers RNN structure N_i – quantity of neural elements in *i-th* layer, *NM=N1* – quantities of neural elements in entrance and target layers are equal

Nonlinear RNNs have shown good results as the detector of anomalies [38, 39]: training RNN is made on normal connections so that input vectors on an output were reconstructed in themselves, thus the connection is more similar on normal, the less reconstruction error is:

$$E^k = \sum_j (\overline{X}_j^k - X_j^k)^2, \qquad (2)$$

where X_j^k – j-th element of k-th input vector, $\overline{X}_j^k$ – j-th element k-th output vector.

Whether $E^k > T$, where T– certain threshold for given RNN connection admits anomaly, or attack, differently – normal connection. Thus there is a problem of a threshold T value determination, providing the most qualitative detection of abnormal connections. It is possible to get threshold value minimizing the sum of false positive (FP) and false negative (FN) errors, basing on cost characteristics of the given errors – FN error seems to be more expensive, than FP error, and its cost should be higher [39].

Private Classifiers
The described technique of definition of an input vector accessory to one of two classes – "normal" or "attacks", that is "not-normal" – it is possible to use in opposite way. If at training the detector of anomalies we used normal vectors which were restored in itself, and the conclusion about their accessory to a class "normal" was made, training the detector on vectors-attacks which should be restored in itself, it is possible to do a conclusion about their accessory to a class of "attack". Thus, if during functioning of this detector the reconstruction error (3) exceeds the certain threshold, given connection it is possible to carry to a class "not-attacks", that is normal connections. As training is conducted on vectors-attacks the given approach realizes technology of misuse detection, and its use together with previous technique is righteous.

Thus, one RNN can be applied to definition of an accessory of input vector to one of two classes – to on what it was trained (class A), or to the second (class $\overline{A}$), to which correspond outliers:

$$\begin{cases} X^k \in A, & if \ E^k \le T, \\ X^k \in \overline{A}, & if \ E^k > T. \end{cases} \qquad (3)$$

Worth to note that it is possible to train RNN in the special way [39] on connections of both classes so that to raise quality of detection on conditions (4).

As already it was mentioned above, database KDD includes normal connections and also attacks of four classes which considerably differ from each other. Therefore it is advisable to train detectors for each of five classes separately, not uniting all classes of attacks in a single whole.

Here again there is a problem of a choice of a threshold T for each concrete detector. If for the anomaly detector it was possible to speak at once, that cost of FN error is higher, than cost of FP error, in case of the detector for a class of attacks R2L it is hard to tell what will be worse – FP error (that is to name "R2L" connection to this class not concerning – attack of other class or normal connection) or FN detection of the given attack (on the contrary).

Many researchers [40] use a cost matrix for definition of cost of errors F (Table 4). Average values of FP and FN errors (Table 5) for each class can be calculated as follows:

$$F_i^{FP} = \frac{\sum\limits_{j,\ j\neq i} F_{ji}}{N-1}, \quad F_i^{FN} = \frac{\sum\limits_{j,\ i\neq j} F_{ij}}{N-1}, \qquad (4)$$

where N – quantity of classes $(N=5)$.

Table 4. The cost matrix F of incorrect classification of attacks and average costs of errors of detectors of each class

Real class	Cost of false prediction					Av. cost	
	normal	dos	probe	r2l	u2r	F_i^{FP}	F_i^{FN}
1 normal	0	2	1	2	2	2,5	1,75
2 dos	2	0	1	2	2	2	1,75
3 probe	1	2	0	2	2	1,5	1,75
4 r2l	4	2	2	0	2	2	2,5
5 u2r	3	2	2	2	0	2	2,25

On the basis of the given costs it is possible to choose value of a threshold which minimizes a total average error on training or validation data base.

Experimental Results

For an estimation of efficiency of the offered approach a number of experiments is lead. Private detectors for each class are trained, and all over again the training set got out of all base KDD, then from connections on concrete services – HTTP, FTP_DATA, TELNET. Nonlinear RNNs were used with one hidden layer with function of activation a hyperbolic tangent and logical sigmoid function of activation in a target layer. Quantity of neural elements in input and target layers according to quantity of parameters of input data – 41, in the hidden layer – 50. The training dataset contained 350 vectors of normalized values for each class. The RNNs were trained with layer-by-layer learning [36].

After each detector was trained the validation on training samples was conducted with the purpose of a finding of value of threshold T at which average cost of an error is minimal. In the further the testing of trained detectors was made on test samples with threshold values received before (Table 5). 10% of KDD database was used for testing purposes.

Table 5. Results of detectors testing

Service	Threshold	FP, %	FN, %	Av. cost	Service	Threshold	FP, %	FN, %	Av. cost
ALL					HTTP				
normal	0,00070	12,56	6,68	0,1844	normal	0,00123	6,8	2,72	0,0841
dos	0,00214	4,33	1,09	0,0542	dos	0,00340	0	0	0
probe	0,00120	7,79	14,21	0,1675	probe	0,00132	0	0	0
r2l	0,00116	2,87	5,38	0,0947	r2l	0,00114	5,17	0,25	0,0463
u2r	0,00112	7,07	5,54	0,1323	u2r	0,00126	0	0,07	0,0009
HTTP					TELNET				
normal	0,00620	2,4	0,17	0,0214	normal	0,00036	44,4	1,31	0,2394
dos	0,00290	1,5	0	0,0098	dos	0,00650	0	0	0
probe	0,00114	0	0	0	probe	0,00162	0	0	0
r2l	0,00110	0	0	0	r2l	0,00136	3,33	0	0,0294
					u2r	0,00076	5,91	2	0,0907

3.1.2   Fusion of Private Classifiers

Joint Functioning
As it was told above the best classification results can be achieved using several independent classifiers of the identical nature, because construction of the general estimation from private can be made by greater number of methods. We shall unite the private detectors trained in the previous section in one general (Figure 6).

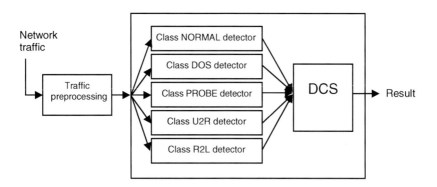

Fig. 6. Fusion of independent private classifiers in one general

The basic problem in construction of such a classifier becomes definition of a cumulative estimation proceeding from estimations of private detectors. In works of various researchers (for example [41]) the set of methods, such as a finding of average value for each class on the basis of indications of all classifiers, the sum of votes for each class, methods of an estimation «a priori» and «a posteriori» is considered. These methods mean that each classifier states a private estimation concerning an opportunity of an accessory of input image to at once several classes, and these classes are identical to all classifiers. However in our case classes, about an accessory to which each classifier judges, first, are various, secondly, are crossed. Therefore all the methods listed above are not applicable.

Dynamic Classifier Selection
The general classifier consists from $N=5$ private detectors, each of which has a threshold T_i. Values of thresholds got out proceeding from minimization of average cost of errors. To make estimation values comparable it is enough to scale reconstruction error on a threshold. Then (4) will be:

$$\begin{cases} X^k \in A_i, & if \quad \delta_i^k \leq 1, \\ X^k \in \overline{A}_i, & if \quad \delta_i^k > 1, \end{cases} \tag{5}$$

where $\delta_i^k = \dfrac{E_i^k}{T_i}$ - a relative reconstruction error. Thus, than less δ_i^k, the probability

of accessory of an input image X^k to a class A_i is higher. Therefore it is possible to allocate the method of determination of a cumulative estimation – by the minimal relative reconstruction error:

$$\begin{cases} X^k \in A_m, \\ \delta_m^k = \min_i \delta_i^k. \end{cases} \tag{6}$$

As the purpose of improvement of efficiency of classification is the minimization of erroneous classification expressed in minimization of average cost of classification, in construction of a cumulative estimation it is possible to act the same as at the choice of a threshold in private detectors – to consider cost of erroneous classification. If δ_i^k - a characteristic of probability of error of classification on i-th detector the estimation of possible average cost of error on each of detectors will be equal:

$$\Omega_i^k = \frac{\displaystyle\sum_{j, j \neq i} \delta_i^k F_{ji}}{N-1}. \tag{7}$$

The estimation (8) shows, what ability of loss in cost if we shall name a vector belonging to j-th class by a vector of i-th class, i. e. i-th classifier instead of j-th will be chosen.

On the basis of the given estimation we shall allocate the second method of a cumulative estimation determination – on the minimal possible cost of false classification:

$$\begin{cases} X^k \in A_m, \\ \Omega_m^k = \min_i \Omega_i^k. \end{cases} \tag{8}$$

Besides it is possible to consider mutual influence of possible errors – to add up an estimation Ω_i^k and an estimation of a prize in cost if i-th classifier instead of wrong j-th will be chosen:

$$\Psi_i^k = -\frac{\sum_{j, j \neq i} (\delta_j^k - \delta_i^k) F_{ij}}{N-1}. \tag{9}$$

Then on the basis of estimations (8) and (10) it is possible to allocate the third rule of winner detector selection – on the minimal possible mutual cost of false classification:

$$\begin{cases} X^k \in A_m, \\ \Omega_m^k + \Psi_m^k = \min_i (\Omega_i^k + \Psi_i^k). \end{cases} \tag{10}$$

Experimental Results

Efficiency of the general classifier functioning we shall check up experimentally using the private detectors trained in section 3.1.1. Results are presented in Tables 6-8.

Table 6. Results of attack detection and recognition by fusion of classifiers with minimal relative reconstruction error DCS

Service	FP, %	FN, %	MC, %	Recognized, %			
				dos	probe	r2l	u2r
ALL	10,80	2,34	3,76	98,17	96,55	91,88	100
HTTP	0	0,08	0,25	99,75	100	100	–
FTP_DATA	0,66	1,09	1,45	100	100	96,66	100

Table 7. Results of attack detection and recognition by fusion of classifiers with the minimal possible cost of false classification DCS

Service	FP, %	FN, %	MC, %	Recognized, %			
				dos	probe	r2l	u2r
ALL	30,80	0,9	30,80	97,8	99,3	92,5	100
HTTP	0	0,08	0	99,8	100	0	–
FTP_DATA	0,70	1,06	0,70	100	100	96,7	100

Table 8. Results of attack detection and recognition by fusion of classifiers with the minimal possible mutual cost of false classification DCS

Service	FP, %	FN, %	MC, %	Recognized, %			
				dos	probe	r2l	u2r
ALL	18,8	0,7	18,8	98,3	98,0	93,1	98,2
HTTP	0	0,08	0	99,8	100	100	–
FTP_DATA	27,3	0,4	27,3	100	77,6	98,7	100

Apparently from results, the unequivocal answer to a question – which method is better – is not present. The method of a choice of a final class with use of mutual cost can minimize a error, but with substantial growth of quantity of false detection, methods with minimal relative reconstruction error and possible cost give basically comparable results, on some service one is better, on some – another.

3.2 Modular Neural Network Detectors

In the following sections, several modular neural network detectors to construct Intrusion Detection Systems (IDS) are discussed. They are based on the integration of different artificial neural networks each of which performs complex classification task. Each neural network is intended for carrying out a specific function in the system. The proposed approaches are results of evolution from a single neural network detectors to multi-agent systems [42, 43, 44].

3.2.1 Basic Element of Intrusion Detection System

Let's examine the basic neural element to construct Intrusion Detection Systems (Fig.7). As input data, the 41 features from KDD-99 dataset will be used, which contain the TCP-connection information. The main goal of IDS is to detect and recognize the type of attack. Therefore, 5-dimensional vectors will be used for output data, because the number of attack classes plus normal connection is five. We propose to use the integration of PCA (principal component analysis neural network) and MLP (multilayer perceptron) as for basic element of IDS. We will name it the first variant of IDS.

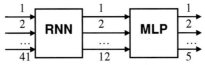

Fig. 7. The first variant of IDS (Model 1)

The PCA network, which is also called a recirculation network (RNN), transforms 41-dimensional input vectors into 12-dimensional output vectors. The MLP processes those given compressed data to recognize type of attacks or normal transactions.

In this section we present two neural networks based on principal component analyses techniques, namely linear and nonlinear RNN networks.

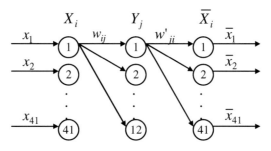

Fig. 8. RNN architecture

Let's consider an auto-encoder, which is also called a recirculation neural network (see Fig. 8). It is represented by MLP, which performs the linear or nonlinear compression of the dataset through a bottleneck in the hidden layer. As shown in the figure, the nodes are partitioned into three layers. The bottleneck layer performs the compression of the input dataset. The output of the j-th hidden unit is given by

$$y_j = F(S_j), \tag{11}$$

$$S_j = \sum_{i=1}^{41} w_{ij} \cdot x_i, \tag{12}$$

where F is activation function; S_j is weighted sum of the output from j-th neuron; w_{ij} is the weight from the i-th input unit to the j-th hidden unit; x_i is the input to the i-th unit.

The output from the i-th unit is given by

$$x_i = F(S_i), \tag{13}$$

$$S_i = \sum_{i=1}^{12} w'_{ji} \cdot y_j. \tag{14}$$

We use two algorithms for RNN training. One is the linear Oja rule and the other is the backpropagation algorithm for nonlinear RNN.

The weights of the linear RNN are updated iteratively in accordance with the Oja rule [45]:

$$w'_{ji}(t+1) = w'_{ji}(t) + \alpha \cdot y_j \cdot (x_i - \overline{x_i}), \tag{15}$$

$$w_{ij} = w'_{ji}.$$

Such a RNN is known to perform a linear dimensionality reduction. In this procedure, the input space is rotated in such a way that the output values are as uncorrelated as possible and the energy or variances of the data is mainly concentrated in a few first principal components.

As already mentioned, the backpropagation approach is used for training nonlinear RNN. The weights are updated iteratively in accordance with the following rule:

$$w_{ij}(t+1) = w_{ij}(t) - \alpha \cdot \gamma_j \cdot F'(S_j) \cdot x_i, \tag{16}$$

$$w'_{ji}(t+1) = w'_{ji}(t) - \alpha \cdot y_j \cdot F'(S_i)(\overline{x_i} - x_i) \tag{17}$$

where γ_j is error of j-th neuron:

$$\gamma_j = \sum_{i=1}^{n}(\overline{x_i} - x_i) \cdot F'(S_i) \cdot w'_{ji}. \tag{18}$$

The weights data in the hidden layer must be re-orthonormalized by using the Gram-Schmidt procedure [44].

Let's consider the mapping of input space data for the normal state and Neptune type of attack on the plane of the two first principal components. As we can see in Fig. 9(a), the data which belong to one type of attack can be located in different areas. The visualization of such data obtained by using only linear RNN will not be satisfactory because of complex relationship between the features. One of the ways to solve this problem is to use the nonlinear RNN network.

As we can see in Fig. 9(b), the nonlinear RNN performs better in visualizing dataset in comparison with a linear RNN.

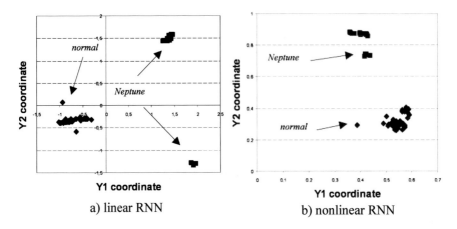

a) linear RNN   b) nonlinear RNN

Fig. 9. Data processed with: a – linear RNN, b – nonlinear RNN

There is a problem in Principal Components Analysis (PCA). We do not know the number of principal components.

Table 9. Recognition Rates for Some Set of Samples Depending on Number of Principal Components

Number of principal components	2	4	5	7	10	12	15	20	41
Recognition rates	39,24%	47,15%	71,84%	78,16%	95,25%	95,70%	96,84%	96,52%	96,84%

We have tried several neural network classifiers with different number of principal components, and analyzed the results of recognition by choosing the number of principal components which gave the best performance in efficiency in each of those classification model.

Our experiments (see Table 9) show that the optimal number of principal components lies near 12.

As already mentioned, the MLP is intended to classify attacks on the basis of components, which are obtained by using RNN. The number of output units depends on number of attack classes. The backpropagation algorithm is used for training MLP. After training of neural networks they are combined together for an intrusion detection system.

3.2.2 Generation of Different Intrusion Detection Structures

Using the results presented in the section 3.2.1, we can suggest several neural network classification models for development of intrusion detection systems.

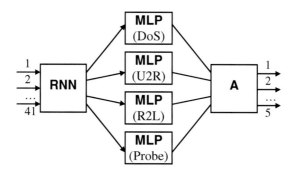

Fig. 10. The second variant of IDS (Model 2)

The second variant of IDS structure is shown in Fig. 10. It consists of four MLP networks. As we can see, every MLP network is intended to recognize the class of attack, that is, DoS, U2R, R2L or Probe. The output data from 4 multilayer perceptrons enter the Arbiter, which accepts the final decision according to the class of attack. A one-layer perceptron can be used as the Arbiter. The training of the Arbiter is performed after leaning of RNN and MLP neural networks. This approach enables to make a hierarchical classification of attacks. In this case, the Arbiter can distinguish one of the 5 attack classes by the corresponding MLP.

Complex computational problems can be solved by dividing them into a number of small and simple tasks. Then the results of each task are integrated for a general conclusion. An appropriate simplicity is achieved by distributing those training tasks to several *experts*. The combination of such experts is known as *Committee Machine*. This integrated knowledge has priority over the opinion taken separately from each expert. We have prepared two modular neural networks for the purpose of intrusion detection.

The third variant of IDS is based on this idea, and is shown in Fig. 11. Expert is represented by a single classification system. We use basic intrusion detection system as an expert (see Fig. 7). Training data sets for each expert are not the same with each other. They are self-organizing during the training process as a result of classification performed by the previous experts. The rule that was chosen for this purpose is Boosting by filtering algorithm [46]. After training, the neural networks have an ability to detect intrusions. In testing mode, every expert is intended for processing the original 41-demensional vectors. The Arbiter performs vote functions and accepts the final joint resolution of three experts. Arbiter is represented by the two-layer perceptron.

1. Train a first expert network using some training set;
2. A training set for a second expert is obtained in the following manner:
   - Toss a fair coin to select a 50% from NEW training set and add this data to the training set for the second expert network;
   - Train the second expert;
3. A third expert is obtained in the following way:
   - pass NEW data through the first two expert networks. If the two experts disagree, then add this data to the training set for the third expert:
   - Train the third expert network.
4. Vote to select output.

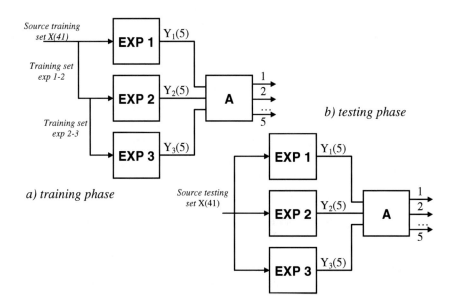

Fig. 11. The third variant of IDS, based on boosting by filtering algorithm (Model 3)

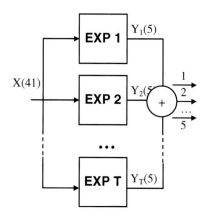

Fig. 12. The fourth variant of IDS, based on AdaBoost algorithm (Model 4)

In the case of AdaBoost algorithm [47] (Fig. 12), Summator performs the functions of the Arbiter. This analog of the Arbiter generates the result of the voting by summarizing private decisions.

3.2.3 Experimental Results

To assess the effectiveness of proposed intrusion detection approaches, a series of experiments were performed. The KDD99 cup network data set was used for training and testing different neural network models, because it is one of only a few publicly available data set of intrusion detection that attracts the researchers' attention due to its well-defined nature.

So we used 10% data selected from KDD dataset (almost 500000 records) to generate a subset for training and testing afterwards. To be more specific, we used 6186 samples for training neural networks, and used all records for testing the system (see Table 10).

Table 10. Training and Testing Samples

	DoS	U2R	R2L	Probe	Normal	Total count
training samples	3571	37	278	800	1500	6186
testing samples	391458	52	1126	4107	97277	494020

The same data sets were applied for model 1 and model 2 as well, so that we can compare the performance of those proposed models here. The approaches proposed are designed to detect 5 classes of attacks from this dataset which includes DoS, U2R, R2L, Probe and Normal.

To evaluate our system, we used three major indicators, that is, the detection rate and recognition rate for each attack class and false positive rate. The detection rate (true attack alarms) is defined as the number of intrusion instances detected by the system divided by the total number of intrusion instances in the test set. The recognition rate is defined in a similar manner. The false positive rate (false attack alarms) represents the total number of normal instances that were classified as intrusions divided by the total number of normal instances.

Let's examine the recognition of attacks using the model 1. This model is quite simple. Table 11 shows statistics of recognition depending on attack class.

Table 11. Attack Classification with Model 1

class	count	detected	recognized
DoS	391458	391441 (99.99%)	370741 (94.71%)
U2R	52	48 (92.31%)	42 (80.77%)
R2L	1126	1113 (98.85%)	658 (58.44%)
Probe	4107	4094 (99.68%)	4081 (99.37%)
Normal	97277	---	50831 (52.25%)

The above results show that the best detection rate and recognition rates were achieved for attacks by DoS and Probe connection. U2R and R2L attack instances were detected a bit worse (80.77% and 58.44%, respectively). Besides, the bottom row in Table 11 shows that some normal instances were (incorrectly) classified as intrusions.

The number of false positives emerged from the first model is considerable. This can be corrected by the second model described above. As shown in table 12, the second model performed quite well in terms of false positives. This is due to the four single multilayer perceptrons corresponding to each of the four attack classes.

Table 12. Attack Classification with Model 2

class	count	detected	recognized
DoS	391458	391063 (99.90%)	370544 (94.66%)
U2R	52	49 (94.23%)	37 (71.15%)
R2L	1126	1088 (96.63%)	1075 (95.47%)
Probe	4107	3749 (91.28%)	3735 (90.94%)
Normal	97277	---	83879 (86.22%)

As mentioned above, each expert in Model 3 and Model 4 is represented by a single classification system. We use model 1 as an expert in the experiments here as shown in Table 13 and 14. But every subsequent expert influences the outputs of other performing aggregated opinions of the several neural networks.

Table 13. Attack Classification with Model 3

class	count	detected	recognized
DoS	391458	391443 (99.99%)	370663 (94.69%)
U2R	52	50 (96.15%)	42 (80.76%)
R2L	1126	1102 (97.87%)	1086 (96.45%)
Probe	4107	3954 (96.27%)	3939 (95.91%)
Normal	97277	---	84728 (87.09%)

Table 14. Attack Classification with Model 4

class	count	detected	recognized
DoS	391458	389917 (99.61%)	369088 (94.29%)
U2R	52	51 (98.08%)	44 (84.62%)
R2L	1126	1119 (99.37%)	636 (56.48%)
Probe	4107	3908 (95.15%)	3668 (89.31%)
Normal	97277	---	77212 (79.37%)

The total results of the detection rates and false positive rates related with each model are shown in Table 15.

Table 15. Total Results for each Model

model	True attack alarms	False attack alarms	Recognized correctly	Total recognized %
Model 1	396696 (99.98%)	46446 (47.75%)	375522 (94,65%)	86.30%
Model 2	395949 (99.80%)	13398 (13.77%)	375391 (94.61%)	92.97%
Model 3	396549 (99.95%)	12549 (12.90%)	375730 (94.70%)	93.21%
Model 4	394995 (99.56%)	20065 (20.62%)	373436 (94.13%)	91.22%

In general, model 3 is shown to achieve the lowest false positive rate and the highest accuracy (93.21%). In fact, it is more accurate than the model 2 (92.97%) and the model 4 (91.22%). So, the three last models can be effectively used for the classification of huge input data set with a complicated structure.

3.2.4 Multiagent Neural Networks

Multiagent neural networks use several detectors that specialize different fields of knowledge.

In our work artificial immune system has been exploited for a development of multiagent IDS. Several important questions that strongly influence the efficiency of the model arise in the course of designing multiagent structures: obtaining of the generalized decision on the basis of the set of detector opinions, selection of detectors, cloning and mutation, destruction of bad and/or irrelevant detectors.

First of all, it is necessary to define what we will use as a detector to classify attacks. As shown in Fig. 13, we offer the model slightly modified the model proposed in the previous sections. See section 3.2.1 for more in detail.

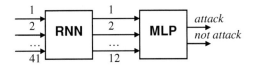

Fig. 13. Modified detector for immune system construction

Each detector is represented by artificial neural network consisted of recirculation neural network and multilayer perceptron, which functions were already discussed above. Such a detector specializes certain type of attack. There are two output values "yes" (when the entrance pattern relates to the given type of attack) and "no" (when the entrance pattern is not attack of the considered type).

The detectors, which represent the same type of attack, are combined in groups from 3 to 10. Generally, all the detectors in the group give the diverse conclusions which is the results of casual processes during the training. Theoretically, the number of detectors in the system is not limited and their number can be easily varied, but in real world problems with computational resources such as operative memory, speed etc..., arise.

Recognition process of an entrance pattern consists of the following sequence of steps:

1. Input pattern is transmitted to the multiagent system.
2. Each detector gives a conclusion about entrance activity.
3. So-called *factor of reliability on each group of the detectors* is formed. This factor reflects percent of voices in the group, given for the type of attack the group is specialized in.
4. The analysis of factors of reliability, obtained from each group, is carried out. A decision of the group with the maximum value of the factor is considered to be the final decision.

The obvious advantage of such an approach is, (i) Training process is made comparatively easily; (ii) Detectors are trained on a smaller number of samples than models considered in the previous sections; (iii) It allows to increase quality of their training and to considerably reduce time spent for preparation of the next detector.

Let's consider how such a multiagent system work from an example of a population of detectors. The population consists of 110 detectors (5 detectors in a group for each attack type from the KDD99 dataset). The results were prepared in the same way as the models in the previous sections (Table 16) so that we can compare them. As we can see, the results are similar to each other.

Table 16. Attack Classification with the Multiagent Neural Network

class	count	detected	Recognized
DoS	391458	383953 (98.08%)	368779 (94.21%)
U2R	52	47 (90.39%)	46 (88.46%)
R2L	1126	1122 (99.67%)	359 (31.88%)
Probe	4107	4105 (99.95%)	2369 (57.68%)
Normal	97277	---	75538 (77.65%)

The second experiment is related with the recognition of new attacks. For this purpose, we prepared a special set of samples for testing and training. The testing samples consist of network connection records that represent some of the most popular network services taken from the KDD99 dataset (http, ftp, ftp data, smtp, pop3, telnet). As dataset for testing, we generated a considerably reducing number of samples for each attack type. Also what is necessary to draw attention is that the records of some scanty attack types were entirely excluded from the training set. Therefore, only 9 types of attacks have been selected here. Accordingly, 9 groups (5 detectors in each) have been generated. So, the quantity of the population has made up 45 detectors.

Table 17. New Attack Detection with the Multiagent Neural Network

type	count	detected	type	count	detected
Normal	75952	71338 (93.93%)	Multihop*	7	7 (100.00%)
Land*	1	1 (100.00%)	Phf*	4	0 (0.00%)
Neptune	901	901 (100.00%)	Spy*	2	1 (50.00%)
Buffer_overflow	30	30 (100.00%)	Warezclient	1015	1003 (98.82%)
Loadmodule	9	9 (100.00%)	Warezmaster	20	20 (100.00%)
Perl*	3	1 (33.33%)	Ipsweep	9	9 (100.00%)
Rootkit*	7	3 (42.86%)	Nmap*	2	2 (100.00%)
ftp_write*	6	6 (100.00%)	Portsweep	15	15 (100.00%)
guess_passwd	53	53 (100.00%)	Satan	10	9 (90.00%)

* - the attacks that were absent in the training set.

The results shown in Table 17 show a lot of records corresponding to new attacks were detected and classified as an "attack". It means that multiagent systems are capable of detecting new attacks and have high generalization capacity.

We have discussed only the prototype of one population. Nevertheless, the results are promising due to the fact that many unknown records were detected. Extension of the proposed approach based on multiagent neural networks with the basic mechanisms of immune system (which exploits selection, mutation, cloning, etc.) will allow us to build a real time intrusion detection system.

4 Conclusion

In this chapter the artificial immune systems and neural network techniques for computer viruses and intrusion detection have been addressed. The AIS allow detecting unknown computer viruses. Integration of AIS and neural networks permits to increase performance of the security system. The IDS structure is based on integration of the different neural networks. As a result fusion classifier, modular neural networks and multiagent systems were proposed. The KDD-99 dataset was used for experiments performing. Experimental results show that the neural intrusion detection system has possibilities for detection and recognition computer attacks.

Proposed techniques have been shown powerful tools with respect to conventional approaches.

References

[1] de Castro, L.N., Timmis, J.I.: Artificial Immune Systems: A New Computational Intelligence Approach. Springer, Heidelberg (2002)
[2] Janeway, C.A.: How the Immune System Recognizers Invaders. Scientific American 269(3), 72–79 (1993)
[3] Dasgupta, D.: Artificial immune systems and their applications. Springer, New York (1999)
[4] Computer virus, http://en.wikipedia.org/wiki/Computer_virus
[5] Traditional antivirus solutions – are they effective against today's threats? (2008), http://www.viruslist.com
[6] Proactive protection: a panacea for Viruses? (2008), http://www.viruslist.com
[7] de Castro, L.N., Timmis, J.I.: Artificial Immune Systems: A New Computational Intelligence Approach. Springer, Heidelberg (2002)
[8] Janeway, C.A.: How the Immune System Recognizers Invaders. Scientific American 269(3), 72–79 (1993)
[9] Handbook of neural network processing. CRC Press LLC, Boca Raton (2002)
[10] Ezhov, A., Shumsky, S.: Neurocomputing and its application in economics and business, Moscow, MIPHI (1998)
[11] Ayara, M., Timmis, J., de Lemos, L., de Castro, R., Duncan, R.: Negative selection: How to generate detectors. In: Timmis, J., Bentley, P.J. (eds.) Proceedings of the 1st International Conference on Artificial Immune Systems (ICARIS), pp. 89–98. University of Kent at Canterbury Printing Unit, Canterbury (2002)
[12] Forrest, S., Hofmeyr, S.A.: Immunology as information processing. In: Segel, L.A., Cohen, I. (eds.) Design principles for the immune system and other distributed autonomous systems, Oxford University Press, New York (2000)
[13] Jerne, N.K.: Clonal Selection in a Lymphocyte Network, pp. 39–48. Raven Press (1974)
[14] Bezobrazov, S., Golovko, V.: Neural Networks for Artificial Immune Systems: LVQ for Detectors Construction. In: Proceedings of the IEEE International Workshop on Intelligent Data Acquisition and Advanced Computing Systems: Technology and Applications (IDAACS 2007), Dortmund, Germany (2007)
[15] Forest, S., Perelson, F., Allen, L., Cherukuri, R.: Self-Nonself Discrimination in a Computer. In: Proceedings IEEE Symposium on Research in Security and Privacy, pp. 202–212. IEEE Computer Society Press, Los Alamitos (1994)

[16] Balthrop, J., Esponda, F., Forrest, S., Glickman, M.: Coverage and Generalization in an Artificial Immune System. In: Proceedings of the Genetic and Evolutionary Computation Conference (GECCO), pp. 3–10. Morgan Kaufmann Publishers, San Francisco (2002)

[17] Hofmeyr, S., Forrest, S.: Architecture for an artificial immune system. EvolutionaryComputation 8(4), 443–473 (2000)

[18] Hofmeyr, S.A.: An interpretative introduction to the immune system. In: Cohen, I., Segel, L. (eds.) Design principles for the immune system and other distributed autonomous systems, Oxford University Press, New York (2000)

[19] Kohonen, T.: Self-organized Formation of Topologically Correct Feature Maps. Biological Cybernetics 43, 59–69 (1982)

[20] Hagan, M.T., Demuth, H.B., Beale, M.H.: Neural Network Design, 1st edn. PWS Pub. Co. (1995)

[21] Golovko, V.: Neural networks: training, organization and application, Moscow, IPRZHR (2001)

[22] Kaspersky Lab: Antivirus software (2008), http://www.kaspersky.com

[23] ESET NOD32 antivirus software (2008), http://www.eset.com

[24] Kumar, S., Spafford, E.H.: A Software architecture to support misuse intrusion detection. In: Proceedings of the 18th National Information Security Conference, pp. 194–204 (1995)

[25] Ilgun, K., Kemmerer, R.A., Porras, P.A.: State transition analysis: A rule-based intrusion detection approach. IEEE Transaction on Software Engineering 21(3), 181–199 (1995)

[26] SNORT, http://www.snort.org

[27] Lunt, T., Tamaru, A., Gilham, F., et al.: A Real-time Intrusion Detection Expert System (IDES) – final technical report. Technical report, Computer Science Laboratory, SRI International, Menlo Park, California (February 1992)

[28] Porras, P.A., Neumann, P.G.: EMERALD: Event monitoring enabling responses to anomalous live disturbances. In: Proceedings of National Information Systems Security Conference, Baltimore, MD (October 1997)

[29] Denning, D.E.: An intrusion-detection model. IEEE Transaction on Software Engineering 13(2), 222–232 (1987)

[30] Lee, W., Stolfo, S., Mok, K.: A data mining framework for adaptive intrusion detection. In: Proceedings of the 1999 IEEE Symposium on Security and Privacy, Los Alamos, CA, pp. 120–132 (1999)

[31] Lee, W., Stolfo, S.: A Framework for constructing features and models for intrusion detection systems. ACM Transactions on Information and System Security 3(4), 227–261 (2000)

[32] Liu, Y., Chen, K., Liao, X., et al.: A genetic clustering method for intrusion detection. Pattern Recognition 37(5), 927–934 (2004)

[33] Eskin, E., Rnold, A., Prerau, M., Portnoy, L., Stolfo, S.: A Geometric framework for unsupervised anomaly detection. In: Applications of Data Mining in Computer Security. Kluwer Academics, Dordrecht (2002)

[34] Shyu, M., Chen, S., Sarinnapakorn, K., Chang, L.: A Novel Anomaly Detection Scheme Based on Principal Component Classifier. In: Proceedings of the IEEE Foundations and New Directions of Data Mining Workshop, in conjunction with the Third IEEE International Conference on Data Mining (ICDM 2003), pp. 172–179 (2003)

[35] Kayacik, H., Zincir-Heywood, A., Heywood, M.: On the capability of an SOM based intrusion detection system. In: Proc. IEEE Int. Joint Conf. Neural Networks (IJCNN 2003), pp. 1808–1813 (2003)

[36] Zhang, Z., Li, J., Manikopoulos, C.N., Jorgenson, J., Ucles, J.: HIDE: a Hierarchical Network Intrusion Detection System Using Statistical Preprocessing and Neural Network Classification. In: Proceedings of the 2001 IEEE Workshop on Information Assurance and Security United States Military Academy, West Point, NY, pp. 85–90 (2001)

[37] 1999 KDD Cup Competition,
http://kdd.ics.uci.edu/databases/kddcup99/kddcup99.html

[38] Golovko, V., Ignatiuk, O., Savitsky, Y., Laopoulos, T., Sachenko, A., Grandinetti, L.: Unsupervised learning for dimensionality reduction. In: Proc. of Second Int. ICSC Symposium on Engineering of Intelligent Systems EIS 2000, University of Paisley, Scotland, pp. 140–144. ICSS Academic Press, Canada (2000)

[39] Hawkins, S., He, H., Williams, G., Baxter, R.: Outlier Detection Using Replicator Neural Networks. In: Kambayashi, Y., Winiwarter, W., Arikawa, M. (eds.) DaWaK 2002. LNCS, vol. 2454, pp. 170–180. Springer, Heidelberg (2002)

[40] Golovko, V., Kochurko, P.: Some Aspects of Neural Network: Approach for Intrusion Detection. In: Kowalik, Janusz, S., Gorski, J., Sachenko, A. (eds.) Cyberspace Security and Defense: Research Issues. NATO Science Series II: Mathematics, Physics and Chemistry, vol. 196, pp. 367–382. Springer, Heidelberg (2005); VIII, p. 382

[41] Kochurko, P., Golovko, V.: Neural Network Approach to Anomaly Detection Improvement. In: Proc. of 8th International Conference on Pattern Recognition and Information Processing (PRIP 2005), Minsk, Belarus, May18-20, pp. 416–419 (2005)

[42] Giacinto, G., Roli, F., Didaci, L.: Fusion of multiple classifiers for intrusion detection in computer networks. Pattern Recognition Letters 24, 1795–1803 (2003)

[43] Giacinto, G., Roli, F., Fumera, G.: Selection of image classifier. Electron 26(5), 420–422 (2000)

[44] Golovko, V., Vaitsekhovich, L.: Neural Network Techniques for Intrusion Detection. In: Proceedings of the International Conference on Neural Networks and Artificial Intelligence (ICNNAI 2006), Brest State Technical University - Brest, pp. 65–69 (2006)

[45] Golovko, V., Kachurka, P., Vaitsekhovich, L.: Neural Network Ensembles for Intrusion Detection. In: Proceedings of the 4th IEEE Workshop on Intelligent Data Acquisition and Advanced Computing Systems: Technology and Applications (IDAACS 2007), Research Institute of Intelligent Computer Systems, Ternopil National Economic University and University of Applied Sciences Fachhochschule Dortmund - Dortmund, Germany, pp. 578–583 (2007)

[46] Golovko, V., Vaitsekhovich, L., Kochurko, P., Rubanau, U.: Dimensionality Reduction and Attack Recognition using Neural Network Approaches. In: Proceedings of the Joint Conference on Neural Networks (IJCNN 2007), Orlando, FL, USA, pp. 2734–2739. IEEE Computer Society, Los Alamitos (2007)

[47] Oja, E.: Principal components, minor components and linear networks. Neural Networks 5, 927–935 (1992)

[48] Drucker, H., Schapire, R., Simard, P.: Improving performance in neural networks using a boosting algorithm. In: Hanson, S.J., Cowan, J.D., Giles, C.L. (eds.) Advanced in Neural Information Processing Systems, Denver, CO, vol. 5, pp. 42–49. Morgan Kaufmann, San Mateo (1993)

[49] Freund, Y., Schapire, R.E.: A short introduction to boosting. Journal of Japanese Society for Artificial Intelligence 14(5), 771–780 (1999)

Immunocomputing for Speaker Recognition

Alexander O. Tarakanov

St. Petersburg Institute for Informatics and Automation, Russian Academy of Sciences,
14-line 39, St. Petersburg, 199178, Russia
tar@iias.spb.su

Abstract. Based on mathematical models of immunocomputing, this chapter proposes an approach to speaker recognition by intelligent signal processing. The approach includes both low-level feature extraction and high-level ("intelligent") pattern recognition. The key model is the formal immune network (FIN) including apoptosis (programmed cell death) and immunization both controlled by cytokines (messenger proteins). Such FIN can be formed from audio signals using discrete tree transform (DTT), singular value decomposition (SVD), and the proposed index of inseparability in comparison with the Renyi entropy. Application is demonstrated on the task of recognizing nine male speakers by their utterances of two Japanese vowels. The obtained results suggest that the proposed approach outperforms state of the art approaches of computational intelligence.

1 Introduction

Artificial immune systems (AISs) (Dasgupta 1999; de Castro and Timmis 2002) and immunocomputing (IC) (Tarakanov et al. 2003; Zhao 2005) are developing with the branches of computational intelligence (Tarakanov and Nicosia 2007; Dasgupta and Nino 2008; Tarakanov 2008) like genetic algorithms (GAs) and artificial neural networks (ANNs) also called neurocomputing. Recent advances in AISs include a stochastic model of immune response (Chao et al. 2004), an aircraft fault detection (Dasgupta et al. 2004), intrusion detection (Dasgupta and Gonzalez 2005) and computational models based on the negative selection process that occurs in the thymus (Dasgupta 2006; Dasgupta and Nino 2008).

Recent advances in IC include a concept of biomolecular immunocomputer as a computer controlled fragment of the natural immune system (Goncharova et al. 2005). A connection of IC with cellular automata (CA) (Adamatzky 1994) leads to encouraging results in three-dimensional (3D) computer graphics (Tarakanov and Adamatzky 2002) and inspires a novel method of identification of CA (Tarakanov and Prokaev 2007). A connection of IC with brain research helps to discover and study at least three deep functional similarities and the fundamental communicative mechanisms the neural and immune system have in common: a) cytokine networks of brain and immunity (Goncharova and Tarakanov 2007), b) nanotubes at neural and immune synapses including ion channels, microtubules and tunneling nanotubes (Goncharova and Tarakanov 2008a), and c) receptor mosaics of neural and immune communication (Agnati et al. 2005ab, 2008; Goncharova and Tarakanov 2008b). Understanding these issues could lead to new therapeutic targets and tools, especially in neurodegenerative disorders related to microglia, like in Parkinson's disease (Fuxe et al. 2008).

Apart from brain research and molecular medicine, recent advances in real-world applications of IC (Tarakanov 2007a) also include intelligent simulation and forecast of hydro-physical fields (Tarakanov et al. 2007b; Tarakanov 2009a), intrusion detection (Tarakanov 2008) and signal processing (Atreas et al. 2003, 2004; Tarakanov et al. 2005b, 2007a). This chapter proposes the IC approach to speaker recognition based on intelligent signal processing.

2   Background

According to (Tarakanov et al. 2003), IC is based on the principles (especially, mathematical models) of information processing by proteins and immune networks. Some similarities and differences between neurocomputing and IC are shown in Tab. 1. Since ANN represents a "hardwired" network of artificial neurons, essential difference of IC is that formal immune network (FIN) represents a network of free bindings between formal proteins. For example, the IC approach to pattern recognition is abstracted from the principle of molecular recognition between proteins, including antibody (also called immunoglobulin: Ig) of natural immune system and any other antigen (including another antibody). Let Ig1 and Ig2 be two antibodies, while Ag be antigen. The strength of biophysical interaction between any pair of proteins can be measured by their binding energy. Let FIN[1] and FIN[2] be values of binding energy between Ag and Ig1, Ig2, correspondingly. Then any protein (including antibody) can be represented and recognized by the corresponding couple of numbers FIN[1] and FIN[2] in such 2D immune network of interactions (formed by two antibodies Ig1, Ig2). Accordingly, any high-dimensional input vector Ag (antigen which can include several Ig-binding sites also called epitopes) can be projected to such low-dimensional space of FIN and recognized by the class of the nearest point of FIN (Fig. 1).

Table 1. Similarities and differences between neuro- and immunocomputing

Approach	Neurocomputing	Immunocomputing
Basic Element	Artificial Neuron (AN)	Formal Protein (FP)
Network	Artificial Neural Network	Formal Immune Network
Plasticity	Rigid connections of ANs	Free bindings of FPs
Learning	Weights of connections	Binding energies
Hardware	Neurochip	Immunochip

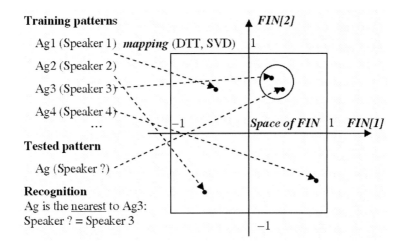

Fig. 1. Immunocomputing approach to intelligent speaker recognition

In such background, the key model of the IC approach to intelligent speaker recognition is the FIN. In the training mode, FIN is formed from audio signals using discrete tree transform (DTT) (Atreas et al. 2003, 2004) and singular value decomposition (SVD) (Horn and Johnson 1986). After the procedures of apoptosis (programmed cell death) and immunization both controlled by cytokines (messenger proteins) (Tarakanov et al. 2005a), the result of such feature extraction by FIN is estimated by the proposed index of inseparability (Tarakanov 2007b) in comparison with the Renyi entropy (Renyi 1961). In the recognition mode, the current audio signals are processed by DTT, mapped to the FIN, and the speaker is recognized by a "cytokine class" of the nearest cell (point) of the FIN.

3 Mathematical Models

3.1 Formal Immune Network

Let vector-matrix transposing be designated by upper stroke ($[]'$). For example, if X is column vector then X' is row vector.

Definition 1. Cell is a pair $V = (f, P)$, where f is real value ("cytokine value") $f \in R$, whereas $P = (p_1, ..., p_q)$ is a point of q-dimensional space: $P \in R^q$ with the restricted coordinates: $\max\{| p_1 |, ..., | p_q |\} \leq 1$.

Let distance ("affinity") $d_{ij} = d(V_i, V_j)$ between cells V_i and V_j is defined by any norm $\|P\|$ (Euclidean, Tchebyshev, Manhattan, etc.) so that $d_{ij} = \|P_i - P_j\|$.

Fix some finite non-empty set of cells ("innate immunity"): $W_0 = (V_1, ..., V_m)$.

Definition 2. FIN is a set of cells: $W \subseteq W_0$.

Definition 3. Cell V_i recognizes cell V_k if the following conditions are satisfied:

$$\left| f_i - f_k \right| < \rho \,, \ d_{ik} < h \,, \ d_{ik} < d_{ij} \,, \ \forall V_j \in W \,, \ j \neq i \,, \ k \neq j \,,$$

where ρ and h are non-negative real values ("recognition threshold" and "affinity threshold").

Let us define the behavior of FIN by the following two rules.

Rule 1 (apoptosis). If cell $V_i \in W$ recognizes cell $V_k \in W$ then remove V_i from FIN.

Rule 2 (immunization). If cell $V_k \in W$ is nearest to cell $V_i \in W_0 \setminus W$ among all cells of FIN: $d_{ik} < d_{ij} \,, \ \forall V_j \in W$, whereas $\left| f_i - f_k \right| \geq \rho$, then add V_i to FIN.

Note that the immunization in Rule 2 is actually "auto-immunization" since the immunizer cell belongs to the set of "innate immunity" W_0. Let W_A be FIN as a consequent of application of apoptosis to all cells of W_0. Let W_I be FIN as a consequence of immunization of all cells of W_A by all cells of W_0. Note that the resulting sets W_A and W_I depend on the ordering of cells in W_0. Further it will be assumed that the ordering is given. It will be also assumed that $d_{ij} \neq 0 \,, \ \forall i \neq j$. Consider some general mathematical properties of FIN. The following Properties 1-3 look obvious while Proposition states more important and less evident feature of FIN.

Property 1. Neither the result of apoptosis W_A nor the result of immunization W_I can overcome W_0 for any innate immunity and both thresholds:

$$W_A \subseteq W_0, \quad W_I \subseteq W_0, \quad \forall W_0, h, \rho \,.$$

Property 2. For any innate immunity W_0 and recognition threshold ρ there exists such affinity threshold h_0 that apoptosis does not change W_0 for any h less than h_0:
$W_A = W_0 \,, \forall h < h_0$.

Property 3. For any innate immunity W_0 and affinity threshold h there exists such recognition threshold ρ_0 that apoptosis does not change W_0 for any ρ less than ρ_0:
$W_A = W_0 \,, \forall \rho < \rho_0$.

Proposition. For any innate immunity W_0 and recognition threshold ρ there exists affinity threshold h_1 such that consequence of apoptosis and immunization $W_1 = W_I(h_1)$ provides the minimal number of cells $|W_1| > 0$ for given W_0 and ρ and any h: $|W_1| \leq |W_I(h)| \,, \ \forall h \,, \ \forall W_I \subseteq W_0$.

The proof of this Proposition can be found in (Tarakanov 2007a). Actually, the Proposition states that the minimal number of cells after apoptosis and immunization is a kind of "inner invariant" of any FIN, which depends on the innate immunity and the recognition threshold but does not depend on the affinity threshold. Practically, it means that such invariant can be found for any FIN by apoptosis and immunization without considering any affinity threshold (in Definition 3) at all.

Now we can define a model of molecular recognition in terms of FIN. Let "epitope" (antigenic determinant) be any point $P = (p_1,..., p_q)$ of q-dimensional space: $P \in R^q$. Note that any cell of FIN also contains an epitope, according to Definition 1.

Definition 4. Cell V_i recognizes epitope P by assigning him value f_i if the distance $d(V_i, P)$ between the cell and the epitope is minimal among all cells of FIN: $d(V_i, P) = \min\{d(V_j, P)\}, \forall V_j \in W$.

If value f (in Definition 1) is natural number $f = c$, $c \in N$ (i.e. "cytokine class"), whereas recognition threshold (in Definition 3) $\rho < 1$, then we obtain a special case of cytokine FIN proposed by Tarakanov et al. (2005a) and applied to pattern recognition. Note that innate immunity W_0 and its ordering of cells are determined by the set and the order of raw data of a particular application. For example, this order is naturally determined by time series in intrusion detection (Tarakanov 2008), spatio-temporal forecast (Tarakanov 2009a), and signal processing (Tarakanov 2009b).

3.2 Singular Value Decomposition

Let pattern ("molecule") be any n-dimensional column-vector $\mathbf{Z} = [z_1,..., z_n]'$, where $z_1,..., z_n$ are real values. Let pattern recognition be mapping $\mathbf{Z} \to P$ of the pattern to a q-dimensional epitope, and recognition of the epitope by the value f of the nearest cell of FIN. Consider a mathematical model of such mapping of any pattern: $R^n \to R^q$. Let $\mathbf{Z}_1,..., \mathbf{Z}_m$ be n-dimensional training patterns with known values $f_1,..., f_m$. Let $\mathbf{A} = [\mathbf{Z}_1,..., \mathbf{Z}_m]'$ be training matrix of dimension $m \times n$. Consider SVD of this matrix (Horn and Johnson 1986):

$$\mathbf{A} = s_1 \mathbf{X}_1 \mathbf{Y}_1' + ... + s_r \mathbf{X}_r \mathbf{Y}_r', \tag{1}$$

where r is rank of the matrix, s_k are singular values and $\mathbf{X}_k, \mathbf{Y}_k$ are left and right singular vectors with the following properties:

$$\mathbf{X}_k' \mathbf{X}_k = 1, \ \mathbf{Y}_k' \mathbf{Y}_k = 1, \ \mathbf{X}_k' \mathbf{X}_i = 0, \ \mathbf{Y}_k' \mathbf{Y}_i = 0, \ i \neq k, \ k = 1,...,r. \tag{2}$$

Consider the following map $P(\mathbf{Z}): R^n \to R^q$, where $\mathbf{Z}$ is any n-dimensional pattern $\mathbf{Z} \in R^n$ and $\mathbf{Y}_1,..., \mathbf{Y}_q$ are left singular vectors of SVD (1):

$$p_k = \frac{1}{s_k} \mathbf{Y}_k' \mathbf{Z}, \ k = 1,...,q. \tag{3}$$

Property 4. Any epitope $P(\mathbf{Z}_i)$ obtained by the application of formula (3) to any training pattern $\mathbf{Z}_i$, $i = 1,...,m$, lies within unit cube of FIN (see Definition 1).

This property can be proved using the properties (2) of singular vectors.

3.3 Discrete Tree Transform

Consider the mathematical model of forming pattern from any one-dimensional signal (time series). Let $T = \{t_1,...,t_n\}$ be a fragment of signal, where $t \in R$ be real value in general case, $n = 2^{N_0}$ and N_0 is some number exponent so that n is a power of 2. Let $u = 2^{N_1}$, $N_1 \leq N_0$. According to (Atreas et al. 2003), the dyadic DTT of T is the following map:

$$T \rightarrow \{a_{u,k}\}, \quad a_{u,k} = \frac{1}{n} \sum_{1+(k-1)u}^{ku} t_i, \quad k = 1,...,2^{N_0-N_1}.$$

Let $l = N_1$ be DTT level: $0 \leq l \leq N_0$. Let us denote the DDT map as follows:

$$T \rightarrow T^{(l)}, \quad T^{(l)} = \{t_1^{(l)},...,t_n^{(l)}\}, \quad t_i^{(l)} = a_{u,k}, \quad 1+(k-1)u \leq i \leq ku. \qquad (4)$$

Consider the values $z_j = t_j^{(l)}$, $j = 1,...,n$, as the pattern (vector) $Z = [z_1,...,z_n]'$ obtained by the processing of any fragment T of the signal.

3.4 Index of Inseparability

According to the above models, the feature extraction method by IC is as follows.

1. Extract m training patterns from the signal.
2. Form q-dimensional FIN1 with $m_1 = m$ cells (using DTT and SVD).
3. Find its inner invariant FIN2 with $m_2 \leq m_1$ cells (using apoptosis and immunization).

As the result, the q-dimensional points of FIN2 $P_1,..., P_{m_2}$ can be considered as the feature vectors that represent the signal.

The following task is to estimate a quality of such feature extraction. This can be done using the special entropy proposed in (Renyi 1961) and proved to be rather useful metric of very large networks regarding the task of intrusion detection (Johnson 2005). According to (Renyi 1961; Johnson 2005), the Renyi entropy of j-th dimension of FIN can be defined as follows:

$$[I_R]_j = -\frac{1}{m} \sum_{i=1}^{m} \log_2(p_{ij}^2), \qquad (5)$$

where $p_{1j},..., p_{mj}$ are the values of j-th coordinate of the points of FIN $P_1,..., P_m$. According to (Tarakanov 2007b), let us consider the maximal entropy as the Renyi entropy of FIN:

$$I_R = \max\{[I_R]_j\}, \quad j = 1,..., q. \qquad (6)$$

Usually, entropy represents a measure of disorder. The lower is entropy the lower is disorder of the system and vice versa. Consider another metric which is more specific to FIN. According to (Tarakanov 2007b), the index of inseparability of FIN2 can be defined as follows:

$$I_2 = \ln\left(\frac{m_2}{m_1 h_{\min}}\right), \tag{7}$$

where m_1 is number of cells in FIN1 and m_2 is number of cells in FIN2 (after apoptosis and immunization), whereas $h_{\min}$ is the minimal distance between any pair of cells of FIN with different values of f :

$$h_{\min} = \min\{d_{ij}\}, \; i \neq j, \; |f_i - f_j| > \rho. \tag{8}$$

Consider that $m_2 = m_1$ for FIN1. Then the index of FIN1 can be derived from (7) as follows:

$$I_1 = -\ln(h_{\min}). \tag{9}$$

Thus, the greater is minimal distance $h_{\min}$ the lower is the index and the better is the separability between different cells of FIN.

4 Speaker Recognition

The IC scheme of intelligent speaker recognition is shown in Fig. 2. In both training and recognition modes, the fragment is extracted from the signal and processed by DTT to form the pattern (antigen). Steps 1-11 below describe the IC algorithm. Steps 1-8 form the training, whereas Steps 9-11 form the recognition.

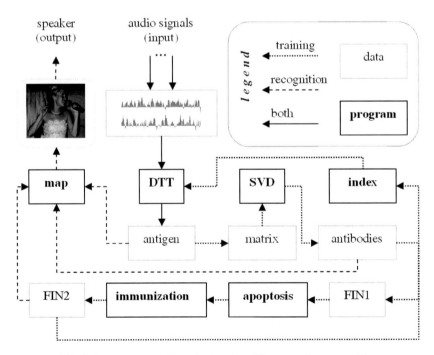

Fig. 2. Immunocomputing scheme of intelligent speaker recognition

Step 1. Fix integers: n, m, u, l, q, where $n \geq 1$ is the number of time points (fragment) of one signal (if speaker should be recognized by one signal) or number of signals (if speaker should be recognized by several signals); $m \geq 1$ is the number of training counts of the signal(s); $u = 2^{N_1}$ is the window size for DTT; $0 \leq l \leq N_1$ is the level of DTT (so that $l = 0$ means no DTT); $q \geq 1$ is the dimension of FIN.

Step 2. Compute DTT (4) of each fragment of the training signal(s): $(t_i,...,t_{i+n+1})$, $i = 1,..., m$.

Step 3. Form training vector $\mathbf{Z}_i = [z_{i1},..., z_{in}]'$, where $z_{ij} = t_{i+j}^{(l)}$ for one signal or $z_{ij} = t_j^l$ for several signals, $j = 1,...,n$, and assign the speaker number (class) for this vector: $f(\mathbf{Z}_i) = c(\mathbf{Z}_i)$.

Step 4. Form training matrix $\mathbf{A} = [\mathbf{Z}_1 ... \mathbf{Z}_m]'$ of the dimension $m \times n$.

Step 5. Compute first q singular values $s_1,..., s_q$ and corresponding right singular vectors $\mathbf{Y}_1,..., \mathbf{Y}_q$ by SVD (1) of the training matrix, where $q \leq r$ and r is rank of the matrix: $r \leq \min\{m,n\}$.

Step 6. Form the points of FIN1 $P_1,..., P_q$ by the mapping (3) of each training vector Z_i, $i = 1,..., m$ to the q-dimensional space of FIN:

$$p_{i1} = \frac{1}{s_1} \mathbf{Y}_1' \mathbf{Z}_i, \ ... \ , \ p_{iq} = \frac{1}{s_q} \mathbf{Y}_q' \mathbf{Z}_i.$$

Step 7. Compute the index of inseparability of FIN1 (9). If $I_1 > 50$ then go to Step 9 without apoptosis and immunization at Step 8. This means that $h_{\min} < 10^{-21}$ and, thus, the training data are conflicting so that at least a couple of actually coincident patterns belongs to different classes (speakers):

$$c(\mathbf{Z}_i) \neq c(\mathbf{Z}_j), \ \mathbf{Z}_i \cong \mathbf{Z}_j, \ i \neq j.$$

Step 8. Using the apoptosis and immunization, reduce $m_1 = m$ training points of FIN1 to $k = m_2$ points of FIN2, where the number of the points k is self-defined by the inner invariant of FIN (see Proposition).

Step 9. Compute DTT of the fragment of test signal, form n-dimensional vector Z and compute mapping (3) to the q-dimensional space of FIN (FIN1 or FIN2, depending on Step 7):

$$p_1 = \frac{1}{s_1} \mathbf{Y}_1' \mathbf{Z}, \ ... \ , \ p_q = \frac{1}{s_q} \mathbf{Y}_q' \mathbf{Z}.$$

Step 10. Among the training points of FIN $P_1,...,P_k$ (where $k = m_1$ or $k = m_2$, depending on Step 7), determine the nearest point to $P(Z)$.

Step 11. The class of this point $c(\mathbf{Z})$ is the recognizing speaker.

Note the following important property of the IC algorithm.

Property 5. If test vector is equal to any training vector: $\mathbf{Z} = \mathbf{Z}_i$, $i = 1,...,m$, then exactly $f(\mathbf{Z}) = f(\mathbf{Z}_i)$.

The proof of this property can be found in (Tarakanov 2007a). Thus, the IC algorithm recognizes exactly any pattern it has been trained. Simply say, the IC approach does not make mistakes on any non-conflicting training set.

5 Numerical Examples

Data for numerical experiments have been taken from KDD archive (KDD 1999). The task is to distinguish 9 male speakers by their utterances of 2 Japanese vowels 'ae' successively (Kudo et al. 1999). For each utterance, Kudo et al. (1999) obtained a discrete-time series with 12 linear predictive coding (LPC) cepstrum coefficients (Rabiner and Juang 1993; Antoniol et al. 2005). This means that one utterance by a speaker forms a time series whose length is in the range 7-29 and each point of a time series is of 12 features (12 coefficients).

The number of the time series is 640 in total. Kudo et al. (1999) used one set of 270 time series for training and the other set of 370 time series for testing. Number of instances (utterances) is 270 for the training file (30 utterances by 9 speakers) and 370 for the test file (24-88 utterances by the same 9 speakers in different opportunities). Each line in the training or test file represents 12 LPC coefficients in the increasing order separated by spaces. Lines are organized into blocks, which are a set of 7-29 lines separated by blank lines and corresponded to a single speech utterance of 'ae'. Each speaker is a set of consecutive blocks. In the training file, there are 30 blocks for each speaker. Blocks 1-30 represent speaker 1, blocks 31-60 represent speaker 2, and so on up to speaker 9. In the test file, speakers 1 to 9 have the corresponding number of blocks (31, 35, 88, 44, 29, 24, 40, 50, 29). Thus, blocks 1-31 represent speaker 1 (31 utterances of 'ae'), blocks 32-66 represent speaker 2 (35 utterances of 'ae'), and so on. For example, first 3 lines (points of time series) of block 1 (speaker 1) are shown below (from the training file):

```
1.86 -0.21 0.26 -0.21 -0.17 -0.12 -0.28 0.03 0.13 -0.31 -0.21 0.09
1.89 -0.19 0.24 -0.25 -0.11 -0.11 -0.31 -0.03 0.17 -0.29 -0.25 0.09
1.94 -0.24 0.26 -0.29 -0.04 -0.10 -0.38 0.02 0.17 -0.31 -0.23 0.07
```

Let $t_i(x_j)$ be values (time series) of LPC coefficients $j = 1,...,12$ in time points $i = 1, ... 4274$. Thus, $m = 4274$, $n = 12$, $t_1(x_1) = 1.86$, $t_2(x_1) = 1.89$, $t_3(x_1) = 1.94$, etc. A fragment of this time series for $i < 520$ is shown in Fig. 3, whereas an example of DTT of this time series is shown in Fig. 4.

According to the test file, number of the testing patterns (time points) is 5687. The obtained results are collected in Tab. 2 for one signal (LPC coefficient 1) and Tab. 3

for all signals (LPC coefficients 1-12). Example of FIN for the best recognition rate (99.1% in Tab. 3) is shown in Fig. 5. The values of parameters of such FIN are as follows (see Step 1): number of signals is $n = 12$, number of time points for the training is $m = 4211$, window size (DTT frame) is $u = 64$, DTT level is $l = 4$, and dimension of FIN is $q = 7$. Actually, Fig. 5 shows 3D projection of 7D FIN, where cells with different degree of gray represent 9 different speakers.

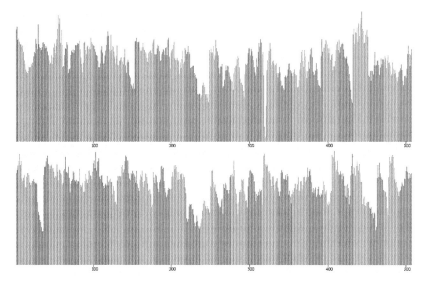

Fig. 3. Time series (LPC coefficient 1) of speaker 1 training (upper) and testing (lower)

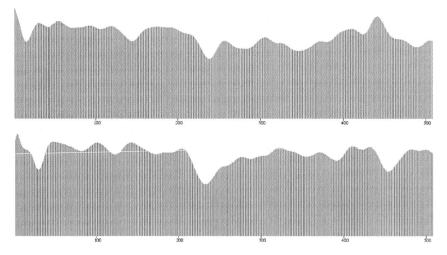

Fig. 4. DTT of time series in Fig. 3

Table 2. Speaker recognition by one signal

window size	DTT level	cells in FIN1	cells in FIN2	minimal distance	entropy	index	test errors	correct recognition	notes
4	0	4271	3454	5.3E-5	12.4	9.6	4145	27.1%	
8	0	4267	3240	1.0E-4	11.0	8.9	3978	30.0%	
16	0	4259	3181	7.9E-5	10.5	9.2	3909	31.1%	
32	0	4243	2990	1.7E-4	10.0	8.4	3684	34.9%	
64	0	4211	2609	1.2E-4	8.8	8.5	3290	41.5%	
	1		3081	1.1E-4	10.4	8.8	3482	38.1%	
	2		3279	7.7E-5	11.0	9.2	3417	39.2%	
	3		3456	9.1E-5	11.6	9.1	3051	45.8%	
	4		3431	1.05E-4	11.8	9.0	2910	48.3%	max%
	5		3431	2.7E-5	11.8	10.3	2982	47.0%	
	6		3100	7.4E-5	10.7	9.2	3123	44.5%	

Table 3. Speaker recognition by 12 signals

Dimension of FIN	Cells in FIN2	minimal distance	entropy	index	test errors	correct recognition	notes
3	3440	8.6E-5	12.3	9.1	984	82.7%	
4	3348	1.4E-4	11.8	8.7	324	94.3%	
5	2805	2.9E-4	10.0	7.7	245	95.7%	
6	2677	3.6E-4	9.5	7.5	120	97.9%	
7	2258	6.3E-4	8.3	6.7	49	99.1%	max%
8	2533	6.5E-4	9.6	6.8	105	98.2%	
9	2079	6.7E-4	7.8	6.6	153	97.3%	
10	2898	6.8E-4	10.6	6.9	157	97.2%	
11	2980	7.7E-4	10.8	6.8	138	97.6%	
12	2894	9.5E-4	10.5	6.6	183	96.8%	

Fig. 5. Example of FIN for speaker recognition by 12 signals

6 Discussion

The classifier proposed by Kudo et al. (1999) showed the classification rate of 94.1%, while a 5-state continuous Hidden Markov Model also considered in their paper and attained up to 96.2%. However, the IC approach shows better classification rate 99.1% (Tab. 3). This advantage of the IC approach has been also confirmed by a row of other comparisons with ANN (Tarakanov A and Tarakanov Y 2004; Tarakanov and Prokaev 2007; Tarakanov et al. 2007b; Tarakanov 2008) and GA (Tarakanov A and Tarakanov Y 2005) as well as nearest neighbor method and support vector machines (Tarakanov 2008, 2009ab).

Main idea of the IC algorithm is the mapping of any high-dimensional vector (antigen) to low-dimensional space of FIN using binding energies between the antigen and antibodies. Apoptosis and immunization also reduce the number of storing cells of FIN without loss of the accuracy of recognition. Due to these mathematically rigorous features, the IC algorithm outperforms state-of-the-art approaches of computational intelligence (Tarakanov 2008, 2009ab).

Another mathematically rigorous property of FIN is the exact recognition of any pattern it has been trained. This property allows using FIN to disclose the ambiguities in any data, e.g., like in the task of identification of cellular automata (Tarakanov and

Prokaev 2007). Such possibility is beyond the capabilities of ANN due to irreducible training errors and the known effect of overtraining when the attempts to reduce the errors may lead to their drastic increase (Tarakanov A and Tarakanov Y 2004). Therefore, FIN demonstrates the combination of faultless recognition with rather low training time. This looks unobtainable for its main competitors in the field of computational intelligence. For example, a comparison in (Tarakanov et al. 2007b) shows that FIN needs just 20 seconds to determine its optimal parameters where ANN trained by the error back propagation needs about 24 hours (!) for the similar purpose.

It is also worth noting that the proposed IC approach to speaker recognition has nothing common with a statistical analysis technique. Moreover, it is well-known that statistical methods (e.g. Markov chains, Bayesian networks etc.) are too slow and inaccurate to cope with real-world signals. For example, our comparison (Tarakanov et al. 2002) showed that the IC worked at least 10 times faster and much more accurate than conventional statistics. At the same time, the IC is able to sharply focus attention at most dangerous situations like the forecast of plague outburst (Sokolova 2003), which is beyond the capabilities of the traditional statistics. No wonder that more and more of modern approaches to signal processing turn to wavelet analysis, where the dyadic DTT (Section 3.3) is also a wavelet-type transform (Kozyrev 2002).

7 Concluding Remarks

The proposed IC approach to speaker recognition belongs to the field of computational intelligence (Tarakanov 2008). This approach is based essentially on the mathematical models of information processing by proteins and immune networks (Tarakanov et al. 2003). On such background, the mathematical model of FIN with apoptosis and immunization controlled by cytokines (Tarakanov et al. 2005a) represents the key model of the approach. On the other hand, it is worth noting that the SVD can also model the binding energy between two proteins (Tarakanov et al. 2003), whereas the dyadic DTT can model an immune-type antigen processing (Atreas et al. 2003, 2004).

The results of comparisons reported by now suggest that the speed and accuracy of the IC approach is probably unobtainable for other robust methods of computational intelligence (in particular, neurocomputing and evolutionary algorithms). These advances of the IC approach together with its biological nature probably mean a further step toward an intelligent immunochip in-silicon.

References

Adamatzky, A.: Identification of cellular automata. Taylor & Francis, London (1994)

Agnati, L.F., Tarakanov, A.O., Guidolin, D.: A simple mathematical model of cooperativity in receptor mosaics based on the "symmetry rule". BioSystems 80, 165–173 (2005a)

Agnati, L.F., Tarakanov, A.O., Ferre, S., Fuxe, K., Guidolin, D.: Receptor-receptor interactions, receptor mosaics, and basic principles of molecular network organization: possible implication for drug development. J. Mol. Neurosci. 26, 193–208 (2005b)

Agnati, L.F., Fuxe, K.G., Goncharova, L.B., Tarakanov, A.O.: Receptor mosaics of neural and immune communication: possible implications for basal ganglia functions. Brain Res. Rev. 58, 400–414 (2008)

Antoniol, G., Rollo, V.F., Venturi, G.: Linear predictive coding and cepstrum coefficients for mining time variant information from software repositories. ACM SIGSOFT 30(4), 1–5 (2005)

Atreas, N., Karanikas, C., Polychronidou, P.: Signal analysis on strings for immune-type pattern recognition. Compar. Func. Genomics 5, 69–74 (2004)

Atreas, N., Karanikas, C., Tarakanov, A.: Signal processing by an immune type tree transform. In: Timmis, J., Bentley, P.J., Hart, E. (eds.) ICARIS 2003. LNCS, vol. 2787, pp. 111–119. Springer, Heidelberg (2003)

Chao, D.L., Davenport, M.P., Forrest, S., Perelson, A.S.: A stochastic model of cytotoxic T cell responses. J. Theor. Biol. 228, 227–240 (2004)

Cover, T.M., Hart, P.E.: Nearest neighbor pattern classification. IEEE Transact Inform Theory 13, 21–27 (1967)

Dasgupta, D. (ed.): Artificial immune systems and their applications. Springer, Berlin (1999)

Dasgupta, D., Krishna-Kumar, K., Wong, D., Berry, M.: Negative selection algorithm for aircraft fault detection. In: Nicosia, G., Cutello, V., Bentley, P.J., Timmis, J. (eds.) ICARIS 2004. LNCS, vol. 3239, pp. 1–13. Springer, Heidelberg (2004)

Dasgupta, D., Gonzalez, F.: Artificial immune systems in intrusion detection. In: Rao Vemuri, V. (ed.) Enhancing computer security with smart technology, pp. 165–208. Auerbach, Boca-Raton FL (2005)

Dasgupta, D.: Advances in artificial immune systems. IEEE Compt. Intell. Mag. 1, 40–49 (2006)

Dasgupta, D., Nino, F.: Immunological computation: theory and applications. Auerbach, Boca-Raton FL (2008)

de Castro, L.N., Timmis, J.: Artificial immune systems: a new computational intelligence approach. Springer, London (2002)

Fuxe, K.G., Tarakanov, A.O., Goncharova, L.B., Agnati, L.F.: A new road to neuroinflammation in Parkinson's disease? Brain Res. Rev. 58, 453–458 (2008)

Goncharova, L.B., Jacques, Y., Martín-Vide, C., Tarakanov, A.O., Timmis, J.I.: Biomolecular immune-computer: Theoretical basis and experimental simulator. In: Jacob, C., Pilat, M.L., Bentley, P.J., Timmis, J.I. (eds.) ICARIS 2005. LNCS, vol. 3627, pp. 72–85. Springer, Heidelberg (2005)

Goncharova, L.B., Tarakanov, A.O.: Molecular networks of brain and immunity. Brain Res. Rev. 55, 155–166 (2007)

Goncharova, L.B., Tarakanov, A.O.: Nanotubes at neural and immune synapses. Curr. Med. Chem. 15, 210–218 (2008a)

Goncharova, L.B., Tarakanov, A.O.: Why chemokines are cytokines while their receptors are not cytokine ones? Curr. Med. Chem. 15, 1297–1304 (2008b)

Horn, R., Johnson, Ch: Matrix analysis. Cambridge University Press, London (1986)

Johnson, J.E.: Networks, Markov Lie monoids, and generalized entropy. In: Gorodetsky, V., Kotenko, I., Skormin, V.A. (eds.) MMM-ACNS 2005. LNCS, vol. 3685, pp. 129–135. Springer, Heidelberg (2005)

KDD, The UCI KDD Archive. University of California, Irvine, CA (1999), http://kdd.ics.uci.edu/

Kozyrev, S.V.: Wavelet theory as p-adic spectral analysis. Izvestia: Mathematics 66, 367–376 (2002)

Kudo, M., Toyama, J., Shimbo, M.: Multidimensional curve classification using passing-through regions. Patt. Rec. Lett. 20, 1103–1111 (1999)

Rabiner, L.R., Juang, B.H.: Fundamental of speech recognition. Prentice Hall, Englewood Cliffs (1993)

Renyi, A.: On measures of entropy and information. In: Fourth Berkeley symposium on mathematics, statistics and probability, vol. 1, pp. 547–561. Cambridge University Press, London (1961)

Sokolova, L.: Index design by immunocomputing. In: Timmis, J., Bentley, P.J., Hart, E. (eds.) ICARIS 2003. LNCS, vol. 2787, pp. 120–127. Springer, Heidelberg (2003)

Tarakanov, A., Goncharova, L., Gupalova, T., Kvachev, S., Sukhorukov, A.: Immunocomputing for bioarrays. In: Timmis, J., Bentley, P. (eds.) Proc. 1st Int. Conf. ICARIS 2002, pp. 32–40. University of Kent at Canterbury, UK (2002)

Tarakanov, A.O.: Formal immune networks: self-organization and real-world applications. In: Prokopenko, M. (ed.) Advances in applied self-organizing systems, pp. 269–288. Springer, Berlin (2007a)

Tarakanov, A.O.: Mathematical models of intrusion detection by an intelligent immunochip. Communicat. Compt. Inform. Sci. 1, 308–319 (2007b)

Tarakanov, A.O.: Immunocomputing for intelligent intrusion detection. IEEE Compt. Intell. Mag. 3, 22–30 (2008)

Tarakanov, A.O.: Immunocomputing for spatio-temporal forecast. In: Mo, H. (ed.) Handbook of research on artificial immune systems and natural computing: applying complex adaptive technologies, pp. 241–261. IGI Global, Hershey (2009a)

Tarakanov, A.O.: Immunocomputing for intelligent signal processing. Neural Compt. Appl. (2009b) (in press)

Tarakanov, A., Adamatzky, A.: Virtual clothing in hybrid cellular automata. Kybernetes 31, 394–405 (2002)

Tarakanov, A., Nicosia, G.: Foundations of immunocomputing. In: First IEEE symposium on foundations of computational intelligence, FOCI 2007, pp. 503–508. Omnipress, Madison (2007)

Tarakanov, A., Prokaev, A.: Identification of cellular automata by immunocomputing. J. Cell Autom. 2, 39–45 (2007)

Tarakanov, A.O., Tarakanov, Y.A.: A comparison of immune and neural computing for two real-life tasks of pattern recognition. In: Nicosia, G., Cutello, V., Bentley, P.J., Timmis, J. (eds.) ICARIS 2004. LNCS, vol. 3239, pp. 236–249. Springer, Heidelberg (2004)

Tarakanov, A.O., Tarakanov, Y.A.: A comparison of immune and genetic algorithms for two real-life tasks of pattern recognition. Int. J. Unconvent. Compt. 1, 357–374 (2005)

Tarakanov, A., Goncharova, L., Gupalova, T., Kvachev, S., Sukhorukov, A.: Immunocomputing for bioarrays. In: Timmis, J., Bentley, P. (eds.) Proc. 1st Int. Conf. Artificial Immune Systems, ICARIS 2002, pp. 32–40. University of Kent at Canterbury, UK (2002)

Tarakanov, A.O., Goncharova, L.B., Tarakanov, O.A.: A cytokine formal immune network. In: Capcarrère, M.S., Freitas, A.A., Bentley, P.J., Johnson, C.G., Timmis, J. (eds.) ECAL 2005. LNCS (LNAI), vol. 3630, pp. 510–519. Springer, Heidelberg (2005)

Tarakanov, A.O., Kvachev, S.V., Sukhorukov, A.V.: A formal immune network and its implementation for on-line intrusion detection. In: Gorodetsky, V., Kotenko, I., Skormin, V.A. (eds.) MMM-ACNS 2005. LNCS, vol. 3685, pp. 394–405. Springer, Heidelberg (2005)

Tarakanov, A., Kryukov, I., Varnavskikh, E., Ivanov, V.: A mathematical model of intrusion detection by immunocomputing for spatially distributed security systems. RadioSystems 106, 90–92 (2007a) (in Russian)

Tarakanov, A., Prokaev, A., Varnavskikh, E.: Immunocomputing of hydroacoustic fields. Int. J. Unconvent. Compt. 3, 123–133 (2007b)

Tarakanov, A.O., Sokolova, L.A., Kvachev, S.V.: Intelligent simulation of hydrophysical fields by immunocomputing. Lect. Notes Geoinf. Cartog. XIV, pp. 252–262 (2007c)

Tarakanov, A.O., Skormin, V.A., Sokolova, S.P.: Immunocomputing: principles and applications. Springer, New York (2003)

Zhao, W.: Review of Immunocomputing: Principles and Applications. ACM SIGACT News 36, 14–17 (2005)

Author Index